PERIODIC INSPECTION OF PRESSURE VESSELS

The Institution of Mechanical Engineers

PERIODIC INSPECTION OF PRESSURE VESSELS

A Conference arranged by the
Applied Mechanics Group
of the Institution of Mechanical Engineers
9–11th May 1972

1 BIRDCAGE WALK · WESTMINSTER · LONDON

ISBN 0 85298 185 6

CONTENTS

Periodic Inspection of Pressure Vessels

This conference was sponsored by the Pressure Vessels section of the Applied Mechanics Group and was held in London. 354 delegates registered to attend. The planning panel consisted of Dr R. W. Nichols (chairman), Mr R. L. J. Hayden, Mr G. P. Smedley and Mr B. Watkins.

Six sessions included: (1) 'Requirements for inspection' (papers C42, C56, C48, C26); chairman, Professor Y. Ando. (2) 'Inspection practice in differing applications' (papers C29, C59, C53, C52, C51); chairman, L. J. Chockie. (3) 'Inspection practice in nuclear plant' (papers C55, C41, C44, C31); chairman, R. D. Wylie. (4) 'Equipment for reactor pressure vessel inspection' (papers C57, C25, C46, C43, C47, C27); chairmen, Dr T. Broome and Dr R. W. Nichols. (5) 'Future developments—acoustic emission and other techniques' (papers C28, C30, C40, C45, C49, C58); chairmen, Dr S. H. Bush and B. Watkins. (6) 'Other developments and general aspects' (papers C50, C54, C60); chairman Dr R. W. Nichols.

A cocktail party was held in the evening of Wednesday 10th May in the ballroom of the St Ermin's Hotel, Caxton St, London, S.W.1.

C25/72 THE NUCLEAR REACTOR VESSEL INSPECTION TOOL USING AN ULTRASONIC METHOD

T. YAMAGUCHI* Y. FUKUSHIMA* S. KIHARA* T. ENDO* Y. YOSHIDA*

This paper describes the outline of an in-service inspection tool using an ultrasonic method for the P.W.R. type of nuclear pressure vessels, which is now under development. This equipment consists of the mechanical components of the tool, the control panel to drive it, and data acquisition apparatus. The tool is designed to be adjustable to reactor vessels of various sizes. Mitsubishi Heavy Industries Ltd (M.H.I.) has now completed the fabrication of most parts of the tool, and are planning to carry out a mock-up test in the near future to check the operational function of the tool and the detectability of the defects.

INTRODUCTION

In the nuclear industry, in order to ensure the protection of the public from any foreseeable accidents, the safety of all components is absolutely essential. In a nuclear power plant especially, periodic inspection is required even after operation commences, to secure the safety of the vessel and to prevent, in advance, the occurrence of any accident.

In operation, a nuclear reactor vessel becomes radioactive after being irradiated by neutrons. Therefore, depending on their positions, some parts of the vessel are inaccessible. For this reason, the inspection of such parts must be remotely performed.

To meet the above requirements, M.H.I. began several years ago to research and develop this inspection method. As a result, development is almost completed of an inspection method and its remote control tool. Furthermore, it was scheduled recently to perform a mock-up test utilizing a model vessel.

OUTLINE OF A NUCLEAR REACTOR VESSEL

The outline of the major dimensions and welded joints of the typical P.W.R.-type reactor vessels which are now under design, or have already been constructed, is shown in Fig. 25.1. The materials used are nickel–chromium–molybdenum forged steel for the upper shell, shell flange, closure head flange, outlet and inlet nozzles, and safety injection nozzle; and manganese–molybdenum–nickel plate for the middle shell, lower shell, closure head, and lower head. In addition, forged austenitic stainless steel (A.S.M.E. SA-182 F316, F304) is used at the safe-end of each nozzle. On the inner surface of the vessel, the overlay cladding equivalent of austenitic stainless steel type 304 is performed and is finished by grinding. The total thickness is approximately 5·5 mm.

In order to secure fully the safety of the nuclear reactor vessel in its fabricating process, extraordinarily strict non-destructive inspection of various sorts are performed. The kind of inspection method used for each component in fabrication is shown in Fig. 25.2, each welded joint being inspected by various methods, such as ultrasonic, radiography, magnetic particle, and liquid penetrant, after the welding is completed or after the overlay welded surface is finished.

GENERAL SYSTEM REQUIREMENT

An inspection method which is applicable to the in-service inspection of a nuclear reactor vessel must satisfy the following conditions (1)†.

(1) It should be capable of non-destructive, volumetric inspection, which can be carried out from one side of a vessel.

(2) It should be capable of automatic and remotely controlled measurement.

(3) It must have appropriate sensitivity resolution and good repeatability of data.

(4) It should be insensitive to irradiation from radioactive materials.

(5) It must have an electrically processible data output.

(6) It should be capable of rapid operation and analysis.

The ultrasonic inspection method is the only one that satisfies all these conditions.

The ultrasonic inspection methods adopted were normal beam, angle beam, and pitch–catch methods, which are shown in Table 25.1. The search unit is composed of lead–zirconate–titanate transducers, which are 20 mm in diameter with a frequency of 1 MHz (for angle beam and pitch–catch methods) and 2·25 MHz (for normal beam method). The effect of radioactive irradiation, for lead–zirconate–titanate transducers, as shown in Fig. 25.3, is hardly observed regarding the loss of sensitivity within the irradiation dosage $<1{\cdot}7\times10^{7}$ r.

Inspection is performed from within a vessel. The inner

The MS. of this paper was received at the Institution on 1st November 1971 and accepted for publication on 13th December 1971.

* *Mitsubishi Heavy Industries Ltd, Kobe Technical Institute.*

† *The reference is given in Appendix 25.1.*

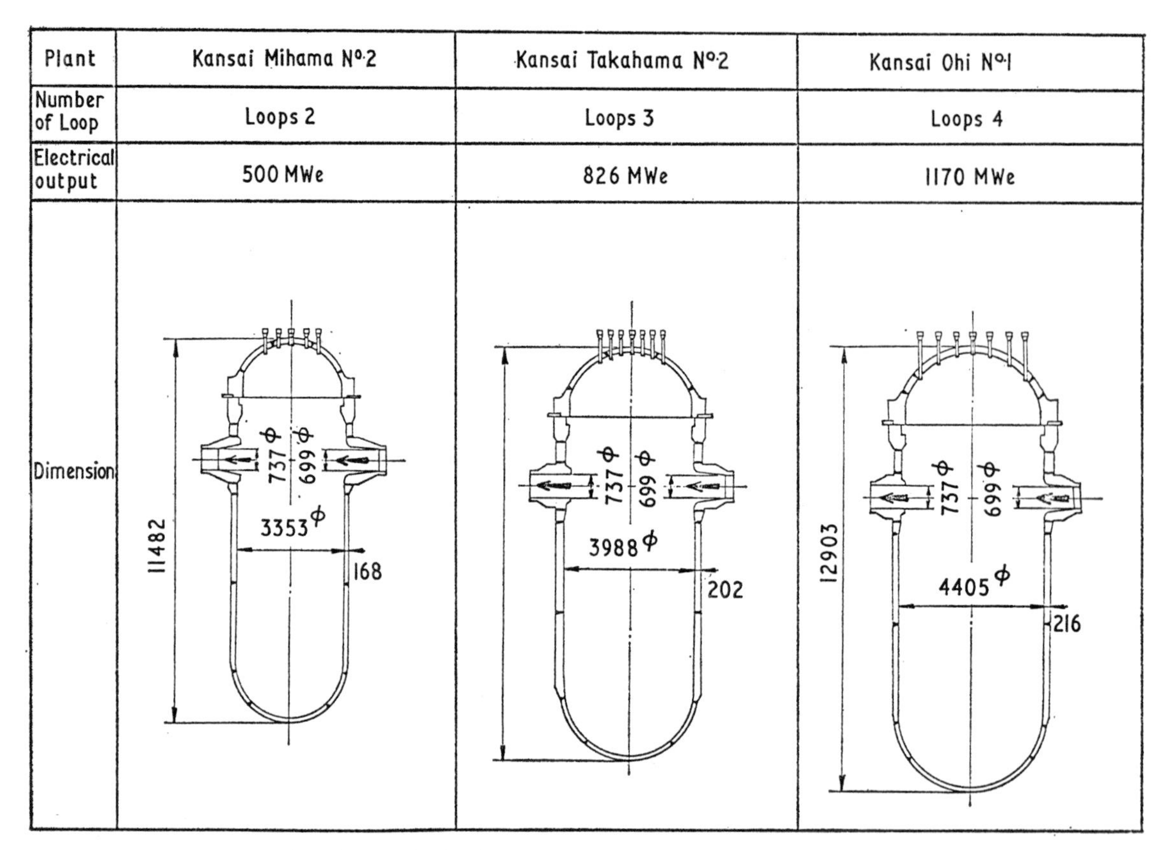

Fig. 25.1. Principal dimensions of typical reactor vessels

surface of a vessel is not entirely smooth. In fact, there exists a valley construction on the bead overlapped section which is a hazard relative to the penetration of an ultrasonic beam. Therefore an immersion method was adopted. This is a method in which an ultrasonic beam is transmitted by keeping a certain distance between a transducer and a surface of the vessel wall to be inspected. One-third of the thickness of the wall to be inspected is usually taken as a water path distance.

Although the cladding surface is finished by grinding, there is still comparative roughness on the bead-overlapped part, i.e. the valley part, as well as a metallurgical change. These items result in scatter of the transmitted energy and, therefore, give an effect on echo height. Consequently, the main endeavour of the M.H.I. researchers' study was to improve the accuracy of inspection by minimizing the effect of cladding relative to the volumetric inspection of a reactor vessel.

Table 25.1. Flaw detecting method

Method	Normal Beam	Angle Beam	Pitch-Catch
Principle	Flaw C.R.T. Display Sound part / Unsound part / Unsound part T S B / T S B₁ / T S F B₂	θ_1 θ_2 Flaw C.R.T. Display Sound part / Unsound part T / T F	Pitch Catch θ_1 θ_1 θ_2 θ_2 Flaw C.R.T. Display Sound part / Unsound part T C / T C
Remarks	T: Transmit echo S: Surface echo F: Flaw echo B: Back echo	θ_1: 19° (Incident angle in water)	θ_2: 45° (Refracted angle in steel) C: (Transmission echo)

Areas to be inspected by this tool are to include those where volumetric inspections are required by the A.S.M.E. Code Section XI. Areas to be inspected and the ultrasonic inspection method which is to be adopted for each area of the vessel are shown in Table 25.2 and Fig. 25.4. In order to promote precision of inspection, normal beam and pitch–catch methods are simultaneously used on the circumferential and longitudinal welds of the vessel (angle beam method is also applicable).

AUTOMATIC REMOTE-CONTROL TOOL

This tool is composed of a mechanical operation component which works underwater in a nuclear reactor vessel, and a control panel to operate this component; this is shown in Fig. 25.5. The construction of the mechanical operation component enables it to be adjusted to reactor vessels of various sizes, and it is remotely operated with high precision and reliability on the areas to be inspected.

The control panel is the operation unit by which the mechanical component is driven, and it is capable of manual and automatic operation. Adequate consideration is given regarding water pressure strength, resistance to radioactive irradiation, and corrosion resistance for the main component. Simultaneously, a special design to maintain the rigidity of the tool has been devised in order to promote driving precision. This tool is also designed to permit transportation in a container, and to give easy disassembly and assembly.

Configuration and function

The tool is composed of the following elements.

Turning table equipment

It consists of rail, turning frame, and driving mechanism. The rail is laid on the upper surface of a pressure-vessel flange. The turning frame with driving mechanism is placed on the rail, and it is rotated on a rail driven by three rollers.

Elevation mechanism

It consists of the elevator, guide frame, elevation driving mechanism, and guide frame support mechanism. The elevator runs vertically along the guide frame whose upper edge is fixed on the turning table.

Manipulator

There are five series of manipulator driven by electric and pneumatic power. Each manipulator is equipped with ultrasonic transducers of various kinds, as shown in Fig. 25.6. The manipulator series 1 is provided on the turning table; series 2, 3, and 4 on the elevator; series 5 at the lower edge of the guide frame. Each of them is operated in accordance with the particular areas to be inspected.

Automatic remote-control equipment

It consists of a control panel and a position indication panel. The control panel is placed on an operation deck. By command from this panel, transducers on a manipulator are moved circumferentially, vertically, and radially to scan the surfaces to be inspected automatically.

Flaw detection

It is important that this tool be able to detect the position of a flaw as well as its presence. By making an initial setting

Table 25.2. Detailed descriptions on inspection areas of the reactor vessel

Areas subject to inspection	Actually inspectable areas		Ultrasonic testing method
	Length (L), mm	Width (B), mm	
(A) Flange ligaments between stud holes	Entire circle	Flange width	Normal beam method (No. 1)
(B) Vessel to flange welds	Entire circle (except stud hole)	Wall thickness	Normal beam method (No. 2)
(C) Shell circumferential welds	Entire circle	Weld metal width $+2x$ (wall thickness)	Normal beam method and pitch–catch method (possible to use angle beam method)
(D) Shell longitudinal welds	Entire length of each weld	Weld metal width $+2x$ (wall thickness)	Normal beam method and pitch–catch method (possible to use angle beam method)
(E) Outlet and inlet nozzle attachment welds to vessel	Entire circle	Upper shell wall thickness (except a part of inlet nozzle)	Normal beam method (No. 1)
(F) Pipe to safe-end welds	Entire circle	Weld metal width $+2x$ (wall thickness)	Normal beam method and angle beam method (No. 2)
(G) Outlet and inlet nozzle to safe-end welds	Entire circle	Weld metal width $+2x$ (wall thickness)	Normal beam method and angle beam method (No. 2)
(H) Inner radius section of the inlet nozzle-to-vessel juncture	Entire circle	Inner radius section	Normal beam method
(I) Shell to lower head welds	Entire circle (except internal support attachment)	Weld metal width $+2x$ (wall thickness)	Normal beam method and angle beam method
(J) Meridional welds in lower vessel head	10% of each longitudinal weld (upper part)	Weld metal width $+2x$ (wall thickness)	Normal beam method

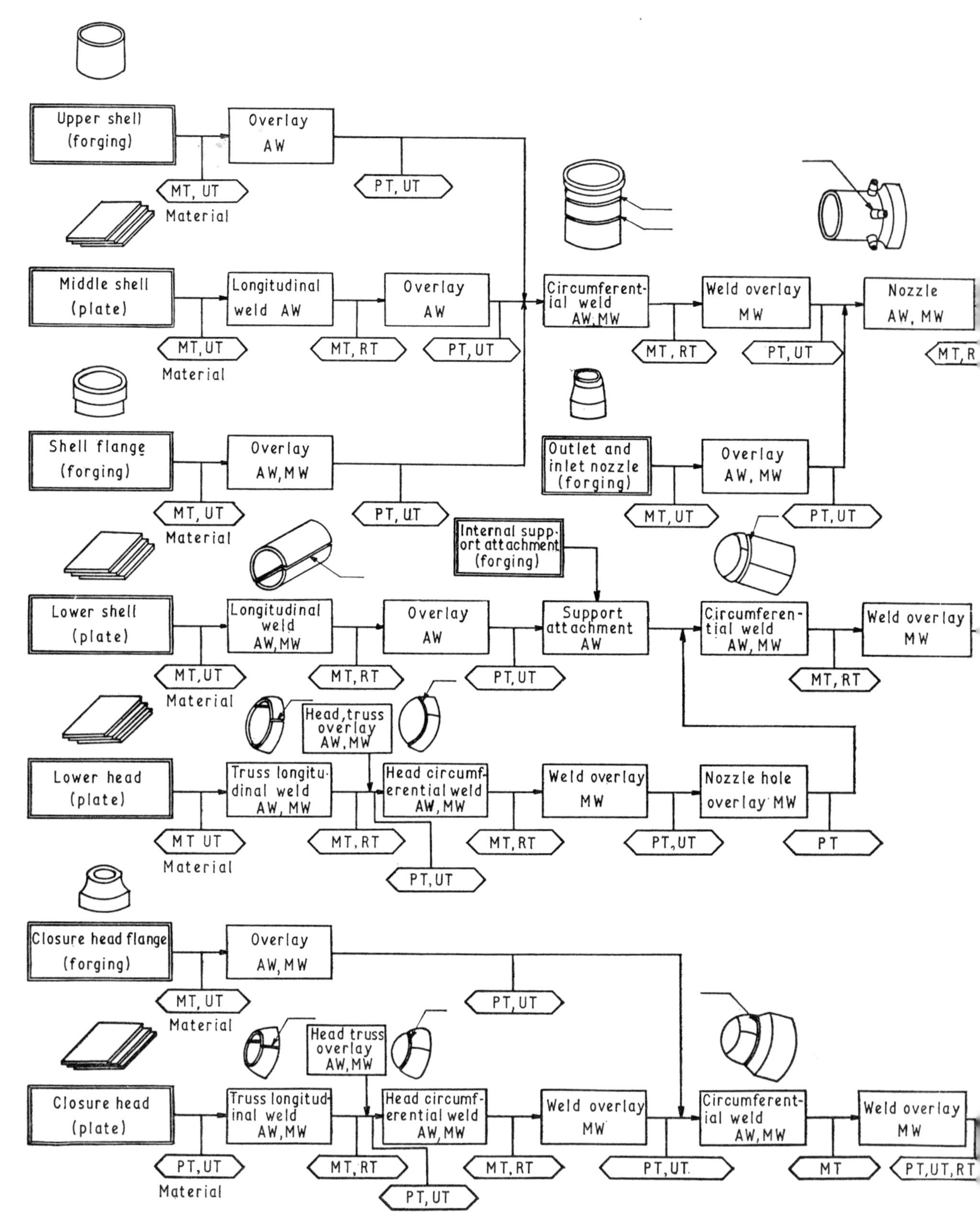

Fig. 25.2. Outline of process for non-destruct

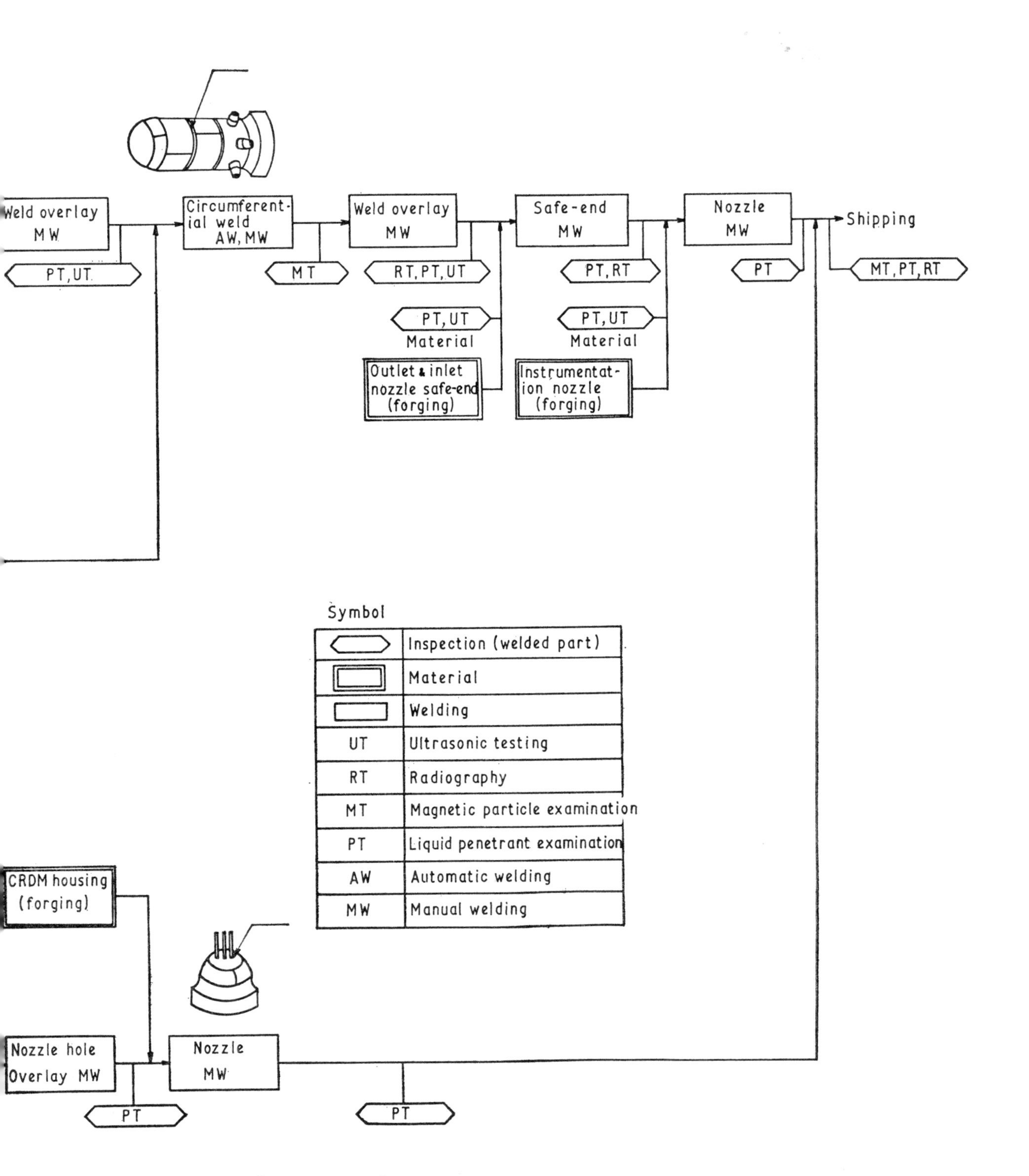

amination during fabrication of reactor vessel

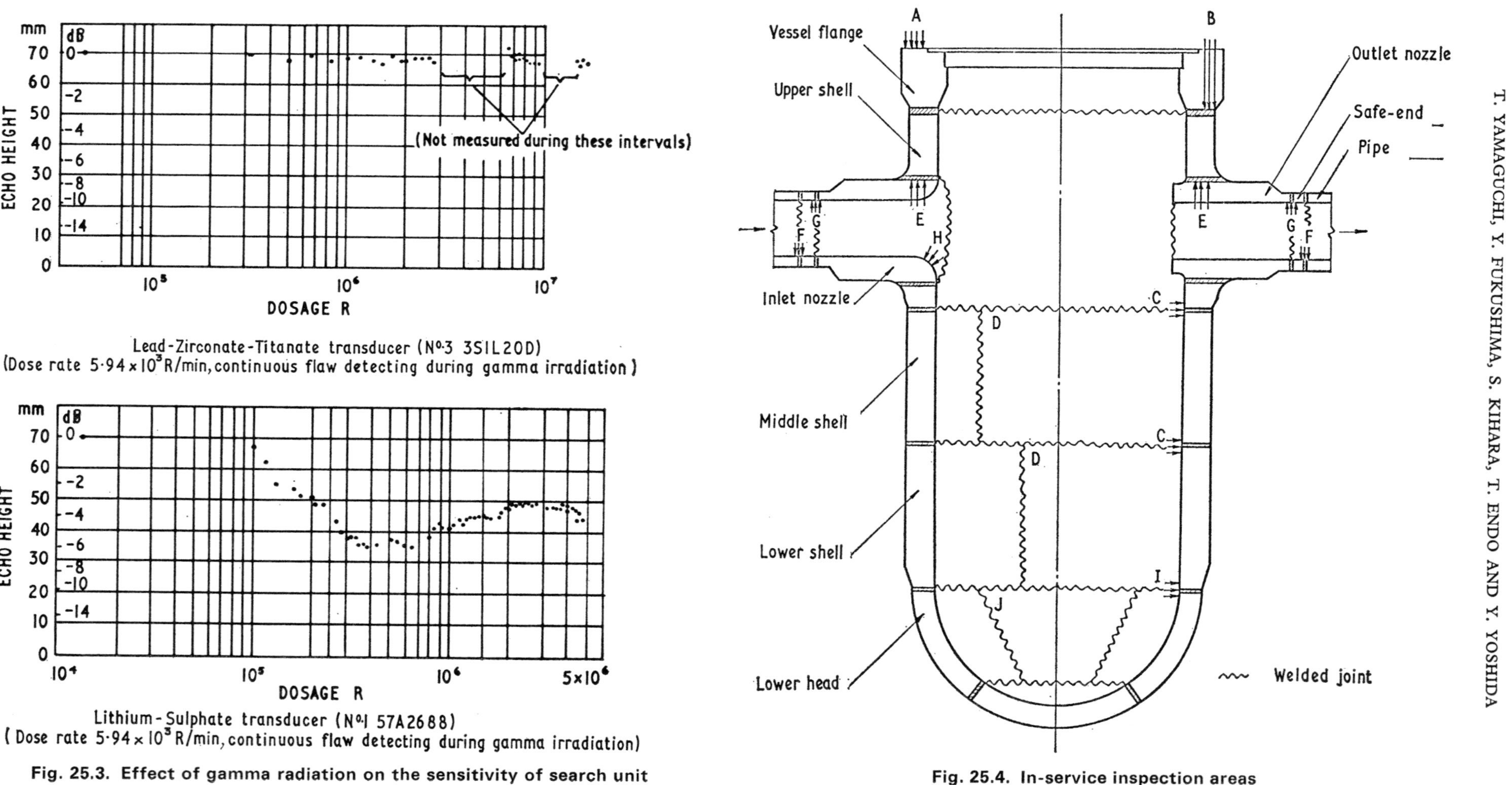

Fig. 25.3. Effect of gamma radiation on the sensitivity of search unit

Fig. 25.4. In-service inspection areas

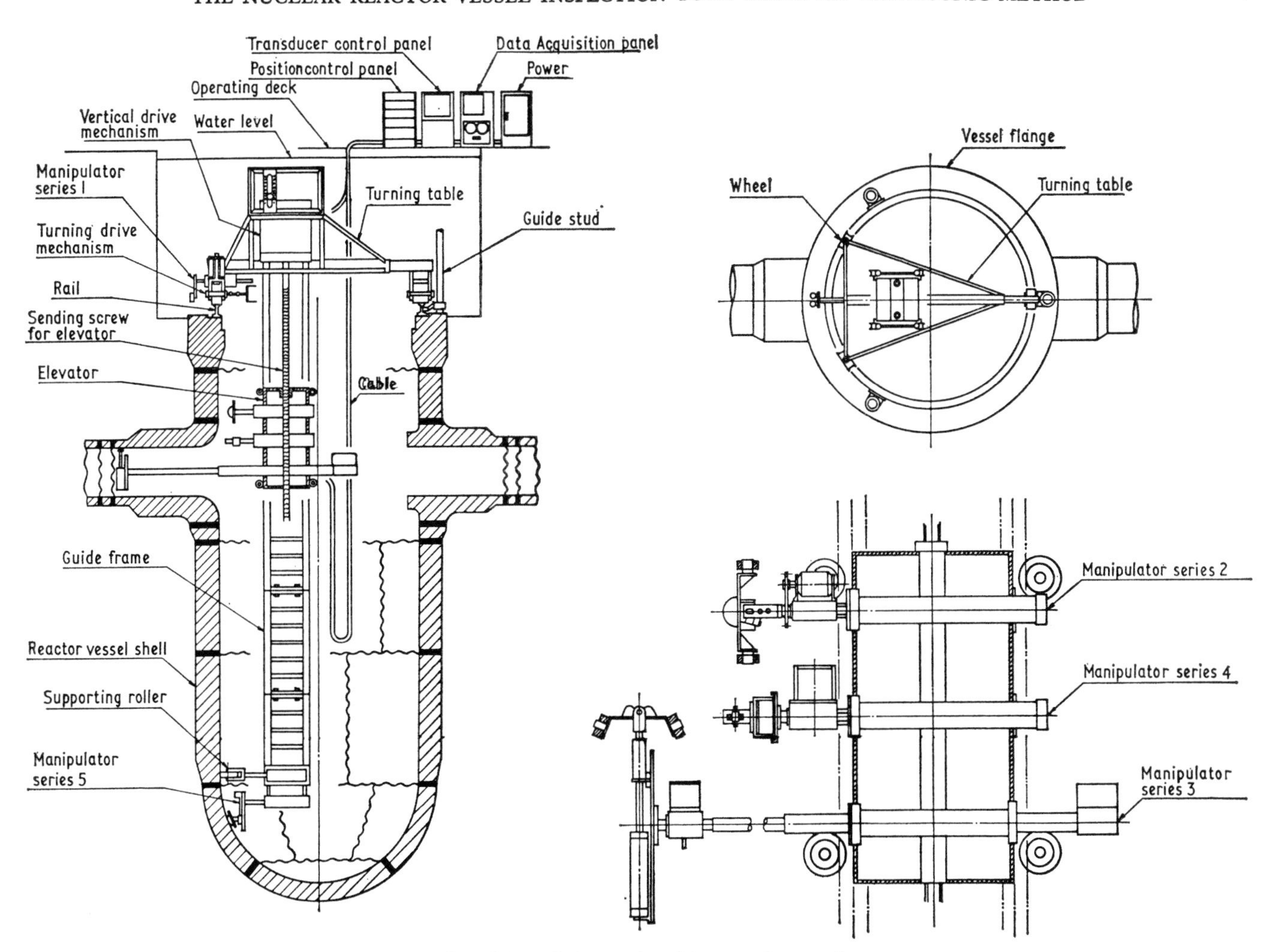

Fig. 25.5. Schematic of inspection tool

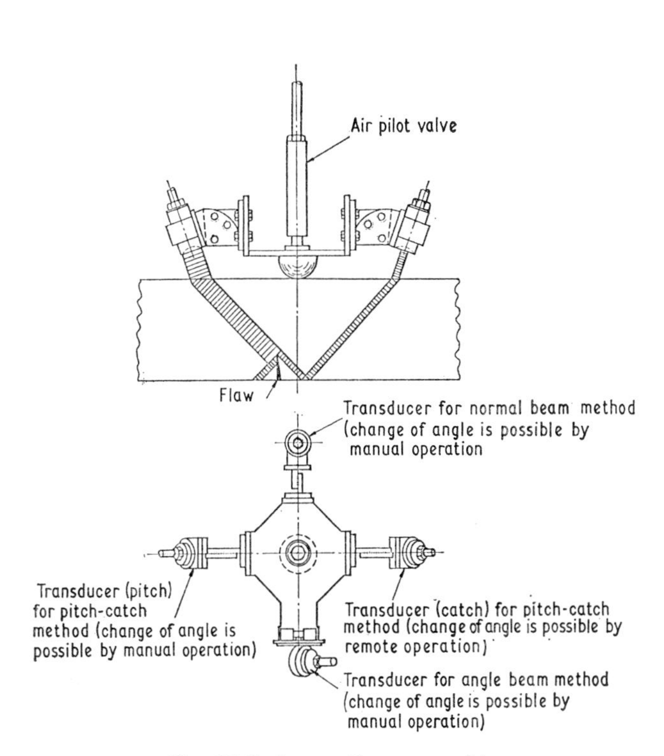

Fig. 25.6. Inspection assembly

position of the tool on a vessel at a reference point, the position of the inspection area is continuously monitored at the control panel. This position-indication signal is simultaneously recorded as an address information in a data acquisition unit.

Dimensions of tool

Because the inner diameter and depth of a reactor vessel differ in accordance with capacity, the method of installing this tool on a vessel changes in each case. This tool is applicable within the following range by adjusting each component element in its assembly process.

Inner diameter of vessel	3300–4420 mm
Maximum depth (from vessel flange surface to shell to bottom head weld)	8400 mm
Nozzle inner diameter	690–900 mm
Nozzle depth (from vessel inner surface)	1130 mm

In addition, the major dimensions and weight of this tool are:

Length × width × height	4430 × 1760 × 10 800 mm
Weight	About 3·5 ton

Scanning precision and speed

The scanning precision and speed of the search units furnished on this tool are:

Angle of circumferential rotation $\pm 0{\cdot}1^\circ$

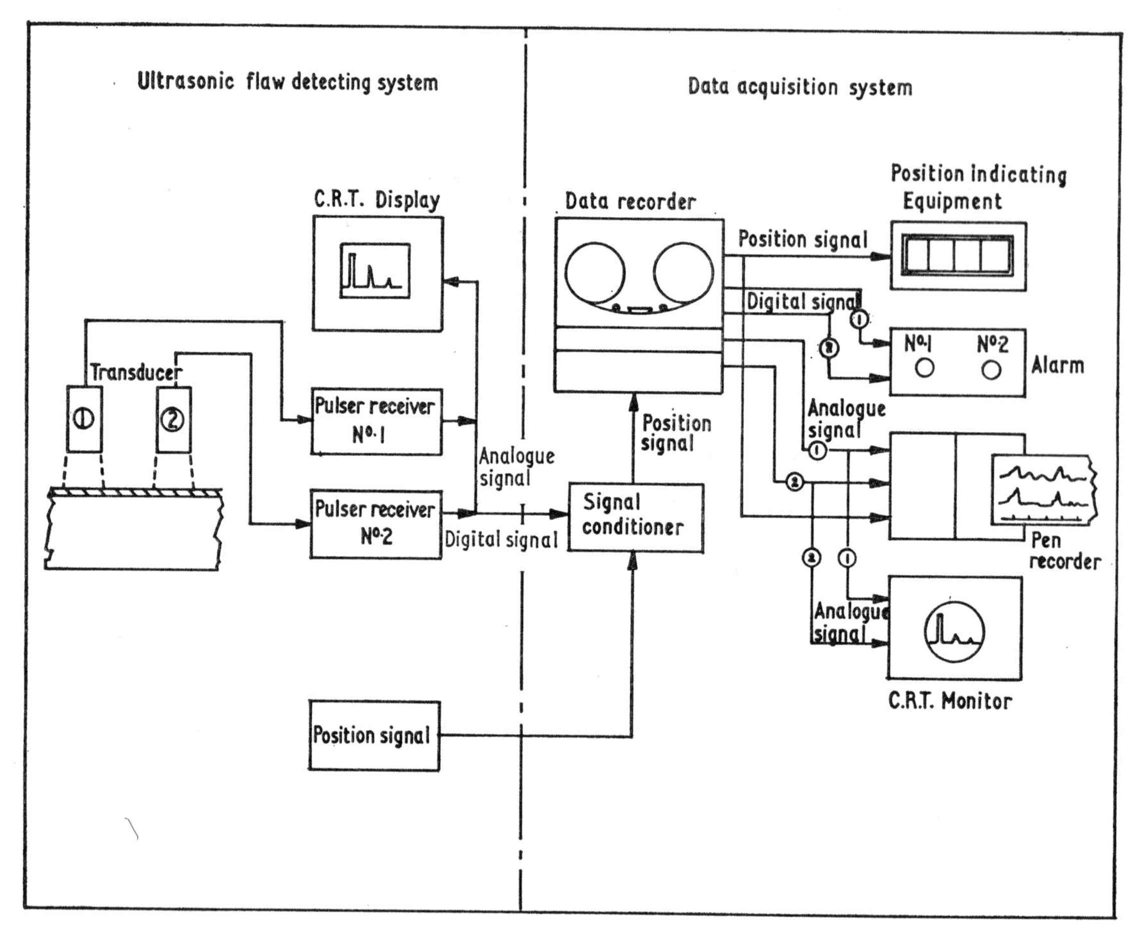

Fig. 25.7. Block diagram of the data acquisition apparatus

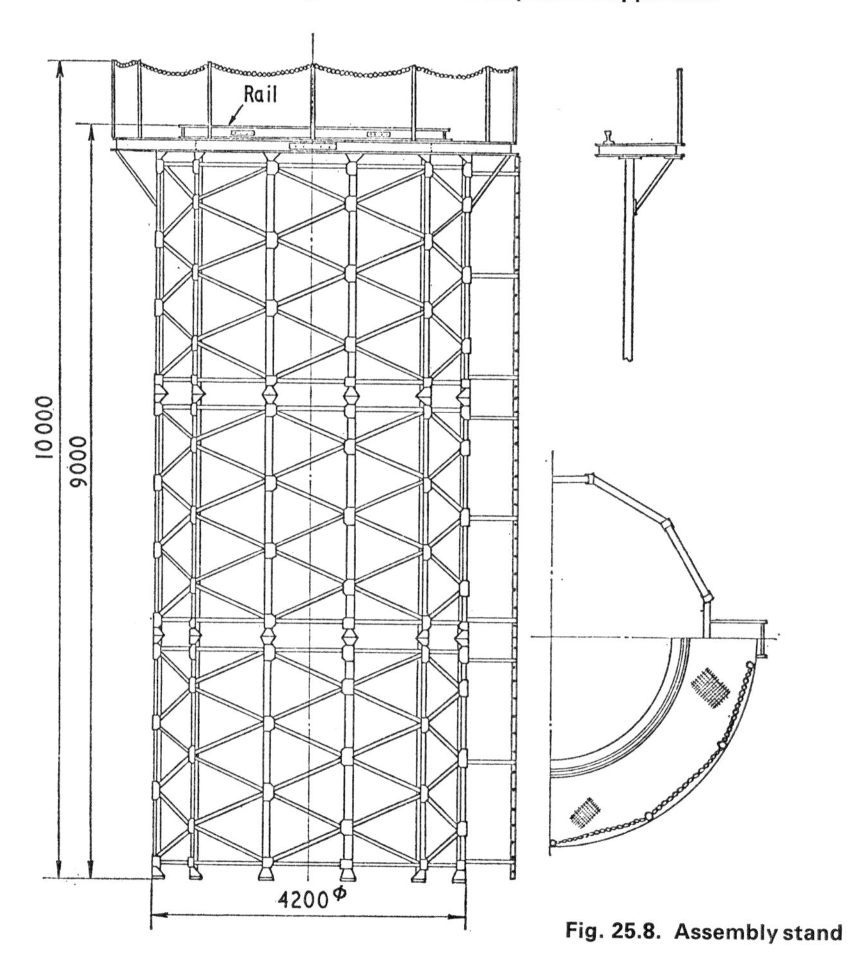

Fig. 25.8. Assembly stand

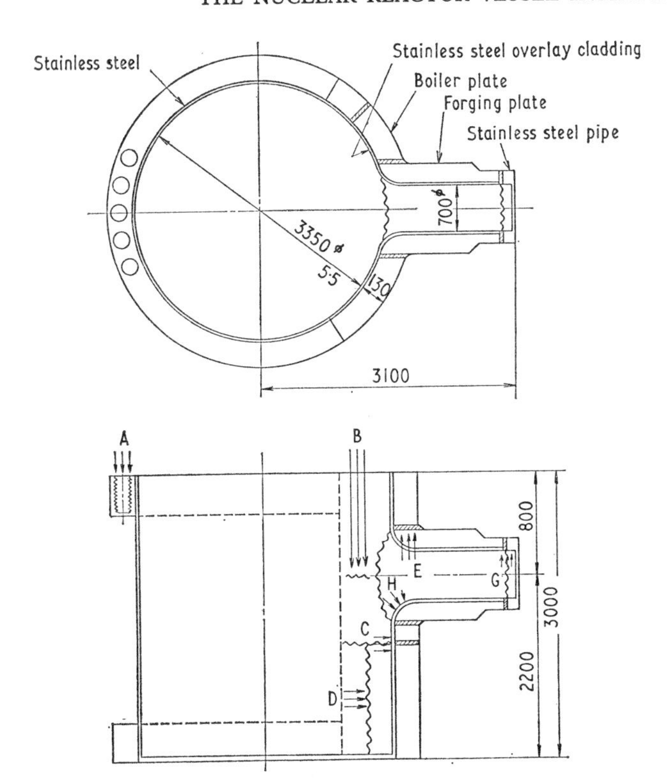

Fig. 25.9. Model vessel for the mock-up test

Vertical position	±3 mm
Lateral position	±2·5 mm
Angle of transducer unit rotation (when transducer is rotated 90°)	±0·1°
Scanning speed	max. 100 mm/s

DATA ACQUISITION APPARATUS

Outline of the system

An overall configuration of the data acquisition apparatus for this tool is shown in Fig. 25.7. Two sets of transmitter/receiver are provided. They are the electronic amplifier to transmit and receive ultrasonic beams. The pulse-echo height received is observed by a cathode-ray tube display. Pre-set voltage is provided at the data recording parts to eliminate the effect of noise echo by a cladding on a surface to be inspected, and to record only a flaw signal which exceeds this pre-set voltage. The flaw signal is recorded by a pen recorder as well as being magnetically recorded using a data recorder. A signal conditioner is provided to adjust the signal voltage level to be recorded by a data recorder. An alarm signal will be emitted if a flaw signal over the pre-set level exceeds an acceptable limit. In addition, the data recorder is recording the presence of a flaw and its position simultaneously.

Position signal

Although the search unit is 20 mm in diameter, the inspecting position will be displayed every 1 mm of its movement. The position is measured as a distance (rotation angle, vertical position, radius) between the reference point and the inspected part, which is recorded and displayed digitally on a control panel. In order to detect the co-ordinates of the moving distance to each direction, the signals are emitted by a shaft-encoder (DECIHI-CODE, Pulcen), which is installed on a driving screw to drive search units.

Recording unit

A data recorder is capable of the simultaneous recording of all signals from the transducers, position, scanning direction, and audio signal for recording. The signals recorded by a data recorder are easy to retrieve. The pen recorder records continuously the signals exceeding the pre-set voltage to inspect visually the presence of flaws macroscopically.

FIELD ASSEMBLY

In order to perform the inspection of a nuclear reactor vessel using this tool, an assembly stand on which this tool is built must be set in advance. Adjustment of the controlling function of each manipulator and the calibration of sensitivity of each transducer are also performed in this state. The stand for a field assembly is constructed of the frames shown in Fig. 25.8, which are easy to assemble and disassemble. The size is approximately 5000 mm in diameter and approximately 9000 mm in height. Therefore a space is required wide enough for the installation of this stand near the nuclear reactor when it is assembled in the field.

MOCK-UP TEST

A mock-up test is planned as a utilization test in the near future. The vessel model for the test has a cylindrical shape with a length of 3000 mm and an inner diameter of 3350 mm, as shown in Fig. 25.9. A part of the wall thickness is constructed of a size comparable to the real vessel, where a nozzle of an actual inner diameter is attached. On the inner surface, cladding of stainless steel is performed. Thus, the model simulates the real vessel. A flaw of proper size (either artificial weld flaws or machined defects) is made on a part of this model which corresponds to the area to be inspected, as shown in Table 25.2. In this manner, the operational function and flaw detectability of this tool is planned to be examined.

CONCLUSIONS

M.H.I. initiated the development of the Mitsubishi nuclear reactor vessel ultrasonic inspection tool several years ago. At present, the greater part of it having been already completed, a utilization test is planned using a vessel model for mock-up testing.

In order to secure the higher safety factors required of a nuclear reactor vessel, the completion of this in-service inspection tool is urgently required, in the development of which M.H.I. is making its greatest effort, with the collaboration of each division of the various related fields.

APPENDIX 25.1

REFERENCE

(1) GROSS, L. B. and JOHNSON, C. R. 'In-service inspection of nuclear reactor vessels using an automated ultrasonic method', *Mater. Eval.* 1970 **28** (No. 7), 162.

C26/72 DEVELOPMENT OF IN-SERVICE INSPECTION SAFETY PHILOSOPHY FOR U.S.A. NUCLEAR POWER PLANTS

S. H. BUSH* R. R. MACCARY†

The American Society of Mechanical Engineers' (A.S.M.E.) Section XI Code on 'In-service inspection of nuclear reactor coolant systems' represents the joint efforts of the United States Atomic Energy Commission's regulatory organization and of the nuclear industry. Efforts initiated in 1967 culminated in a code in 1970. Significant features of this code are: (1) the concept of designing the system to permit inspection and possible repairs; (2) the requirement of a complete examination prior to start-up to serve as a base line for future examinations; (3) the acceptance of new inspection systems or techniques more amenable to remote applications, provided such systems can be validated; and (4) the establishment of inspection periods and level of inspection for given components or sections of components based on the concepts of relative probability of degradation of the various portions of the systems and the significance of such degradation to the safety of the reactor system.

The general safety philosophy that led to the development of the Section XI Code is described, together with a general description of the code coverage. This code represents a significant departure from A.S.M.E. policy in that it deals with the operational phase of a pressure system compared to the design, fabrication, and construction stage.

Some consideration is given to future revisions in the code and their significance to safety and to reactor operation. It is felt that this code corrects one of the major reservations inherent in nuclear systems where great care was given to design and fabrication but no analogue existed to the regular inspections typical of fossil-fuel plants during their operation.

INTRODUCTION

THE PRACTICE of periodic inspections of pressure-retaining components of fossil-fuelled power stations is well established as essential to a high degree of availability of the power station. Although the nature and extent of the inspections have, in a large measure, evolved from several decades of operating experiences of many power stations, no formalized codes for conducting these inspections had been developed in the United States. Rather, a variety of rules have been imposed by the enforcement authorities having jurisdiction at the power station sites. Industry codes focused principally upon the development of recommended rules for the care of power boilers (1)‡, with relatively limited attention being given to other pressure-retaining components such as piping, pumps, and valves.

The design arrangements of pressure-retaining components in fossil-fuelled power stations have always considered the need for ready access to all components to facilitate inspections, maintenance of equipment, repairs, and replacement of components as required. However, with the development of nuclear power plant designs, system designers initially believed that periodic inspections of the reactor coolant pressure boundary would be impracticable due to the radioactivity of the system. Consequently, very limited attention was given to the needs for in-service inspection in early nuclear power plant designs, so systems were not provided with adequate access for inspection of many components.

The system designers assumed that in-service inspections would be unnecessary, provided the nuclear power system components were designed and constructed to higher quality standards than those applied to fossil-fuelled power stations. The development and publication of the first edition (1963) of the A.S.M.E. Boiler and Pressure Vessel Code, Section III, 'Nuclear Vessels' (2) was a first step in response to this need for high-quality standards of vessel components in nuclear power stations.

DEVELOPMENT OF AN IN-SERVICE INSPECTION CODE

As the number of nuclear power stations in service increased, operationally induced defects of components requiring repairs also increased. The U.S. Atomic Energy Commission (A.E.C.) recognized as early as 1966 that the enhanced quality standards applied in the design and construction of components of the reactor coolant pressure boundary did not justify the omission of a planned programme of periodic in-service inspections. Therefore, the regulatory division of U.S. A.E.C. began to develop criteria for the in-service inspection of nuclear reactor

The MS. of this paper was received at the Institution on 11th November 1971 and accepted for publication on 16th December 1971. 32

* *U.S. A.E.C. Advisory Committee on Reactor Safeguards, Battelle Memorial Institute, Pacific Northwest Laboratory.*

† *U.S. Atomic Energy Commission, Washington, D.C., U.S.A.*

‡ *References are given in Appendix 26.1.*

coolant systems; the U.S. A.E.C. also encouraged nuclear industry code-writing organizations to further up-grade the level of quality standards for all pressure-retaining components of nuclear power stations.

A comparable effort on development of an in-service inspection code by the U.S. nuclear industry began about the same time, eventually leading in late 1967 to a joint A.E.C.–industry co-operative code development programme under the auspices of the American National Standards Institute (A.N.S.I.) N-45 Committee, and with the sponsorship of the A.S.M.E. An initial 'Draft code for in-service inspection of nuclear reactor coolant systems' was published in October 1968, followed by the publication of the 1970 edition of the A.S.M.E. Boiler and Pressure Vessel Code, Section XI, 'In-service inspection of nuclear reactor coolant systems' (**3**). In addition, the A.S.M.E. appointed a standing committee, the A.S.M.E. Subgroup on In-service Inspection, responsible to the A.S.M.E. Subcommittee on Nuclear Power (Section III Code). Participation by members of the U.S. A.E.C. Advisory Committee on Reactor Safeguards and the Division of Reactor Standards on these committees should assure the continued joint co-operation with industry in further developments and improvements of the rules for in-service inspection.

A.E.C. ADOPTION OF IN-SERVICE INSPECTION CODE

The U.S. A.E.C. formally accepted several A.S.M.E. codes, including Section XI, in its recently published (1971) 'Codes and standards for nuclear power plants' (**4**), which includes adoption of the rules of the in-service inspection code as mandatory requirements to be met by applicants for licences to build and operate nuclear power stations in the United States. The rules of the A.S.M.E. Section XI Code also fulfil the requirements of A.E.C. General Design Criterion 32 (**5**) with respect to periodic inspection and testing of important areas and features of components which are part of the reactor coolant pressure boundary.

LOSS-OF-COOLANT ACCIDENT 'ENVELOPE'

In the early stages of development of the rules for in-service inspection, the regulatory division of the U.S. A.E.C. was faced with the major task of identifying those pressure-retaining components of light-water cooled and moderated nuclear power plant systems [i.e. pressurized water (P.W.R.) and boiling water reactor (B.W.R.) plants] most important to safety. Components whose malfunction or structural failure could potentially impair the safe continued operation of the nuclear power system merited continued surveillance throughout the service lifetime of the plant. In-service inspection should determine the prevailing structural integrity of the components so that necessary corrective measures could be taken on a timely basis.

Although the structural integrity of reactor primary coolant systems of P.W.R. and B.W.R. plants was generally accepted as essential to the operation of power plants without undue risk to the health and safety of the public, criteria specifically defining the boundary had not yet been formulated at that time.

REACTOR COOLANT PRESSURE BOUNDARY

A comparative study was undertaken to delineate the reactor coolant pressure boundary of typical P.W.R. and B.W.R. plants. The system boundary criteria which evolved from this study identified for the first time the loss of reactor coolant accident 'envelope' of the system. More specifically, the reactor coolant pressure boundary envelope included all those piping runs interconnecting components of the system whose postulated failure could result in loss-of-coolant accidents which would require the isolation of all systems penetrating the primary reactor containment and the operation of engineered safety feature systems to mitigate the consequences of such postulated failures.

The boundary definition, as currently specified in paragraph IS-120 of the A.S.M.E. Section XI Code (**3**), identifies all components of the system (e.g. pressure vessels, piping, pumps, and valves) which must be designed to the highest quality standards (**6**), and the number of such components subject to in-service inspection under the rules of the articles of the A.S.M.E. Section XI Code. Figs 26.1 to 26.4 of typical P.W.R. and B.W.R. plants prepared by the U.S. A.E.C., which graphically delineate the reactor system pressure boundary as well as associated systems, were included in the code for the purpose of general guidance. In effect, the boundary encompasses the reactor coolant system, and portions of systems connected thereto, located within the confines of the primary reactor containment structure up to and including the outermost isolation valves in system piping penetrating the containment.

Identification of the components (and their overall system arrangement) within the reactor coolant pressure boundary provided the initial bases for the development of appropriate in-service inspections. The primary problem associated with the inspection of reactor coolant pressure systems was the need to examine components in radiation fields where access by personnel would not always be practical. This difficulty, totally unlike conditions in fossil-fuelled power stations, posed the greatest challenge in the development of in-service inspection rules.

INSPECTION ACCESSIBILITY

During the initial efforts to develop a meaningful programme of in-service inspection, both the U.S. A.E.C. staff members and the A.N.S.I. N-45 Committee recognized the restrictive space allowances and encumbrances, such as the structural and concrete members surrounding components of nuclear power stations, which would seriously interfere with any contemplated inspection, and, in some areas, preclude any examination other than with special mechanized and remotely operated equipment developed for such purposes.

To remedy this situation, the designers of nuclear power stations had to be made aware of the need to change the prevailing design philosophy from one of 'restricted accessibility' to one of adequate 'in-service inspection accessiblity'. This intent is now clearly reflected in the general provisions for access requirements specified under Article IS-140 of the A.S.M.E. Section XI Code. These accessibility requirements, as applied to recently designed power plants, have substantially influenced the current design of components, the component supporting arrangements in the reactor coolant systems, and the spatial

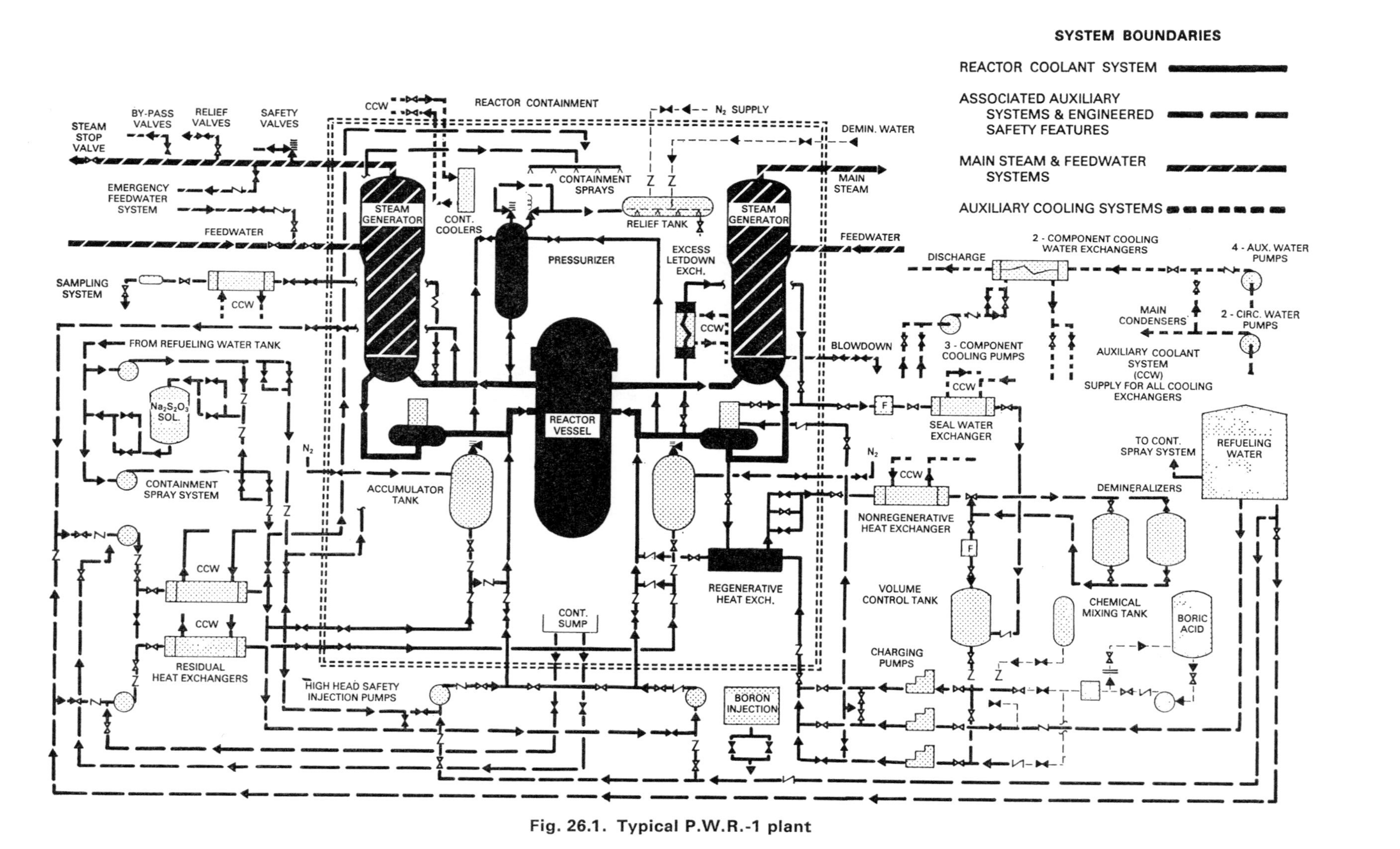

Fig. 26.1. Typical P.W.R.-1 plant

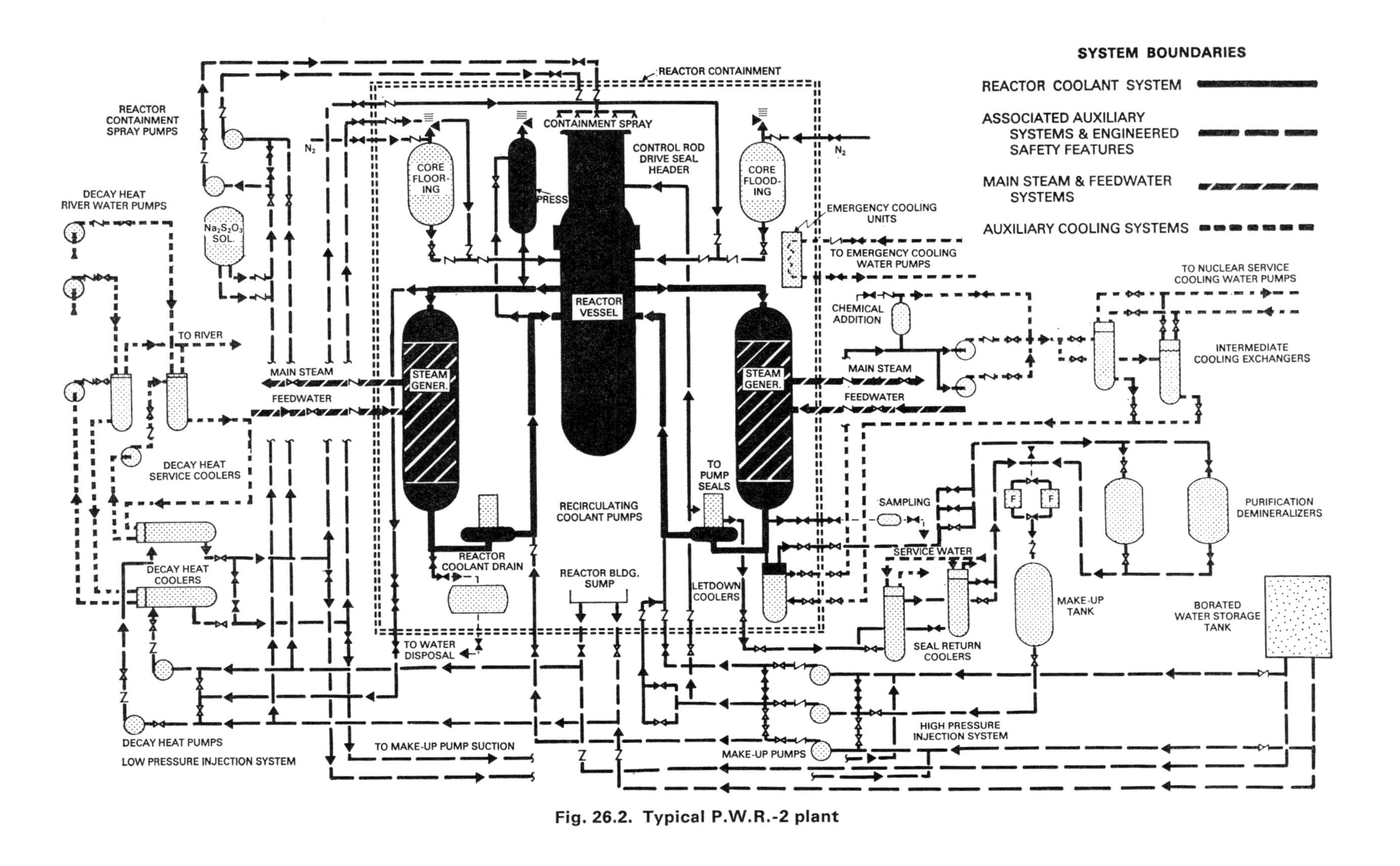

Fig. 26.2. Typical P.W.R.-2 plant

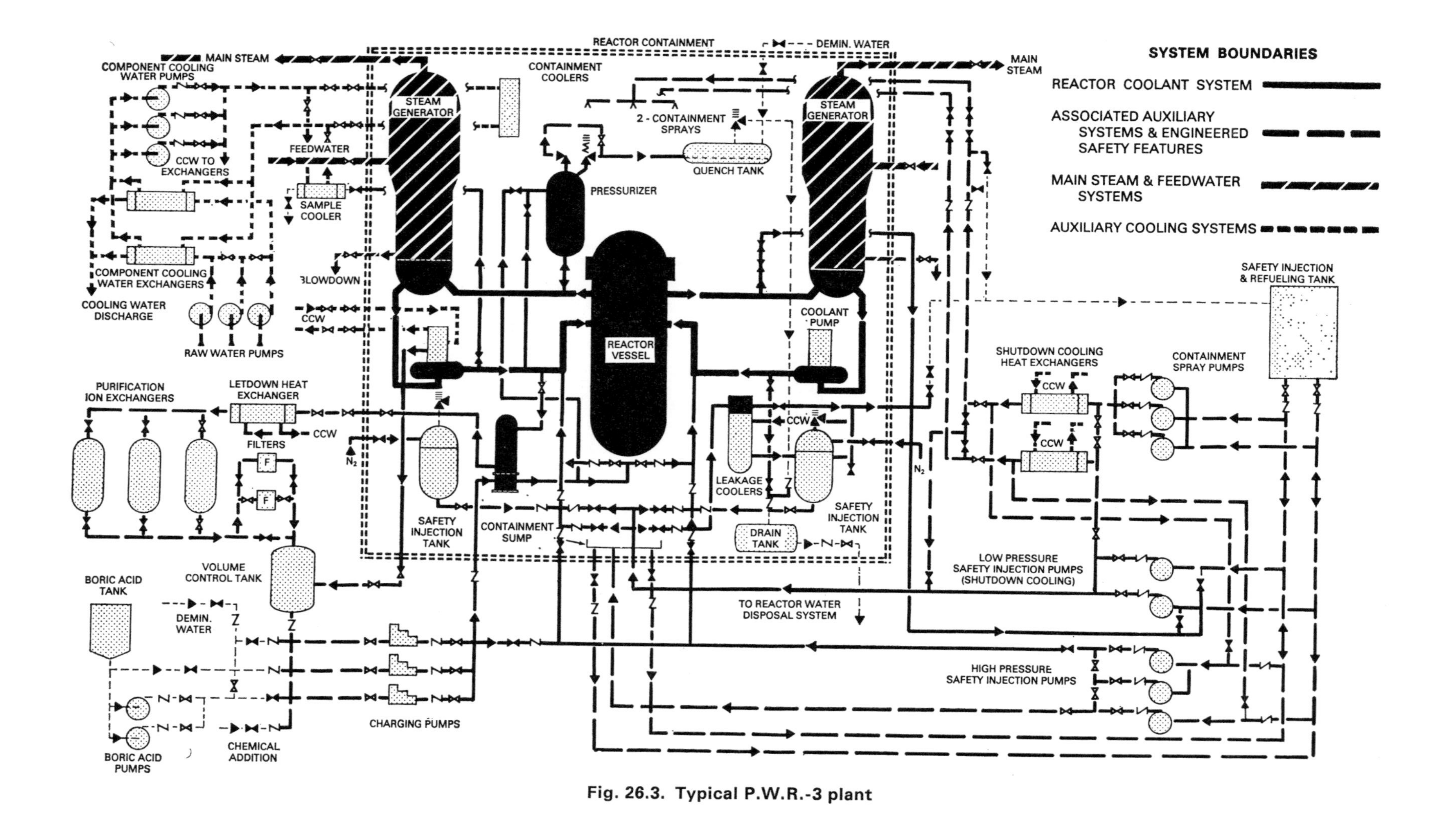

Fig. 26.3. Typical P.W.R.-3 plant

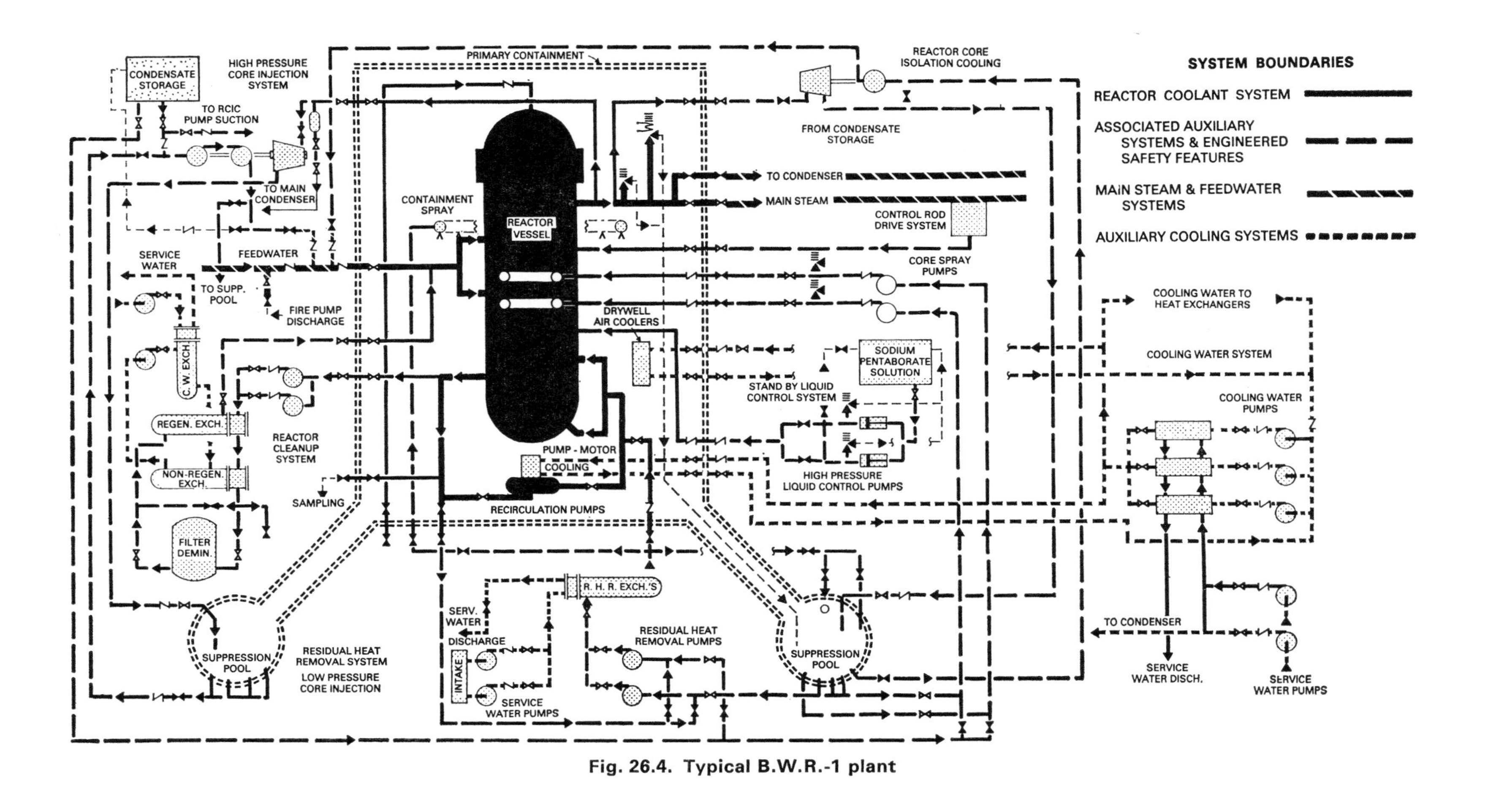

Fig. 26.4. Typical B.W.R.-1 plant

allowances included for the conduct of examination. Many practical measures and design considerations have been taken to facilitate insulation removal, to simplify the introduction and erection of examining equipment, and to enable the performance of in-service examinations in a straightforward and expeditious manner.

Surprisingly, such design changes and measures made to conform with the requirements for adequate inspection accessibility, as specified in the code, have neither significantly affected the system layouts within the reactor containment structures nor resulted in an increase in the size of the containment structures. Instead, designers have emphasized the development and application of special, remotely operable examining equipment which would permit inspections to be performed in limited spaces, rather than the inclusion of more liberal spatial allowances, and shielding protection (where necessary) to facilitate more direct examinations. Whether such an approach is ultimately acceptable will depend on the experience gained with the current generation of examining equipment designed and developed to permit inspections in limited spaces.

REPAIR AND REPLACEMENT OF COMPONENTS

Despite the admonition inherent in A.S.M.E. Section XI Code's intentional reference to the need to consider accessibility to perform necessary operations associated with repair or replacement in the event that structural defects or indications are revealed by in-service examinations, power plant designers have not yet fully incorporated such considerations into system arrangements. Admittedly, the designer is beset with the difficulties associated with anticipating conditions of structural degradation which may never occur in service. Nevertheless, the question still remains whether the nuclear power plant owner's interest is adequately protected in the event that unacceptable defects are discovered during an in-service inspection. If adequate working space is not available to permit repairs or replacements, the difficulties encountered may significantly affect the nuclear power plant downtime. In the event that acceptable repairs cannot be performed, restoration of service may not be readily justified. In all cases, the plant owner is faced with satisfying relevant safety criteria of the U.S. A.E.C.

DESIGN FOR 'REPAIRABILITY'

The associated risks of inordinate delays in effecting repairs, or justification of continuation of service with defects in components that can only be removed with great difficulty, should be considered seriously by the owner as well as by the plant designer. A more prudent approach might be a broadening of the design philosophy to include 'repairability' in addition to 'inspection accessibility', at least for those components where service experience in operational power stations has demonstrated that they are more likely to suffer structural degradation. This approach might encourage the system and components designers to improve their designs to minimize the likelihood of structural failures in service rather than to redesign plants with additional space for the purposes of 'repairability' considerations. The economics and risks associated with these alternative design approaches are factors which the power plant owner must consider in achieving the difficult balance between safe operation and plant availability.

DEVELOPMENT OF AN INSPECTION BASIS

In the selection of appropriate areas of components of the reactor coolant pressure boundary which should be periodically examined, it became obvious that there were practical inspection considerations which would result in a departure from the inspection practices normally applied to components in fossil-fuelled power stations.

THE NEED FOR REMOTELY OPERABLE EQUIPMENT

The principal consideration was a recognition that examinations must be conducted on pressure-retaining components whose materials of construction would become radioactive in service (e.g. the reactor vessel), while other components of the system would develop radiation fields as a consequence of accumulations of radioactive products circulating within the reactor coolant system (e.g. P.W.R. steam generators). In the case of reactor vessels, the most direct and practical approach chosen to protect personnel from radiation exposure involved acceptance of remote means of examination. Such means would require the development of examination equipment specifically engineered for remote operability and adapted to use techniques which would yield meaningful inspections.

A significant requirement was the complete base-line inspection prior to reactor start-up, using the same non-destructive testing techniques planned for in-service inspection. Both visual and ultrasonic examination techniques could be used in performing examinations with remotely operable equipment similar in design to devices and mechanisms developed for inspection purposes in other than the nuclear power industry. However, it was recognized that special designs would be needed to examine some nuclear power plant components. In the United States, industry has successfully met this challenge not only by developing examination tools and equipment for in-service inspection of nuclear power plant components, but also by offering the power stations the services of qualified operators to conduct the examination at the site.

ACCEPTABLE EXAMINATION TECHNIQUES

Although the A.S.M.E. Section XI Code rules identify three acceptable methods of examination (namely, visual, surface, and volumetric), alternative methods, which may develop as the state of examination technology advances, are acceptable, provided their application leads to equivalent or superior inspection results. Several promising new examination techniques, such as acoustic emission and acoustic holography, are currently being developed in the United States, and may complement the presently used techniques.

REDUCING RADIATION FIELDS

With respect to system components other than the reactor vessel, examinations would be substantially simplified if appropriate cleaning and decontamination processes were available to reduce or remove the radioactive products built up within the components. However, acceptable cleaning procedures are not fully developed for reactor coolant pressure systems constructed primarily of austenitic stainless steels and ferritic steels. In addition, such cleaning processes would need to be investigated to assure that the cleaning compounds used would not become contaminants in the reactor coolant system. For these reasons the A.S.M.E. Section XI Code includes a precautionary note pertaining to the use of decontamination processes prior to the performance of the in-service examination. To date, those nuclear power stations in the United States which have completed in-service inspection have not employed overall system cleaning processes but, instead, have adopted special techniques to flush and clean local areas in high radiation fields or have added temporary shielding to permit the performance of the examination with a minimum of exposure to personnel. Nuclear station operators have also recognized that surface finish of materials, the absence of build-up pockets and design crevices, and the avoidance of stagnant areas in steam flows within the components are significant factors to be considered in reducing radiation fields and examination time.

INSPECTION BY REPRESENTATIVE SAMPLING

Aside from the problems identified with the examination of components in radiation fields, an equally important consideration was the limited permissible access of personnel to the primary reactor containment building during normal reactor operation. Consequently, examinations were virtually limited to plant shutdown periods. Examinations of components in fossil-fuelled power stations suffer no such limitation in that there is daily access to system components, providing a timely detection of any evidence of structural distress.

These considerations led initially to the suggestion of a 100 per cent examination of the most critical areas of all components comprising the reactor coolant pressure boundary during each scheduled shutdown period. Because the time required to complete such examinations proved to be excessive and prohibitively costly in downtime, a more practical approach was taken, based on the frequency of anticipated refuelling outages and an inspection programme which relied on a representative sampling of components to determine the prevailing conditions of structural integrity.

INSPECTION INTERVALS

The decision to require some in-service inspections during each second (or third) refuelling outage (i.e. approximately a 36-month period, based on refuelling outages of 12–18 months) reflected a judgement based on the relatively limited service experience of operating nuclear power stations. A three-year interval between inspections was considered an acceptable frequency to detect evidence of any deteriorative process which could affect the structural integrity of components.

The degree of examination at three-year intervals of the reactor vessel was limited because of the more difficult problems associated with radiation exposures, limited accessibility, the use of special examination equipment and, in some cases, the need to remove the entire fuel inventory and the reactor core structure from the vessel in order to facilitate examinations.

In recognition of the high quality standards applied in the construction of the vessel, the improved design techniques employed which consider metal fatigue damage as a consequence of operational transients, and the more stringent quality assurance requirement imposed during the fabrication of the vessel, an extension of the inspection interval from three years to 10 years was considered justifiable for the more difficult examinations. An inspection interval of 10 years was thereby selected as the base period, within which time all required examinations would have been completed. These examinations would be repeated during each successive 10-year service period over the design lifetime of 40 years generally adopted for nuclear power stations.

EXAMINATION PERIODS

Inherent in the acceptance of the 10-year-inspection interval was the realization that the complete base-line inspection required prior to start-up should detect flaws developed during construction, plus furnishing a benchmark for future inspections. The A.S.M.E. Section XI Code requires at least three inspections during each 10-year service interval. The purpose of this inspection programme is to assure periodic examination of components at approximately uniform intervals of $3\frac{1}{3}$ years over each 10-year service period (inspection interval). The code percentage of any examination required for the 10-year inspection interval which may be credited to each nominal interval of $3\frac{1}{3}$ years is shown in Table 26.1.

ORDER OF EXAMINATION

A subject of considerable study was the establishment of the preferred order of examinations (or re-examinations) over successive inspection intervals (i.e. 2nd 10 years; 3rd 10 years; and 4th 10 years). The question which required resolution was whether the same areas of a component examined (such as vessel seam welds) during the first inspection interval should be repetitively examined during successive inspection intervals to monitor the changes which might occur in service over a 40-year period, or whether additional areas of the component (not previously examined) should be examined during successive inspection intervals to obtain a better sampling of the structural condition of the components over the service lifetime. A significant conclusion of the study was that the latter statistical sampling approach to inspection was preferred, since it increased the likelihood of detecting the development of flaws in welds in a component subjected to the same service conditions compared to the case where the same areas of a component received more examinations over the service lifetime. This conclusion provided the basis for establishing the code requirements on the order of examination for the various categories.

In those cases where the extent of examination during an inspection interval for a given category was 100 per

Table 26.1. Code percentage

Fraction of inspection interval	Elapsed time	Percentage of examinations credited
1st inspection ($\frac{1}{3}$) .	$3\frac{1}{3}$ years	25 min. to $33\frac{1}{3}$ max.
2nd inspection ($\frac{2}{3}$) .	$6\frac{2}{3}$ years	50 min. to $66\frac{2}{3}$ max.
3rd inspection ($\frac{3}{3}$) .	10 years	100

cent, an identical examination sequence was adopted during each successive inspection interval. This approach ensured that such components would be re-examined completely over each inspection interval, and that the same part of the component would be re-examined during the same time fraction of the 10-year period.

BASIS FOR SELECTION OF EXAMINATION AREAS

The selection of those areas of the reactor coolant pressure boundary whose examination would provide representative assessment of the structural integrity existing at the time of inspection required consideration of design and service conditions as well as the operating history of nuclear and non-nuclear power stations. These considerations led to the examination categories specified in the A.S.M.E. Section XI Code.

Particular emphasis was given to the examination of the pressure-retaining welds in each component (i.e. vessels, piping, pumps, and valves). Service experiences confirm that weld joints, and the weld heat-affected zones in base metal, are the most likely areas for flaws to occur. In addition, the weld joints, due to their design, often are structural discontinuities in the component. Such stress raisers may act as points of initiation of cracks that often propagate into the base metal. The weld examination areas specified in the Section XI Code include the base metal for one plate thickness on each side of the weld joint to ensure examination of the weld, heat-affected zone, and adjacent base metal. About 50 per cent of the examination categories listed in the code pertain to such pressure-retaining welds.

The selection of the specific weld areas to be examined in each component category was based on such factors as:

(1) Environmental conditions, such as irradiation embrittlement of the reactor vessel belt-line.

(2) Operational transients, such as system start-ups and shutdowns, which induce metal fatigue as a result of cyclic strains.

(3) Component design configurations identified with higher stress fields, such as vessel nozzles, and weld-joint structural discontinuities between piping, pumps, and valves.

(4) Material properties of dissimilar metal joints, such as weld joints between austenitic stainless steel and ferritic steels that may be subject to additional thermal strain in service.

EXTENT OF WELD EXAMINATIONS

Some idea of the differences in level of examination of weld areas during each inspection interval and cumulatively completed over a 40-year design service lifetime is given in Table 26.2.

Table 26.2. Extent of weld examinations

Examination category	10-year inspection interval, per cent	40-year service lifetime,* per cent
A Welds in reactor vessel belt-line region		
(*a*) Circumferential welds	5	20
(*b*) Longitudinal welds	10	40
When neutron-fluence exceeds 10^{19} nvt		
(*a*) Circumferential welds	50	> 100
(*b*) Longitudinal welds	50	> 100
B Welds in vessels		
(*a*) Circumferential welds	5	20
(*b*) Longitudinal welds	10	40
C Welds—vessel-to-flange, head-to-flange	100	400
D Nozzle welds in vessels	100	400
E-1 Vessel penetration welds	25	100
F Dissimilar metal welds	25	100
J-1 Welds in piping	25	100
L-1 Welds in pump casings (One pump in each group performing same system function)	100	400
M-1 Welds in valve bodies (One valve in each group performing same system function)	100	400

* The percentages listed are based on the assumption of completion of the fourth 10-year inspection. In the event, the nuclear power station is retired from service following 40 years of operation, the extent of examination is reduced for those components whose examination is normally performed at or near the end of an inspection interval.

EXAMINATION CATEGORIES (OTHER THAN WELDS)

Interior weld-clad surfaces in vessels were included as a separate examination category (I-1 and I-2) to serve as a monitor of conditions such as micro-fissuring which have occurred in some earlier reactor vessels. Weld cladding in operating reactors in the United States has generally been free from problems. However, because of the limited experience in systems where weld-cladding is applied by different deposition methods, it is believed that these examinations will serve a meaningful purpose.

The examination categories L-2 and M-2 were added to acquire a visual examination of the internal surfaces or cast components at locations where flaws are likely to develop in service. These examinations are to be performed in conjunction with the scheduled disassembly of components to verify the structural integrity of internal parts (e.g. pump impeller, valve discs, etc.).

Pressure-retaining bolting in vessel flange connections, pump casing flanges, and valve body bonnet flanges are covered in examination categories G-1 and G-2; there has been a history of bolting failures in operating power stations.

Support members and structures which serve as the primary restraints and supports for components of the reactor coolant pressure boundary are vital structural elements whose failure could jeopardize the safe operation of the system and the structural integrity of the components. The structural elements included in examination categories H, K-1, and K-2 encompass:

(1) Vessel supports and attachments to foundations which are designed to accommodate operating loads and seismic-induced motions in the event of an earthquake loading.

(2) Piping supports, snubbers, and shock absorbers designed to carry operating loads, seismic loads, and

pipe whip and jet forces in the event of a postulated severance of piping.

(3) Rigid anchors for piping, pumps, and valves to prevent the transmission of reaction loads into any components which could impair the operability of valves and pumps.

The special examination category N covers the examination of the interior surfaces and internal components of the reactor vessel; it is considered one of the most critical examination requirements in the A.S.M.E. Section XI Code. Among the considerations contributing to the development of this examination category were the reported experiences and difficulties encountered in operating facilities. These interior examination areas should assure:

(1) Inspection of all internal support attachments welded to the reactor vessel whose failure could result in reactor core disarrangement.

(2) Discovery of any loose parts which might have accumulated at the bottom of the reactor vessel during service.

(3) Detection of undue wear as a result of flow-induced vibrations of components of the reactor core structure.

(4) Verification of the overall structural integrity of the core structure, including supplementary internal components such as moisture separators, material surveillance capsules, instrumentation, and reactor control rod assembly guides.

In recognition of the obvious difficulties in the examination of the interior of the reactor vessel and of the internals, the code rules permit performance of examinations at or near the end of each 10-year inspection interval. These examination requirements have prompted the U.S. designers of reactor vessels, particularly for P.W.Rs, to provide for the complete removal of reactor internals. Current designs of the reactor internals of B.W.Rs do not yet incorporate features which permit removal of core structures; therefore, highly specialized mechanized non-destructive equipment has been developed to perform the required inspections.

LEAK DETECTION SYSTEMS

Although U.S. nuclear power stations include systems for the detection of unidentifiable leakage which might develop in the pressure-retaining boundary of the system during normal reactor operation, such leak detection systems have not yet been developed to the point where unidentifiable small leakage sources can be readily distinguished from the numerous identifiable leakage sources (e.g. valve stems, component seals, leak-off connections, etc.) which normally prevail in an operating plant.

SYSTEM HYDROSTATIC TESTS

A system hydrostatic test during a shutdown period permits a planned examination for evidence of any leakages that might originate from through-wall cracks of the pressure boundary. Such a pressure test enhances the possibility of timely discovery of small through-wall flaws which, because of leak size, might not be readily detected by the installed leak-detection systems. Leakage should be determinable without removal of insulation by dripping from the insulation joints. The code rules specify that the examination be conducted after the system has been at test pressure for at least four hours to ensure that water from leaks will reach the insulation joints.

The selection of an appropriate test pressure for the examination was based on the maximum test pressure compatible with the practical limitations imposed by system design and operational considerations; the test temperature used must satisfy applicable material fracture toughness criteria established for the ferritic materials of the system components. For example, as the years of service increase, the system test will be conducted at higher temperatures, dictated by the fracture toughness test results obtained from the reactor material surveillance programme.

Since any system test pressure selected could not duplicate the membrane strain levels in pressure-retaining components during reactor operation (because of the absence of thermal gradients, and the higher yield strengths and modified modulus of elasticity of each material at test temperature as compared to operating temperature), a compromise was necessary in setting an appropriate test pressure. The system test pressure (P_t) selected for in-service inspection is derived from the system nominal operating pressure (P_o) at rated reactor power modified by the minimum ratio of the respective yield strengths of any material used in the construction of the system at test and operating temperatures (S_{yt}/S_{yo}) to compensate for the temperature-dependent difference in the material's physical properties. This approach recognized the inherent limitations in setting test pressures in systems constructed of materials having a wide range of physical properties. As an example, the difference between the maximum and minimum ratio of (S_{yt}/S_{yo}) among all materials used in a typical reactor coolant system may be as much as 30 per cent.

The in-service system hydrostatic test required by the A.S.M.E. Section XI Code reflects the acceptance of the pressure test as, primarily, a means to enhance leakage detection during the examination of components under pressure, rather than solely as a measure to determine the structural integrity of the components. This test differs from the shop and pre-operational system hydrostatic tests required by the rules of the A.S.M.E. Section III Code (under which rules the components were fabricated and tested), in that substantially longer test pressures are required.

THE BROADER SAFETY ASPECTS OF IN-SERVICE INSPECTION

The U.S. A.E.C. views the in-service inspection of nuclear power stations in a much broader safety sense than a periodic assessment of the structural integrity of components to assure that the system may continue to operate safely. As part of its regulatory responsibilities, with respect to operating nuclear power stations, the U.S. A.E.C. is equally interested in a continuing assessment of the adequacy and maintenance of safety margins incorporated in the design and operation of the individual nuclear power stations. The reporting of defects detected by an in-service inspection of components may influence the operation and in-service inspection of other similarly designed power plants as well as dictate the need to modify the design of future plants.

Periodic in-service inspections that uncover unexpected flaws or structural degradation may serve to identify design and material deficiencies which, in turn, may dictate the need to revise design rules in construction codes. For example, if undue fatigue damage was discovered during in-service inspection of a class of plants, design safety margins might have to be increased, or design details improved in future plants.

FUTURE EXTENSIONS OF IN-SERVICE INSPECTIONS

The 1971 Section XI Code does not cover safety-related auxiliary systems such as parts of the emergency core cooling systems and main steam and feedwater systems. In the future, the code may require inspection of all components whose failure may affect the safety of the nuclear power station or result in a significant release of radioactivity.

Another change being considered in the code is the inclusion of sections on in-service testing of pumps and of valves covering the functional and performance testing of pumps and valves. Such code rules should prove valuable in that the extensive testing required could detect the onset of degradation earlier than would be true with routine operation.

SUMMARY

The development of an in-service inspection code for components of nuclear power stations represents a major step toward the formulation of a set of inspection rules by which the nuclear power station operators and regulatory authorities may periodically gain confidence in the conditions of structural integrity of the system components for continued operation.

The experience that will be gained by in-service inspection should guide the plant designers toward the improvements in the design, construction, and quality standards incorporated into new nuclear power stations as well as provide assurance to regulatory authorities that the safety margins applied in the design of the plant components are adequate and can be maintained throughout the service lifetime of the nuclear power plant.

APPENDIX 26.1

REFERENCES

(1) A.S.M.E. Boiler and Pressure Vessel Code, 'Section VII: Recommended rules for care of power boilers' (Am. Soc. Mech. Engrs, New York).

(2) A.S.M.E. Boiler and Pressure Vessel Code, 'Section III: Nuclear vessels' (Am. Soc. Mech. Engrs, New York).

(3) A.S.M.E. Boiler and Pressure Vessel Code, 'Section XI: In-service inspection of nuclear reactor coolant systems' (Am. Soc. Mech. Engrs, New York).

(4) A.E.C. Codes and Standards for Nuclear Power Plants, 'Title 10: Code of Federal regulations', Pt 50, Sect. 50.55a (U.S. Atomic Energy Commission).

(5) A.E.C. General Design Criteria for Nuclear Power Plants, 'Title 10: Code of Federal regulations', Pt 50, Appendix A (U.S. Atomic Energy Commission).

(6) A.S.M.E. Boiler and Pressure Vessel Code, 'Nuclear power plant components', 1971 edn (Am. Soc. Mech. Engrs, New York).

C27/72 TECHNIQUE FOR INSPECTION OF LIGHT WATER REACTOR PRESSURE VESSELS

B. WATKINS* H. JACKSON*

Application of the measures advocated in the A.S.M.E. Section XI Code has led to an increased need for non-destructive testing (N.D.T.) development. The methods employed in the main consist of conventional N.D.T. techniques engineered to withstand the particular environment. Thus, the equipment must be capable of operating up to a depth of 100 ft of water in a high radiation background to a high standard of sensitivity and reproducibility. As the daily outage cost for a power reactor is several orders of magnitude greater than the daily cost of inspection, it is essential that the inspection time be kept to a minimum. Additionally, the aim of any inspection is to compare the current results with earlier and base-line data, and for this reason all the records must be immediately available and comparable for assessment. The paper deals with N.D.T. techniques evolved to meet these requirements.

INTRODUCTION

IT IS MANDATORY that conventional pressure vessels are periodically inspected using both visual and other non-destructive testing (N.D.T.) techniques. As a result, the number of unplanned shutdown periods in conventional power stations has been minimized and vessels have been repaired as necessary (1)†. This has resulted in an enviable safety record.

In addition to the conventional engineering considerations, i.e. stress, temperature effects, etc., designers of nuclear plants must take into consideration the effects of neutron irradiation on mechanical and corrosion behaviour. The high radiation background makes the type of inspection carried out on conventional plant impracticable, and this has led to greater attention being paid to the selection of materials and the development of improved fabrication and inspection techniques (2). Additionally, large programmes of work have been undertaken in many countries to apply fracture mechanics techniques to determine the critical defect sizes of engineering concern (3) (4).

Nevertheless, because of the consequence of failure and the high cost of outage time it is becoming mandatory to inspect nuclear circuits at periodic intervals. Due to access difficulties, manual non-destructive techniques often cannot be applied once the vessel has gone into service, and it is difficult to relate the results of an in-service inspection to the 'acceptance' tests usually carried out during manufacture. For these reasons it has become customary to carry out a 'finger-printing' exercise using the same or similar techniques as those used during subsequent in-service inspections, thus facilitating comparison between this and future tests.

Access to the pressure vessel for periodic inspection is usually limited by the radiation level and by the reactor operator's programme. The latter may limit the amount of inspection which can be carried out at any one time, but the intention of current codes of practice for in-service inspection is that the pressure vessel should be completely inspected in any given 10-year cycle (5) (6).

The outage costs on a large nuclear plant are extremely costly, typically £20,000 per day for a 800-MW plant, and for this reason it is necessary that the 'on-line time' taken up by the inspection be kept to a minimum. This needs careful programming, and spare equipment must be available so as not to increase the inspection period. Furthermore, the results of the inspection must be presented in such a form that they can be assessed quickly, so that any remedial action necessary is not delayed.

This paper sets out to indicate those areas in light water reactors which require post-operational inspection, the methods used to determine the standard of inspection, and the techniques used to carry out that inspection. It considers the improvements in design which could be applied to reactors for which periodic inspection was considered at the design stage, and the problems posed by reactors already constructed in which no provision for periodic inspection was made. In many of these instances an initial finger-printing may not have been carried out.

AREAS REQUIRING INSPECTION

The reactor vessel of a typical light water reactor is a cylindrical shell with a hemispherical bottom, and top closure head. The latter is joined to the reactor vessel by mating flanges with large studs and nuts. The cylindrical shell contains inlet and outlet water nozzles. The top and bottom domes contain numerous penetrations for instrumentation and control rods.

The wall thickness varies from 4 in to 10 in but the section at the nozzle–vessel weld can be up to 18 in. Some vessels are made from plate and contain both longitudinal and circumferential weld, but in other cases forged rings are used for the barrel sections of the vessel. The vessel is generally made from quenched and tempered low-alloy steel, and all the inner surfaces are clad with weld-deposited stainless steel.

The MS. of this paper was received at the Institution on 15th November 1971 and accepted for publication on 16th December 1971. 33

* *Advisory and Reactor Inspection Dept, R.E.M.L., Culcheth.*

† *References are given in Appendix 27.1.*

The head and vessel are joined by means of studs located in threaded holes in the lower flange, passing through clearance holes in the head, and held down by means of nuts. Fig. 27.1 shows a typical P.W.R. vessel, and Table 27.1 gives the code requirements and suggested techniques for inspecting critical areas.

Flaws and defects are inherent in all engineering structures at all stages of manufacture and fabrication. Fracture mechanics provides a quantitative assessment so that for any specific combination of material, geometry, and applied stress there is a critical size of stress concentration, e.g. a crack-like defect from which an unstable crack can grow. This approach is therefore used to quantify the permissible defect size. Various fracture mechanics approaches can be used depending on material ductility (**4**), but in the case of the thicker sections and complex geometries it is often assumed that the more brittle plane-strain conditions apply (**7**).

Using this linear elastic fracture mechanics approach an analysis has been done for the critical defect in a typical light water reactor pressure vessel. Thus in the nozzle area and the barrel section the critical defect size

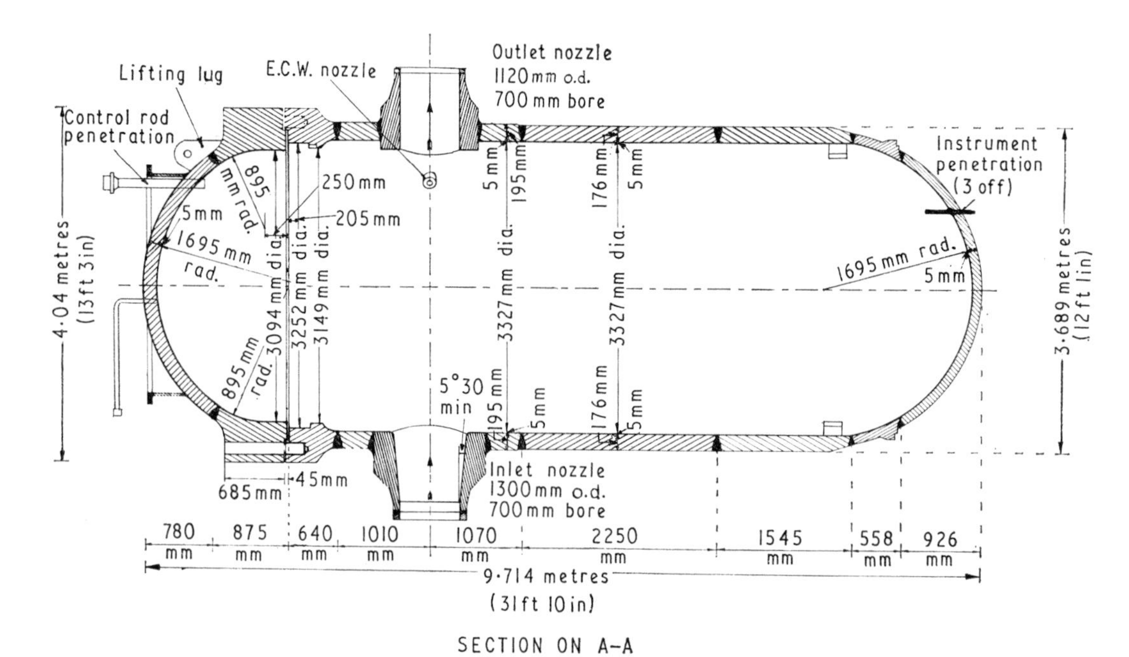

Fig. 27.1. Typical P.W.R. vessel

Table 27.1. Code requirements and suggested techniques

Components and parts to be examined	A.S.M.E. XI method (rec.)	N.D.T. method proposed by R.E.M.L.
Longitudinal and circumferential shell welds in core region	Volumetric	Ultrasonic B-scan, transverse and longitudinal. Visual C.C.T.V.
Vessel-to-flange and head-to-flange circumferential welds	Volumetric	Ultrasonic A- and B-scan
Primary nozzle-to-vessel welds and nozzle-to-vessel inside radiused section	Volumetric	Ultrasonic B-scan. Visual C.C.T.V.
Vessel penetrations	Volumetric	Automatic A-scan or fixed-angle B-scan
Primary nozzle to safe end welds	Visual, surface and volumetric	Ultrasonic B-scan
Closure nuts and studs	Volumetric and visual or surface	Automated ultrasonic and eddy current
Ligaments between threaded stud holes . . .	Volumetric	Automatic A-scan or B-scan
Integrally welded vessel supports	Volumetric	Ultrasonic B-scan
Closure head cladding	Visual and surface or volumetric	Ultrasonic C-scan. Visual C.C.T.V.
Vessel cladding	Visual	Visual C.C.T.V. Ultrasonic C-scan
Interior surfaces and internals and integrally welded internal supports	Visual	Visual C.C.T.V. and coupled introscope

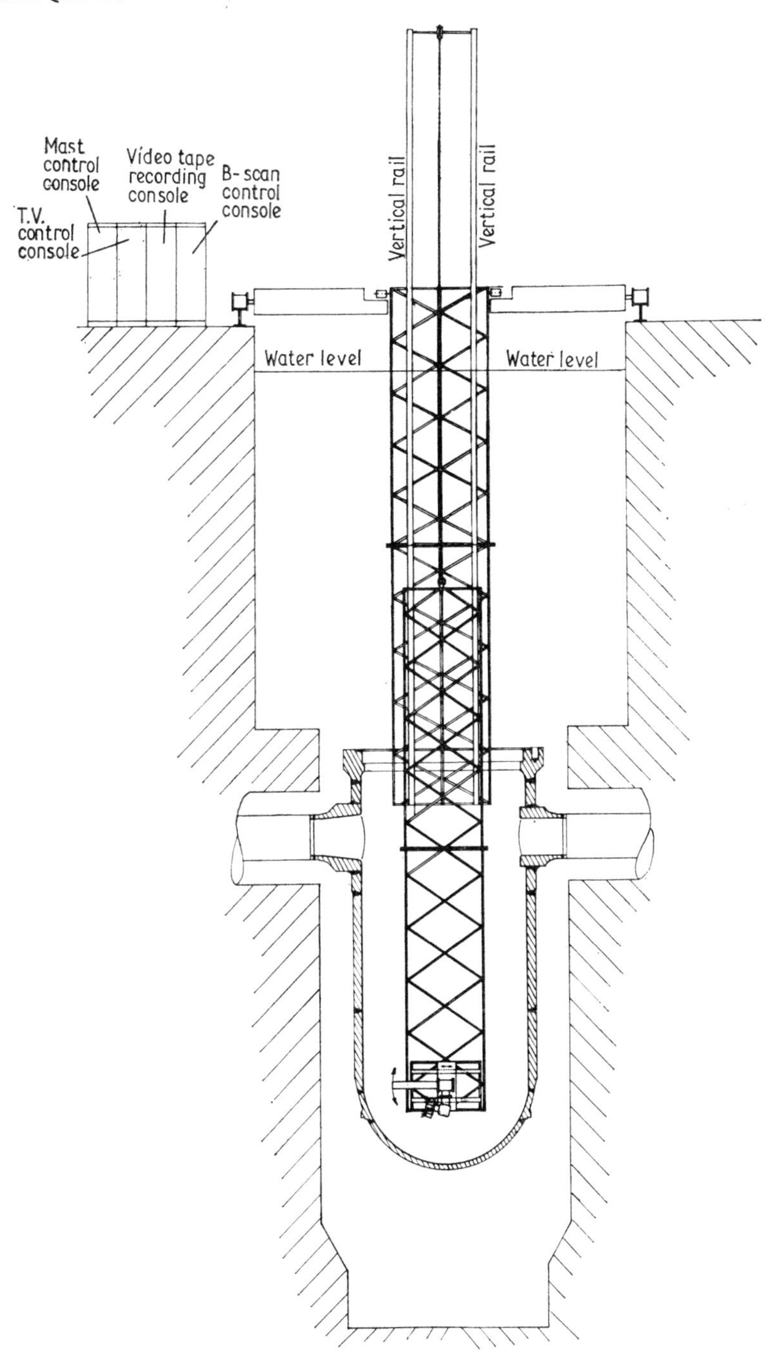

Fig. 27.2. Mast arrangement

is of the order of 2–2½ in, and that this may be as low as 1½ in after irradiation (**8**). This approach allows a judgement to be made of the level of sensitivity required for the inspection of pressure vessels.

It is suggested that the inspection techniques should be capable of detecting linear defects of the order ½ in long by ½ in thick. In order to ensure that defects of this size can be detected it may be necessary to calibrate the technique on test pieces in which the joint geometry is simulated. Furthermore, it is preferable that test pieces contain both natural and artificial defects. In some instances the ultrasonic response from a natural crack is greater than that from an artificial defect of the same size: in other cases the reverse can apply. Thus, in order to get maximum information, the angle of the incident beam needs to be variable over a wide range, so as to optimize defect response.

METHOD OF INSPECTION

Any inspection process must, of necessity, be non-destructive. The range of N.D.T. techniques available at the construction stage becomes severely restrictive once the reactor has been commissioned. Due to the hostile radiation environment, most of the inspection has to be carried out underwater, typically 50 ft deep, but possibly up to 100 ft deep. Under these conditions it is essential that some form of mast or manipulator is used in order to carry out the inspection. This mast must locate the inspection instruments and be capable of reproducing this location within precise limits, typically 4 mm at any repeat inspection. In order to do this a form of indexing device is necessary, and it is preferable that the readout of this device be recorded simultaneously with the inspection records.

A typical mast arrangement is shown in Fig. 27.2; the mast support arrangement can be modified to accommodate vessels of varying diameter. With this particular arrangement the mast can be rotated in a circumferential direction and additionally can be moved in the vertical and X and Y directions. All the inspection equipment is carried on the mast by means of manipulator devices which are capable of independent movement.

Thus, for example, in the case of a nozzle examination the ultrasonic head is rotated circumferentially around the bore of the nozzle, whilst at the same time a general surveillance closed-circuit TV camera will enable the operator to view the operation. Similarly in the case of the main-vessel welds, a closed-circuit TV camera will survey the weld area being examined ultrasonically.

With the arrangement described the results are presented on video tape using a split-image technique. This allows the ultrasonic results, a TV picture of the area being examined, and the indexing system to be presented as one record. All relevant information is thus presented as a unit and simplifies the comparison of the results obtained over a series of inspection intervals.

INSPECTION TECHNIQUES

Surface examination

Many of the proven and accepted conventional techniques for surface inspection cannot be used in post-operational examination. Penetrant inspection, for instance, cannot be considered because it is incapable of underwater application. Similar considerations apply to magnetic-flaw detection and, additionally, the austenitic layer of cladding on the inside surface proves a major disadvantage.

The main method of surface examination is closed-circuit TV. In recent years major advances have been made in the use of this technique in the presence of ionizing radiation; but loss of equipment due to accidental damage or contamination is expensive. After service the cladding surface is covered with a matt black magnetic layer which significantly reduces the reflectivity. If light intensity is increased to light the dark areas, highlights may develop on the edges of weld beads causing local burn-out of the television picture.

Pictures of adequate resolution can be obtained by using good-quality commercial cameras, albeit fitted with irradiation-stabilized lenses, in custom-built water-tight cases. In the event of serious contamination the cases are disposable. Fig. 27.3 shows a typical surveillance camera in a stainless-steel underwater case. The case is mounted on an immersion pan-and-tilt arrangement, thus facilitating viewing of a wide area from one camera position. A series of 100-W quartz–iodine lamps surround the camera, and these can be angled to obtain optimum lighting and can be dimmed to soften the lighting or switched to provide general or accentuated lighting. The lighting, pan-and-tilt, and camera controls are all located at the viewing point on the charge floor.

For detailed work, miniature cameras are available for inspection at close quarters, with a basic size 6 in long × $1\frac{1}{2}$ in diameter. Fig. 27.4 shows a typical camera mounted in a 9-in diameter machined nozzle with its own lighting arrangement and rotating mirror mounted at 45° in front of the lens. The resolution of both types of camera is better than that required by the design codes.

For areas of limited access, custom-built introscopes are available up to 6 m long, completely waterproofed with a maximum outside diameter of 22 mm. The optical components are all made of stabilized glass and have been able to withstand 3 mega-rods without any blackening or deterioration in resolution. This instrument has 100-W quartz–iodine lighting, which in fact can be increased to 200 W if required, the view heads can look forward, backward, or at 90°; all the optical glass in the introscope can be readily changed. Introscopes of this type can give magnifications up to 30, and this coupled with closed-circuit TV is frequently the best technique for examination of drain rings and other areas where chloride build-up can lead to a risk of stress-corrosion cracking.

Deciding whether surface indications are surface marks or cracks can lead to difficulties. Ultrasonic examination of cladding even in the ground state may give a multiplicity of reflections, and even when the cladding is well ground, reflection from the back surface of the vessel is necessary to ensure confident interpretation.

Frequently, cladding cracks which penetrate the base metal cannot be detected by a non-chromatic TV system but can be detected using colour television, as the reddish-brown stain seeping from the crack can be detected against the black, matt background. Positive quantitative information can be obtained using an ultra-high-frequency conductivity device developed at R.E.M.L. A high-frequency current is passed at 90° across the crack through

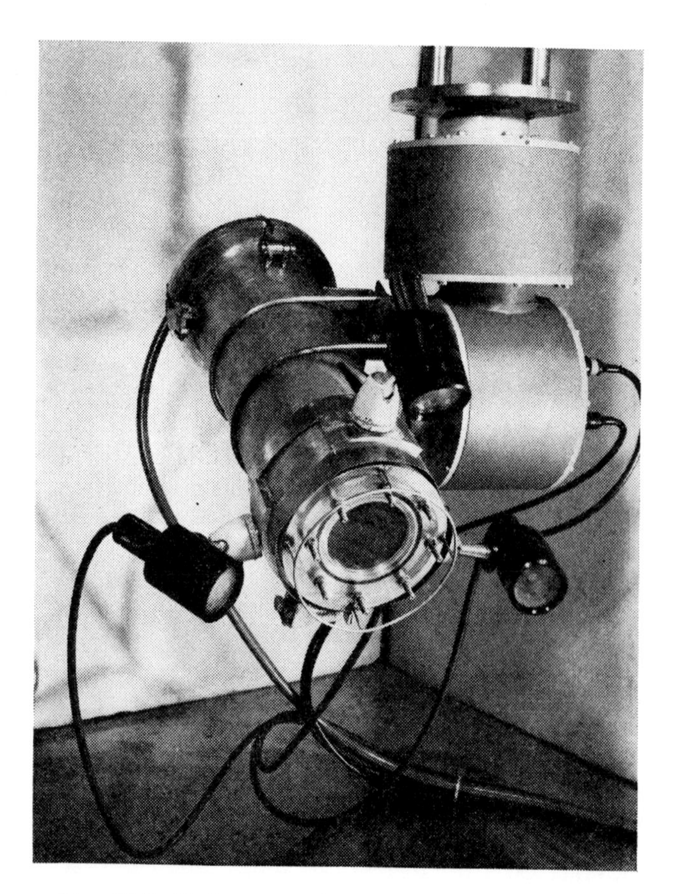

Fig. 27.3. Underwater TV camera and pan and tilt

Fig. 27.4. Small TV camera in nozzle

two contacts and a voltage collected across the crack by two inter-contacts. Due to the high permeability of the material the current travels in the skin of ferritic materials. If no crack is present the instrument records the resistance between the inner probes, but if a crack is present the increased resistance due to the longer path the current takes is indicative of crack depth. It has been found that if a crack is contained within the cladding no increased reading is obtained, but if the crack penetrates the ferritic base it can be measured.

VOLUMETRIC EXAMINATION

The two main methods available to us for volumetric examination of welds are ultrasonic inspection and radiographic inspection. There is a tendency to discount the use of radiography if the subject is itself radioactive; but the head penetrations, which often include dissimilar weld metals, can be accommodated by this method. To overcome fogging from background radiation, the film used is one which has a much greater response to X than to γ radiation. Fogging due to radiation from the surroundings can be minimized by using cassettes backed with either lead or brass shield, whilst other elementary precautions, such as applying the film as late as possible to removing it immediately after the exposure is completed, all contribute to minimize background fogging.

Although radiography is feasible in post-operational inspection it is not practical for the bulk of the welds to be examined, and the most useful volumetric method of examination for in-service inspection is ultrasonic inspection.

ULTRASONIC EXAMINATION

Ultrasonic inspection is not affected unduly by the thickness of the material so the technique is dependent on crack area rather than crack width. By varying the technique the method can be changed so that it becomes more sensitive to defects lying parallel to or inclined to the plate surface.

One major obstacle to the use of ultrasonic techniques on reactor vessels is the weld-deposited cladding on the vessel inside surfaces and head. Depending on the method of deposition and the chemical composition of the cladding and the base material, the acoustic attenuation of the cladding can vary considerably. In the extreme case, e.g. thick, cast stainless steel, materials can be opaque to ultrasound. Cladding finish presents another problem. The beam of ultrasound enters the cladding via a water coupling, and only where the surface is correctly aligned to the beam of sound will a fully sensitive examination be achieved. If the cladding is deposited as a series of weld beads and no weld dressing is done, the useful part of the beam can in fact be a relatively small proportion of the whole. To combat this problem, specifications for the surface finish required on cladding in the weld inspection areas are being written in terms of ultrasonic attenuation (**9**).

There are two very different methods so far developed for the post-operational inspection of reactor welds. One is based on stationary, multiple, fixed-angle probes in tandem; the other relies on a single probe which traverses the weld and can assume any angle in steel between 0, i.e. the compression wave mode, and 70° shear.

Variations of the first technique have been widely advocated for reactor inspection (**10**) (**11**), and these are typified by the Tazelwurm system (**12**). The second technique based on the principle demonstrated by Videoson in the U.K. has been further developed and refined, and

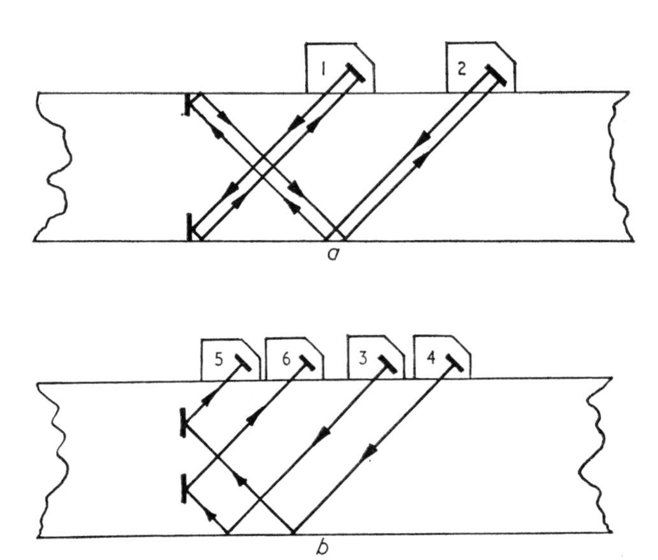

Fig. 27.5. Composite of multiprobe systems

extended into a viable proposition for reactor inspection at R.E.M.L.

The multiprobe system relies on a series of shearwave tandem probes capable of injecting a beam of ultrasound into the material at an angle of 45°. Cracks breaking either surface and substantially parallel to the weld axis will be detected by single probes (see Fig. 27.5*a*), and defects lying in the weld which do not break the surface are detected by pairs of probes (see Fig. 27.5*b*). Probes 1 and 2 act as transceivers and scan the surfaces, whilst probes 3 and 4 are transmitters to be received by 5 and 6. The switching sequence can be made effectively at the rate of 10/s. The inevitable variations due to beam-path length are compensated by adjusting the sensitivity levels of each pair of probes. The coupling efficiency is checked by a crystal in each probe working in the compression wave mode as a monitor. The system is designed to look for cracks with smooth wall surfaces. The smoother the crack, the more likely it is to be detected. The results of the examination are fed on to paper charts and are often difficult to interpret quickly. This difficulty is recognized in the RTD Sonolog 70, which provides the records in a 'plural analogue' or in a facsimile form (**10**).

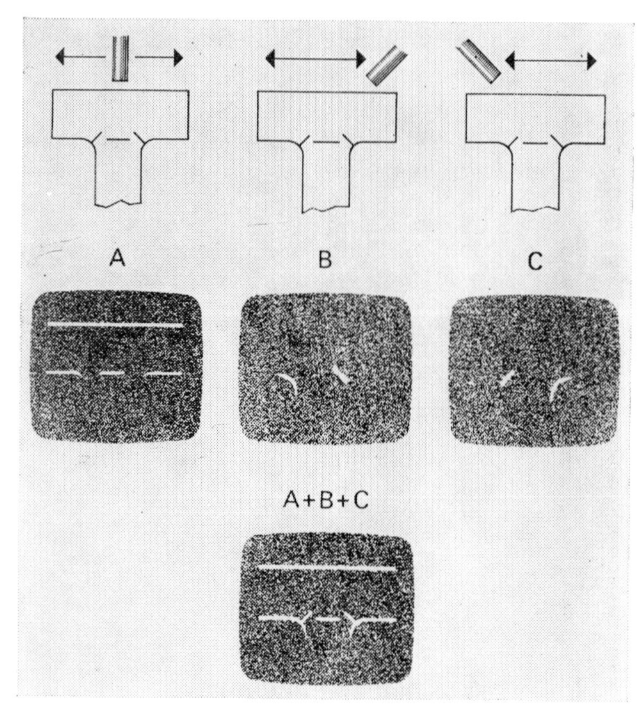

Fig. 27.6. B-scan tee-diagram

B-SCAN TECHNIQUE

The B-scan system uses an immersion probe that moves across the weld so that a beam of ultrasound sweeps the weld and one wall thickness either side of it. The probe can be remotely controlled to move the entry angle of the sound beam from 70° on one side of the weld through the normal position through to 70° on the outer side of the weld. The ultrasonic system is driven by Krautkramer USE 1, and the A-scan presentation is always available.

The data provided by the conventional A-scan are processed together with information fed back to the control unit about the position of the probe with respect to the weld and the angle of the beam in the material under test. All this information is built up to provide a cross-sectional image of the material swept by the sound beam.

Thus on a tee-weld containing cracks examined from the top surface of the tee the picture from a compression wave scan would appear as shown in Fig. 27.6 (A). The picture for a 45° scan from the left side would appear as shown in Fig. 27.6 (B), and the picture for the 45° scan from the other side would appear as shown in Fig. 27.6 (C). The presentation is made on a storage oscilloscope and the scans at any number of angles can be presented together so that the finished picture for the tee-examination would appear as shown in Fig. 27.6 (A+B+C).

As the angle is changed from compression to shear wave mode the instrument automatically changes all the time constants and additionally can be programmed to change receiver attenuation to maintain a constant gain. All the conventional controls on the most sophisticated A-scan equipment are available to the operator using B-scan. A defect can be found either by a defect echo or loss of bottom echo, save that in B-scan presentation loss of bottom echo is denoted by a hole in the backwall image.

In practice the weld examination sequence is predetermined and the weld is scanned at two or three angles, or more if required, automatically and the results recorded on video tape. If required, any defect can be further explored by stopping the automatic scanning sequence and using a manual override to move the probe precisely to any position angle and to carry out what is effectively a manual examination with an unlimited range of probe angles with automatically maintained gain. The defect response can in this way be optimized and the maximum information obtained, thus simplifying interpretation.

It would be expected that the method would suffer from the deficiencies of manual, single-probe systems and be unable to detect crack-like defects normal to the surface but not breaking the surface. In practice, this has not proved to be the case. Natural defects are faceted and once an indication either of a small reflection or hole in the bottom rail has been received, then the ability of the equipment to change to any angle to carry out a detailed exploration has meant that reflections coming back at any angle can be utilized to build up a complete picture of the defect.

The ultrasonic probe is located in a mechanical head which allows both axial and rotation movement of the probe (Fig. 27.7). This box is carried on a manipulator which in turn is located on the inspection mast. The manipulator holds the mechanical B-scan head against the wall of the reactor in register with the weld being examined. The digital readout system on the mast platform indicates the position of the box axially to the reactor and radially round the reactor. In the case of a nozzle examination the indexing system also shows the axial position of the probe inside the nozzle and the position of the probe around the bore of the nozzle.

To inspect the nozzle the B-scan head is placed in contact with the bore of the nozzle and is rotated around the centre-line of the nozzle with the sound beam pointing outward and towards the centre-line of the reactor. The head is progressed so that one compression wave and any number of angle scans are made, and the head then

Fig. 27.7. B–scan box

moved forward by the distance of half the probe diameter. Fig. 27.8 shows the B-scan picture of a nozzle examination on a nozzle containing porosity and a natural weld crack.

For each sequence the B-scan picture is recorded by a television camera, and by means of TV viewing devices the index details are added in one corner. Additionally, either the A-scan response on the optimum angle or a picture from a closed-circuit surveillance camera can also be included. The composite picture is then fed into a video-tape recorder, when it can be replayed for examination as required.

STUD AND NUT EXAMINATION

The principal concern relating to the closure nuts and studs are cracks at the root of the threads. The inspection can be carried out either by means of an ultrasonic examination or by means of a surface examination of the thread roots.

Some reactor studs have a $\frac{1}{2}$–1 in diameter axial hole for a heater element, and in these cases the most effective way of carrying out an ultrasonic examination is by means of a shear-wave probe operating from this hole. The angle of the probe must be carefully calculated so that a fatigue crack at the root of the thread will reflect the beam back into the probe but echoes from the threads are minimized. This method has been shown to give a full-scale deflection from a defect with a side area of 6 mm^2.

Studs which do not contain a centre hole can be examined from the ends, provided the geometry is such that the beam of ultrasound can be focused on all the threads. Again, this method has been shown to be effective.

At R.E.M.L. the preferred method of surface examination of the thread roots is by use of the Amlec flaw detector. This is an eddy current device operating at 30 kHz which is capable of detecting surface cracks in high-permeability materials. A specially profiled probe is moved along the thread and responds well to tight cracks but has little reaction to the thread roots. This device can detect cracks at the roots of threads 2 mm long × 1 mm deep.

In order to facilitate examination of the studs, the R.E.M.L. manipulator rotates the stud, the probes remaining stationary; this avoids entanglement of the probe lead. To provide acoustic coupling an external irrigator is used which pumps demineralized water to the probe face. The ultrasonic and eddy current checks can be made at the same time, the head being so designed that the ultrasonic beam and the eddy current probe are at the same point on the thread during the inspection.

This technique of examining the stud threads from the heater hole can be used to carry out a full ultrasonic examination without removing the stud from the top flange of the vessel. The manipulator will naturally have to be designed for each application, but the technique can offer economic advantages.

The nuts are inspected from the top surface using a compression wave probe, and the Amlec eddy current probe is used to examine the thread roots. The same order of sensitivity can be obtained as in the stud.

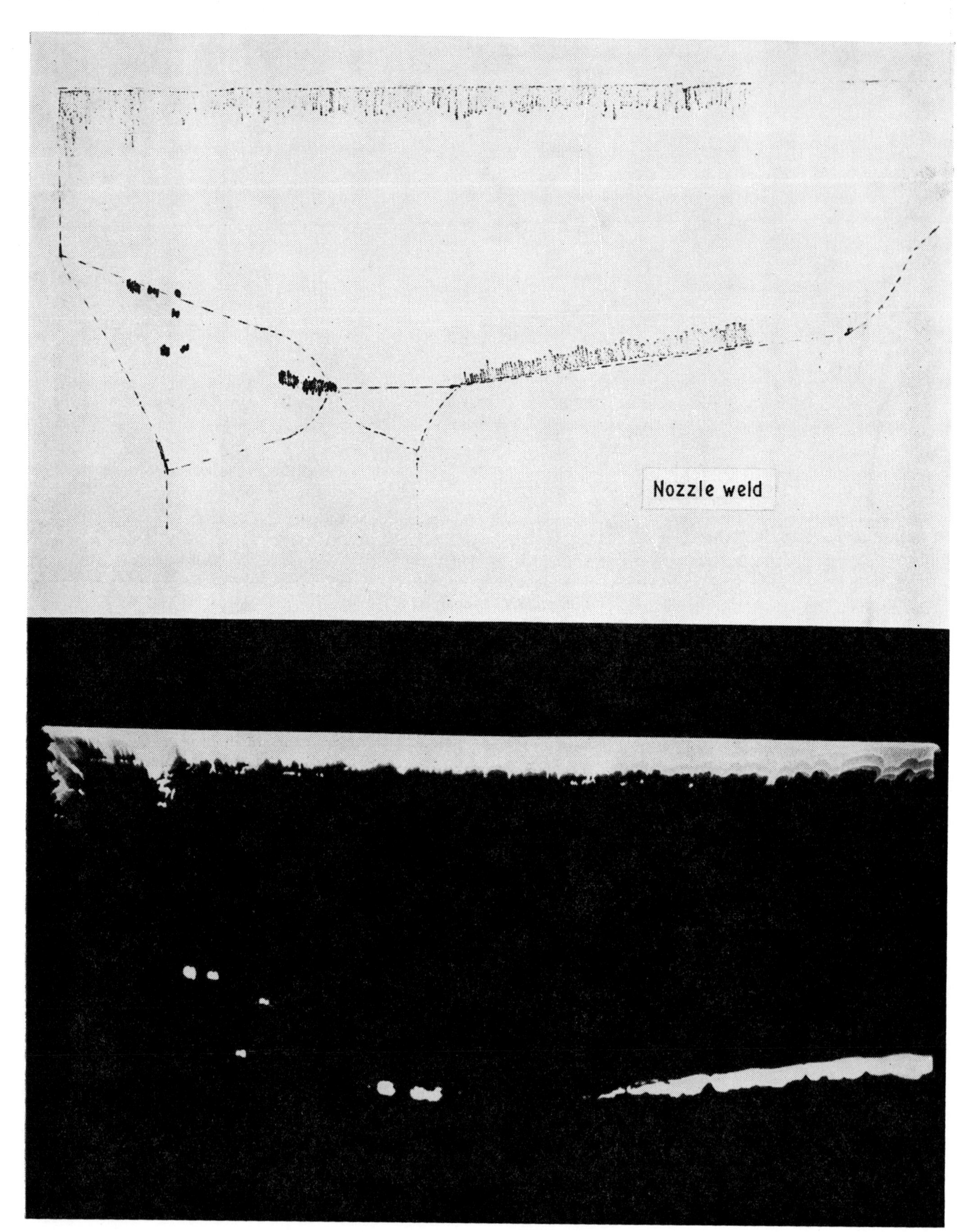

Fig. 27.8. B–scan nozzle display

CONCLUSIONS

This paper has described some of the techniques and the methods currently available for reactor post-commissioning inspection. Emphasis has been placed on the need to present the results of the inspection in a form such that they can be immediately assessed. This has influenced the development work and is aimed at giving reactor operators and licensing authorities information that will permit continued operation.

Any factors which reduce inspection time will influence the final choice of technique as both inspector and operator will be under continual pressure to reduce the cost of 'on-line' time. This attitude is already influencing reactor designs in that some designers are considering leaving a gap between the reactor vessel and the insulation, to permit inspection from the outside of the vessel (13). Rails would be welded to the vessel during construction or fixed by means of suction devices on a temporary basis to carry the inspection devices. Any welded fixture must be applied before stress relief of the vessel but offers the advantage that inspection can be carried out without removal of the reactor head. Inspection from the outside offers advantages in that it reduces many anomalies caused by the S.S. cladding and significant improvement in the ability of the equipment to detect small micro-cracks in the band area. Development work is already in hand to allow a B-scan presentation of the ultrasonic signals from such a device.

The use of acoustic emission techniques to augment and supplement N.D.T. surveillance offers significant advantages for the future. Current work has shown the feasibility of applying such techniques to large pressure vessels. The method can be used during initial pressurization to locate defect areas for confirmation by other N.D.T. techniques (14). Further work is in hand that will permit surveillance of vessels during operation and during periodic inspections to characterize the signals to the extent where they can be interpreted in terms of defect size and shape (15). Coupled with the type of inspection described in the paper, these techniques offer an important step forward in the overall inspectability of reactor pressure vessels in that they will permit 100 per cent volumetric inspection of the vessel in one operation.

APPENDIX 27.1

REFERENCES

(1) PHILIPS, C. A. G. and WARWICK, R. G. 'A survey of defects in pressure vessels built to high standards of construction and its relevance to nuclear primary circuit envelopes', U.K.A.E.A. Rep. AHSB(S)R162 1968 (H.M.S.O., London).

(2) POULTER, D. R. (Ed.) *The design of gas-cooled graphite-moderated reactors* 1963 (Oxford University Press, London).

(3) WHITMAN, G. D. and WITT, F. J. *Proc. Symp. Technology of Pressure-retaining Steel Components, Nucl. Metallurgy* 1970 **16** (A.I.M.M.P.E. and A.S.M.E.).

(4) *Practical fracture mechanics for structural steel, Proc. Symp. at U.K.A.E.A. Reactor Materials Lab.* 1969 (April) (Chapman and Hall, London).

(5) 'Rules for in-service inspection of nuclear reactor coolant systems', A.S.M.E. Boiler and Pressure Vessel Code, Section XI.

(6) 'Draft proposal for in-service inspection of reactor pressure vessels', by Dr Kellerman, RS-TUV.

(7) WESSEL, E. T. and MAGER, T. R. 'Fracture mechanics technology as applied to thick-walled nuclear pressure vessels', *Conf. Practical Application of Fracture Mechanics to Pressure-vessel Technology* 1971, 17 (Instn Mech. Engrs, London).

(8) NICHOLS, R. W. and COWAN, A. *First Int. Conf. Structural Mechanics in Reactor Technology* Berlin, 1971, LG/1. Preprints of papers (Bundesanstalt für Materialprüfung (B.A.M.), Berlin; Commission of the European Communities, Brussels).

(9) CLAYDEN U.K.A.E.A. Private communication.

(10) STERKE, A. de. 'Ultrasonic inspection of welds in nuclear reactor pressure vessels', *Int. Symp. Non-destructive Testing of Nuclear Power Reactor Components* Rotterdam, 1970 (February).

(11) MYER, H. J. 'Ultrasonic in-service inspection of reactor pressure vessels', *Kerntechnik, Isotofentechnik Chemie* 1971 (13th January) Bk 2, 5, 56–68.

(12) MEYER, H. J. 'Ultrasonic inspection of pressure vessels and walls mechanical and automated techniques', *Sanderdvick aug Materialprüfung* 1970 **12** (No. 10), 329.

(13) WYLIE, R. D., LAUTZENHEISER, C. E. and WORTMAN, O. 'Development of an inspection system for the Atucha nuclear reactor', *Proc. Second Int. Conf. Material Technology* Mexico City 1970 (August), 678.

(14) BENTLEY, P. G., BURNUP, T. E., BURTON, E. J., COWAN, A. and KIRBY, N. 'Acoustic emission as an aid to pressure vessel inspection', Paper C30 at this Conference.

(15) 'In-service inspection programme for nuclear reactor vessels', Progress Rep. No. 5, 1971 (Southwest Research Inst.).

C28/72 EVALUATION OF ACOUSTICAL HOLOGRAPHY FOR THE INSPECTION OF PRESSURE VESSEL SECTIONS

G. J. DAU* D. C. WORLTON† H. DALE COLLINS‡

This paper describes work that demonstrates the feasibility of using acoustical holography for the detection and measurement of flaws in thick-walled reactor pressure vessels. A brief discussion of the holographic imaging system and hologram reconstruction is given. Results presented include those which demonstrate porosity imaging in weld zones, flaw detection in rotor sections, and resolution capability. The resolution measurements indicate that the system will perform at the predicted value. Perhaps the most significant and exciting results are those which demonstrate that one hologram contains all the information necessary to image flaws at many different depths within an object. Design criteria are given for a prototype system for field evaluation.

INTRODUCTION

THE DEPENDENCE OF material serviceability on size, shape, orientation, and location of internal flaws heightens the need for improved methods of measuring detail flaw characteristics non-destructively. Unfortunately, existing ultrasonic inspection procedures are limited in this respect. This limitation is particularly felt in the nuclear industry where integrity of critical structures is assured on the bases of ultrasonic tests periodically performed to monitor and reveal flaw growth with service life.

At this time, the holographic process appears to offer one of the most promising approaches to improve non-destructive methods of delineating flaw characteristics. The same concepts which produce remarkably realistic images from film recordings apply, in principle, to ultrasonic as well as light waves. This suggests that with ultrasound used as the illuminating source, holographic techniques would enable the production of visible images of flaws located in opaque materials. Theoretically, these images would quantitatively reveal dimensions, orientation, etc., of the flaws imaged.

Over the past 18 months the Edison Electric Institute has supported a research programme aimed at verifying these theories experimentally and evaluating the possibility of reducing them to practical use for examination of pressure vessels and other nuclear structures. The work has been conducted at Battelle-Northwest Laboratories under contracts to the Jersey Nuclear Company and Southwest Research Institute. In brief, this work has demonstrated the validity of holographic concepts as applied to ultrasonic waves, and from it has evolved equipment and application techniques which appear compatible with practical inspection practices. The paper presented here highlights results of this programme.

The MS. of this paper was received at the Institution on 8th November 1971 and accepted for publication on 8th December 1971. 22

* *Battelle-Northwest, Richland, Washington.*
† *Jersey Nuclear Company, Richland, Washington.*
‡ *Formerly with Battelle-Northwest: now with Holosonics Corporation, Richland, Washington.*

DESCRIPTION OF HOLOGRAPHIC IMAGING SYSTEM

A variety of possible methods of producing ultrasonic holograms were investigated during the course of this work. The following discussion is restricted to the experimental arrangements shown in Figs 28.1 and 28.2, which together were used to form and reconstruct the images presented in this paper. This system has evolved from the exploration of many avenues of approach and appears at the time to incorporate an optimum balance of experimental parameters for this application. Further improvement, of course, would be expected to follow from longer range developments that have been suggested by this work.

Outwardly, the system appears in operation to be similar to a conventional ultrasonic C-scan. A single, focused ultrasonic transducer is scanned (under water) over a prescribed area, called the aperture, in a rectangular raster. The transducer is driven in a pulse-echo mode, with time gating employed to isolate flaw echoes from front- and back-surface echoes.

Signal processing techniques, however, differ markedly from conventional testing practices. Within the electronic system (Fig. 28.1) are circuits which mix and compare the echo signals with a reference signal obtained from the pulse transmitter. This reference wave allows retention of phase information in the received signal, and in this respect serves the same purposes as the reference light beam employed in optical holography.

The phase shifter also plays a vital role in the process, but does not have a direct counterpart to the optical case. Its purpose is to put a linear grating on the hologram which, during the reconstruction process, displaces the images away from the zero-order light. If this were not done it would be almost impossible to view the recon-

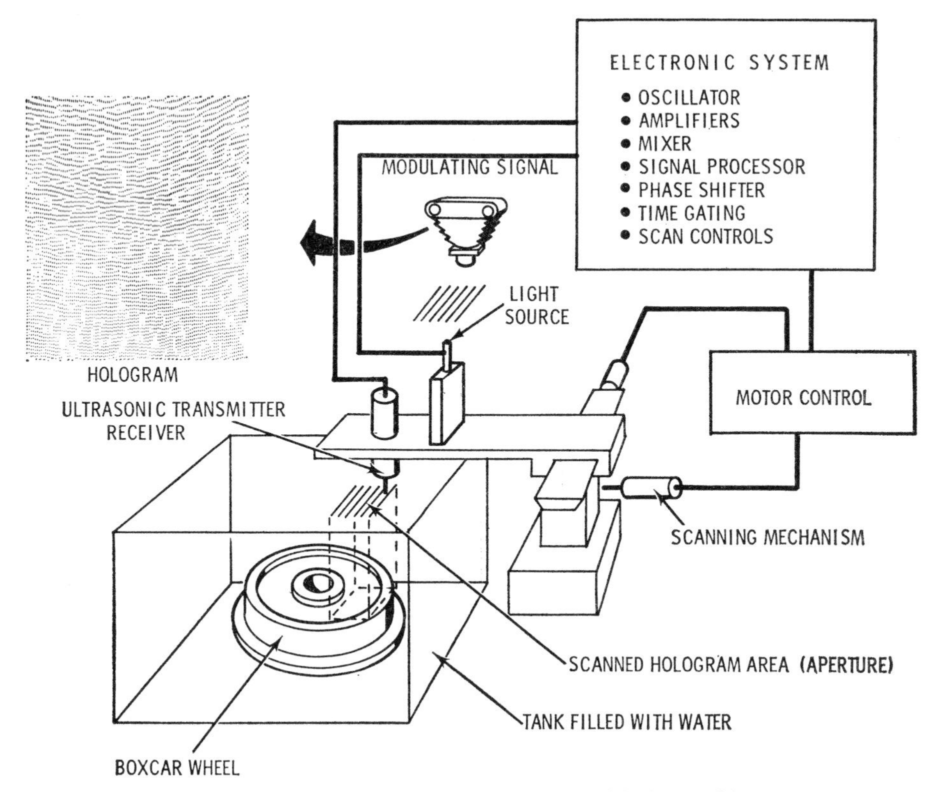

Fig. 28.1. Schematic of a laboratory acoustical holographic system

structed images because it would be masked by the undiffracted laser beam. From practical considerations the phase shifter was a most important development because it permits a single transducer to be used in both the send and receive mode and it allows the ultrasonic beam to be directed perpendicular to the surface of the part being examined. An analytical treatment of this system is given in Appendix 28.1.

The output from this operation modulates a light source which moves in synchronism with the transducer. The resulting light field is then recorded on Polaroid transparency film. This transparency is the hologram; a typical example is shown in Fig. 28.1.

The focused transducer plays an important role in this system, which goes beyond a simple concentration of beam intensity. In the first place, waves emanating from the transducer have a spherical wave front. If a flat transducer were used, plane waves would result. The area illuminated by the beam would then be limited to the size of transducer employed. If a plane wave is incident upon the surface at some angle, the exact volume illuminated within the slab is dependent upon metal characteristics, because of beam diffraction. Thus, the scan aperture must be carefully situated as determined by the angle of incidence of the source wave, distance to the interface, and type of metal. In practice, a very large transducer would be required.

The spherical wave front produced by the focused transducer, on the other hand, illuminates a large volume within the slab. The spherical wave front brings an added advantage in that the focal point, which is the apparent centre of the waves, can be selected at will. Since the plane of the resulting hologram corresponds to the plane of the wave centre, the hologram plane can be different from the plane of the transducer. Therefore, adjusting the beam focal point to coincide with the metal surface eliminates the need to consider the water path in calculating the position of the flaw from the image position. Thus, only one set of parameters (those of metal) need be involved in the data reduction. The result is to effectively bring the flaw closer to the hologram plane (by the amount of the water path), thereby enlarging its reconstructed image.

Another important advantage follows from the use of a single transducer as both sender and receiver. Analysis has shown that if both sender and receiver are scanned coincidentally, resolution can be increased by two and the image appears closer to the hologram by two. Time-sharing a single transducer in both send and receive mode, of course, satisfies this requirement.

The experimental system used to obtain the data described here provides the following capabilities:

Aperture size: 15×15 cm
Line density: up to 88 lines/cm
Scan rates: up to 6 cm/s.
Pulse length: nominally 10–20 μs adjustable
Pulse repetition rate: nominally 1 kHz adjustable.
Transducer (typical): 2·54 cm, ground to provide a focus at 10 cm; transducer can be operated at 1–10 MHz.

Hologram reconstruction procedure

To produce the optical image from an acoustical hologram, it is necessary to place the hologram (i.e. the Polaroid transparency) in an optical bench arrangement. The optical system needed is straightforward, as indicated in Fig. 28.2*a*. A laser provides the needed coherent light and the spatial filter shapes the beam so that it approaches a point source. A mechanical shutter provides

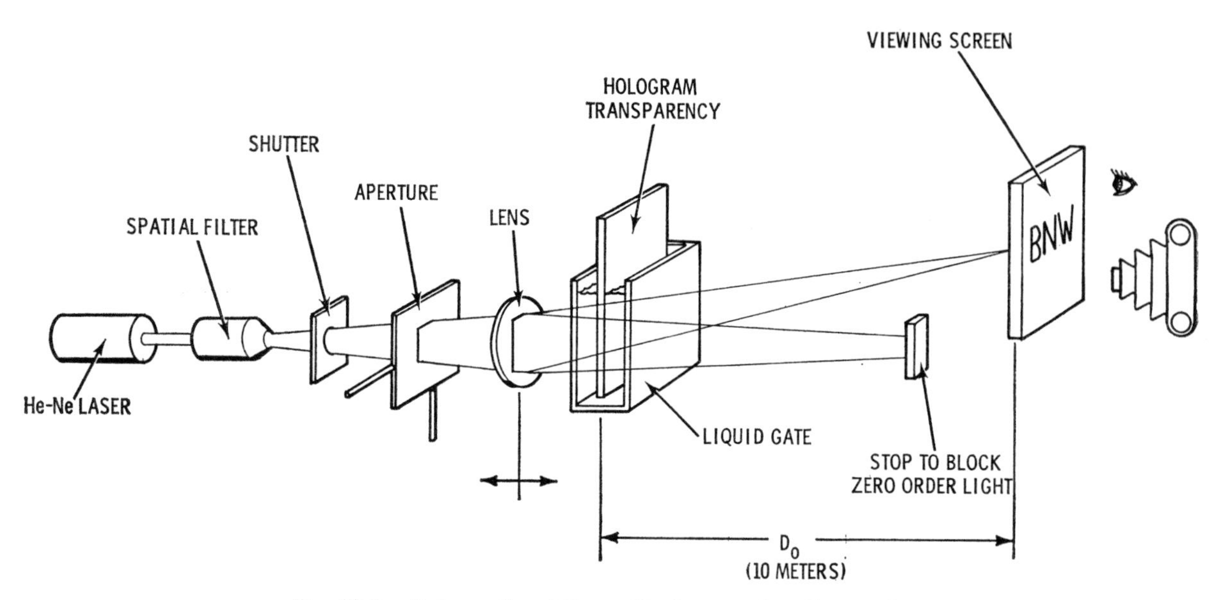

Fig. 28.2*a*. Schematic of the optical reconstruction system

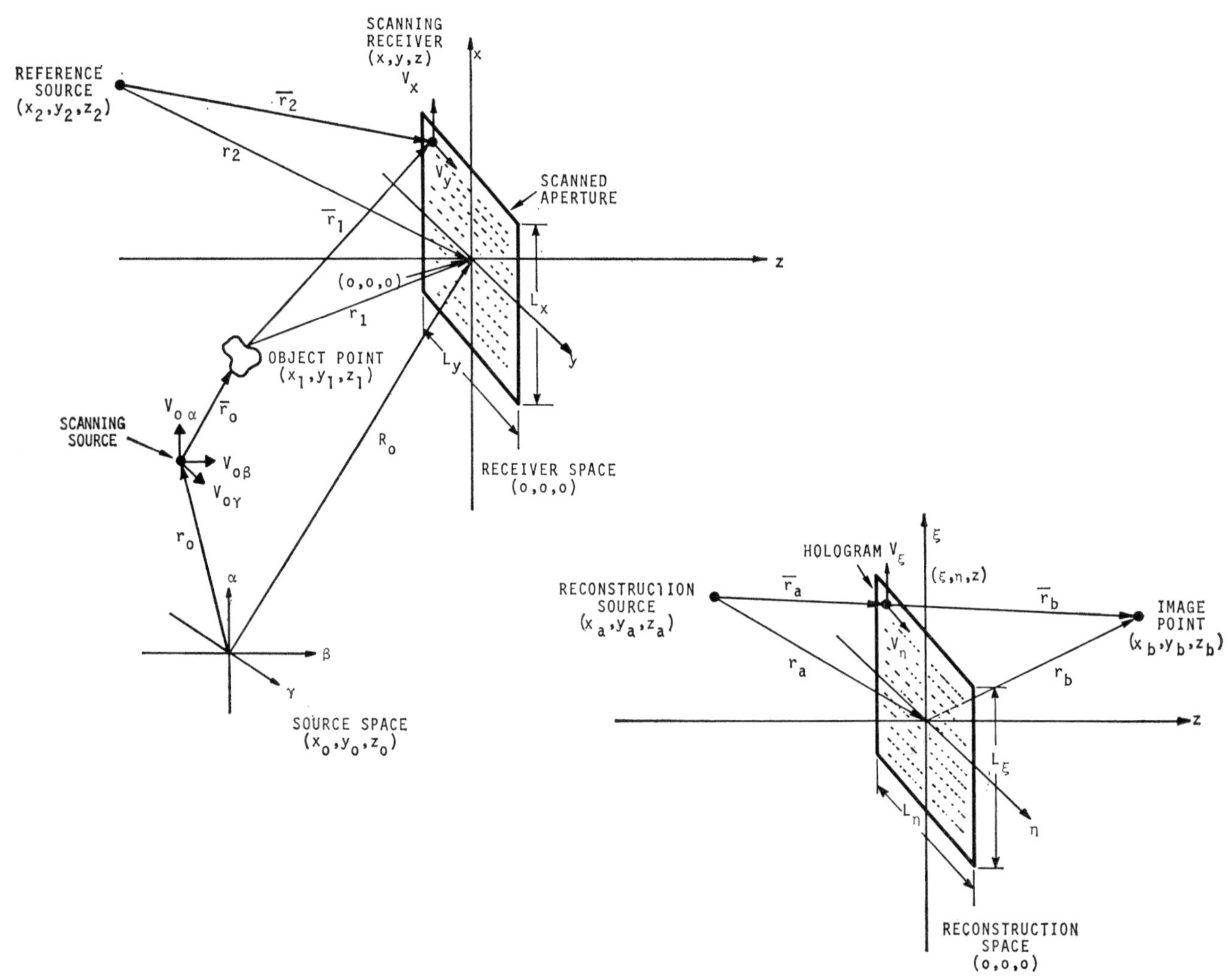

Fig. 28.2*b*. Schematic for the recording and reconstruction geometry used to derive image parameters

a precisely timed amount of light needed for photographing the images. The mechanical aperture limits the light beam so that it covers the desired area on the hologram. The movable lens is used to focus the beam on the viewing screen. Movement of this lens brings different depth (or longitudinal) planes of information into focus. Under controlled conditions, movement of the lens solves the image location equations (see Appendix 28.1). Thus, this set-up may be considered as a simple optical computer.

A liquid gate is used to hold the hologram. As the main function of this apparatus is to provide a fluid to surround the film and match indices of refraction, film flatness is not of concern.

A ground glass viewing screen is located downrange at a specified and fixed distance (for this work the distance was set at 10 m). This screen is used for focusing and viewing the reconstructed image. Permanent records are produced by substituting a camera for the viewing screen.

As indicated earlier, this device can be calibrated by fixing components at specific points given by the appropriate optical equations. When this is done, movement of the lens to the focal point of an image effectively solves the image location equation. Thus, properly calibrated, the lens position can be used to indicate the image location in the component being inspected.

As indicated, use of a single, focused transducer adjusted so as to put the focal point at the metal surface simplifies formation and interpretation of the reconstructed images. Fig. 28.2*b* illustrates key parameters of the process when applied under those conditions. The transducer is scanned over the aperture area, whose dimensions are L_x by L_y. The field generated by the light source is the hologram and it is recorded on the film, whose dimensions are L_ξ and L_η. The distance from the flaw to the metal surface is r_μ.

In the reconstruction process, shown in the lower part of Fig. 28.2*b*, $\bar{r}_b$ is the distance from the hologram to the reconstructed image and $\bar{r}_a$ is the distance from the movable lens to the hologram. When a hologram is viewed in this system, an image will be in focus at the distance $\bar{r}_b$ which corresponds to a particular depth in the metal. The parameters are related by the expression

$$r_\mu = \frac{\bar{r}_b r_a}{[(\lambda_{SL}\lambda_{\mu w})/2m^2](r_a+\bar{r}_b)} \quad . \quad (28.1)$$

where λ_{SL} is the ratio of the wavelength of the ultrasonic wave in water to wavelength of the reconstruction light wave, $\lambda_{\mu w}$ is the ratio of the wavelength of ultrasound in the metal to the wavelength of ultrasound in water, and m is a magnification ratio equal to L_x/L_ξ and to L_y/L_η.

Thus, any given plane throughout the thickness of the metal can be viewed in the reconstruction process by making the appropriate adjustments to r_a and $\bar{r}_b$. In practice, it is convenient to hold $\bar{r}_b$ fixed. For this work $\bar{r}_b$ was fixed at 10 m and is designated as D_0 in Fig. 28.2*a*. Successive planes can then be viewed from top to bottom of the test piece by incrementally changing r_a by the appropriate amount. Alternatively, one may examine a hologram by adjusting r_a until an image comes into focus on the viewing screen. When this is done, the actual depth of the discontinuity in the metal, r_μ, may be determined from the above equation.

It is of interest to also determine from the visible image the actual dimensions of the discontinuity being viewed. Dimensions lying in the plane under observation are related by a lateral magnification ratio, M_2:

$$M_2 = \frac{\text{image size}}{\text{actual size}} = \frac{2\lambda_L m\bar{r}_b}{\lambda_u r_u} \quad . \quad (28.2)$$

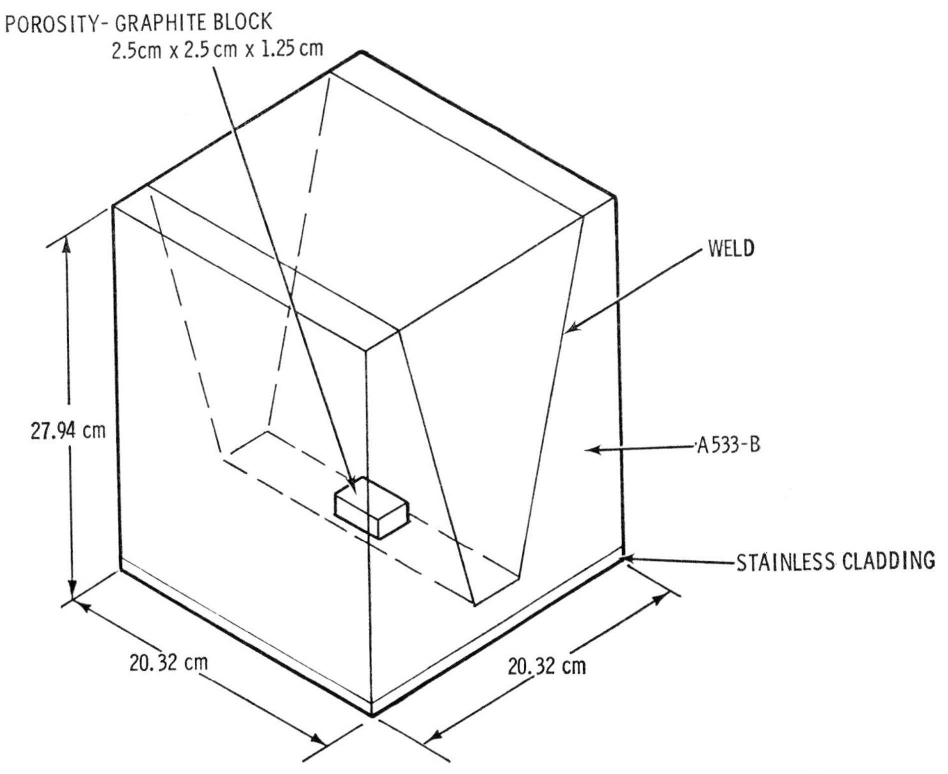

Fig. 28.3. Test specimen made to simulate a flaw in a pressure vessel weld

where the focal plane of the transducer was placed at the metal surface. At ultrasonic frequencies near 5 MHz, M_2 may be typically around 0·1.

Resolution is a convenient parameter for evaluating the quality of any imaging system. The limiting theoretical lateral resolution of this system is determined by the transducer and is given by:

$$\Delta X_1 \simeq 1{\cdot}22\lambda_s(t/a)$$

where a is the diameter of the transducer. For the transducer used in this work, this equation indicates that flaws separated 1·46 mm in the lateral direction can be resolved at an ultrasound frequency of 5 MHz.

EXPERIMENTAL RESULTS

Development and evaluation of equipment and the application technique have involved the examination of many objects containing both natural and artificial flaws. Presented here are the results of a holographic examination of the following conditions:

Weld flaws in nuclear pressure vessel stub—imaged from both clad and unclad surfaces

a Test specimen clearly showing the clad surface.
b Image of the flaw taken through the smooth surface (bottom side in photograph).
c Image of flaw taken through clad surface (top side in photograph).

Fig. 28.4. Photograph and acoustical images of the simulated flaw

Natural flaw in rotor section
Simulated line flaws
Drilled letters 'BNW'
Milled letter 'F'
Point reflector at different depths
Resolution measurements.

The results are presented by first showing a sketch of the geometry of the defect, followed by a photograph of the image. In some cases a photograph of the object itself is included. For those interested in complete detail, the image parameters are also given.

Weld defect

In any large pressure vessel, weld areas are always of concern; thus, inspection of a weldment is very desirable. To evaluate the selected system, a large weld was made in a steel block. The weld was prepared according to appropriate code procedures in a section of pressure vessel steel. During welding, a 2·54 cm (1 in) by 2·54 cm (1 in) by 1·25 cm (0·5 in) graphite block was inserted to produce an actual defect. The flaw was located 20 cm (8 in) below the top or unclad surface of the block and 5 cm (2 in) up from the stainless steel clad surface. Fig. 28.3 is a sketch of the test specimen. Fig. 28.4*b* shows the results obtained from inspecting the weld from the smooth, unclad side. For this test, the transducer was scanned through the test aperture about 10 cm (4 in) above the block surface. Fig. 28.4*c* shows results obtained by imaging the same flaw through the stainless steel clad surface. The clad surface consists of a series of wave-like ripples spaced about 2·5 cm (1 in) apart with a height of about 1 mm (0·04 in). In spite of this non-uniformity, the two results shown in Fig. 28.4 are very good. (Note that in Fig. 28.4*c* the image is magnified by a greater amount than the image in Fig. 28.4*b* because the distance the sound propagates through steel is less.) These results are dramatic evidence that flaws can be imaged deep within a thick weldment via ultrasonic holography with adequate resolution, whether inspecting through the clad or unclad surface. Resolution is about 1·2 mm (0·048 in) at depths exceeding 20 cm (8 in) when using an ultrasonic frequency of 5·1 MHz.

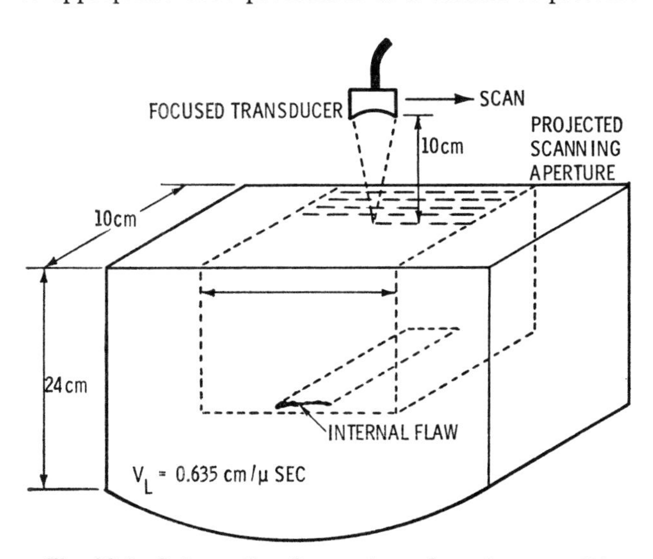

Fig. 28.5. Schematic of a section of steel rotor with natural flaw

Rotor section flaw

Fig. 28.5 shows the hologram construction geometry used to image an actual flaw in a section of a large steel rotor block. The narrow width of the section caused multiple internal reflections which decreased the signal-to-noise ratio, thus somewhat degrading the image.

Fig. 28.6 shows the image of the flaw. The image is about 2·7 mm (0·106 in) wide and 19 mm (0·75 in) long. Using the calculated demagnification factor of 0·186, it is possible to show that the actual flaw is 1·45 cm (0·606 in) wide and 10 cm (4 in) long. The flaw does not lie in a plane parallel to the rotor section top surface. Thus, the ends are out of focus because this picture was taken at the averaged computed depth of the flaw. Resolution was

Fig. 28.6. Image of the flaw obtained at a frequency of 3·15 MHz yielding a resolution of 2 mm at a magnification of 0·186

limited to 2 mm (0·09 in) by the focused transducer (i.e. focal length, diameter, and frequency). Image quality is very good and demonstrates the ease of displaying complex information via the holographic process.

Simulated line flaw

During the course of this work it appeared worthwhile to attempt imaging line flaws that were parallel to the scanning aperture, and at different depths. Simulated flaws were produced by drilling various-sized holes and then chiselling grooves and ridges on their top surfaces to make them as rough as possible. Hole arrangements were positioned so that some holes were directly above others, thereby providing a shadowing effect. Fig. 28.7 shows the hole locations.

In theory, one hologram should contain all of the information necessary to reconstruct images of each flaw. (All of this information could not be obtained in one scan using conventional ultrasonic inspection techniques.) The images, Fig. 28.8, that were obtained from this hologram are dramatic evidence that the theory is correct. Since the magnification factor is different for each flaw, the images appear to be vastly different in size. In reality, all have about the same dimensions.

The photographs are arranged in order of increasing depth within the metal block. Fig. 28.8*a* shows a sharp image of the uppermost hole (No. 1). Note that all the other holes are so out of focus that they do not appear. Fig. 28.8*b* is taken at the plane of hole No. 2. As predicted, it appears smaller than the uppermost hole. Fig. 28.8*c* is perhaps the most interesting in that it shows hole No. 3 very clearly, even though it was shadowed by hole No. 2. Fig. 28.8*d* is focused on the plane of hole No. 5 which is 22·7 cm (9 in) below the surface.

This series of photographs is perhaps the most exciting of our results. It must be emphasized that all were obtained from one hologram, simply by moving the camera through the three-dimensional image and taking photographs at the planes of interest. The images are precise enough so that, knowing the magnification factors, we can make fairly accurate measurements of the flaws.

Drilled letters 'BNW'

In another series of tests, the letters 'BNW' were formed in an aluminium test block by drilling holes up from the bottom surface. The hole tops were milled flat and were located 11·4 cm (4·5 in) below the top surface. The hole diameter is 6·35 mm (0·25 in) and separation varies from 1·5 to 2 mm (0·059 to 0·08 in). Fig. 28.9 shows the test specimen details and Fig. 28.10 illustrates the acoustical images obtained. The edges of the 'B' and 'W' are not as sharp as the 'N' because the 'BNW' configuration exceeded the aperture length, thus resulting in inadequate illumination of the edges. The image in Fig. 28.10*a* was made at 5·1 MHz, and thus shows greater resolving capability than the image in Fig. 28.10*b*, which was made at 3·15 MHz. Although both figures show very good image quality, comparison shows increased resolving power at the higher frequency. Note particularly that the holes are better separated and edges more sharply defined in the higher frequency image.

Milled letter 'F'

We performed experiments to compare the results of the selected scanning systems with another variation of the source–receiver system. Our results show the superiority of the selected system.

A hologram was made using the optimum configuration, i.e. a single focused source–receiver transducer with electronic phase shifting. The object was the letter 'F' situated as shown in Fig. 28.11. Fig. 28.12*a* shows the resultant image formed with a frequency of 3·4 MHz using a pulse repetition rate of about 500 Hz and a pulse length of 30 μs. The resolution is 1·2 mm (0·079 in). Another hologram of the same object was formed using a frequency of 5·1 MHz. It is apparent that the higher frequency produces greater resolution: 1·2 mm (0·047 in).

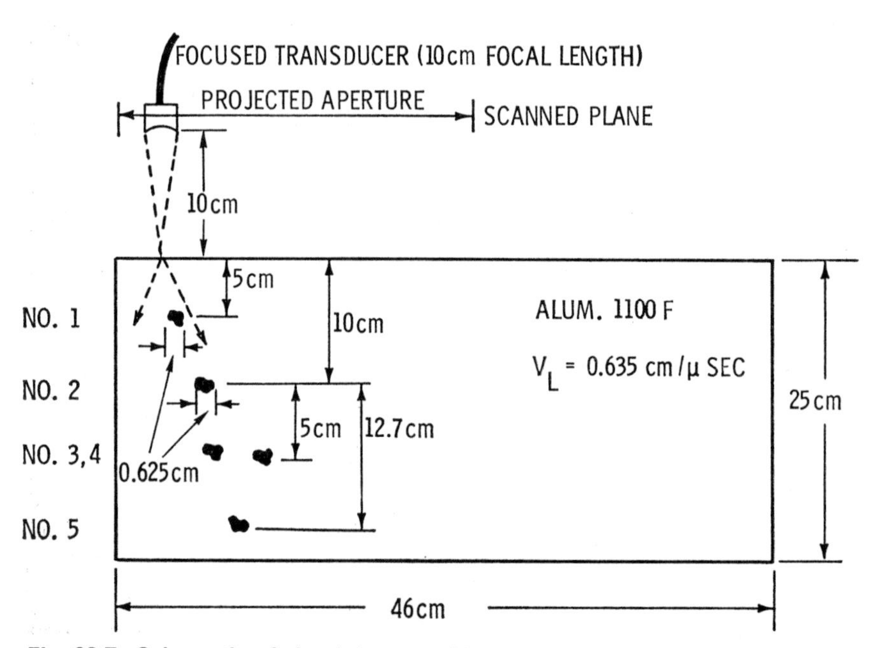

Fig. 28.7. Schematic of aluminium test block containing simulated flaws at varying depths

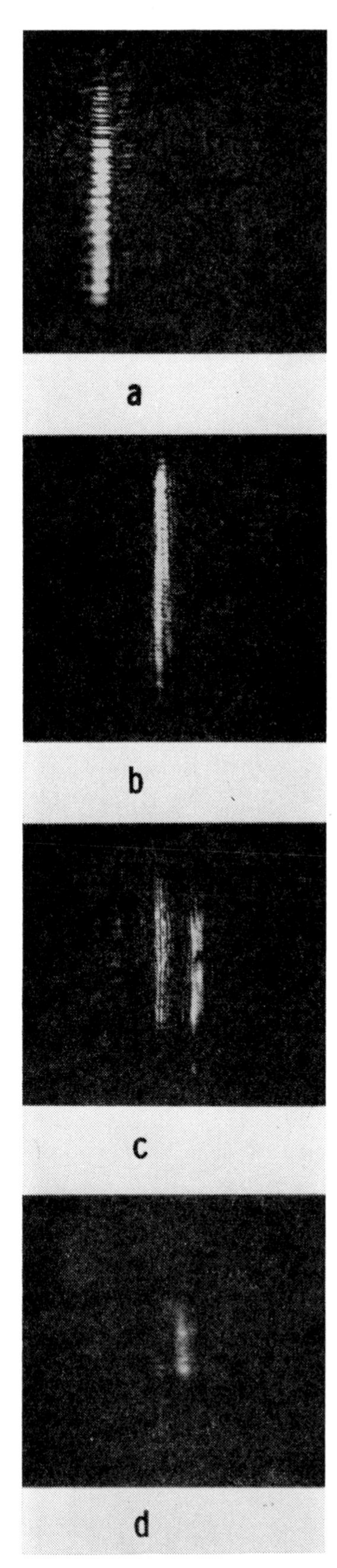

Fig. 28.8. Images of simulated flaws obtained at a frequency of 3·25 MHz yielding a resolution of 2 mm and magnifications of 0·574, 0·278, 0·186, and 0·123 respectively

The image shown in Fig. 28.12*b* also has more sharply defined edges and greater lateral magnification.

Using the same object another hologram was constructed; however, an inclined source (plane wave) transducer and a point receiver were used. The source was inclined approximately 5° with respect to the block surface, imposing a linear grating on the hologram necessary for image separation. In essence, the inclined source was used in place of electronically shifting the reference signal. Fig. 28.13 shows that the image is of rather poor quality compared with Fig. 28.12. One reason is that the source distance for this geometry is infinite, which lowers the resolution by a factor of 2 compared to the simultaneous source–receiver configuration.

Point reflectors at different depths

One of the unique features of holographic imaging is the ability to record all depth (longitudinal) information about an object in one hologram. This was already demonstrated in Fig. 28.8. However, to quantify this capability a special test was performed. Two 1·27-cm (0·5-in) diameter holes were drilled into the bottom of a large aluminium block. The tops of the holes were spherically shaped to simulate point reflectors, as shown in Fig. 28.14. One hole top was 8·9 cm (3·6 in) and the other was 10 cm (4 in) below the top surface. Fig. 28.15 shows the reconstructed images.

Since the reflectors are at different depths, the images also appear at different distances from the hologram. Consequently, in Fig. 28.15*a* the bottom image is in focus but the deeper reflector is out of focus. By the simple movement of a lens the shallower reflector is brought into focus and the deeper reflector now appears out of focus, as shown in Fig. 28.15*b* (see Appendix 28.1 for discussion). Measurement of the lens movement, which is required to go from one focal point to the other focal point, permits easy calculation of the distance between the reflectors. The hole depths were chosen to be separated by about the depth resolution limit. If they were any closer together (in depth) it would not be obvious that one was in better focus than the other. This test then was essentially a depth resolution test.

Verification of resolution ability

Because scanned holography offers much promise as a non-destructive evaluation tool, it was appropriate to experimentally verify its resolution ability. Consequently, two test blocks were constructed to verify resolution calculations. Fig. 28.16 shows the geometry of one block with 15 holes drilled from the bottom surface. The holes are 3·18 mm (0·12 in) in diameter with the tops milled flat, and are placed in three rows. In one row the holes are separated 1 mm (0·04 in), in the second row the spacing is 2 mm (0·08 in), and in the third row the spacing is 4 mm (0·16 in). The tops of the holes are 10 cm (4 in) below the block surface. A 10-cm (4-in) focal length transducer of 2·54 cm (1 in) diameter was used to construct the holograms.

The maximum obtainable holographic lateral resolution was calculated using the following expression:

$$\Delta X_{\max} = \frac{\lambda_{\mu} f_{\mu}}{d}$$

where λ_μ is the acoustical wavelength in metal, f_μ the transducer focal length, and d the transducer diameter.

Using a frequency of 3·12 MHz (theoretical resolution of 2 mm), we constructed a hologram of the test specimen. Fig. 28.17*a* shows the resultant image. The left row of holes have merged together, showing that the resolving capability of the system is less than 1 mm. The middle row of holes are separated, which shows that the system resolution equals or slightly exceeds the calculated limit of 2 mm. Thus, we concluded that the experimental system resolution agrees with the calculated value.

Another hologram was made of the same object but using a frequency of 5·1 MHz. The results of Fig. 28.17*b* show that the increased frequency has also increased the resolution limit to 1 mm. The middle row of holes is clearly separated here.

ACKNOWLEDGEMENTS

The authors gratefully acknowledge the contribution of R. P. Gribble for the very significant effort he made in performing the experimental work. Dr B. Percy Hildebrand also deserves credit for making significant contributions during the preparation of the manuscript.

APPENDIX 28.1

ANALYTICAL DISCUSSION

Image position calculation

A general schematic layout for both hologram generation and reconstruction is shown in Fig. 28.2*b*, mentioned previously. The image positions of the internal flaws were calculated theoretically from the image location equations given below (1)*.

The following discussion shows how these equations are used for the conditions of this project.

$$\frac{x_b}{r_b} = \pm\frac{\lambda_L}{\lambda_S}\cdot\frac{V_x}{V_\xi}\left\{\frac{x_1}{r_1}+(x_1-x_0)\frac{V_{0\alpha}}{V_x}\cdot\frac{1}{R_0}-\frac{x_2}{r_2}\right\}-\frac{x_a}{r_a} \quad (28.3)$$

* *The reference is given in Appendix 28.2.*

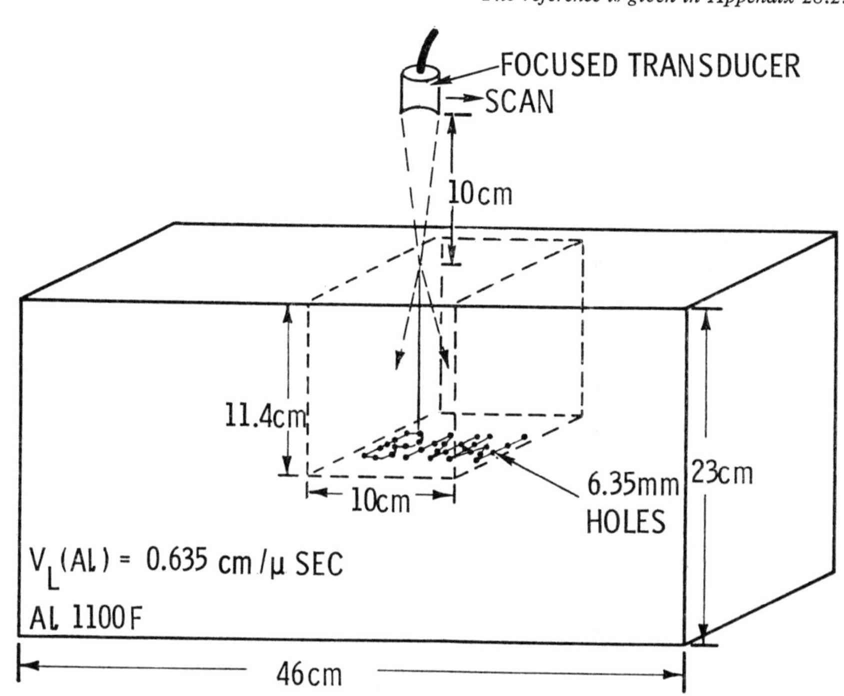

Fig. 28.9. Schematic of aluminium test block with holes drilled in 'BNW' pattern

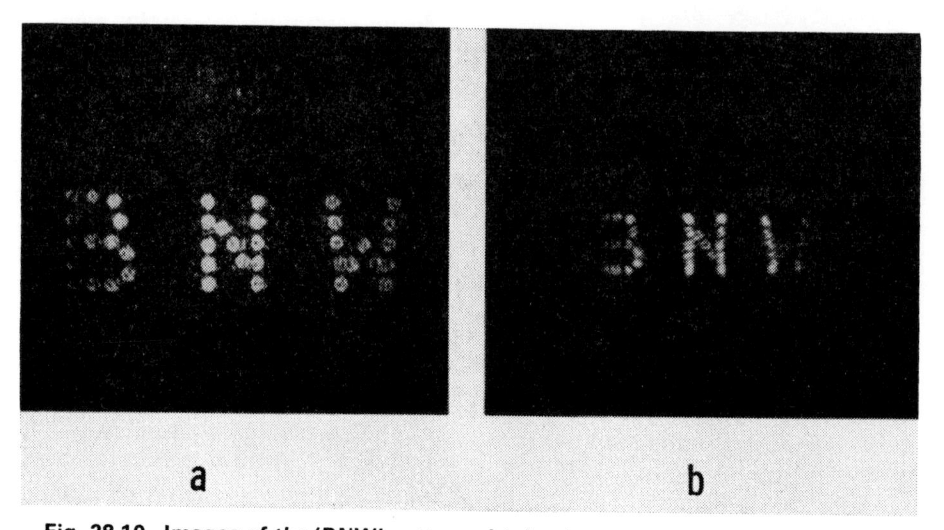

Fig. 28.10. Images of the 'BNW' pattern obtained at two different frequencies

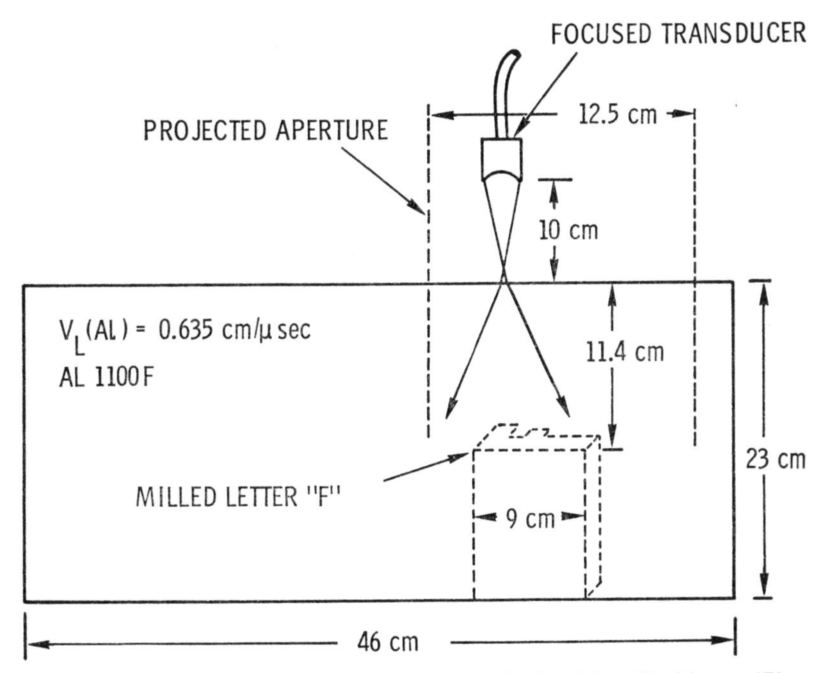

Fig. 28.11. Schematic of aluminium test block with milled letter 'F'

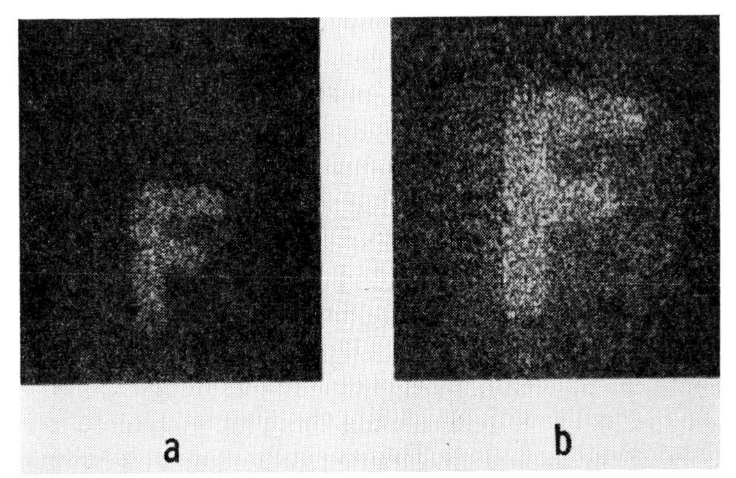

Fig. 28.12. Images of milled letter 'F' obtained at two different frequencies

Fig. 28.13. Image of milled letter 'F' taken with a stationary plane wave illumination with no electronic phase shifting

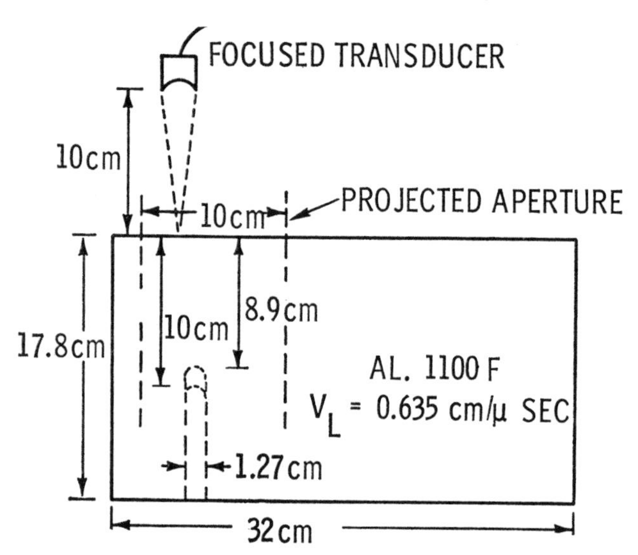

Fig. 28.14. Schematic of aluminium test block designed to measure depth resolution

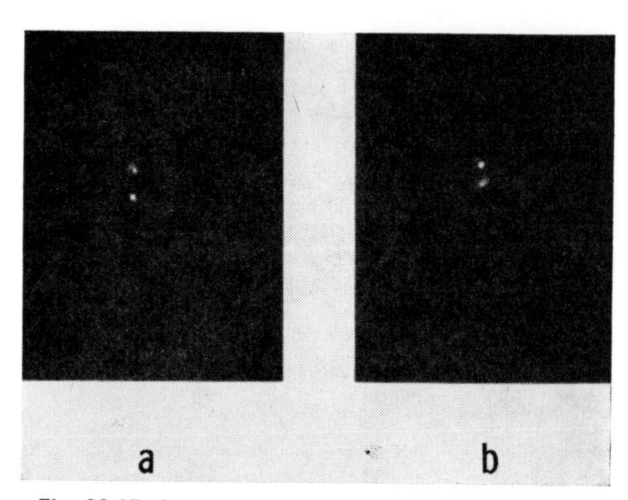

Fig. 28.15. Images obtained from the depth resolution test block at 3·15 MHz

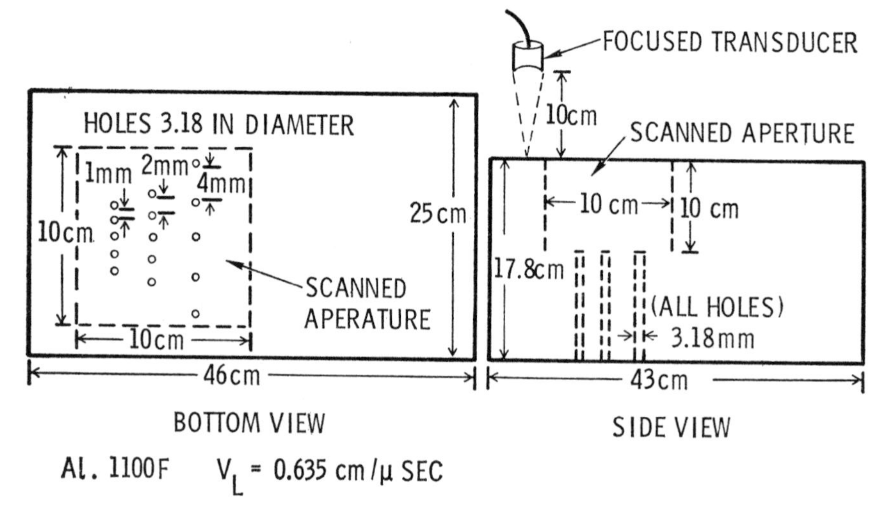

Fig. 28.16. Schematic of aluminium test block made to measure lateral resolution images of the resolution test block

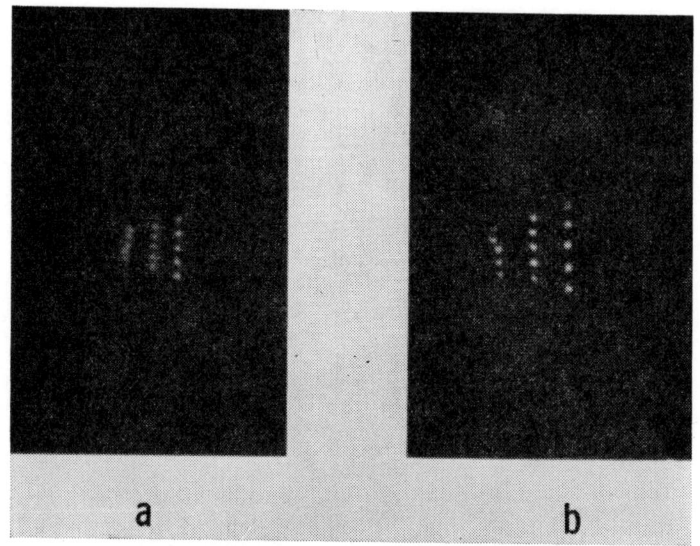

a Image taken at 3·12 MHz; magnification, 0·274.
b Image taken at 5·1 MHz; magnification, 0·435.

Fig. 28.17. Images of the resolution test block

$$\frac{y_b}{r_b} = \pm\frac{\lambda_L}{\lambda_S}\cdot\frac{V_y}{V_\eta}\left\{\frac{y_1}{r_1}+(y_1-y_0)\frac{V_{0\beta}}{V_y}\cdot\frac{1}{R_0}-\frac{y_2}{r_2}\right\}-\frac{y_a}{r_a} \quad (28.4)$$

$$\frac{1}{r_b} = \pm\frac{\lambda_L}{\lambda_S}\left(\frac{V_x}{V_\xi}\right)^2\left\{\frac{1}{r_1}+\left(\frac{V_{0\alpha}}{V_x}\right)^2\frac{1}{R_0}-\frac{1}{r_2}\right\}-\frac{1}{r_a} \quad (28.5)$$

$$\frac{1}{r_b} = \pm\frac{\lambda_L}{\lambda_S}\left(\frac{V_y}{V_\eta}\right)^2\left\{\frac{1}{r_1}+\left(\frac{V_{0\beta}}{V_y}\right)^2\frac{1}{R_0}-\frac{1}{r_2}\right\}-\frac{1}{r_a} \quad (28.6)$$

where λ_L is the wavelength of the reconstruction source, λ_S the wavelength of the construction source, a plus sign refers to the conjugate flaw image, and a minus sign refers to the true flaw image.

In order for r_b to be the same for each co-ordinate (i.e. stigmatic), the following conditions must be satisfied:

$$\frac{V_x}{V_\xi} = \frac{V_y}{V_\eta} \quad . \quad . \quad . \quad . \quad (28.7)$$

and

$$\frac{V_{0\alpha}}{V_x} = \frac{V_{0\beta}}{V_y} \quad . \quad . \quad . \quad . \quad (28.8)$$

The actual and simulated velocity components of the receiver (V_x, V_y, V_ξ, V_η) are related to the scanning aperture and hologram dimensions by the following equations:

$$\frac{V_x}{V_\xi} = \frac{L_x}{L_\xi} = m_x, \qquad \frac{V_y}{V_\eta} = \frac{L_y}{L_\eta} = m_y \quad (28.9)$$

The velocity ratios define the hologram magnification in the x and y directions (i.e. lateral magnifications). If the point source and receiver occupy the same position and are scanned simultaneously, then the flaw image location equations simplify to the following expressions:

$$\frac{1}{r_b} = \pm\frac{\lambda_L}{\lambda_S}m^2\left\{\frac{2}{r_1}-\frac{1}{r_2}\right\}-\frac{1}{r_a} \quad . \quad (28.10)$$

$$\frac{x_b}{r_b} = \pm\frac{\lambda_L}{\lambda_S}m\left\{\frac{2x_1}{r_1}-\frac{x_2}{r_2}\right\}-\frac{x_a}{r_a} \quad . \quad (28.11)$$

$$\frac{y_b}{r_b} = \pm\frac{\lambda_L}{\lambda_S}m\left\{\frac{2y_1}{r_1}-\frac{y_2}{r_2}\right\}-\frac{y_a}{r_a} \quad . \quad (28.12)$$

where

$$\mathbf{m} = V_x/V_\xi = V_y/V_\eta = V_{0\alpha}/V_x = V_{0\beta}/V_y$$

and

$$R_0 = r_1$$

To obtain the image location within the block, it is necessary to solve the above equations for r_1. Once this quantity is determined, the distance from the surface of the inspected block to the image location is calculated by solving the following equation for r_μ:

$$r_1 = r_w+\frac{\lambda_\mu}{\lambda_w}r_\mu \quad . \quad . \quad . \quad (28.13)$$

where r_w is the distance from the receiving transducer to the upper block surface, r_μ the distance from the block surface to the internal flaw, λ_w the wavelength of sound in water, and λ_μ the wavelength of sound in the metal.

The wavelength ratio (i.e. λ_μ/λ_w) is approximately 4, thus the distance from the inspected component surface to the point being imaged appears to be magnified by a factor of 4.

Our experimental work showed excellent agreement between the calculated flaw location and the actual location. The above equations would be cumbersome to work with in a field situation. However, it can be shown that a properly constructed and calibrated image reconstruction arrangement functions as a simple optical computer. By moving the lens position in this apparatus (see Fig. 28.2) these questions are solved and the position of the lens indicates image position. Since we have chosen the focused transducer configuration, we can move the effective hologram surface to coincide with the metal surface ($r_w = 0$), thus removing this factor from the equation.

Flaw lateral magnification

The flaw lateral hologram magnifications for point source–receiver scanning are defined as:

$$M_L(x) = \frac{\partial x_b}{\partial x_1} = \pm\frac{\lambda_L}{\lambda_S}\cdot\frac{V_x}{V_\xi}\cdot\frac{r_b}{r_1}\left(1+\frac{r_1}{r_0}\cdot\frac{V_{0\alpha}}{V_x}\right) \quad (28.14)$$

$$M_L(y) = \frac{\partial y_b}{\partial y_1} = \pm\frac{\lambda_L}{\lambda_S}\cdot\frac{V_y}{V_\eta}\cdot\frac{r_b}{r_1}\left(1+\frac{r_1}{r_0}\cdot\frac{V_{0\beta}}{V_y}\right) \quad (28.15)$$

where $z_1 \gg x_1$ or y_1

and $z_1-z_0 \gg x_1-x_0$ or y_1-y_0

If the source and receiver are located at the same position, i.e. ($r_1 = R_0$) and $V_{0\alpha}/V_x = V_{0\beta}/V_y = V_\xi/V_x = V_\eta/V_y$, then the flaw lateral magnification is simplified to:

$$M_L(x) = M_L(y) = \pm 2\frac{\lambda_L}{\lambda_S}\cdot\frac{r_b}{r_1}m \quad (28.16)$$

For focused transducer scanning (pulse echo operation), the lateral flaw magnification can be expressed by the following equation:

$$M_1 = \pm 2\frac{\lambda_L}{\lambda_S}\cdot\frac{r_b m}{(r_1-f)} \quad . \quad . \quad (28.17)$$

where f is the focal length of the transducer.

Flaw radial magnification

The radial or depth magnification for simultaneous source–receiver scanning is

$$M_R = \pm 2\frac{\lambda_L}{\lambda_S}\left(\frac{r_b m}{r_1}\right)^2 \quad . \quad . \quad (28.18)$$

If a focused transducer is scanned, then the radial magnification is

$$M_R = \pm 2\frac{\lambda_L}{\lambda_S}\left(\frac{r_b m}{r_1-f}\right)^2 \quad . \quad . \quad (28.19)$$

The flaw images will have magnification distortion if the ratio of the radial to lateral magnification is equal to or less than unity. The images will be stretched in depth (i.e. a cubic flaw in the metal will appear as a hexahedral flaw in the reconstructed image). This restriction and a limited aperture eliminate the possibility of viewing the flaw images in three dimensions as in optical holography. The flaw image can be viewed in depth by focusing on different planes through the flaw.

We now proceed to use these expressions to illustrate the appearance of a holographic image and to elaborate on our previous statement that the optical bench may be calibrated to act as a simple computer. To facilitate this objective, we repeat the particular equations we must use:

$$\frac{1}{r_b} = \pm\frac{\lambda_L}{\lambda_S} m^2 \left\{\frac{2}{r_1} - \frac{1}{r_2}\right\} - \frac{1}{r_a} \quad . \quad (28.10)$$

$$\frac{x_b}{r_b} = \pm\frac{\lambda_L}{\lambda_S} m \left\{\frac{2x_1}{r_1} - \frac{x_2}{r_2}\right\} - \frac{x_a}{r_a} \quad . \quad (28.11)$$

$$M_L = \pm 2\frac{\lambda_L}{\lambda_S} m \frac{r_b}{r_1} \quad . \quad . \quad (28.20)$$

$$M_R = \pm 2\frac{\lambda_L}{\lambda_S} m^2 \left(\frac{r_b}{r_1}\right)^2 \quad . \quad . \quad (28.21)$$

The phase shifter generates a simulated plane reference beam. This means that the equivalent reference source is at $r_2 = \infty$. Hence, equation (28.10) becomes

$$\frac{1}{r_b} = \pm\frac{2\lambda_L}{\lambda_S} \cdot \frac{m^2}{r_1} - \frac{1}{r_a} \quad . \quad . \quad (28.22)$$

If, in addition, we reconstruct with a plane wave ($r_a = \infty$) we have a further simplification of equation (28.22) to

$$\frac{1}{r_b} = \pm\frac{2\lambda_L}{\lambda_S} \cdot \frac{m^2}{r_1} \quad . \quad . \quad . \quad (28.23)$$

or

$$r_b = \pm\frac{\lambda_S}{\lambda_L} \cdot \frac{r_1}{2m^2}$$

Substituting this expression into equations (28.20) and (28.21) yields

$$M_L = \frac{1}{m} \quad . \quad . \quad . \quad . \quad (28.24)$$

$$M_R = \frac{\lambda_S}{2\lambda_L m^2} \quad . \quad . \quad . \quad (28.25)$$

These two equations yield the primary reason why the image is not viewable in three dimensions as we have come to expect of holography. As an example, let $m = 3$, $\lambda_S/\lambda_L = 500$ as would be the case of 5 MHz sound in water. If we are imaging in metal, $\lambda_S/\lambda_L \simeq 2000$. Thus, a 1 cm cubic flaw will appear as a hexahedral with a cross-section $\frac{1}{3}$ cm square and length of approximately 100 cm for metal and 25 cm for water. Furthermore, from equation (28.23) we see that the image will be 100 times as far from the hologram as the flaw. Therefore, to image a relatively large volume, say 15 cm×15 cm×25 cm (6 in×6 in×10 in), we should need to search an image space of 5 cm×5 cm×2500 cm (2 in×2 in×100 in) volume with a movable screen or camera. Since this is obviously not a desirable situation, we choose equation (28.22) as our guiding expression and allow r_a to change. This is accomplished by moving the lens shown in Fig. 28.2. Rearranging equation (28.22), we then have

$$r_b = r_a \left[\frac{\frac{\lambda_S}{\lambda_L} \cdot \frac{r_1}{2m^2}}{r_a - \left(\frac{\lambda_S}{\lambda_L} \cdot \frac{r_1}{2m^2}\right)}\right] \quad . \quad (28.26)$$

Now we see that the image distance depends not only on the flaw distance, r_1, but also on the reconstruction source distance, r_a. Thus, we can set the camera or viewing screen a fixed distance from the hologram and manipulate the lens position (r_a) to bring the desired cross-section of the image into focus. That is, we set r_b, the image distance, equal to a constant $\bar{r}_b$, in our case 10 m, and solve the equation for r_a, yielding

$$r_a = \bar{r}_b \left[\frac{\frac{\lambda_S}{\lambda_L} \cdot \frac{r_1}{2m^2}}{\bar{r}_b - \frac{\lambda_S}{\lambda_L} \cdot \frac{r_1}{2m^2}}\right] \quad . \quad . \quad (28.27)$$

As an example, suppose the 1 cm cubic flaw was located at $r_1 = 15$ cm, then the required reconstruction distance to focus the near surface of the flaw ($r_1 = 15$ cm) is $r_a = -10{\cdot}7$ m, and for the far surface ($r_1 = 16$ cm), $r_a = -10{\cdot}67$ m. The negative sign merely means that the light source must be converging, as shown in Fig. 28.2. A suitable calibration on the lens motion can then be used to calculate the distance to the flaw.

Unfortunately, this type of reconstruction scheme also brings about a variation in the magnification so that various cross-sections of the cubic flaw, for example, will appear different in size. For our example, using equation (28.20), the near face will present an image $\frac{1}{5}$ cm square, whereas the far surface will be shown as $\frac{3}{16}$ cm square. Consequently, we must also have a calibration chart handy for converting image size to flaw size. However, this is not difficult, so the reconstruction system we chose is quite practical.

Resolution

The most convenient parameter for evaluating the quality of any imaging system is resolution, both in depth and in cross-section. Holography is no different than optics in this regard. The traditional expression for lateral or cross-sectional resolution in optics is

$$\Delta x \simeq 1{\cdot}22\frac{\lambda r}{L} \quad . \quad . \quad . \quad (28.28)$$

where λ is the wavelength, r the distance to image from the lens, and L the diameter of the lens.

The depth resolution is defined as

$$\Delta r \simeq 2\lambda\left(\frac{r}{L}\right)^2 \quad . \quad . \quad . \quad (28.29)$$

Thus, we note that no conventional imaging system, including holography, has as good depth resolution as lateral resolution.

For the acoustical holography system we have chosen, we have the following resolution expressions:

$$\Delta x_1 \simeq \frac{\lambda_S r_1}{2L_x} \quad \text{and} \quad \Delta r_1 \simeq \frac{\lambda_S}{2}\left(\frac{r_1}{L_x}\right)^2 \quad (28.30)$$

where λ_S is the wavelength of sound in the medium, r_1 the distance to flaw, and L_x the aperture of hologram.

The improvement in lateral resolution by 2 and radial resolution by 4 is a consequence of the simultaneous source and receiver scan mode we have chosen (1). These expressions assume that the transducer produces a perfect equivalent point source. This, of course, is not true. Thus, as in any imaging system using a cascade of components, the weakest link dominates. Therefore, the best resolution we can attain is that of the transducer, namely,

$$\left.\begin{aligned} \Delta x_1 &\simeq 1{\cdot}22\lambda_S\frac{f}{a} \\ \Delta r_1 &\simeq 2\lambda_S\left(\frac{f}{a}\right)^2 \end{aligned}\right\} \quad . \quad . \quad (28.31)$$

where f is the focal length of the transducer and a the diameter of the transducer.

For our transducer this amounts to

$$\left.\begin{aligned} \Delta x_1 &\simeq 4{\cdot}88\lambda_S \\ \Delta r_1 &\simeq 32\lambda_S \end{aligned}\right\} \quad . \quad . \quad . \quad (28.32)$$

At 5 MHz in a metal this means that we can resolve flaws separated by 1·46 mm laterally and 1 cm in depth.

Note from equations (28.30) and (28.31) that for large depths it is possible that the hologram resolution becomes larger than the transducer resolution. In this case the hologram resolution dominates. If we equate equations (28.30) to equations (28.31), we find that the depth at which crossover occurs is

$$r_1 = 2{\cdot}44\frac{fL_x}{a} \simeq 10L_x \quad . \quad . \quad (28.33)$$

Thus, for depths greater than 10 aperture widths we must use equations (28.30) for our resolution calculations. This point has never been reached in our experiments.

Electronic simulated off-axis reference

The reference beam in acoustical scanned holography can be simulated with respect to any inclination angle. This provides the necessary conditions for imaging directly in the projected scanning aperture with separation of the images (true and conjugate) and the undiffracted light. The acoustic reference beam can be expressed in two dimensions as

$$S_R(x, y) = P_R(x, y) \cos(\omega_S t - \beta_S x \sin a_{RS}) \quad (28.34)$$

where $\beta_S = 2\pi/\lambda_S$ and α_{RS} is the inclination angle with respect to the z-axis. The signal contribution to the scanning acoustic receiver by the reference beam is

$$P_R(x, y) \cos(\omega_S - \omega_{RS})t \quad . \quad . \quad (28.35)$$

where $\omega_{RS} = (2\pi/\lambda_S)V_x \sin \alpha_{RS}$ in which V_x is the scanning velocity of the receiver. Equation (28.35) represents a sinusoidal wave whose phase is a function of the scanning velocity and inclination angle of the reference transducer. This shows that the acoustic reference beam can be simulated with an electrical signal of this form and combined with the flaw signal in a balanced mixer or multiplier. If the simulated plane wave is inclined with respect to the plane (i.e. $\alpha_{RS} > 0$), then a linear diffraction grating will be imposed on the hologram. The grating spacing is a function of scanning velocity, V_x, phase shifter frequency, ω_{RS}, and the magnification, m. The grating spacing on the hologram is

$$d = \frac{\lambda_S}{m \sin \alpha_{RS}} \quad . \quad . \quad . \quad (28.36)$$

The finest grating imposed on the hologram, according to equation (28.36), has a spacing of λ_S/m and the maximum frequency is $\omega_{RS} = 2\pi V_x/m\lambda_S$. This is indeed the finest grating that can be achieved with a physical reference beam. It should be obvious, however, that any grating spacing can be imposed on the hologram by proper adjustment of the phase shift control voltage frequency if an electronic reference is used. Therefore, the grating spacing can be less than a wavelength, which indicates that electronic simulation is more versatile than a physical acoustic reference beam. Simply by driving the phase shifter with a signal $\omega_p > \omega_{RS}$, we can impose a finer grating on the hologram to diffract the image as far as we please. In this case

$$d = \frac{2\pi V_x}{m\omega_p} \quad . \quad . \quad . \quad (28.37)$$

An upper limit is imposed by the pulse repetition rate, as mentioned earlier.

The image displacement due to the diffraction grating obeys the grating formula. The angle of diffraction by a grating of spacing d may be calculated by the formula

$$\sin\theta = \frac{\lambda_L}{d} = \frac{\lambda_L m\omega_p}{2\pi V_x} \quad . \quad . \quad (28.38)$$

If the image is reconstructed at the distance $\bar{r}_b$ and θ is small (usually the case), we have an image displacement of

$$\delta \simeq \bar{r}_b\theta = \frac{\lambda_L m\omega_p \bar{r}_b}{2\pi V_x} \quad . \quad . \quad (28.39)$$

A typical example with $m = 3$, $V_x = 6$ cm/s, $\omega_p/2\pi =$ 100 Hz and $\bar{r}_b = 10$ m, yields an image offset of 3·2 cm. This is usually ample for separating the image from the central light spot.

APPENDIX 28.2

REFERENCE

(1) HILDEBRAND, B. P. and BRENDEN, B. B. 'Introduction to acoustical holography', BNWL-SA-3467 1971 (Battelle, Richland, Washington).

BIBLIOGRAPHY

ALDRIDGE, E. E. 'Ultrasonic holography', in *Research techniques in nondestructive testing* 1970, chapter 5 (Academic Press, London).

COLLINS, H. D. 'Holographic scanning techniques for imaging flaws in metals', BNWL-SA-3587 1971 (Battelle, Richland, Washington).

HILDEBRAND, B. P. 'Simultaneous object illumination scanning and detector scanning in holography', *Phys. Lett.* 1968 **27A**, 376.

HILDEBRAND, B. P. 'Holography by scanning', *J. opt. Soc. Am.* 1969 **59**, 1.

HILDEBRAND, B. P. 'The effect of high scanning velocities on the holographic image', *J. opt. Soc. Am.* 1970 **60**, 1166.

C29/72 IN-SERVICE INSPECTION OF PRESSURE VESSELS—STATUTORY INSPECTIONS

J. C. BROWN* P. J. NICKELS* R. G. WARWICK*

For every nuclear reactor to which the very highest standards of monitoring and inspection techniques are unquestioningly applied, there are many hundreds of thousands of items of lesser plant in daily use throughout the country. Examination of Boiler Explosion Acts reports shows it is the latter classes of plant which overall constitute the greater hazard to life and property. Several incidents involving explosion of pressure plant are examined and the conclusion is drawn that periodical inspection alone cannot prevent every type of accident, that in the vast majority of instances the traditional visual methods developed over the years and conducted by trained engineer surveyors provide adequate safeguard, and that the value of modern techniques is recognized by the inspecting authorities and employed by them in those instances which are felt to warrant these methods. The inspection approach to specific types of pressure plant is considered in greater detail, followed by an examination of the dangers involved and inspection requirements for safety devices whose function it is to protect boilers and pressure vessels from shortage of water, overpressure, etc. Finally, the problems of present-day process plants in which corrosion rates greatly exceed the order of figure anticipated in more conventional plant, which is inspected at intervals measured in years, are considered and a brief outline is given of techniques adopted by the inspecting authorities to give warning of a developing dangerous situation.

INTRODUCTION

THROUGHOUT THE WORLD the realization that pressure vessels are potentially dangerous and need to be constructed to satisfactory designs and maintained in a safe condition has, in the majority of cases, been recognized by legislation. In the United Kingdom, for example, the Factories Act governs the safety requirements for boilers and pressure vessels used widely throughout industry. Unlike many other countries, the Factories Act does not lay down detailed specifications for the design of boilers and pressure vessels but requires that they shall be of good construction, sound material, adequate strength, and free from patent defect. In the limited cases where these phrases have been tested in the courts, these words have been held to give absolute liability, but the fact that detailed rules are not laid down enables a greater flexibility of design to be used. The Factories Act also calls for regular inspection of steam boilers and pressure vessels, and the period between these examinations varies with their size and type. In practice, the majority of these inspections are carried out by engineer surveyors employed by the specialist engineering insurance companies.

Firstly, we intend to indicate the distribution of various types of pressure plant in the U.K. and the spread of accidents that do occur. The general question of the place of various types of testing procedure in boiler and pressure vessel inspections will be considered, as will be the extent of damage which can result from the explosion of 'conventional' boilers or pressure vessels. This will be followed by a study of the inspection approach to a wide range of different pressure vessels.

Nuclear pressure vessels, very thick boiler drums, high-pressure chemical reactor vessels all contain very large amounts of energy when operating. This potential is recognized by the operators, and the necessary resources are made available for ensuring that the risks of a catastrophic explosion are reduced as far as is practicable; however, this plant only represents a minor part of that in use in industry. Looking at the figures of items inspected by one of the larger specialist engineering insurers, it should be noted that for every nuclear pressure vessel or nuclear heat exchanger under inspection there are 35 water-tube boilers (of which 11 are large, high-pressure units in power generation), 290 shell boilers of all types, 1500 steam receivers of all types, and 2000 air receivers and other gas pressure vessels of all types.

As a requirement of the Boiler Explosions Acts 1882 and 1890, explosions occurring to all steam boilers and pressure vessels, and other vessels used for heating liquids, have to be reported to the Department of Trade and Industry. The only exceptions to these are boilers or vessels in government or railway service and those on domestic premises. The Department produces an annual report (1)† on the investigations carried out under these Acts, and Table 29.1 summarizes the 45 explosions reported in 1968 (the latest report available). Forty-five per cent (i.e. 20) of the reported explosions were due to the failure of tubes in water-tube boilers, and this will be dealt with at length later in the paper, as will the failure to provide adequate safeguarding against overpressure, which accounted for six of the explosions.

The MS. of this paper was received at the Institution on 10th November 1971 and accepted for publication on 30th December 1971. 33

* *Associated Offices Technical Committee, St Mary's Parsonage, Manchester M60 9AP.*

† *References are given in Appendix 29.1.*

Table 29.1. Analysis of explosions reported under Boiler Explosions Acts during 1968

Cause	Shell boilers	Water-tube boilers		Bakers' steam-tube ovens	Vessels, pipes	Fittings
		Drums and headers	Tubes			
Overheating or shortage of water	6	nil	9	4	nil	nil
Overpressure	nil	nil	nil	nil	6	nil
Corrosion, erosion failure of seams, etc., at normal working pressure	nil	nil	11	nil	3	1
Water hammer	nil	nil	nil	nil	nil	4
Blowing out of manhole door or joint	1	nil	nil	nil	nil	nil

Table 29.2. Analysis of explosions reported under the Boiler Explosions Acts for the five years 1964–68

Cause	Shell boilers	Water-tube boilers		Bakers' steam-tube ovens	Vessels, pipes	Fittings
		Drums and headers	Tubes			
Overheating or shortage of water	22	nil	45	50	nil	nil
Overpressure	1	nil	1	nil	28	nil
Corrosion, erosion failure of seams, etc., at normal working pressure	5	nil	33	7	23	5
Water hammer	nil	nil	nil	nil	5	10
Blowing out of manhole door or joint	3	nil	nil	nil	1	nil

The dangers of and problems associated with bakers' steam-tube ovens have been fully considered by Moorhouse (**2**) in an earlier paper. The six explosions of shell boilers all resulted from the collapse of furnaces, followed by rupture due to failure to maintain the feed-water supply to the boiler. It is interesting to note that none of the three vessels which failed at normal working pressure had been subject to independent inspection. Table 29.2, which gives the figures for five years (**3**), shows a similar pattern with water-tube boiler failure and bakers' steam-tube ovens a major cause of accident.

We would like to consider in more detail two of the explosions (**4**) which occurred in 1968. The first relates to a shell boiler which exploded as a result of a shortage of water, due principally to either the closing of a valve or to insufficient maintenance of the controls rather than inadequate inspection of the boiler, and illustrates the damage that can be done by a boiler explosion. In the second case an explosion occurred because of the inadequacy of the design and inspection of a new vessel.

The boiler was a 13 ft 6 in diameter economic type designed for 150 lb/in^2, fired with pulverized fuel, and was one of a battery of seven boilers of various types supplying steam for textile processing. About an hour before the accident, complaints were received from the process departments that the steam temperature was too high for processing, and shortly afterwards there was a further complaint that a joint in the steam main was probably faulty. On removing the lagging it was found that the pipe was excessively hot and that the lagging was smouldering. The full-time boiler attendant was instructed to check the boilers and was reported to have just completed this check when the furnace failure occurred. Fortunately in this case, even though the explosion occurred during the working day, no one was killed or injured, but the extent of the damage to the factory, shown in Fig. 29.1, involved direct repair costs in excess of £100,000.

Although a very full investigation was carried out after the explosion, it was never finally established whether the shortage of water which caused it was due to sludge deposits in the controls and water gauges, a defect in operational maintenance, or the inadvertent closing of a feed valve or operation failure. However, this case illustrates the devastation which can result from a boiler explosion.

In the second case, an oil-heated mixing vessel exploded, killing the engineer who was commissioning the plant and injuring another man. The vessel concerned was a jacketed mixer which was heated by using one of the heat-transfer oils. The vessel was to be used at a pressure of 30 lb/in^2 and a temperature of 400°F. It was being commissioned at the time of the explosion, which resulted from the complete failure of the seam connecting the cylindrical part of the jacket to the annular closing ring on the shell of the mixing vessel. The report of the investigation states:

> 'During construction the welding procedure adopted conformed to well-accepted practice, excepting for the welding of the flange plate to the outer shell. On completion of welding, the circumferential seam which failed was finished flush with the shell and flange plate in order to present a good appearance.'

Fig. 29.2 is a photomacrograph of part of the seam in question and shows the poor quality of the weld concerned. Clearly this weld detail was not in accordance with any recognized pressure vessel standard, and there

Fig. 29.1

was no inspection of any weld preparation. The report on the accident has considered very fully the possibility of the vessel being subjected to overpressure and has concluded that failure occurred at a pressure of approximately 28 lb/in^2 and temperature of 400°F. After the accident a replacement vessel was designed to B.S. 1500 and subjected to independent inspection during manufacture.

It would be easy to call for non-destructive testing (N.D.T.) by ultrasonics or radiography of all pressure vessels at prescribed intervals. However, when one looks carefully at the present effectiveness of inspection methods—which are basically a detailed visual inspection backed up by N.D.T. or destructive testing as may be appropriate where the engineer's suspicions are aroused—it is considered that the vast majority of boilers and pressure vessels in ordinary use do not require the regular application of sophisticated N.D.T. examinations. In this connection, it is worth remembering that the majority of air receivers and smaller pressure vessels sold in this country are only subject to a hydraulic test on completion. Indeed, of the hundreds of thousands of air receivers in use in this country, we can only recall one where an explosion occurred due to a failure that could have been found by N.D.T. and was not found by the normal visual examination, although many hundreds of air receivers have been repaired following the finding of defects by visual methods; Fig. 29.3 illustrates this case.

In the light of this experience, should the use of N.D.T. at high cost be extended to this type of vessel on a regular basis? The fact that there are so few instances of explosions which can be attributed to original manufacturing faults, or defects attributable to working conditions, is a tribute to the general standard of inspection by the engineering inspecting authorities using standard visual inspection techniques developed over the years. These are backed up by ultrasonic, radiographic, and other methods of flaw detection as considered necessary.

We have tried to illustrate the general philosophy behind the inspection of pressure plant and will now consider in more detail the actual approach used over a wide spectrum of plant. We make no apologies for concentrating on the more common plant—this is where the greatest hazard to life and property lies—or for concentrating on the finding of defects and not discussing the application of fracture mechanics techniques to defects once found. One has first to find one's defect, and in the majority of cases a repair will in practice be more economic than carrying out a detailed defect assessment with very limited or non-existent material data.

WATER-TUBE BOILERS

Water-tube boilers must be considered as two separate problems: the drums and headers, and the tubes. Boiler drums and headers present similar inspection problems to other pressure vessels, and the general approach is to carry out a thorough visual examination of these components. This will be backed up by surface flaw detection as experience demands. To carry this out effectively, regular removal of the drum internals must take place. In addition to these examinations, ultrasonic tests are carried out on a regular basis at discontinuities such as branch

a Junction between the annular ring and the outer shell.

b Junction between the annular ring and the outer shell

Fig. 29.2

welds, particularly where certain high-yield steels have been used in the construction of the parts.

Now turning to tubes, the splitting of a water-tube boiler tube and the consequent ejection of contents presents a serious hazard, but even the more minor failure can result in serious downtime and consequent production losses. The inspecting authority is concerned with three main areas: external corrosion or erosion, internal corrosion, and the build-up of scale. Erosion of the external surfaces of tubes by soot blower or fly ash is relatively easy to detect with experience as it is localized. Gauging of the tubes is always carried out, and this can be supplemented by ultrasonic thickness testing. However, these techniques are limited to those tubes which are accessible.

Corrosion of the internal surfaces of tubes, the majority of which are not straight and frequently of considerable length, is much more difficult to detect or measure, but an indication of the general state can be gained from the tube ends as seen from drums and headers. Usually when pitting or wasting of the bores of tubes is suspected it is the practice to remove a percentage for sampling in order to determine the condition of the surfaces remote from the ends. This is expensive both from the point of view of outage and materials, and there is always the possibility that the selection may not be representative. Miniaturized film or TV cameras with integral illumination are now available, and these can be manoeuvred through bent tubes to photograph the complete circumference of the bore at any desired interval. Fig. 29.4 illustrates the clarity of photographs obtained at intervals along the bore of a water tube.

In general, water-tube boilers will only tolerate quite thin scales before the tubes start to overheat; but with modern water treatment methods, trouble due to scale build-up is rare. Equipment is available and used for measuring the scale in a boiler tube by means of a head which can be manoeuvred in the same way as a miniature camera and measures the variation in its inductance as it is separated from the tube wall by scale. The regular proving of a clear waterway through normal tubes will detect debris or massive scale build-up. Nevertheless, experience indicates that tube sampling is essential as the boiler ages, especially in areas where access for gauging or ultrasonic examination is impossible and where scale build-up problems are known or suspected.

SHELL BOILERS

The genetic term shell boilers covers a wide range of types, but the main method of inspection is a thorough visual examination of the internal and external parts.

Approximately full size.
a Detail of fracture face showing the ruptured weld, (top) unwelded plate edge, and intermittent reinforcement of internal weld.

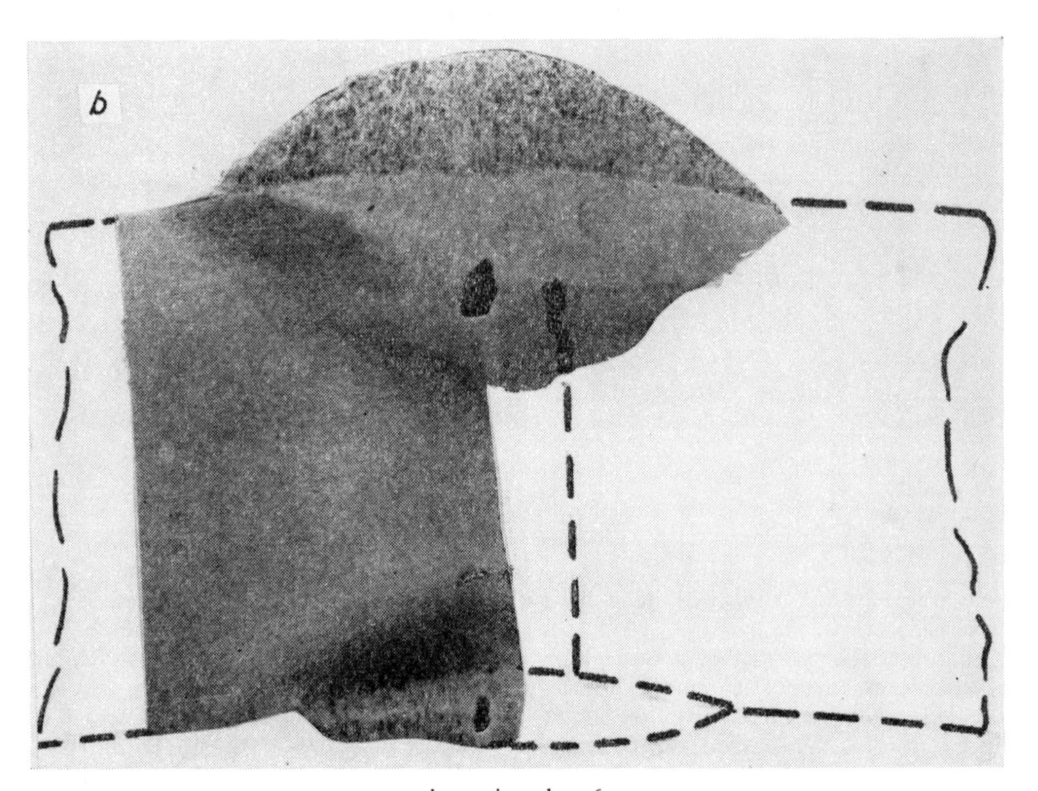

Approximately ×6.
b Macrosection through fractured longitudinal seam showing undesirable features: (1) partial penetration, (2) unwelded plate edge, (3) single-run internal weld.

Fig. 29.3

Although this is normally carried out after the boiler has been thoroughly cleaned and descaled, either mechanically or chemically, there can be advantages in carrying out a preliminary internal examination before any descaling has been performed. With the scale still in position, it is possible to ascertain the extent of scale build-up which can lead to troubles, particularly in boilers with a high evaporative capacity, and the absence of scale in areas can be a tell-tale sign for indicating excessive localized straining.

Where conditions indicate that general corrosion or erosion of areas is taking place, the normal examination will be supplemented by thickness measurements, probably using ultrasonic techniques. Although welded boilers have superseded riveted boilers, the majority of shell boilers in service are still of riveted construction. These present particular problems for inspection in that slight leakage of a riveted seam can lead to serious corrosion under lagging and persistent leakage can lead to severe cracking in the seam itself. The necessity of regular baring of seams had been recognized by the engineering insurance companies for many years, but this practice was made a requirement for boiler examinations at specified intervals in the Examinations of Steam Boiler Regulations 1964 (**5**).

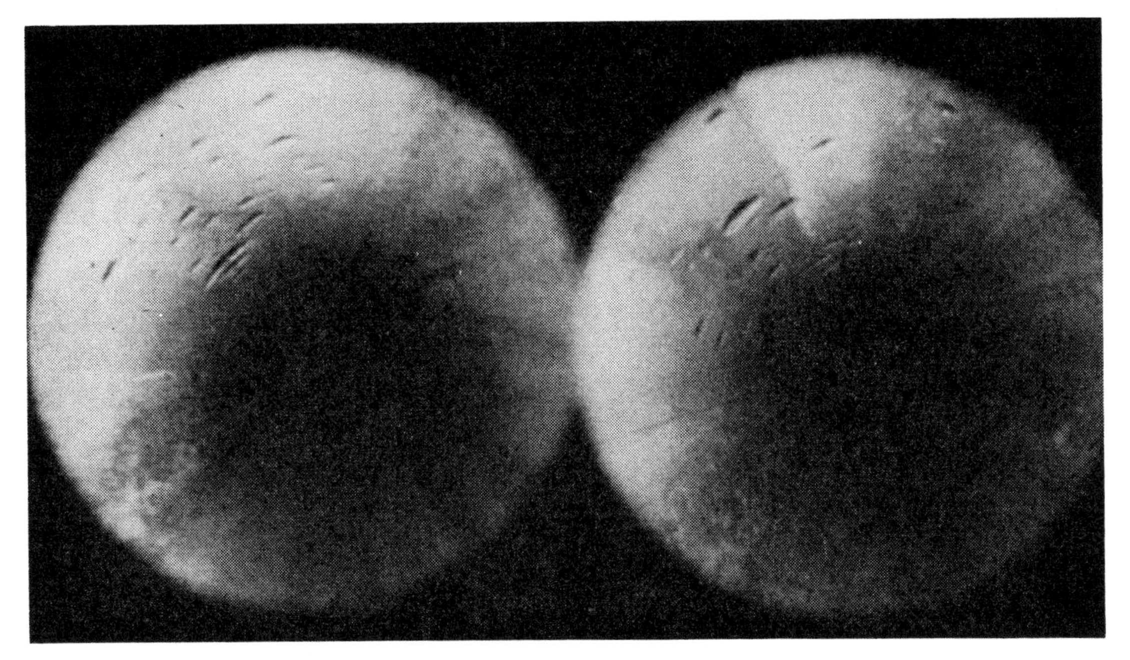

Fig. 29.4

Where regular leakage is taking place from seams it is modern practice to call for an ultrasonic inspection of the area to ascertain the presence and extent of any cracking. This eliminates the method of rivet removal, and in serious cases butt strap removal, which was the practice before ultrasonics were developed.

An increasing percentage of shell boilers in service are fitted with smoke tubes, and these must be considered carefully from the inspection point of view. Although the failure of a smoke tube in a shell boiler is unlikely to result in a serious explosion, their failure in service due to waterside pitting or corrosion, or to fireside corrosion or erosion, can lead to operational embarrassments. The access to these tubes within the boiler shell is normally extremely limited, and when one considers that the tubes only have a limited useful life, which probably normally ranges from 5 to 15 years, the removal of tubes for sampling as they get older is necessary to prevent unexpected outage.

Experience with welded multitubular shell boilers indicates that fatigue cracking does occur in areas such as the attachment welds of the furnace tubes or shell to the end plates and at the tube ends. These defects are normally found initially during a visual examination or manifest themselves by slight leakage. However, increasing use is being made of magnetic and ultrasonic inspection methods to locate defects in areas where experience of particular types of boiler or the particular usage has indicated this to be desirable. Fig. 29.5 shows a macrosection through a shell to end plate weld of a boiler which was

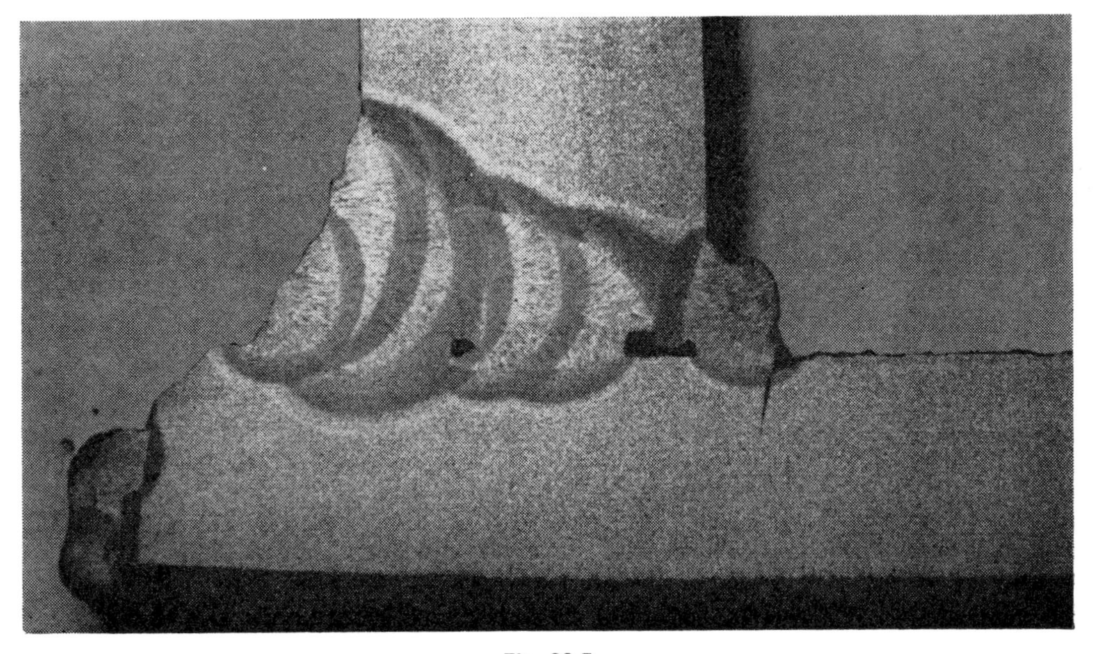

Fig. 29.5

only nine years old. The defect was discovered due to slight leakage from under the covering in the area of the seam, which, when an ultrasonic inspection was carried out of the defect, was found to extend round the shell with a depth varying from 5 mm to 1 mm. A detailed metallurgical investigation was carried out which indicated that the crack was smoothly perpendicular to the shell surface and had the characteristic progressive striation marks typical of a fatigue failure. The crack had started from the toe of the inner seal weld and there was evidence of undercutting and crevice corrosion within this region. The fact that the crack extended round the circumference, however, seemed to eliminate the possibility of a local defect being responsible; in addition, the general condition of the internal boiler surface was good with no sign of any other significant corrosion. Mechanical tests and an analysis of the shell material indicated that it was suitable for the application. The fact that the end plates on this type of boiler are restrained by the shell, the furnace tube, and the smoke tubes, which will all be operating at different temperatures, probably led to the growth of this defect but also prevented a sudden rupture. After the finding of this defect, steps were taken to ensure that particular note was taken of this area on similar boilers. The normal visual examination was backed up by surface crack detection; or where due to the size of the boiler and difficulties of access a sufficiently detailed examination could not be made, ultrasonic examination was called for.

LARGE FLAT SURFACES

A flat plate is a most unsatisfactory shape from a pressure-resisting point of view. When it is necessary to use this design, stays must be provided to limit the bending moment and to prevent distortion. This type of construction is normally employed as little as possible, but when, as in the case of locomotive-type boilers which tend also to use unsatisfactory feed water, it is unavoidable, the engineer surveyor has special problems. However, a hammer in trained hands is effective in detecting broken stays, and the application of regular hydraulic tests with observations of deflections will detect serious local corrosion of the plate material and stays.

STEAM-JACKETED VESSELS

Steam-jacketed vessels from the basic open-topped pan to those designed for special trades present difficulties of inspection owing to the surfaces within the jackets being inaccessible. This is particularly important when ladles or power-operated agitators are employed for mixing, when abrasive action results in an even reduction in the thickness of the plates of the inner vessel.

With this type of vessel, visual inspection must be supplemented by regular thickness and hydraulic tests; but with the increasing problem of obtaining replacements at short notice it is desirable to anticipate the point at which repairs or replacement might be expected, and thickness testing is being more widely used. Often this is carried out by ultrasonic thickness measurement, but frequently firms find the time-honoured drill test to be more economic.

Because grey cast iron will invariably fail with practically no elongation (less than 1 per cent), tensile design stresses of not more than one-tenth of the minimum specified tensile strength are usually employed. In addition, a generous allowance is necessary to compensate for any possible misalignment of the core in the mould, which may be present but not detectable.

As a result of this extra thickness, straightforward corrosion is not usually a significant factor as the plant is often made obsolescent before minimum thicknesses are reached. One particular mechanism, however, that known as graphitic wastage, must be checked for whenever the items are opened up for inspection. There is a fairly well-known pattern related to the incidence of this form of attack, being more severe along flange faces and other parts where crevice conditions and higher stresses exist. A close examination and testing with a sharp probe will indicate the presence of the weak spongy mass of corrosion products.

Defects are present in iron castings in a number of forms, including blow-holes, porosity, or core misalignment—all of which can give rise to failures under certain conditions where stresses over and above the design level are imposed. This situation can arise when flange bolts are overtightened, temperature gradients are imposed, and it is here that the use of hydraulic tests and judicious hammer testing reveals the hypercritical defects, and so forestalls the occurrence of a failure.

It is unusual, however, to have to resort to more sophisticated inspection techniques for this class of pressure vessel, except for steam-heated rolls used in the paper industry requiring a high-quality surface finish that can only be maintained by frequent machining. Here, ultrasonic methods probably represent the best and quickest method of ensuring that sufficient metal remains to withstand the applied loads. Although it is now usual for vessels in the catering trade to be fabricated from stainless steel and for laundry equipment to be of fabricated steel or cast-aluminium construction, cast-iron vessels were previously employed extensively in both trades and large numbers of them are still in use.

AIR RECEIVERS

As previously explained, air receivers constitute the bulk of pressure vessels examined under the various Acts and Regulations, and the majority, being of simple design, do not require sophisticated techniques to determine their condition. In the main, largely due to the vigilance of engineer surveyors, failures as a result of corrosion are remote, but cracking as a result of vibration at the point of attachment at the supports is frequent. Therefore, parts subject to defects of this nature are examined carefully. Magnetic crack detection for preference or dye-penetrant testing enable this type of defect to be detected before the cracks propagate to dangerous proportions.

A number of explosions are brought about by the ignition of oil vapour within air receivers. Although this type of explosion is not associated with the condition of a vessel, obvious signs of contamination by oil carried over from the compressor, coupled with an excessive temperature at the point of discharge, will alert the engineer surveyor and will prompt him to suggest precautions such as the provision of fusible plugs, oil separators, and after-coolers.

LIQUEFIED PETROLEUM GAS VESSELS

The hydrocarbon fluids normally contained in liquid petroleum gas vessels are non-corrosive. Nevertheless, an examination sequence at five- and ten-yearly intervals is recommended. A code of practice issued by the industry sets minimum standards for the examinations made, and these may call for ultrasonic testing to supplement visual examinations. Ultrasonic testing is considered particularly necessary for vessels installed underground which are not being hydraulically tested. As an example of the advisability of regular inspection it should be mentioned that occasionally contamination of liquid petroleum gases does occur. In one case, blistering of a segment of a spherical storage sphere was reported which subsequent investigation established was brought about by the presence of small quantities of water and minute quantities of hydrogen sulphide.

SAFETY FITTINGS

We have dealt with the current inspections approach to the actual physical condition of pressure plant. Equally important for the safe operation of plant is the condition of the safety fittings, and any engineer surveyor of pressure plant must also satisfy himself that the safety devices in connection with that plant are in a satisfactory condition. The Factories Act, British Standards, and other codes lay down which fittings shall be provided in different circumstances. However, in general, safety fittings for boilers and pressure vessels are of four types: (*a*) safety or relief valves, (*b*) reducing valves, (*c*) pressure gauges, and (*d*) water-level control or indicating devices on boilers.

The inspection of safety and relief valves is to ensure, firstly, that the valve is free to operate and, secondly, that the valve will, in fact, operate at the correct pressure. The safety valve is generally dismantled at each thorough examination to ensure freedom of movement and the correct setting, and operation checked by pressure test under operating conditions. These inspections of the item under working conditions are just as important in the overall examination as the inspection of the plant cold.

Reducing valves are normally performance tested to ensure that the pressure on the reduced side of the valve cannot exceed the safe working pressure in the most adverse conditions, i.e. with the supply pressure at its operating maximum and with the fluid demands on the low-pressure side at their minimum. The reducing valve should always be fitted with a suitably sized relief valve and pressure gauge on the downsteam side to act as protection in the event of failure of the reducing valve. Alternatively, a suitable appliance for cutting off automatically the supply of steam as soon as the safe working pressure is exceeded may take the place of the safety valve. When inspecting any steam or pressure vessel the engineer surveyor must ensure that suitable safety fittings are provided whenever the pressure at the source of supply exceeds the maximum permissible working pressure of any item concerned. It is not sufficient to rely on process controls, such as temperature regulators, which govern the flow of steam to the vessel to protect it against overpressure.

The pressure gauge is an important instrument. It is checked at each examination either by removing it from the plant and checking it over its range against a standard test gauge or by fitting a calibrated test gauge to the plant whilst it is operating, and comparing the readings.

The loss of water in steam boilers is a major cause of serious accidents, and particular care must be taken to ensure that water-gauge and water-level controls are working satisfactorily. However, in this connection it is extremely important to realize that the correct operation of these fittings requires regular testing. It is not sufficient to rely on an annaul or bi-annual thorough examination to ensure that they are working satisfactorily.

The problem has become much more acute since the wholesale introduction of automatically controlled boilers with minimum attendance and the decline in full-time manning. The A.O.T.C. published requirements for the insurance of automatically controlled steam boilers as long ago as 1958 which drew particular attention to the necessity of regular tests. The same philosophy is now incorporated in H.M. Factories Inspectorate's Technical Data Note 25 (**6**).

The philosophy of these recommendations on the subject of regularly testing controls can be summarized as follows:

(*a*) The steam and water legs of the water gauge should be blown through separately at least once every 8 h of normal steaming.

(*b*) The water-level feed pump control, the firing controls, and the independent overriding water-level control should be blown down and given a functional test at least once a day or once a shift under normal operating conditions by a trained boiler attendant or a technician familiar with boiler controls. This functional test should ensure that the controls actually operate the feed pump, sound the alarms, and cut off fuel and/or air supplies as may be appropriate.

(*c*) At least once a week the water controls should be checked by manually interrupting the feed supply and lowering the water level by evaporation until the alarms and cut-outs operate.

During the regular annual thorough examination of the boiler the fittings are opened up to ensure that there is no sludge or scale build-up in the chambers or the lines, and the freedom of operation of all moving parts is checked. Functional tests are, of course, witnessed by the engineer surveyor at each visit to the plant whilst it is in operation.

PLANT OPERATING IN ARDUOUS CONDITIONS

Finally, we should like to make reference to the problems associated with the operation of chemical plant using highly corrosive materials where, even when using the best available materials, the corrosion rate on the surface of vessels and pipework and the development and propagation of cracks are such that these limit the time between overhauls and extensive repair or replacement of parts of the plant. In the practical case this period may be less than six months, and a totally different approach to pressure vessel inspection is involved.

When problems such as these are involved, a full-time monitoring team will be provided which is trained in a wide range of N.D.T. techniques. Before the plant is operated, an ultrasonic thickness survey is carried out of all the vessels and pipework at literally thousands of marked

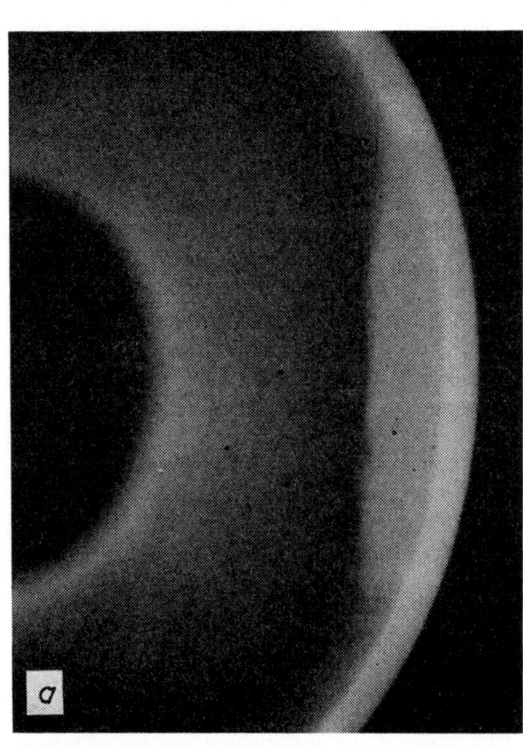

a Profile radiograph of superheater element bend taken to determine if priming and chemical scale carry-over has occurred. A build-up of scale in the outer radius is clearly shown.

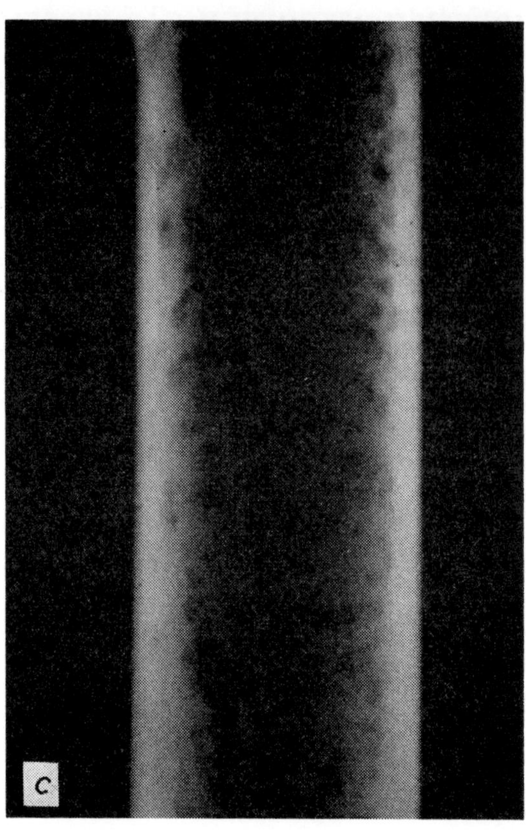

c Profile of radiograph of pipe section to determine the extent of internal pitting.

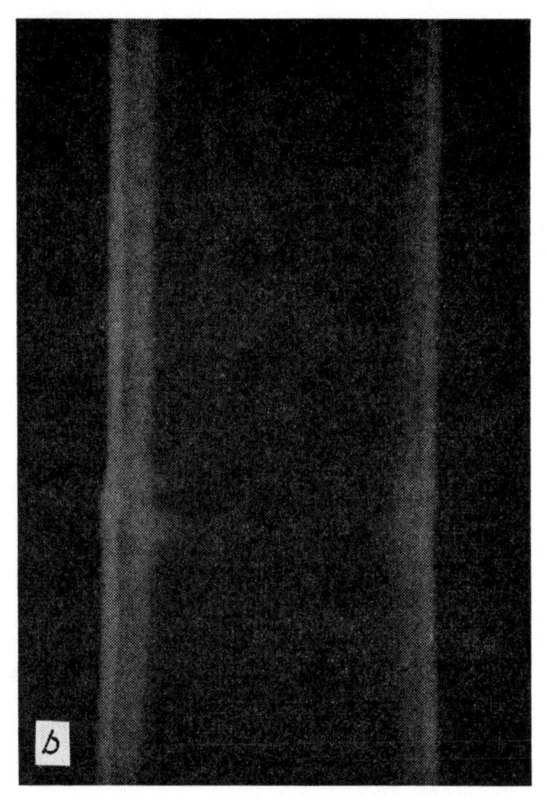

b Profile radiograph of pipe section in gas reforming plant to determine if carbon build-up present. Carbon deposit approximately twice the wall thickness is clearly shown.

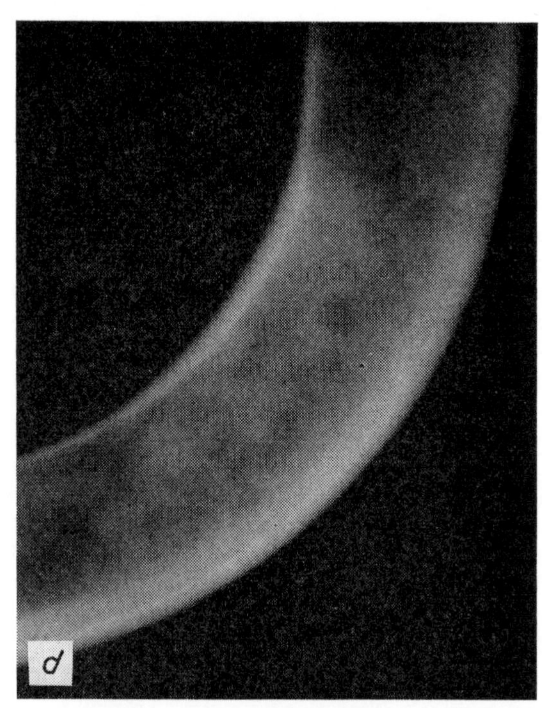

d Profile radiograph of pipe bend. Normally corrosion and thinning of the outer radius of the bend is expected. In this case the radiograph clearly shows a peculiar form of corrosion in which the inner radius has been mostly affected.

Fig. 29.6

points, and all the readings recorded. In addition, areas of high stress concentration and discontinuity, and, in particular, welds in these areas, are surveyed by ultrasonic or radiographic flaw-detection methods.

During operation of the plant, regular thickness measurements are taken and recorded to ascertain the corrosion rate in various parts of the plant, normally on the basis of very frequent checks on a limited number of key points throughout the whole plant to ascertain the general level of material loss. The checking and study of all the points are carried out at extended frequencies on a regular basis, modified as necessary by the results from the key points. From this information the plant is shut down before a minimum safe thickness of the critical components is reached.

Parallel with this, ultrasonic and radiographic examinations are made of areas of stress concentration to locate any cracks which develop and to study their growth. When cracks are detected they are kept under regular study and the plant shut down and repairs carried out before the critical defect size is reached. In addition to the normal radiography designed for flaw detection, profile radiography is proving a useful tool in this type of detailed inspection work, particularly as it can be carried out on hot pipework; Fig. 29.6 illustrates four examples of its use.

CONCLUSIONS

In this brief study of the current practice used by the specialist engineering insurance companies in the periodic inspection of pressure vessels and boilers we have indicated the very wide range of plant in use, and the diversity of techniques is recognized. Without them the more sophisticated plants could not run, but their indiscriminate use on simple plant would not only increase the direct cost of inspection but would also increase the downtime of the plant. In our view, the experienced engineer surveyor's visual examination continues to be the best basis for in-service pressure vessel inspection, but he must be backed up by the necessary N.D.T. services which he can call upon and must be trained to appreciate their full scope and potentialities.

ACKNOWLEDGEMENTS

Acknowledgement is made to the Controller of Her Majesty's Stationery Office for permission to use material published in Annual Reports on the working of the Boiler Explosions Acts and Reports of Preliminary Enquiries. Also to Commercial Union Assurance Company Limited and National Vulcan Engineering Insurance Group Limited for the use of photographic material, and to these two companies and British Engine, Boiler and Electrical Insurance Company Limited for help with the text.

APPENDIX 29.1

REFERENCES

(1) DEPARTMENT OF TRADE AND INDUSTRY. *Boiler explosions 1968* 1971 (H.M.S.O., London).

(2) MOORHOUSE, W. E. 'Investigation into the reasons causing explosion in steam tubes of bakers' ovens', *Proc. Instn mech. Engrs* 1960 **174**, 561.

(3) BOARD OF TRADE. *Boiler explosions 1964*; *Boiler explosions 1965*; *Boiler explosions 1966*; *Boiler explosions 1967*; and also reference (1) (H.M.S.O., London).

(4) BOARD OF TRADE. 'Boiler Explosions Acts 1882 and 1890', Reports of preliminary enquiries, Nos. 3458 and 3459 (H.M.S.O., London).

(5) 'The examination of steam boiler regulations 1964', Statutory Instruments 1964, No. 781 (H.M.S.O., London).

(6) DEPARTMENT OF EMPLOYMENT. 'Safe operation of automatically controlled steam and hot-water boilers', Technical Data Note 25 1971 (H.M.S.O., London).

C30/72 ACOUSTIC EMISSION AS AN AID TO PRESSURE VESSEL INSPECTION

P. G. BENTLEY* T. E. BURNUP* E. J. BURTON* A. COWAN* N. KIRBY*

The development by the U.K.A.E.A. of acoustic-emission monitoring as a non-destructive tool to aid pressure vessel inspection has progressed in three main ways. Firstly, an experimental facility, S.W.E.L., has been built and tested which is capable of detecting and locating small flaws from the acoustic signals received by up to 10 sensors mounted on the pressure vessel during a pressure test. Secondly, measurements have been made using early components of this facility to gain experience on large pressure vessels at manufacturers' works. Thirdly, a large number of measurements have been made in the laboratory of emission rates from artificial defects in test pieces and pressure vessels of low- and medium-strength steels. Thus, the capability is now available to make a 100 per cent inspection of pressure vessels over a wide range of engineering steels; and there is the prospect, with further development, of monitoring during operation the severity of any defects relative to vessel failure.

INTRODUCTION

When a metal is stressed, plastic deformation and micro-cracking can occur before fracture; these processes emit high-frequency stress waves which can be detected by acoustic instrumentation. If a metal structure contains significant defects, cracks, inclusions, lamellar tears, etc., these will give stress concentrations such that plastic flow or cracking will occur at the tips of these defects while the bulk of the structure is at a low-stress level. Thus by measurement of the stress-wave emission, it is possible to detect defects in a pressure vessel when it is stressed well below its failure point. In principle, the stress waves may be monitored by a single remote acoustic sensor; the defects may be located by triangulation methods using an array of sensors.

A metal once stressed gives little or no emission during a subsequent stressing, unless the previous stress is exceeded. This is true also of a metal containing a defect, provided there has been no growth of the defect between the loadings and that the defect does not contain a rapidly growing fatigue crack. This so-called Kaiser effect (1)† can give considerable confidence in the operation of a structure or vessel if observed, for example, in a repeat pressure test.

Conversely, the release of acoustic emission could imply that new defects had been introduced or that existing defects had grown in the intervening period. However, in some steels recovery can occur, so that emission in a repeat test does not necessarily imply the growth or appearance of a new defect.

Acoustic-emission techniques thus offer the possibility of inspection of the whole pressure vessel in one operation, together with the specification of the position of any defects. Such a capability has important applications to structures requiring a high degree of confidence in their integrity, e.g. nuclear reactors, rockets, and to structures where access for conventional non-destructive testing (N.D.T.) methods is limited either by engineering features or by cost.

The development of this capability demands first of all a thoroughly reliable acoustic system for the detection and location of the signals. Much work has been carried out in this field during the past five years, particularly in the U.S.A., including, for example, Idaho Nuclear Corporation (2) and Battelle North-West (3). In the U.K. an acoustic instrumentation has been constructed by the U.K.A.E.A., Risley Engineering and Materials Laboratory (R.E.M.L.), and used for tests ranging from small laboratory test pieces to a 24-m (80-ft) nuclear pressure vessel. As confidence is steadily growing in the acoustic technology, more attention is being paid to the originating signal and the extent to which a quantitative relation might be achievable between the signal and the defects, and this work is also described in outline.

The original American investigations tended to concentrate on high-strength brittle materials where large-amplitude acoustic emissions were obtained with a large increase in the rate of emission just before failure. The U.K.A.E.A. work has been more on low- and medium-strength steels of greater relevance to the bulk of industrial engineering structures. Signals were found to be much weaker and the relationships between emission rate and failure were less clear cut. However, as the work has progressed an understanding is developing to relate the acoustic emission of the mechanics of failure.

ACOUSTIC-EMISSION TECHNIQUE

When a metal is stressed, deformation first occurs at the tips of any defects. This deformation, which may be plastic flow, micro-cracking, or large-scale cracking, produces stress waves that contain very-high-frequency

The MS. of this paper was received at the Institution on 15th November 1971 and accepted for publication on 3rd January 1972. 22

* *Acoustic Dept., R.E.M.L., U.K.A.E.A., Risley, Warrington, Lancs.*

† *References are given in Appendix 30.1.*

Fig. 30.1. Transducer (70 kHz variety)

components, often above 10 MHz. These stress waves are not normally audible (a well-known exception is the 'cry' of tin), and the waves propagate as an elastic disturbance or pulse taking the form of compression, shear, Rayleigh, and Lamb waves. Attenuation occurs with distance and increases with frequency. Shadowing by highly absorbing areas and reflection both occur.

Design and operating principles

The acoustic pulses may be detected at the metal surface by means of piezo-electric transducers which convert the stress to an electric signal, a typical device being shown in Fig. 30.1. The signal is then amplified, generally by using a pre-amplifier near to the transducer. Since the pulses are often small, the signal-to-noise ratio is important and a

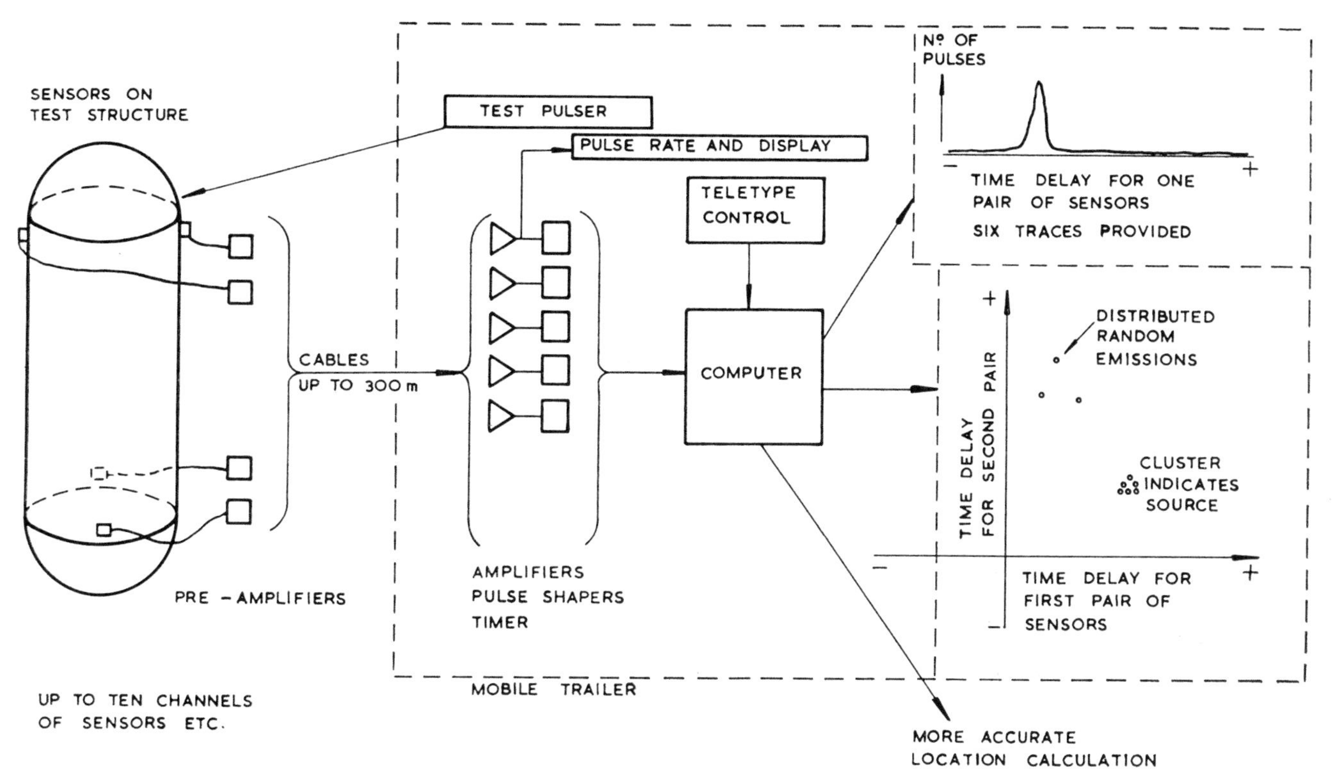

Fig. 30.2. Block diagram of equipment

restricted bandwidth is used to minimize noise. Mechanical noise in the test structure caused by the movement of supports, operating machinery, and fluid flow is high in the audio-frequency range but decreases with frequency, while at much higher frequencies the attenuation of the pulses with distance is serious. The detector pass band is usually in the region 50–500 kHz, and the precise frequency to get a reasonable optimum between noise, attenuation, and sufficient precision in the delay time measurement may be adjusted to suit the specific application.

The detected pulses should then be treated in two ways. The first method is to count the pulses and from the number of counts per second, or the total number of counts, to gain an indication of the importance of the emitting defect. This method is used as a general monitor, especially for characterizing material specimens and for estimating the risk of gross failure of a complex structure.

The second method is to use several sensors appropriately spaced across the vessel surface and then to time the arrival of the same pulse at each sensor. These time differences can be used to determine by triangulation the location of the emitting defect with respect to the sensors. The principles of such location methods have been applied in several technological fields but are complicated in this application by the geometry of the test structure. A few simple cases do exist, e.g. long pipes or defects in welds where a single dimension is sufficient to position the defect, but for structures like pressure vessels, multiple paths and attenuating areas tend to confuse the location calculations. Methods are necessary to assist the operator to sort out the information.

The system has to provide accurate measurement of delays which is presented either externally to the operator or internally to a data-handling system with a technique for associating corresponding delays. This then will allow the source to be located, for example, by the operator from the use of tables calculated before the test or from comparison with the results of test pulses injected artificially; or by a data-handling system from direct computation.

The more important results from both methods at least should be available quickly to allow monitoring of the test, but accurate location of suspect areas can usually wait until after the test.

Fig. 30.3. Photograph of mobile trailer, S.W.E.L.

Summary of equipment

The U.K.A.E.A. (R.E.M.L.) has assembled equipment for experimental use to carry out the functions described above, and this is shown diagrammatically in Fig. 30.2. The equipment is located physically in a single trailer known as S.W.E.L., and is shown in Fig. 30.3. The trailer can be towed to the test site and only needs connecting to the mains to become operational. The equipment of the system is summarized below.

The piezo-electric ceramic sensors used are generally in the form of flat discs less than 2·5 cm in diameter, contained in a simple housing that may be bonded to the material surface with glue or with a film of grease. These sensors resonate mechanically when they are excited by the stress-wave emissions, and this resonant frequency gives the best signal-to-noise ratio. Typical frequencies used so far are 70 kHz and 200 kHz. The sensors can be used up to 250°C if suitable coupling is provided.

The pulse-amplifying channel uses a low-noise pre-amplifier near the sensor to feed a long cable to the main amplifier and discriminator in the trailer. There are 10 such channels available.

A single channel is selected for pulse-rate analysis, and any four channels can be selected for on-line location analysis. Any seven channels can be recorded on magnetic tape for later analysis. The single-channel pulse-rate system uses standard equipment with a chart recorder and displays of rate and total count.

The location analyser is based on a pulse timer that generates information on the delays seen by three sensors following the detection of a particular pulse by the first sensor. The information goes to a PDP-8L computer where the validity of the group of pulse signals, as seen by the four sensors, is tested to remove as many spurious signals as possible. The output of the computer is available in permanent form on punched tape or plots on a recorder,

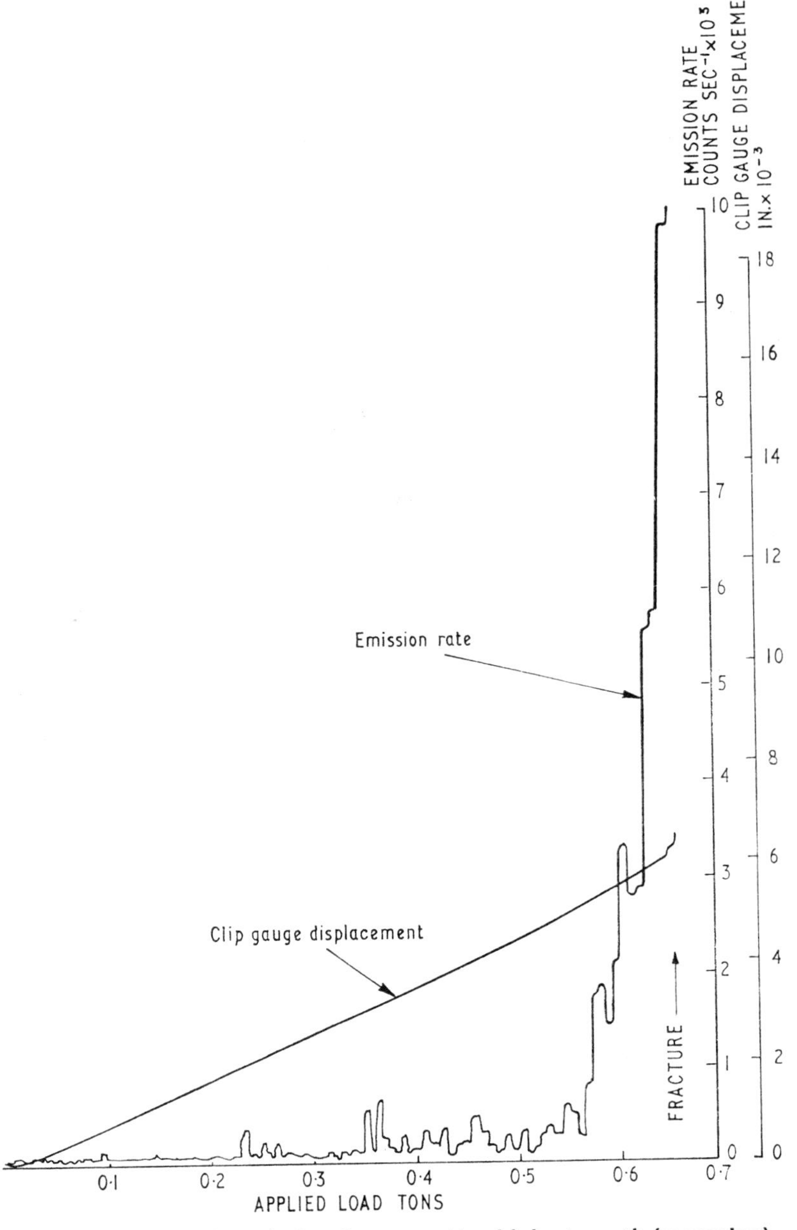

Fig. 30.4. Acoustic emission from an ultra-high-strength (maraging) steel (notched bend specimen)

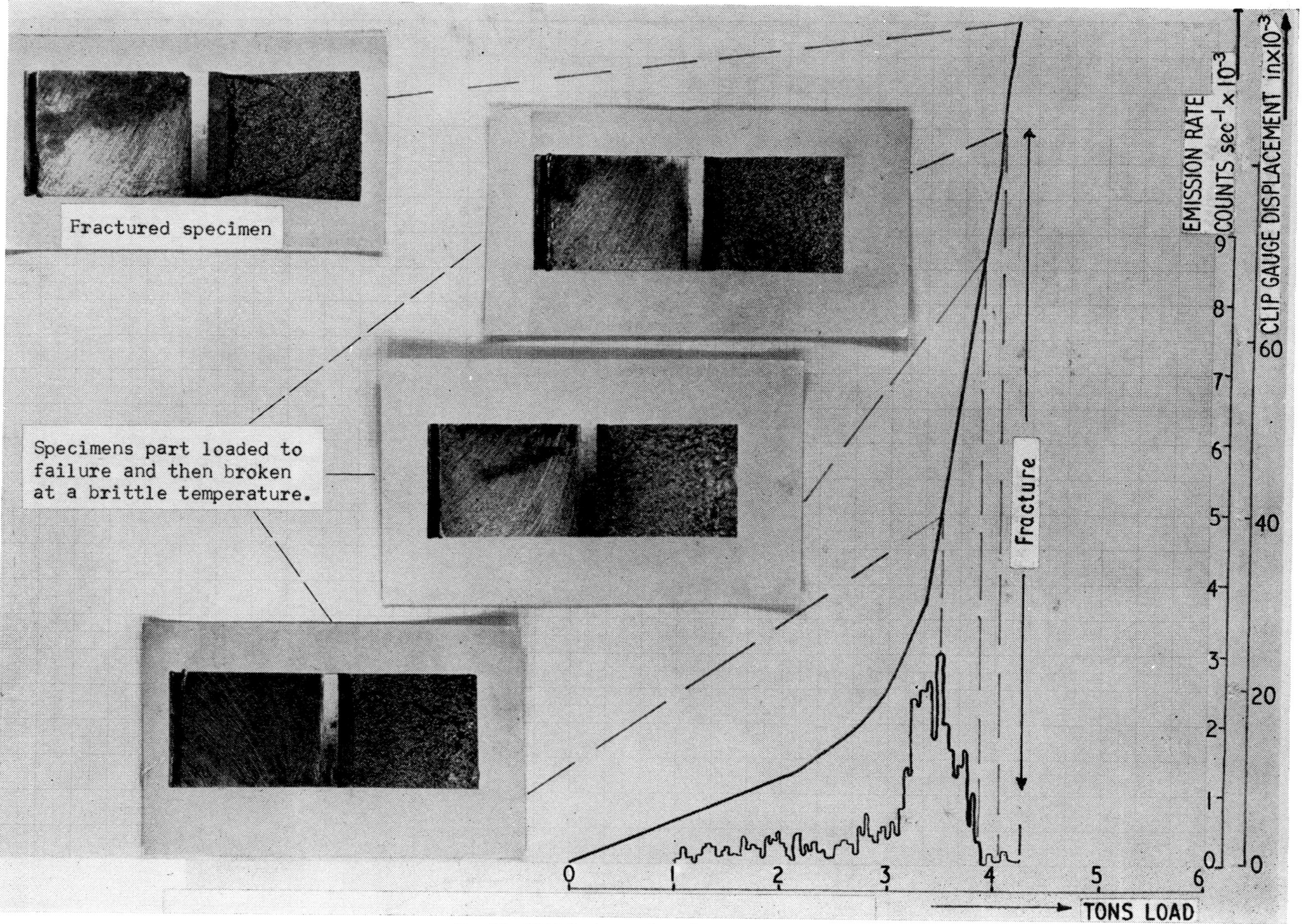

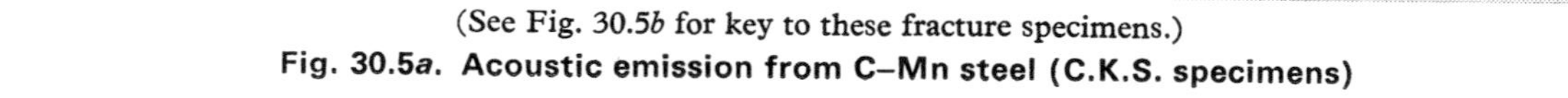
(See Fig. 30.5*b* for key to these fracture specimens.)
Fig. 30.5*a*. Acoustic emission from C–Mn steel (C.K.S. specimens)

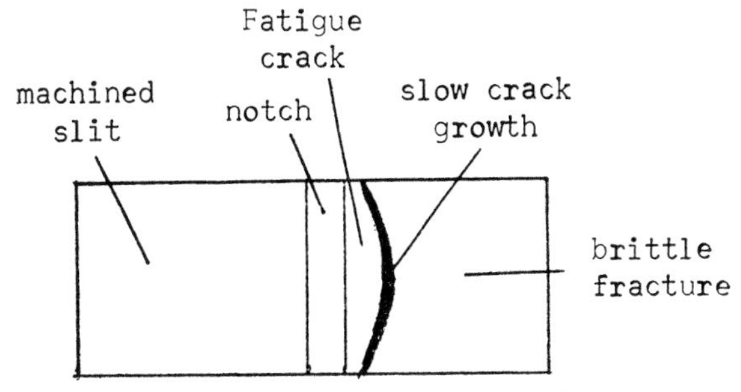

Fig. 30.5*b*. Key to fracture specimens

but for conducting a test it is made available immediately as an oscilloscope display. The system is operated from the teletype mounted in the trailer.

Although three sensors allow the location of a source, the use of four allows considerable redundancy in location so that spurious indications due to reflections and unusual acoustic paths, e.g. round a vessel, can be rejected. The group of four sensors can be combined in six pairs and the results from each pair plotted as a number of pulses received against a base of delay times. One of the six plots is shown in Fig. 30.2, and they are presented during the test on an oscilloscope and later in permanent form. If the number of sources is small, then the peaks can be associated visually and the time delays read off accurately by a strobe system.

A second display is also available and uses a storage oscilloscope. It is a two-dimensional (2D) plot where the co-ordinates are pulse delay times, and there are four separate plots arranged in a square. Each pair of sensors can give rise to a pulse delay measurement and are referred to as a base-line. Any pair of base-lines can be chosen and their delay times used as co-ordinates on the 2D plot. Each emission will then appear as a spot on the oscilloscope in each of the four plots. As the emissions are plotted, those from particular sources will tend to coalesce but those from random sources will be scattered. These plots look something like a map of the source locations but are distorted because the plot is linear instead of the correct hyperbolic.

Location of sources can be achieved in the trailer by reading the delays associated with the several base-lines from either of the displays and calculated either directly or by reference to previously calculated tables. Alternatively, the test structure can be mapped out by using an artificial source which is a piezo-electric element driven by pulses. In addition to the on-line analysis, detailed analysis of the recorded channels can be carried out in the trailer and accurate location of sources achieved by access to a larger computer for part of the work.

TEST PROGRAMME AND RESULTS

The test programme has ranged widely from small laboratory specimens and vessels to large reactor pressure vessels. The specimen tests were aimed at obtaining data which would be relevant to the failure modes of low- and medium-strength engineering steels. Experimental pressure vessels were used to investigate further the effect of pressure cycling on acoustic emission.

Large laboratory vessels were used to develop the S.W.E.L. trailer system whilst the nuclear reactor tests gave experience in field trials as well as direct information on the levels and frequency of background noise.

Specimen tests

Although the potential of acoustic emission has been known for many years, there is relatively little data on the emission characterization of defects in large-scale structures. Ideally, such data should be obtained from tests on actual structures, but a first appreciation of the parameters controlling emission characteristics is possible from tests on small specimens which are notched to simulate the effects of a defect. In some exploratory studies, tests have been made to measure the acoustic emission in notched fracture toughness specimens and relate it to the deformation and fracture occurring at the tip of the notch.

The steels were selected to cover a range of yield strengths (15–30 ton/in^2) found in the more commonly used structural and pressure vessel steels, although one ultra-high-strength steel ($\sim$100 ton/in u.t.s.) was also tested. A.S.T.M. Standard specimens (**4**), 1-in-thick compact K specimens (C.K.S.) for the low- to medium-strength steels and $\frac{1}{2}$-in-thick notched bend specimens for the ultra-high-strength steel, were fatigue cracked before test to accord with the recommended procedure for fracture toughness measurement. A clip gauge was attached to measure the opening of the notch at the surface of the specimen and so provide an indication of the strain occurring at the root of the notch. The acoustic emission was recorded as the number of rings of the transducer above a set threshold amplitude. With this method the larger emissions contribute a greater number of counts than the smaller emissions, and so the measurements are a function of the acoustic energy release.

Two extremes of behaviour are represented by the results from the high-strength steel and a C–Mn steel of 15 ton/in^2 yield strength, and these are shown in Figs 30.4 and 30.5*a* and *b*.

Failure of the high-strength steel occurred under linear elastic plane-strain conditions. Large emissions were detected, some of which were directly audible, and clear warning of failure was obtained from a dramatic rise in the emission count which began at about 80 per cent of the failure load. In contrast, the C–Mn steel gave emissions of a much lower amplitude; no sharp rise in emission occurred in the elastic range, but the emission rate increased appreciably when a marked increase in clip gauge displacement occurred and indicated gross yielding at the notch. In a series of tests in which specimens of the C–Mn steel were part loaded to failure and then broken open at brittle temperatures, it was possible to show fibrous crack extension (dark area beneath the fatigue crack, Fig. 30.5*a*) had occurred after yielding spread across the test piece and that this gave no significant acoustic activity. In other test pieces, this 'quiet' period of slow tearing was followed by an increased emission rate.

The difference between the two types of steel may be caused by the different fracture mechanisms. In a high-strength steel showing negligible plasticity before failure the emission is probably caused by minor instabilities at the leading edge of the crack. Emission in the low-strength steel, on the other hand, appears to be a function of the amount of yielding at the tip of the crack. On this basis, the reduction in emission rate in low-strength steels shown shortly before failure could be due to the effects of work hardening which are known to reduce the emission occurring in a metal deforming plastically (5).

A direct comparison of the cumulative emission count $\sum E$ with the crack-opening displacement, δ (6), at the root of the notch has shown that up to the stage at which the emission begins to decrease, the results approximate reasonably well to a relationship of the form $\sum E = A\delta^n$, where A and n are constants. The relationship is similar to that based on the linear elastic stress intensity factor (i.e. $\sum E = AK^n$) (5) (7), but has the advantage that conditions of higher plasticity can be encompassed corresponding to the failure modes of low- and medium-strength engineering materials in the thicknesses in which they are commonly used.

Experimental pressure vessels

A series of 5-ft-diameter pressure vessels containing artificial defects to study crack initiation conditions have been used to develop acoustic emission methods preparatory to tests on operational vessels. Typical of these was a 3 in thick, 5 ft diameter × 18 ft long cylindrical vessel containing a 0·36 per cent carbon steel test panel in the cylindrical section. The steel was specifically developed for test purposes but has similar mechanical properties to several mild steels commonly used. After welding, the vessel had been stress relieved and pressure tested. The vessel was monitored for acoustic emission during the proof test and shown to be free from major natural defects. A 12-in-long axial slit was then cut through the wall thickness of the test panel and sealed against leakage. To induce growth of fatigue cracks at the slit the vessel was pressure cycled at 50°C and a rate of 4 cycles/h; the cyclic hoop stress (13 500 lb/in²) was approximately one-third the yield strength of the steel.

Acoustic sources were identified by the S.W.E.L. analysis system. Transducers were positioned in two groups, the first to monitor the whole vessel and the second to cover an area around the artificial defect in more detail.

Chart recordings showing the variation in emission rate with applied pressure for a number of cycles are shown in Fig. 30.6. High-amplitude avalanches of signals, overlapping in time, and discrete signals were observed. The avalanches produced the spikes shown in the figure and

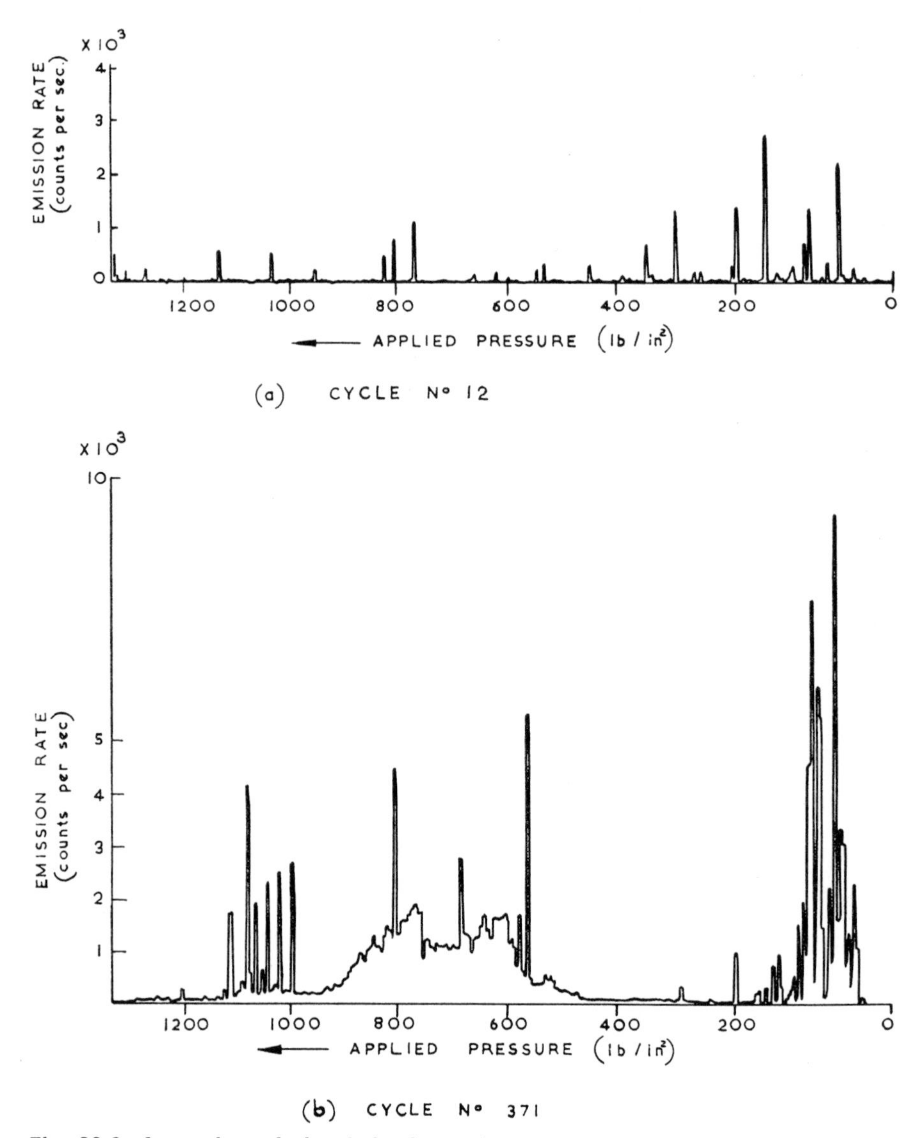

Fig. 30.6. Acoustic emission behaviour of an artificially flawed pressure vessel

were responsible for a continuously high count during the first pressure cycle. The discrete signals were characteristic of those from structural steels and were the cause of the general increase in count rate apparent in the later chart recordings, at intermediate pressure, as shown especially in Fig. 30.6*b*. It will be noted that as the cycling proceeds, the count rate may be divided into three distinctive pressure intervals, i.e. 0–200 lb/in^2, 200–900 lb/in^2, and 900–1350 lb/in^2. However, the interpretation of Fig. 30.6 requires both an identification of the acoustic sources and an appreciation of the mechanism of the growth of fatigue cracks.

The distinctive pressure intervals noted by the emission rate monitor were analysed separately. The analyses are best demonstrated by the two-dimensional plots previously described.

For the 0–200 lb/in^2 interval, Fig. 30.7 shows a result indicating acoustic sources spread apparently randomly over the entire vessel. This is characteristic of mechanical and hydraulic noises which have slow rise times, cannot be timed accurately, and hence give a spurious spread on both time axes. This suggested bedding-down noise from the defect sealing patch and pumping system, an interpretation which was consistent with the nature of the signals observed on an oscilloscope, large in amplitude and overlapping in time to give an almost continuous noise.

The 200–900 lb/in^2 region gives a result which is the opposite extreme. Fig. 30.8*a* shows a very localized source, so small in dimensions that several hundred emissions are superimposed. The position of this source was shown to be very close to that of the artificial defect. This was obtained with the widely spaced group of transducers which monitor the whole vessel, and it is worth noting that there were no other major sources.

Fig. 30.8*b* shows the same source examined with the smaller spaced group of transducers and with much finer time resolution. The source was clearly associated with only one end of the defect and there was an apparent spread in the source around the defect end. The emissions analysed in this pressure region were, therefore, of the greatest interest.

In the highest pressure interval, 900–1350 lb/in^2, diffuse source was found centred around the artificial defect. Some signals came from the defect itself, but a quantitative one-dimensional analysis showed that these formed only a small proportion of the total. The remainder were probably from the region of the outside of the rubber-and-metal patch which sealed the artificial defect and were not true emissions, an interpretation supported again by the signals being much larger than normally experienced for vessel steel of this ductility. Thus it was clear that the significant emissions occurred in the pressure cycling range 200–900 lb/in^2, and this led to the enquiry whether the increase in the rate of emission could be correlated with the development of the fatigue crack.

Even in the presence of a notch a number of stress cycles are required to initiate a fatigue crack. During this incubation period significant emission would be expected to occur only during the first stress application. Emission occurring during crack growth would increase as the crack spread from the point(s) of initiation to cover the entire section (vessel wall thickness) of the notch. In a vessel with a through thickness slit initiation tends to occur at mid-thickness so that a period of growth may be needed

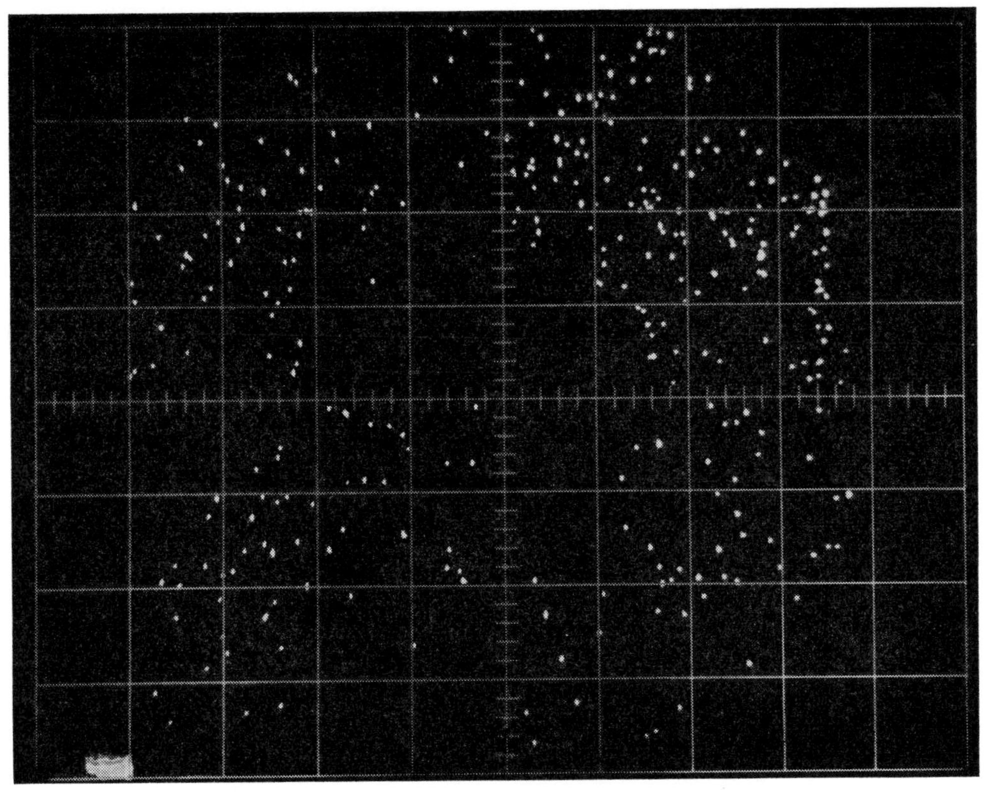

NOTE: Each quadrant shows a different 'view' of the vessel.

Fig. 30.7. Source location analysis—2D plot for 3-in vessel, 0–200 lb/in^2

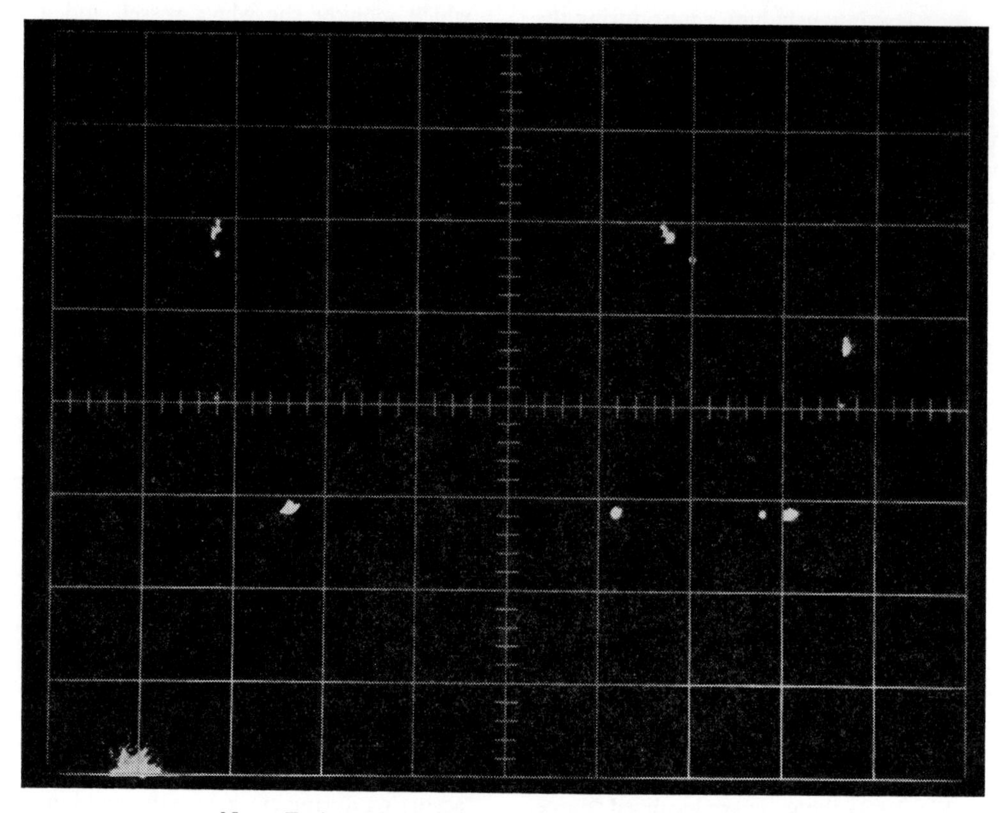

NOTE: Each quadrant shows a different 'view' of the vessel.

Fig. 30.8*a*. Source location analysis—2D plot for 3-in vessel, 200–900 lb/in^2 (coarse time scale)

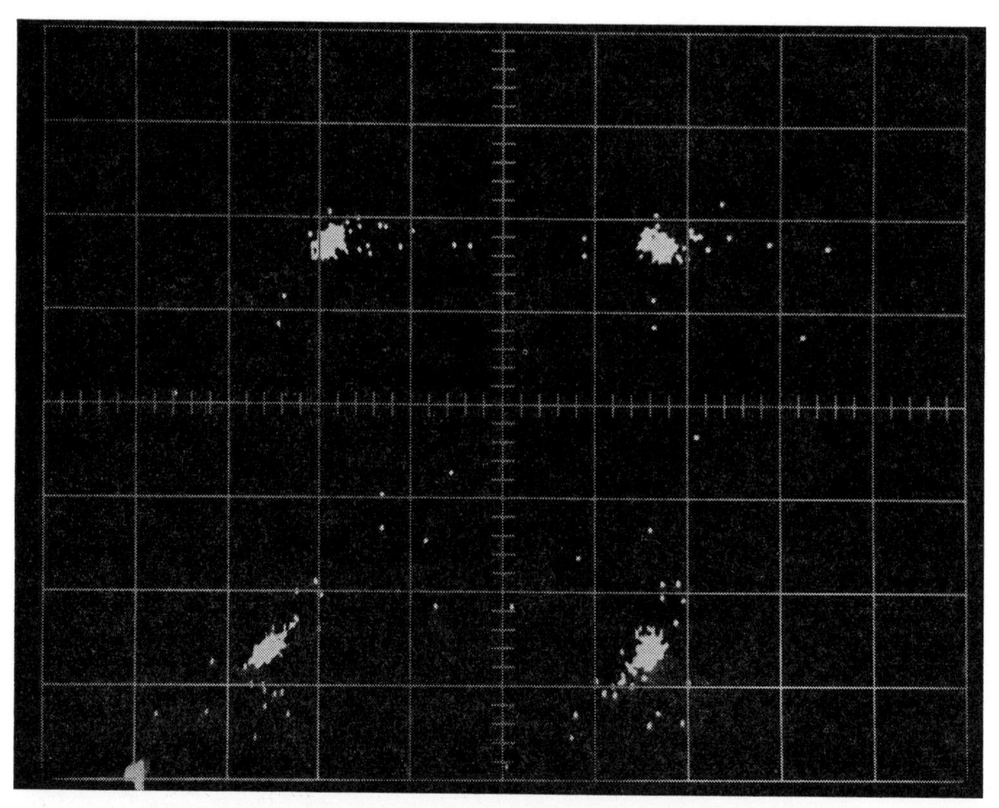

NOTE: Each quadrant shows a different 'view' of the vessel.

Fig. 30.8*b*. Source location analysis—2D plot for 3-in vessel, 200–900 lb/in^2 (fine time scale)

before the crack can be seen at the surface. Thus emission may be detected before a crack is visible.

The discrete emissions identified as coming from the slit were consistent with this pattern. They were first seen in the 121st cycle, and the number per cycle increased steadily before apparently levelling off at about the 350th cycle. Visual confirmation of a crack was obtained some 50 cycles after the first appearance of the emission. The pressure cycling of the vessel is continuing but the data are as yet insufficient to relate the emission quantitatively to the extension of the crack.

Pressure test on nuclear (thick-walled) vessel

Opportunity was taken to make measurements on a thick-walled pressure vessel both to gain field experience and to develop location techniques.

The steel pressure vessel measured about 3 m in length by 2 m in diameter and contained four nozzles. It had been designed in accordance with A.S.M.E. Section I. The vessel had undergone an acceptance hydrostatic test at the manufacturer's works, some six months earlier. The test described was a proof test in which the vessel was stressed to 3500 lb/in^2 (although it had been previously subjected to a hydrostatic test at 3750 lb/in^2) and 50°C approximately; it consisted of three parts: bolt-down, heat-up, and hydro-test.

The instrumentation system used to detect, process, and record the acoustic emissions was based on units described in the above paragraphs. A typical channel comprised a piezo-electric sensor, a low-noise amplifier, a variable bandwidth amplifier, and a pulse shaper. A total of 17 sensors were attached by an Araldite adhesive in positions shown in Fig. 30.9. An X–Y plotter was connected to plot count rate against test pressure during the hydro-test. Because at the time the mobile trailer was not yet constructed, 12 channels were recorded on magnetic tape for subsequent analysis.

Test equipment included a unit for injecting acoustic pulses into the vessel wall. This enabled checks to be made that the sensors and associated electronics were functioning correctly. It also enabled the velocity of propagation of the acoustic impulses to be measured directly from the time delay between the injection of a pulse and its receipt by several sensors. The velocity of transmission was subsequently calculated to be $1{\cdot}15 \times 10^5$ in/s.

During the measurements and recording relatively little electrical interference was observed, but a number of spurious emissions occurred from operators touching the vessel and by disturbances such as air and water blow-offs.

From the subsequent analysis of the magnetic-tape recordings, a source was identified during bolt-down on the top surface of the core barrel on which the closure head was sitting. From the hydro-test analysis, isolated emissions were located as coming from all parts of the vessel. Several sources were located, mostly at one side of the barrel section of the vessel. These are shown in Fig. 30.9, which gives the transformed outline of the vessel.

The test demonstrated the capability of the techniques to take measurements of acoustic-emission activity and to locate sources of emission during a major pressure vessel test when no work was conducted directly on the vessel shell. Acoustic activity was low during the test. This was consistent with pre-test measurements on small vessels, which had shown no appreciable recovery for this type of

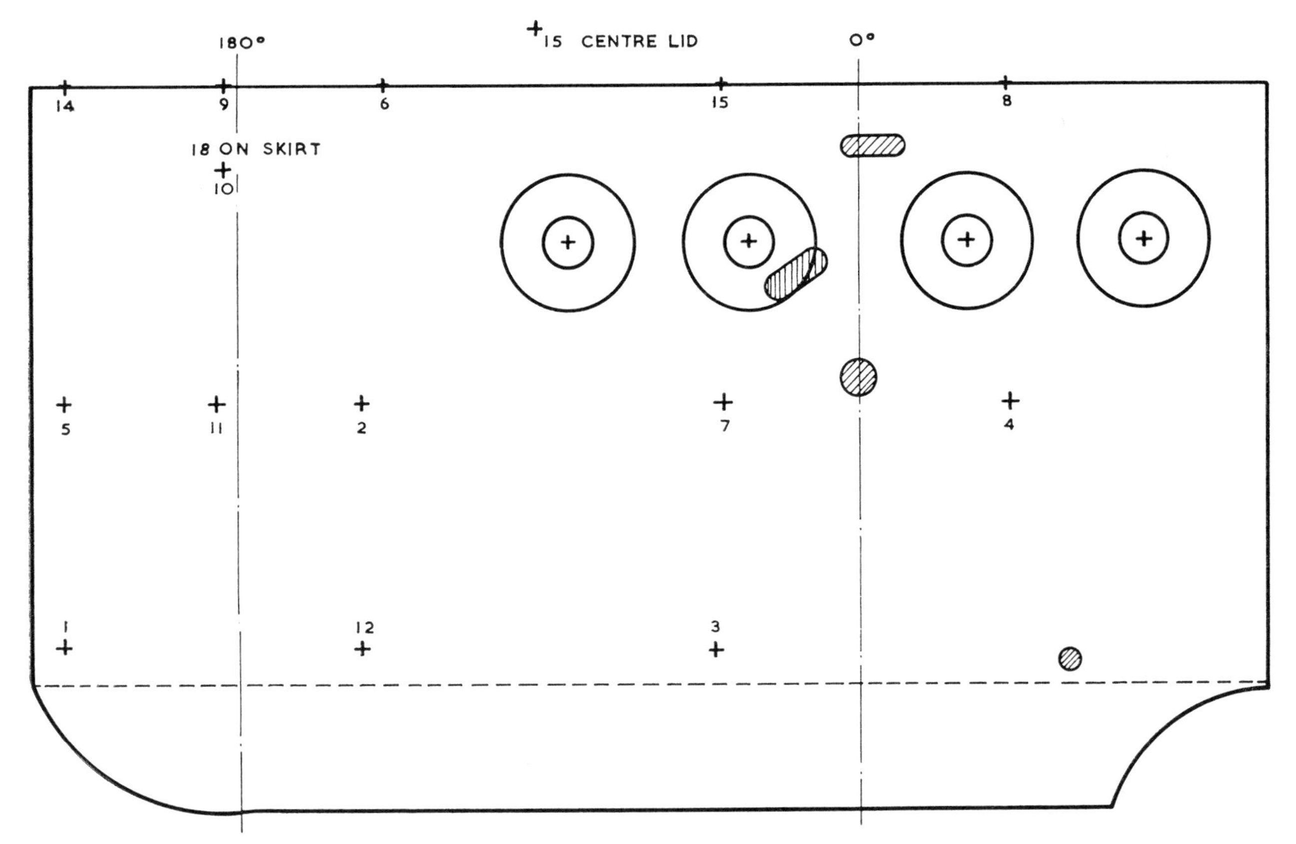

Fig. 30.9. Development of vessel showing sensor locations and sources located during hydro-test

steel. Therefore the previous proof test ensured that there would be minimal activity during this test. However, if any significant defects had developed between the manufacturer's proof test and this restress test, there would have been very much greater emission rates than those actually measured.

Noise analysis on pressure vessel of gas-cooled nuclear reactor

Measurements were made during June 1971 on the Reactor No. 2 at Berkeley Nuclear Power Station during the maintenance shutdown period and subsequent start-up. The pressure vessel is approximately 15 m (50 ft) diameter by 24 m (80 ft) high and operating conditions are approximately 340°C and 110 lb/in^2. This reactor has been operating since November 1962. The object of the measurements was to determine whether background levels were sufficiently small to allow acoustic-emission monitoring on such reactor vessels in the future, either during repressurization following shutdown or during reactor operation.

Because of problems of access, made more difficult by radiation levels, in these initial experiments the transducers could only be fixed to extensions to the pressure vessel and not to the membrane itself. Thus 12 transducers were fitted to fuel standpipes, sight tubes, Wigner tubes, and inlet ducts as indicated in Fig. 30.10. The count level and root mean square (r.m.s.) noise level from one transducer was monitored during the tests. In addition, six transducer inputs were recorded on magnetic tape for subsequent analysis.

Background-noise measurements were made on each transducer during 'quiet' conditions and up to full reactor

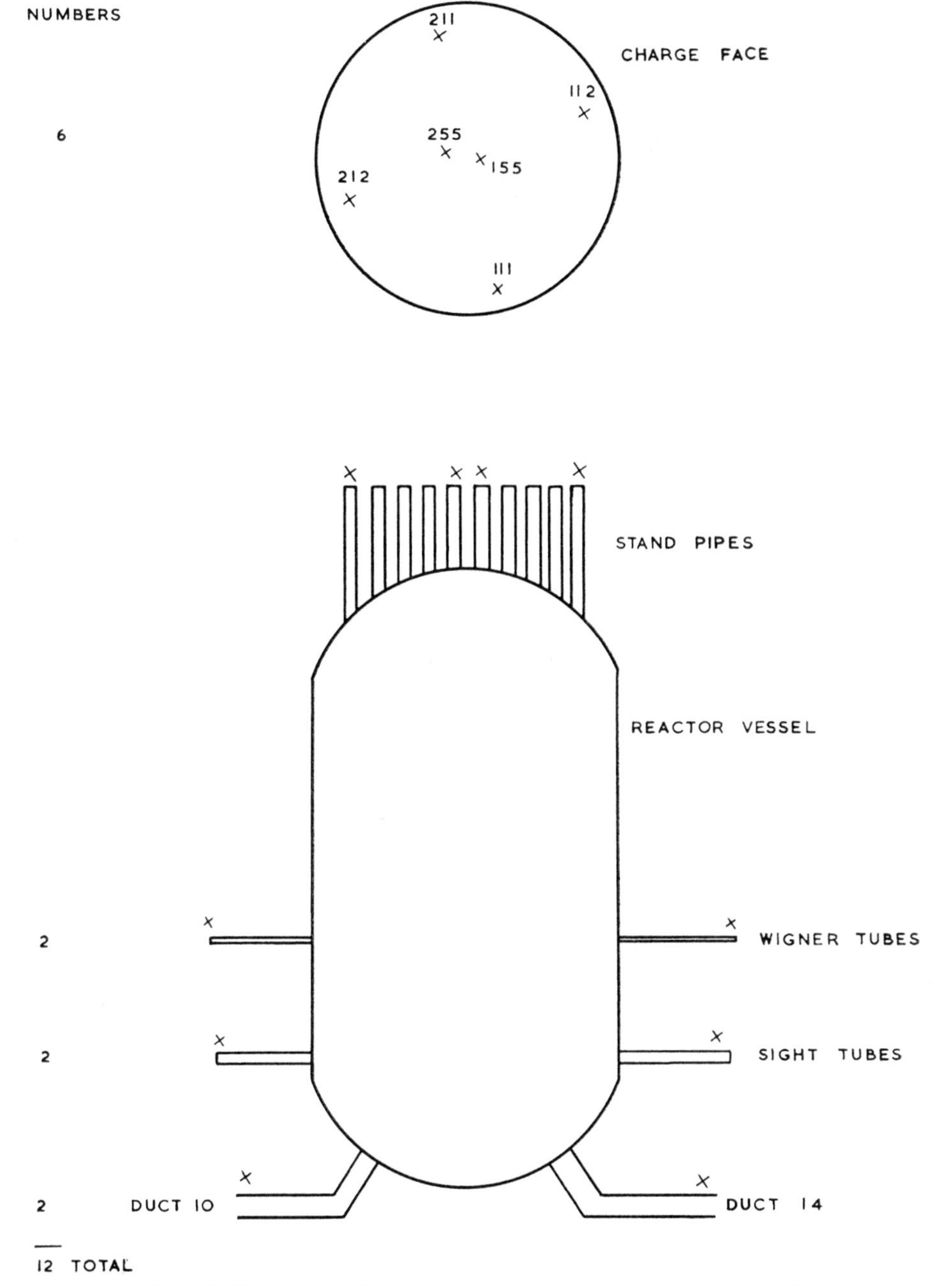

Fig. 30.10. Layout of transducers on Magnox reactor vessel

operation. The noise was generally impulsive rather than random, as indicated by peak/r.m.s. ratio of 4–6 rather than the expected value of 3 for random signals. To aid the subsequent interpretation attenuation measurements were also taken by pulsing electrically a chosen transducer and measuring the peak pulses received on other transducers.

Further measurements were taken by recording transducer outputs whilst the reactor was being pressurized. Subsequent examination showed a large amount of activity on the Wigner tube measuring position, but the signals were of slow rise time, characteristic of the vessel moving on its supports. Signals similar to emissions were recorded on the fuel standpipe transducers, but location analysis has indicated that any possible source of emission must lie within the reactor and is not in the vessel itself.

From the present experiments it may be concluded that tests during repressurization should give useful information under normal (nuclear) shutdown conditions, provided that significant emission occurs from the defects, i.e. that recovery occurs to a useful extent so that cyclic growth gives similar phenomena to that found in the 3-in-thick experimental vessel. On existing magnox reactors measurements would require exploitation of all available shield penetrations; attenuation in fuel standpipes is too high for them to be usable. Location analysis is essential to eliminate spurious signals as the vessel moves on its supports. Monitoring by acoustic emission on an operating nuclear gas-cooled reactor is not feasible with presently available techniques because of the level of background noise, and further development is required.

ACOUSTIC EMISSION AND INSPECTION

The outstanding value of acoustic emission is its ability to give inspection of 100 per cent of a pressure vessel in one operation. Any other conventional N.D.T. technique relies on incremental inspection of the vessel by detailed examination of specific areas and is limited to the unstressed condition. This may lead to doubts as to the overall inspection which has been given to a vessel, whilst examination under the stressed condition relates more closely to the operational condition. Thus, for example, a coarse-grained austenitic cladding in a thick-walled vessel for a light reactor may prevent adequate ultrasonic examination of underlying ferritic material, while tightly held cracks may not give adequate deflection for positive identification. Neither occurrence detracts from acoustic-emission measurements, and indeed it may be possible for the measurements to yield a direct measure of the severity of the defect relative to failure.

Ideally, acoustic emission should be applied at the first proof test of the pressure vessel. Under these conditions, background-noise problems can be reduced and defects located more readily than may be possible with the restricted access after operation. Conventional N.D.T. methods can be used to verify the location and quantify the size of defects—a necessary safeguard until confidence in the technique is increased by increasing application. Measurements obtained at this stage serve as a datum for repeat tests, which would preferably be made during commissioning and at subsequent start-up or periodic proof tests if they are applied. Ideally, the design should include provision for acoustic monitoring throughout life; and this may require the use of wave-guides and/or permanent location of sensors. Continuous monitoring for defect growth in the vessel in service introduces problems of stability of equipment, of continuous noise of varying levels, and represents probably the most complex aspect of the application of acoustic emission. Considerable development of techniques is taking place in this area especially related to thick-walled reactor vessels (**8**).

In applying data from acoustic measurements it is essential to consider the response of the vessel to such defects. The signal from a given defect will probably be related to the severity of the defect, i.e. some function of the applied stress and the size and shape of the defect. As has been demonstrated above, low-strength steels failing under non-plane-strain conditions may show marked emission as yielding spreads across the section, but the emission rate may decrease as the strain is increased to the point of fracture. In contrast, high-strength steels failing under plane-strain conditions generally give a rising signal to the point of failure (**7**) (**9**). The probable acoustic behaviour of the different materials in the vessel can be estimated from prior 'characterization' tests on small specimens taking into account plate, forgings, heat-affected zone, weld metal, etc. Such tests could lead to a direct relationship being established between acoustic signal and the proximity to failure, e.g. as a function of the stress intensity K or crack opening displacement (C.O.D.). However, conditions in a test piece, for instance, the finite volume of yielded material, the possibility of signal reflection from geometries of necessarily limited size, could produce signals not necessarily directly related to the behaviour of a defect in a vessel.

The significance of defects detected by acoustic-emission methods would normally be assessed from a fracture mechanics analysis. The method of analysis would depend on the type of pressure vessel, linear elastic fracture mechanics (L.E.F.M.) being appropriate to the thick-walled vessels of light water reactors (**10**) and general yield fracture mechanics to the thinner vessels, e.g. of gas-cooled reactors (**11**). The thinner-section vessels probably involve less problems in that the critical size of defect is very much larger and the initial proof test usually gives more severe conditions than in operation. In thick-section nuclear reactor vessels the initial proof test is often less severe than the service condition in which thermal stresses can cause the critical condition (**12**) (**13**). The difference in critical defect size may be even more pronounced when neutron damage in the thick-walled vessel produces significant irradiation embrittlement, and hence a continuously decreasing critical size with increasing life. Under these circumstances any fracture mechanics treatment demands that defects which could approach criticality in service should be found, but also that there is a high degree of confidence that all such defects have been found prior to service. This necessary degree of confidence is only likely to be achieved by the use of a technique which is capable of guaranteeing inspection of 100 per cent of the vessel both at the commissioning stage and at periodic post-irradiation inspections.

CONCLUSIONS

Acoustic-emission monitoring offers the potential of a new non-destructive method of confirming the validity of large pressure vessels and engineering structures. It has the great advantage of detecting flaws anywhere in the structure from a surveillance at a limited number of sensor positions. In particular:

(1) The acoustic technology has the capability of detecting very small flaws in complex, thick-walled vessels during hydrostatic pressure testing of pressure vessels without interfering with the works production schedules.

(2) These flaws may be located to within a few per cent of the sensor spacing.

(3) Presentation techniques, using computer analysis, enable the experienced operator to locate the position of the flaws whilst the stress test is actually taking place.

(4) The technique is now being extended from high-strength brittle materials to low- and medium-strength materials. A relationship is being developed between plastic deformation and acoustic emission, but more work is required to substantiate this.

(5) The technique has some limitations which require further investigation, viz.:

(*a*) There is not yet available sufficient experience to characterize a defect from its emission. Thus at present, faults which have been detected and located need to be identified by existing N.D.T. methods.

(*b*) The technique cannot yet be applied to some operating plants because the background noise, especially from fluid flow, limits the detectable signals so that further development is required, especially to optimize the acoustic frequency at which the measurements are made.

(6) The acoustic emission technique is the only known method of non-destructive inspection capable of examination of all parts of a vessel in one operation. Moreover, when applied under pressure-test conditions the acoustic output is related to the severity of the defects relative to vessel failure. Its potential therefore lies in the ability to give 100 per cent inspection and hence greater confidence in vessel integrity than is possible with conventional N.D.T. methods.

ACKNOWLEDGEMENTS

The measurements at Berkeley Nuclear Power Station were made co-operatively with staff of Berkeley Nuclear Laboratories, notably Dr A. E. Souch and Dr A. C. E. Sinclair. The assistance provided by the Station Superintendent and the Reactor Operating Staff was of the greatest value.

APPENDIX 30.1

REFERENCES

(1) KAISER, J. 'Untersuchungen über das Aufreten von Gerauschen biem Zugversuch', A doctoral dissertation presented to Fakultet für Waschinenwesen und Elektrotechnik der Technischen Hochschule München, München, Germany, 1950.

(2) PARRY, D. L. and ROBINSON, D. L. 'Incipient failure detection by acoustic emission—a development and status report', IN-1398, 1970 (August).

(3) HUTTON, P. H. 'Acoustic emission in metals as an N.D.T. tool', *Mater. Eval.* 1968 (July), 125.

(4) 'Proposed recommended practice for plane-strain fracture toughness testing of metallic materials', *A.S.T.M. Stand.* 1969 (Pt 31), 1099 (A.S.T.M., Philadelphia).

(5) DUNEGHAN, H. L., HARRIS, D. O. and TATRO, C. A. 'Fracture analysis by acoustic emission engineering', *Fracture Mech.* 1968 (June) **1** (No. 1), 105.

(6) BURDEKIN, F. M. 'Crack opening displacement—a review of principles and methods', *Practical fracture mechanics for structural steel, Proc. Symp. Fracture Toughness Concepts for Weldable Structural Steel* 1969 (April) (U.K.A.E.A. in association with Chapman and Hall, London).

(7) HARTBOWER, C. E., REATER, W. G. and CRIMMINGS, P. P. 'Stress wave characteristics of fracture instability in constructional alloys', 1970, I.I.W. Paper.

(8) 'In-service inspection program for nuclear reactor vessels', Bi-annual Progress Rep. No. 5, 1971 (28th May) (Southwest Research Inst., U.S.A.).

(9) TETELMAN, A. S. 'Acoustic testing and testing process', *Metall. Res. Stand.* 1971, 13.

(10) WESSEL, E. T. and MAGER, T. R. 'Fracture mechanics technology as applied to thick-walled nuclear pressure vessels', *Conf. Practical Application of Fracture Mechanics to Pressure-vessel Technology* 1971, 17 (Instn Mech. Engrs, London).

(11) FORMBY, C. L. and CHARNOCK, W. 'The effect of dynamic strain-ageing embrittlement on pressure-vessel integrity', *Conf. Practical Application of Fracture Mechanics to Pressure-vessel Technology* 1971, 1 (Instn Mech. Engrs, London).

(12) MAGER, T. R. and RICCARDELLA, P. G. 'Use of linear-elastic fracture mechanics in safety analysis of heavy-section nuclear reactor pressure vessels', *Conf. Practical Application of Fracture Mechanics to Pressure-vessel Technology* 1971, 56 (Instn Mech. Engrs, London).

(13) YUKAWA, S. 'Evaluation of periodic proof testing and warm prestressing procedures of nuclear vessels', Rep. HSSTP-TR-1, 1969 (General Electric, U.S.A.); also *Nucl. Metall.* 1970 **16**, 250.

C31/72

INSPECTION OF PRESSURE PARTS IN PROTOTYPE NUCLEAR REACTOR CIRCUITS

J. M. CARSON* F. TURNER†

This paper describes the periodic inspection programmes used in U.K.A.E.A. reactor installations producing nuclear power. These cover the Calder and Chapelcross reactors, the Windscale advanced gas cooled reactor, and the Winfrith Heath steam generating heavy water reactor. The inspection methods used include irradiation monitoring of material properties (surveillance), temperature and deflection measurements, creep strain measurement, non-destructive testing, and visual and TV observations. An account is given of some of the inspections carried out and comments are made on future procedure.

INTRODUCTION

Prior to operation of the first Calder Hall gas cooled reactor a committee was set up to consider requirements and procedures for periodic inspection of those parts of the plant not covered by statutory legislation. These included the reactor vessel, heat exchangers, main ducting with gas circulators and isolating valves and essential ancillary gas circuit and equipment.

The committee initially included representatives from the Design and Operations Groups of the U.K.A.E.A., Lloyd's Register of Shipping, the Eagle Star Insurance Company and subsequently from the Authority Health and Safety Branch.

Preliminary proposals for inspection were issued in May 1956, followed by a further note on inspection in August 1956, and by a revised steel irradiation programme in July 1958. The proposals were extended to include the Chapelcross reactors.

A similar procedure was adopted for the Windscale advanced gas cooled reactor where the initial proposals for periodic inspection were issued in April 1961.

Similarly proposals for periodic inspection of the Winfrith steam generating heavy water reactor were issued in March 1967.

In each case government departments concerned with the safety aspects of boilers and other pressure vessels were advised of the proposals and their representatives have been kept informed of experience gained.

CALDER AND CHAPELCROSS

Reactor vessels

The items listed below were included in the original schedule of inspection.

The MS. of this paper was received at the Institution on 2nd November 1971 and accepted for publication on 3rd January 1972. 3

* *Lloyd's Register Industrial Services, Norfolk House, Croydon CR9 2DT.*

† *U.K.A.E.A. Reactor Group, R.E.M.L., Wigshaw Lane, Culcheth, Warrington, Lancs.*

Shell leakage measurement

The intention was to monitor the air in the void between the reactor vessel and the biological shield for CO_2 content each shift. However, this was not implemented because operating experience showed that an early indication of leakage came from the CO_2 make-up required.

At one period the CO_2 usage in Calder No. 2 rose from some 2·5 ton/day to over 4 ton/day, and there was concern at the time that there might be a reactor vessel leak. However, a concentrated effort by Calder operations staff resulted in the major sources of leakage being traced to circulator casing joints. Concurrently, they introduced a deliberate CO_2 leak into a reactor void and established that it was possible to detect leakage in this region of the order of 500 lb/day.

The tendency has been for CO_2 make-up to reduce over recent years consequent on the use of improved jointing materials.

Shell temperature

Regular monitoring of reactor vessel shell and diagrid temperatures has been carried out. However, it is now considered that the original provision of thermocouples was deficient in two regions: (1) the shell at the diagrid support area; (2) the reinforcing plates in way of the main gas outlet ducts. Additional thermocouples to give a clearer indication of temperature differentials between the shell and diagrid at start-up and shutdown would have been advantageous.

The operating metal temperature limit on the reactor vessels is 345°C. From strain gauge readings during pressure tests, the reinforcing plate region is known to be an area subject to relatively high strain. Although additional thermocouples have been fitted in the upper part of this region, further information on the actual metal temperature in the area would be of value.

Stability

The original requirement was that the general movement

of the vessel should be checked at shutdown with suitable pointer gauges and other indicators.

The intention was to check the general position of the reactor vessel relative to the biological shield. Measurements were to be taken on the outer ends of the main gas ducts and other pipes radiating from the cylindrical part of the vessels.

It became obvious that, owing to temperature variations between one shutdown and another, successive readings could not be correlated. The emphasis then changed to readings under steady full power conditions, necessitating modification to indicators that were installed in high-radiation regions.

Readings taken since then have been satisfactory and do not indicate any unusual vessel movement.

Movement of charge tubes

Radial and tangential positions of charge tubes relative to the biological shield were to be checked by direct measurement during operation. The vertical position relative to the shield was to be checked at shutdown, but later the requirement changed to on-load measurements.

The intention was to detect creep of the vessel top head either by tilt of the charge tube standpipes or by vertical movement. Measurements of tilt have been inconclusive as the standpipes are long and flexible. Vertical measurements of movement to date give no cause for concern.

Noise—vibration

At the design stage of Calder and Chapelcross reactors structural damage in the pressure circuit due to high-frequency vibration was not considered probable. Vibration checks during commissioning confirmed this assumption. However, due to the many uncertainties associated with fatigue damage assessment and the implications of pressure circuit failure, work was initiated to establish vibration levels.

The work programme consisted of three stages:

(1) a vibration study carried out in conjunction with Lloyd's on Calder reactors Nos 1 and 2;

(2) collection of acceleration data in conjunction with Aston University, Birmingham;

(3) confirmation of sound pressure level predictions by installation of measuring devices inside the pressure circuit.

In the first stage of the programme external displacement measurements were carried out in the following positions:

(*a*) tops of standpipes;
(*b*) top centre of reactor shell;
(*c*) top bend of an outlet duct;
(*d*) blower regions;
(*e*) bottom 54 in valve mechanism;
(*f*) flange of a Wigner tube.

Measurements on about one-quarter of the standpipes were made with the blowers at maximum speed. Vertical vibration was negligible and lateral displacements led to pessimistic estimates of stress of the order of ± 200 lb/in^2 maximum.

The vertical displacements measured at the vessel top dome were very small, as were those at the outlet ducts, being less than 0·003 in. Localized vibration with a maximum of the order of 0·008 in were measured in the vicinity of a blower motor and ducting. Data collected at the base of the control mechanism of a bottom gas valve showed low vibration levels, $\not> 0{\cdot}010$ in, and measurements at the end flange of a Wigner tube indicated negligible movement. There was no significant difference between reactors 1 and 2. It was concluded that vibration levels were low and the predicted stresses at the base of standpipes were unlikely to initiate fatigue failure.

In the second stage of the investigation a preliminary check confirmed the generally low order of vibration, and as it indicated that the highest values were in the blower region, measurements were concentrated here. External measurements were made at the following points:

(*a*) bellows restraint downstream of blower;
(*b*) vertical duct downstream of blower;
(*c*) horizontal duct under blower;
(*d*) blower inlet bellows restraint;
(*e*) blower inlet duct upstream of bellows restraint;
(*f*) cascade bend below heat exchanger; and
(*g*) bottom of heat exchanger.

The analyses of signals at the University of Aston led to an estimate of an r.m.s. stress of ± 133 lb/in^2 and a peak stress of ± 400 lb/in^2 in the duct wall. Vibration levels were similar in reactors 1 and 2. A statistical assessment of the sound pressure in the gas over the whole frequency bandwidth gave a value of 156 dB. In the third stage of the programme an attempt was made to check the accuracy of duct wall stresses and gas sound pressure level derived in stage 2. Measurements of sound pressure level and sound power level were compared with the theoretical predictions for the Peistrup–Wesler equation for a range of blower speeds. Measured values of sound power level were consistently several decibels higher than the calculated values, but the measured value of duct wall sound pressure for maximum blower speed was in reasonable agreement with the predicted value derived in the second stage of the investigation, viz. 158 dB compared to 156 dB. Attempts to measure vibratory stresses in duct walls using strain gauges have so far been unsuccessful due to low signal/noise ratio.

It is concluded from the above work that noise levels in Calder reactors are low and induced stresses are unlikely to lead to structural damage. Collaborative work (1)* carried out by the 'Fatigue Discussion Group of the Joint Nuclear Power Committee' has indicated that at mean stresses applicable to Calder pressure circuits, vibratory stresses of up to ± 100 lb/in^2 are acceptable for ground fillet welds in the frequency range 100–2500 Hz, temperature range RT–350°C, steel thickness $\frac{3}{8}$–1 in, and a life of 10^{12} cycles. These parameters generally apply to the Calder pressure circuit, i.e. main ducting and associated items.

Permanent strain gauges (creep)

At the design stage of the Calder and Chapelcross reactor pressure vessels consideration was given to constructional problems and various possible modes of failure during service. The requirements for resistance to brittle fracture during construction and to irradiation embrittlement led to the choice of an aluminium grain refined C–Mn steel

* *References are given in Appendix 31.1.*

for the pressure vessel. The outlet-gas design temperature was in excess of 300°C and, therefore, creep properties were also important. As little information was available on creep properties of carbon manganese steels in the early 1950s, a conservative approach to design was dictated. The design was equated to a low deformation creep limit of around 0·2 per cent in 20 years (5 ton/in^2 at 360°C) to avoid the risk of undue standpipe movement with consequent difficulties in charge/discharge. Creep rupture failure at areas of stress concentration was considered improbable with the estimated extrapolated creep rupture strength of 15 ton/in^2 at 345°C in 20 years compared with an expected actual stress value of around 10 ton/in^2. Although air cooling of the standpipe area of the top dome was built into the design, thus limiting the temperature to 320°C, it was decided to monitor low deformation creep in this and other areas to check the design assumption and give warning of excessive movement. Accordingly, Wayne Kerr type capacitance strain gauges were installed on Chapelcross No. 4 pressure vessel. In all, 34 gauges were fitted, as follows:

Top dome	8
Outlet ducts—upper	8
Outlet ducts—lower	8
Central barrel	4
Saddle—bottom dome weld	4
Forged ring—bottom	2

A description of these gauges, and the results of the first three years' operation, has already been described (**2**). Creep strains of up to 0·07 per cent were measured after the first nine months' operation since when no significant creep ($<1\times10^{-8}$ in/in h) has been recorded within the gauge accuracy of $\pm$0·01 per cent strain. Random gauge variations are within this accuracy limit.

Creep and rupture test data of up to 10 years' duration have confirmed the original material properties assumption, so it is considered that under the current operational conditions of top dome temperature $\ngtr$320°C and duct/shell temperature of $\ngtr$345°C, embarrassment due to creep or failure due to rupture is extremely unlikely.

Strain cycling at refuelling shutdowns has been measured, giving a maximum value of 0·15 per cent initially and subsequently 0·11 per cent. Consideration of this and basic fatigue data indicates that at the current rate of cycling the pressure vessels are free from a fatigue type of failure.

Shell corrosion

The original schedule referred to examination for corrosion effects of irradiated steel specimens and to examination of the shell as far as practicable with TV camera or introscopes inserted through charge tubes.

The requirement for TV or introscope examination results, from Calder No. 4 onwards, in TV and still photography of the interior of the vessel top heads prior to start-up, in order to have reference photographs for comparison with those obtained at periodic inspection.

To date there has been no conclusive additional evidence of shell condition by the use of TV or introscopes. Heat shimmer affects viewing, there is difficulty in cooling viewing equipment, and considerable obstruction from the fuel element leak detection piping.

Still photography has been attempted over the years with some success, and with improved lighting systems and camera manipulators the standard of photography can be expected to improve in the future.

Irradiation monitoring of pressure vessel steel

Specimens of steel plate, forging, and weld metal in the form of tensile, bend, and Charpy test pieces have been exposed to representative pressure vessel temperatures and radiation levels since reactor start-up.

During the earlier years withdrawals of specimens were on an annual basis from three different areas of exposure within the pressure vessel:

(*a*) in the top head below shield plugs;
(*b*) opposite bottom of core;
(*c*) under diagrid.

Those under the diagrid receive a neutron dose of 5–10 times that on the pressure vessel.

Additional specimens were given an accelerated radiation exposure in core channels so as to obtain advanced warning of damage to be expected after several years' service. It was very soon apparent from neutron flux measurements that the pressure vessel was subject to a very low dose rate, and the experimental irradiations reached an unrealistically high neutron dose in only a few months. Over the first 10 years no radiation damage was detectable in monitoring specimens with doses of up to $4{\cdot}4\times10^{16}$ nvt (diagrid specimens). This, coupled with the results of a planned series of accelerated dose irradiation experiments and data from the literature, suggested that neutron radiation damage was never likely to be embarrassing to the Calder and Chapelcross pressure vessels in upwards of 30 years. Consequently, with the agreement of the U.K.A.E.A. Inspection Committee, the monitoring withdrawals were reduced to one in five years from each reactor site. Monitoring results to date (15 years) show that mechanical properties of all materials are still within the appropriate specification limits and within the scatter band of control specimen properties. The above comments apply also to irradiation specimens stressed to 5 ton/in^2, equivalent to the pressure vessel stress. Furthermore, there is no evidence of graphitization of weld heat affected zone specimens at the top reactor temperature of 345°C, nor of enhanced corrosion of specimens exposed to reactor coolant.

Since the neutron fluxes are now well established and the assessed maximum fast neutron dose in 30 years is only $1{\cdot}5\times10^{17}$, it is considered that Calder and Chapelcross reactors' pressure vessel steel will not suffer significant deterioration in this timescale. Nevertheless, adequate numbers of specimens are still available to monitor the vessel steel during this period.

Heat exchangers

The original proposals are given below.

Survey on a two-year cycle

All units to be subjected to a complete examination of shell drums, headers, elements, and fittings. Two units to be examined after twelve months, thereafter one every six months.

Early examination of the first units was required to establish if there was any systematic fault which might affect all heat exchangers, e.g. tube vibration. No such fault was found. Examinations continued for some years on a two-year cycle, but when sufficient experience had been obtained agreement was reached on extension of the period between examinations to four years. This extension applied to the main shell of the heat exchangers and the boiler tubes within the shell, but not to the boiler drums, headers, and external pipework which continued to be examined on a two-year cycle.

Defects found in heat exchangers have been few. A small number of pinhole leaks have occurred in the butt welds of bends in boiler tubes. These have either been ground out, rewelded and tested or the affected element blanked off. One laminated plate has been found in the bottom head of a main shell. This was sufficiently close to the inside plate surface to permit cutting out and blending smooth without reducing the plate below acceptable thickness.

In the early years of operation some cracking was noted in intermittent fillet welds attaching the internal gas casing to the main shell. None of the cracks extended into the shell and no additional cracking has been noted in later years.

Corrosion of exposed external boiler pipework has occurred, in the main, on the low-temperature sections and has been made good by a planned maintenance programme.

Hydraulic test on a two-year cycle

This related to drums, headers, and boiler elements. Periodic hydraulic tests have been carried out, but without rigid adherence to the two-year cycle. Some of these tests have, of course, coincided with boiler tube or external pipework repairs. Results have been satisfactory.

Removal of boiler element after four years' operation

This requirement was not implemented. Sufficient evidence was forthcoming by introscope examination of boiler elements when repairs to corroded external pipework were undertaken.

Main ducting and associated equipment

Original requirements were for an internal survey of the ducting system, an internal survey of the bellows, a survey of valves and fittings, and of the main circulators. These items are detailed below.

Survey of ducting system. Five-year cycle for the four circuits. An internal survey to be carried out on one section of the inlet and outlet ducts within the specified cycle; the first circuit to be surveyed after two years.

This requirement was later extended considerably to include internal examination of all ducting accessible through manholes within the five-year cycle. Not all sections are so accessible, but the ducting condition generally has given no cause for concern to date.

Until recently the same remark could have been applied to the cascade corners which form an integral part of the ducting. However, in 1969 at Chapelcross cracks were noted in the fillet attachment welds of cascade vanes to cascade corners. These were short defects, at about mid-length of the weld, and were found mainly in the second and third cascade corners on the discharge side of the circulators. These were repaired and the cracks were detected satisfactorily.

A general investigation of cascade vane welds followed at Calder and Chapelcross, but no further defects of this nature were found.

Owing to the difficulty of access for inspection, the use of ultrasonic techniques applied from the outside was explored.

A full-scale model of a cascade section was made and artificial defects machined in weld areas 1 in long by 0·050 in and 0·150 in deep. Compression and shear wave probes were used to explore weld geometry and defects. As it was found possible to obtain a good signal from the 0·050-in deep defect, this was used as a reference standard for routine inspection. Probes with angles of 45°, 60°, or 70° were used depending on vane/duct geometry. Unfortunately, due to the cascade stiffening ribs, only 40 per cent of any weld is accessible for ultrasonic examination, but this includes the most critically stressed area where the cracks had been discovered. However, critical crack length assessment suggests that cracks, if hidden, would extend beyond the rib zone before approaching a significant size and would therefore be detectable. The poor weld profile and variable fillet size presents some difficulty in interpretation and leads to a pessimistic assessment, but in the hands of experienced personnel the ultrasonic technique is considered to be practical. A critical visual and ultrasonic examination of a cascade bend in Calder reactor No. 1 in June 1970 showed good correlation, but no cracks were found. Ultrasonic indications of $\frac{1}{4}$–2 in were established visually as undercut, poor profile, and areas of negligible weld deposit. The use of ultrasonic inspection is continuing since it is easier to carry out than internal visual inspection and should give adequate warning of significantly sized cracks. Meanwhile, development of introscope/TV techniques is continuing to assist internal visual inspection.

Internal survey of bellows. A selected unit on the inlet and outlet ducting of all circuits to be dismantled for examination of the convolutions within the specified cycle. Units on the first circuit to be examined after two years.

The design of the Calder and Chapelcross bellows units is such that axial loads are carried by a heavy hinged external casing. It was therefore decided that, in order to examine the convolutions, the internal fairing welded to the bellows end pieces should be cut in way of the first convolution and removed. The first units examined showed signs of instability, i.e. opening out of one convolution and closing of that adjacent. The scope of this requirement was then extended to all bellows units within the five-year cycle. Further, to ease subsequent inspection, modified fairings were designed with removable bolted panels to be positioned over the areas thought to be unstable.

The modified fairings in themselves introduced complications. There was extreme difficulty in achieving satisfactory alignment of the sections, which was not helped by the fact that the work was carried out in a confined space and in full protective clothing. In some cases lack of alignment resulted in vibration under gas flow with loosening of bolts and failure of attachment welds.

The position now reached is that since instability does not appear to be progressive and bellows convolutions so far examined show no other sign of deterioration, exposure of convolutions by removal of fairings is not carried out on a periodic basis.

Other requirements. Survey of valves and fittings. Inspection of position indicators on constant load spring support. Monitoring of the system for CO_2 leakage.

These items do not call for comment since no troubles have been experienced.

Main circulators. Survey—two to three years. One circulator to be opened up after 12 months' operation, the remaining three within a two- to three-year operating period. The interval for subsequent inspections to be reviewed after all circulators have been examined.

The impeller of one of the first circulators to be examined had become overheated, apparently due to insufficient clearance in the gas seal region. This impeller was replaced and no other impeller deterioration has been seen.

The periods between circulator inspections have increased with experience, first to four years, then to five. It is now agreed that circulator top half casings need not be removed on a regular periodic basis, but only when necessary for maintenance.

Present position and future procedures

Periodic inspection of the accessible regions of the primary circuits of the Calder and Chapelcross nuclear stations, extending over 15 years, indicates that these continue to be in satisfactory condition. It has been possible, with increasing experience, to relax on a number of the original inspection requirements.

The results of the reactor vessel steel irradiation programme are reassuring and the readings from permanent strain gauges fitted in the gas outlet duct reinforcement region of Chapelcross reactor No. 4 are satisfactory. However, owing to the radiation problem it has not been possible to carry out a significant inspection of the reactor vessels—particularly those regions known to be subject to high strain—apart from the photography mentioned previously.

In view of the limited information on the condition of the vessels, the periodic inspection committee in 1968 decided to set up a working party to consider the possibility of proof testing.

The basic reasons for proof testing were agreed as follows:

(1) Comprehensive examination by orthodox techniques is not possible due to difficulties of access and radiation hazards.

(2) In the absence of periodic inspection and/or proof pressure testing it is not possible to relate the probability of catastrophic failure to failure statistics from conventional plants which have been subject to periodic inspection and/or proof testing.

(3) During a proof test the hazards to personnel on site and in the district can be reduced compared to the hazards if pressure circuit failure occurs during normal operation. A reduction in hazard by a factor of 100 is possible.

(4) Catastrophic failure cannot be prevented by a minimum temperature limit and consideration of a critical crack length approach is necessary.

(5) Unless a proof pressure test is carried out the need to revise operational limits in order to derate the pressure circuit should be considered.

The members of the working party agreed that the Calder and Chapelcross pressure circuits should be proof tested at a pressure of 112·5 lb/in^2 (gauge) at a temperature not exceeding 140°C. The normal working pressure of the circuits is 100 lb/in^2 (gauge), the normal safety valve setting 107·5 lb/in^2 (gauge), thus the proof test pressure is similar to that which would be reached under safety valve accumulation tests.

The general feeling of the working party was that pressure systems which could not withstand this modest amount of excess pressure should not be in operation.

A fracture mechanics approach to the problem, carried out at the time by the Authority Health and Safety Branch, indicated that a successful test at 112·5 lb/in^2 (gauge) would give protection against failure during 10–20 subsequent pressure cycles.

The majority of the Calder and Chapelcross pressure circuits have since been tested successfully as outlined above, and the present recommendation of the periodic inspection committee is that these tests should be repeated at two-year intervals.

WINDSCALE A.G.R.

Comments are limited to the reactor vessel and heat exchangers. The original inspection proposals followed a similar pattern to that established for Calder and Chapelcross.

Reactor vessel

For shell leakage, shell temperature, reactor vessel and internal movement, and air clearance round the vessel the original provisions were adequate; no difficulties have been experienced to date and no further comment is proposed.

Shell corrosion is dealt with in conjunction with two items additional to the Calder–Chapelcross pattern; namely, internal examination of reactor vessel top dome above neutron shield, and external examination of reactor vessel top dome at two-year intervals.

Two entries have been made to the interior of the top dome, and examinations made where accessible of the hot box, refuelling tubes, and gas baffle plates. Sections of the gas baffle plates have been removed and the top dome shell and standpipe penetrations examined in the regions exposed. No defects have been seen and no corrosion noted.

External examination of the top dome has been carried out according to schedule. The vessel supports have been examined and found to be satisfactory. Some re-examination has been undertaken in way of the main gas duct connections, including ultrasonic examination of butt welds which were originally radiographed, and no significant defects have been found. Other areas of the top dome have been examined, including accessible top standpipe penetrations, and no defects noted.

An unscheduled internal inspection has been carried out in the region of the large central nozzle in the bottom dome. This inspection was, of necessity, brief due to a high radiation level but no significant defects were seen.

The present position in the inspection of the reactor

vessel is thus satisfactory, even though there are considerable areas which cannot be inspected by orthodox means.

Irradiation monitoring of pressure vessel steel

The monitoring programme of Windscale A.G.R. is similar in content to that for Calder and Chapelcross, but vessel and specimens are all at 260°C. Design flux calculations suggested that neutron irradiation of the pressure vessel could result in a measurable effect on properties in two to three years, therefore a withdrawal of specimens was made after this period. Since all specimens are exposed beneath the diagrid above the tundish, they receive a somewhat higher fast neutron dose than the bottom of the vessel—i.e. about 5:1—and about 25 per cent more than the vessel barrel at mid-core level. These relationships were established by flux measurement during and since commissioning, and they also showed that the calculated neutron fluxes were excessive.

The results of the two withdrawals of specimens to date after $2\frac{3}{4}$ and $6\frac{1}{2}$ years have shown that the changes in steel properties are small, the tensile results of plate forging and weld metal being within the controls' scatter bands and still within specification limits. This applies also to tensile tests carried out at the operating temperature of 260°C. Charpy tests have shown only marginal shifts in the energy and fracture transition curves with no fall-off in maximum ductile energy. Tests have been carried out up to 300°C. Accumulated fast neutron doses are $4{\cdot}1\times10^{16}$ for the bottom dome and $1{\cdot}6\times10^{17}$ for the barrel, the monitoring specimens having had a dose of 2×10^{17}. The fast neutron dose over 30 years should not exceed 8×10^{17} under current operational conditions. A significant and measurable change in material properties could occur over this period, but data from experimental irradiations and the literature show that this should not be operationally embarrassing.

Heat exchangers

The original requirements were:

(*a*) A survey at two-year intervals.

(*b*) Removal of selected bolts from each main flange to check for permanent stretch.

(*c*) Shell and tube corrosion samples to be removed at two-year intervals.

(*d*) Hydraulic test (on steam side) when required by Inspecting Authority.

(*e*) Steam leakage monitoring continuously.

(*f*) Gas leakage monitoring continuously.

(*g*) Check on shell movement and supports.

All units have been surveyed to schedule. No defects have been found in the main shells and heads, nor is there evidence of corrosion in these parts. The steam drums, external pipework, and headers are in satisfactory condition.

Measurements of bolts removed from the main flanges indicate no permanent stretch, and these have been refitted.

One evaporator tube leak has been detected. The defective tube section was cut out and the live ends of the tubes plugged. The steam side of this heat exchanger was then hydraulic tested satisfactorily.

In early service, vibration of fairings attached to the lower end of the main isolating valve at the entry to the circulators resulted in one becoming detached and grooving the lower head of the exchanger. The damage was not such as to affect the strength of the head significantly. These fairings were removed.

Cracking of the main circulator diffuser was also experienced. One of these has been replaced but the remaining three are now giving satisfactory service following the addition of stiffeners and careful rewelding and blending of the regions affected.

There have been indications of pitting/corrosion on the gas side of the superheater tubes (2·5% Cr–1% Mo) which have led to a decision to reverse the steam flow in the superheater bank and so reduce the maximum metal temperature of the tubes. Consequently, the headers now operating at superheater outlet temperature are of carbon steel, which are working in the creep range. Replacement outlet headers in Cr–Mo steel are ready for installation when necessary. Meanwhile, regular non-destructive examination and creep measurements are being undertaken.

Creep monitoring and non-destructive testing of superheater headers

On reversal of steam flow the headers, 10·75 in diameter ×0·625 in wall, were subject to a possible maximum temperature of 430°C and a pressure of 700 lb/in^2. Since the header material was not originally chosen for use in the creep range, a careful appraisal of its likely behaviour was undertaken using creep data and vessel failure data from the literature. The most likely area of failure under creep conditions was considered to be at nozzle/shell welds. Using a pessimistic assumption of 2 per cent creep ductility to failure in weld metal under multi-axial loading, and a strain concentration factor of 5:1 relevant to the header shell, a limiting shell strain of 0·4 per cent was derived. Creep strains in the header membrane region have been monitored by the use of a replica technique with orthogonal and spiral grids over a period of five years to September 1970. Although difficulties have arisen in the accurate measurements of some of these grids, average values of strain of up to 0·12 per cent were assessed during the 1970 shutdown. Extrapolation of these data indicates that 0·2 per cent strain should not be reached before 1973 and 0·45 per cent not before 1980. Ultrasonic inspection of the headers has been carried out, assisted by a full-scale weld mock-up with artificial defects, revealing intermittent root defects at all 5·5 in outlet pipe welds. These are possibly lack of fusion and original defects not found on works inspection since ultrasonic techniques were not then employed. There is no evidence of the defects propagating or extending into the header shell, but inspection is being continued until the headers can be replaced. Fracture mechanics assessment indicates a critical crack size of about 8 in in the shell, so the growth of a significant sized crack should not go undetected. Incidentally, the operating conditions have been less severe than assumed, viz. 370–430°C and 560–640 lb/in^2.

Heat exchanger ball supports

Each heat exchanger is supported on six sets of three balls, $2\frac{1}{2}$ in diameter, resting on wear plates. The balls and supports were originally made from similar material, 12% Cr–2% C, the high chromium content being intended to resist corrosion under the expected ambient conditions at around 200°C. During a precommissioning hot run a number of the balls cracked, so all were replaced by the only balls available, i.e. 1% Cr–1% C. The original batch were examined and cracks were found to be associated with wear tracks on the balls. All cracked balls were of the 12% Cr–2% C variety, but surprisingly some of the uncracked balls were of 1% Cr–1% C and exhibited no wear tracks. After a detailed examination it was found that the 1% Cr–1% C balls were harder than the 12% Cr–2% C balls and the temperature attained by the balls in use was $\not> 100$°C. Therefore, high chromium material was not essential and 1% Cr–1% C could be safely used. The 12% Cr–2% C wear support plates, being somewhat softer than the balls, were retained on the basis that wear (tracking) was preferable in the plates. Since a subsequent hot run, followed by inspection, showed no damage to the replacement balls, they were retained. Further inspections at shutdown have shown no evidence of ball wear or cracking, but ball and wear plate inspection is now scheduled.

Present position and future procedures

Period inspection of the heat exchangers of the Windscale A.G.R. indicates that the main shells, steam drums, headers, and external pipework are in satisfactory condition. Early troubles within the heat exchangers due to vibration appear to have been overcome. Regular examination of the superheater tubes will continue, as will non-destructive examination and creep measurements of the superheater outlet headers.

Periodic inspection of the reactor vessel top head has been satisfactory to date. Nevertheless, the limited amount of inspection of the vessel makes it desirable to consider other measures. Concurrently with the decision to carry out proof tests on the Calder and Chapelcross reactor vessels, as previously outlined, the periodic inspection committee considered the case of Windscale A.G.R. and decided that a proof test should be carried out for the same reasons. This was completed successfully in July 1968, the maximum pressure at the reactor vessel top dome during the test being 323 lb/in^2 (gauge), i.e. 13 per cent above the normal operating pressure temperature of 170°C.

WINFRITH STEAM GENERATING HEAVY WATER REACTOR

To date periodic inspection experience is limited. It is proposed to outline experience related to the steam drums, channel tubes, primary circuit headers, manifolds, feeders and risers, and the emergency cooling water vessel. As previously, the original inspection requirements are summarized and modifications described.

Steam drums

Internal and external survey, the first drum after one year and the second after two years. Thereafter, the period between inspections subject to review.

It was anticipated at the design and construction stage that a thorough periodic internal examination would be possible and drum internal fittings were designed for removal. However, radiation levels within the drums have restricted the time available for internal inspection, and internal fittings have not been removed. The internal surfaces have so far appeared satisfactory.

In view of the limited internal examination, efforts have been directed towards a searching examination of the drums from the external surfaces, using ultrasonics, magnetic particle and dye penetrant examinations, and eddy current testing.

Both drums have been examined as follows:

By ultrasonics: The main welded seams, longitudinal and circumferential; bonding of the internal austenitic cladding in way of the main seams; other selected areas containing imperfections in bonding of the cladding known to exist from the construction records; in way of downcomer and other selected nozzle welds; ligaments, where accessible, between adjacent nozzles.

Ultrasonic examination was carried out using USM2 and USK5 Flaw Detectors (Krautkramer) in conjunction with 2 MHz and 4 MHz 24-mm diameter compression wave probes, 2 MHz × 35° and 45° shear wave probes, and a 4 MHz × 45° miniature shear wave probe. Prior to inspection, full-scale sections of set on and set through nozzle welds were made, containing spark eroded defects in welds and adjacent areas. Defects included simulated lamellar tears in drum and nozzle material. These mock-ups were used to optimize the inspection technique and assist in location of defect position.

Main weld seams plus 9 in each side in the drum shell were examined with compression probes prior to shear wave examination, to explore the stainless steel cladding interface. A similar cladding examination was carried out around nozzles.

By magnetic particle and/or eddy current methods: Nozzle welds to drum, where accessible; main support bracket welds to drum.

Magnetic crack detection of nozzle and attachment welds was carried out using portable a.c./d.c. 1500 A Videosons equipment, but due to difficulty in avoiding arc strikes on painted areas this technique was abandoned in favour of the Amlec crack detector Mk VI.

The main welded seams were examined by radiography during fabrication. The re-examination by ultrasonics gave good correlation with the original radiographic records. There was no indication of change in areas of cladding containing known imperfections in bonding. The condition of the drums to the extent established by the above external examination and by close visual examination was satisfactory.

Headers, manifolds, feeders, and risers

Inspection to date has been external only, consisting of visual and dye penetrant examinations, mainly in weld areas and regions of stress concentration. Some calibration of the external diameter of risers has been carried out for future reference and a number of ultrasonic thickness checks have been undertaken. Results to date have been satisfactory.

Monitoring of pressure tubes

Creep. The bore diameters of selected tubes are measured to an accuracy of $0{\cdot}2 \times 10^{-3}$ in at each shutdown by means of an English Electric Co. gauge. This gauge, which operates automatically, can measure the diameter at six radial positions and in 1 in vertical steps over a length of 15 ft. The diameter, radial position, vertical position, and water temperature are recorded on a chart and a complete channel measurement can be accomplished in 3 h. Measurements to date show a maximum increase in bore diameter of 0·013 in on an original of 5·142 in. Failure of pressure tubes by creep rupture is not expected at less than 10 per cent diametric strain. A safety factor of 3:1—equivalent to, say 3 per cent strain—has been adopted, and it is estimated that this value will not be reached in less than 15 full power years.

Waterside corrosion—Zircalloy 2. Corrosion has been measured on specimens exposed to reactor coolant in a special fuel cluster. Results to date indicate extrapolated metal losses of 0·0025–0·005 in over a period of 200 000 h, which is well within the corrosion allowance for the 0·2-in thick tubes. Hydrogen pickup, resulting from corrosion, is assessed at around 150 p.p.m. in 200 000 h.

Interspace vault gas corrosion is also being monitored. Current data suggest similar corrosion behaviour and hydrogen pickup to that on the waterside.

Embrittlement. Monitoring specimens in the form of COD test pieces have been exposed to channel conditions as for waterside corrosion. Some of these specimens have been pre-hydrided to 200 and 500 p.p.m. hydrogen to give advance warning of the combined effects of irradiation and hydrogen pickup. Tests, after exposure, have been made at 20 and 300°C. Although hydrogen has an embrittling effect at low temperatures, up to 500 p.p.m. has no obvious effect on COD values at 300°C; therefore, it should have no effect on critical crack length for fast fracture. Calculations of critical crack length for 500 p.p.m. hydrogen in Zr 2 (much more than expected in 200 000 h), after a saturation irradiation dose of around 4×10^{20} neutrons/cm^2, give values of not less than 2·5 in at 300°C. Experimental work indicates that a crack of less than 0·75 in would result in leakage, and therefore detection before reaching a critical size.

Pressure tube surface. Visual inspection of channel walls is made using introscopes or TV cameras. Photographic or video-tape records are made of any surface marking found. Replication techniques have been used to determine the depths of fretting marks. Local surface damage not exceeding 0·004 in has been so measured.

Hydraulic test

Hot overpressure hydraulic test. Will include all of primary circuit. Subject to review after second drum inspection —may be used to extend period between internal surveys.

No specific time interval was set originally for the first hydraulic test, nor was the test temperature specified. In view of the limited internal examination of the steam drums, and in order to prove the primary circuits as a whole, it was decided to test each circuit individually early in 1971.

The test pressure was 1575 lb/in^2 (gauge), i.e. the same as the initial cold hydraulic test pressure, and the temperature during test was of the order of 50–60°C. These tests were satisfactory.

It should be noted that non-destructive examination of each drum, as outlined previously, was undertaken before and after the hydraulic tests.

Emergency cooling water vessel

Internal and external examination after two years' operation, then as required.

This vessel, which is fabricated in material similar to B.S. 1501:271 (Mn–Cr–Mo–V), is maintained at a pressure of 1160 lb/in^2 (gauge) during reactor operation. In order to reduce thermal shock to S/P, the shell temperature is maintained at 70°C by means of external electric heating elements and the shell is lagged.

Access for internal inspection presents no difficulty, but external inspection requires removal of lagging and heating elements.

Prior to the first inspection consideration was given to the possibility of carrying this out from the internal surface only, but it was decided that owing to incidence of cracking in service with this type of material a full external and internal inspection should be undertaken.

Apart from close visual examination, non-destructive testing was carried out using ultrasonic, eddy current, and dye penetrant methods. Ultrasonic techniques were employed on all butt welds in the shell, areas adjacent to saddle fillet welds and lifting lugs, the weld and hinge on the 18-in manway, and all accessible branch welds. Eddy current testing was used to scan all branch welds and fillet welds on the outside of the vessel. Fillet welds on the inside platform of the vessel and a repaired area in one head were checked by dye penetrant.

Apart from evidence of segregation or inclusions in one plate, known to exist from the original records, and an indication of a defect at the original root face of a 6-in branch to shell weld (consistent with a slag line or lack of root fusion), no other significant defects were observed. Examination by a normal probe of the above weld showed that the defect did not extend into either the shell or the nozzle. This area will be re-examined on an annual basis. Non-destructive testing techniques used were similar to those for the steam drums.

Present position and future procedures

Although the original expectation of a thorough periodic internal examination of the steam drums has not been realized, the non-destructive examinations carried out from the external surfaces of the drums are reassuring, as no significant defects have been found nor any appreciable deviation from the original fabrication records. It is felt that the examinations completed at the first inspection of these drums compare favourably with the requirements of A.S.M.E. Code, Section XI, 'In Service Inspection'.

The examination of headers, manifolds, feeders and risers is considered satisfactory at this stage. Present monitoring evidence on the Zr 2 pressure tubes suggests that corrosion, including resulting hydrogen pickup plus irradiation effects, will not limit reactor life up to 200 000 h. The successful hydraulic overpressure tests of both primary circuits in 1961 add further reassurance.

The thorough examination of the emergency cooling water vessel has not revealed defects that give cause for concern, although the defect indication in the root of the 6-in diameter nozzle to shell weld will be re-examined regularly.

Future procedures are under discussion by the periodic inspection committee, but at present no detailed statement can be made other than the fact that consideration is being given to the development of television and photographic equipment to make possible a more detailed examination of the steam drums.

ACKNOWLEDGEMENT

The assistance and information supplied by staff at Windscale, Calder, Chapelcross, S.G.H.W. Winfrith, and R.E.M.L. is gratefully acknowledged.

APPENDIX 31.1

REFERENCES

(1) *Conf. Fatigue of Welded Structures* 1970 (July), Paper 21 (The Welding Institute).

(2) Wood, D. S. and Anderson, D. 'The time-dependent deformation behaviour of Chapelcross No. 4 reactor pressure vessel', *Conf. Thermal Loading and Creep in Structures and Components, Proc. Instn mech. Engrs* 1963–64 **178** (Pt 3L), 157.

C40/72 UTILIZATION OF ACOUSTIC EMISSION FOR IN-SERVICE INSPECTION

B. H. SCHOFIELD*

The utilization of acoustic emission for reactor vessel in-service inspection shows considerable promise. System development is currently quite active among several commercial sources. Notable is the support given by the utility industry in the U.S.A. to the development of acoustic emission, holography, and the refinement of ultrasonic non-destructive testing methods through the Edison Electric Institute and its Steering Committee. This paper describes the emission characteristics significant to pressure vessel surveillance and their utilization in precluding catastrophic failure, and the detection and locating of defects in the structure. The typical components of an emission surveillance system are discussed, followed by a review of the current state of the art.

INTRODUCTION

THE APPLICATION of acoustic emission as a practical tool has been pursued by various investigators in the U.S.A. over the past six or seven years at a relatively mild pace. Innovations have ranged from production inspection of jet engine turbine blading to the measurement of concrete runway deterioration during landings of Boeing 747 aircraft. The most active and undoubtedly the most enthusiastic efforts have been the relatively recent efforts within the commercial nuclear reactor area as expressed by the public utility concerns. The motivations within this industry are unique and varied, with the technical complexities compounded by the perplexities of the sociopolitical community. Notwithstanding these many difficulties, the utility industry has moved with decisiveness and would appear to be on the threshold of a major advancement in the in-service surveillance of nuclear reactors through the utilization of acoustic emission.

A discussion of the desire and needs for safety in the design, fabrication, and operation of nuclear reactors would be superfluous at this point in time and certainly at this particular conference. Suffice it to say that the spirited interest in acoustic emission technology represents merely one of many undertakings in conformity with such goals and clearly evidences the commitment of the utility industry to meeting their responsibilities.

The utilization of the acoustic emission phenomenon as a non-destructive tool for in-service inspection of nuclear reactors shows substantial promise both technically and economically. Acoustic emission as a technology is new, less than 20 years old. The 1971 issue of Section XI of the A.S.M.E. Code ('In-service inspection of nuclear reactor coolant systems') notes this developing area and makes provision for its utilization within the Code philosophy.

In this paper the significant technical parameters of acoustic emission are reviewed in the light of their utilization for in-service inspection, followed by a discussion of system hardware developments and the current state of the art.

The MS. of this paper was received at the Institution on 25th October 1971 and accepted for publication on 3rd January 1972. 34
* *Teledyne Materials Research, 303 Bear Hill Rd, Waltham, Massachusetts 02154, U.S.A.*

ACOUSTIC EMISSION PARAMETERS

The acoustic emission observed from all metals, including the carbon steels, chrome–molybdenum steels, and other pressure vessel materials, has often been characterized as of two types: 'burst' and 'continuous' emission. Although their appearance to a greater or lesser degree in different metals suggests the presence or absence of a unique property or characteristic, present thinking is that the 'burst' and 'continuous' emissions are produced by the same micro-deformation mechanisms. Whether or not the emission is primarily of the 'burst' character or of the 'continuous' form is entirely dependent on the time–energy activation of the deformation mechanism. For example, in the case of twinning deformation, a mechanism which produces high energy 'burst' type pulses, there is a relatively long period during which elastic energy is being absorbed and a much shorter time period over which the deformation interchange occurs in forming the twin. Consequently, the emission is observed as a discontinuous process in time, relatively long quiescent periods interrupted by the short period of the emission burst. Fig. 40.1 is a representation of the 'burst' type emission response.

Conversely, in a material such as aluminium, energy absorption and release each occur during comparable time periods resulting in lower energy emission signals, but signals emanating at a much more rapid rate and giving the appearance of a continuous form of emission. An example of the latter is shown in Fig. 40.2. Progressing from top to bottom (1–9), left to right, the increase in the amplitude of 'continuous' emission observed during a tensile test can be readily observed. The bottom right photo is typical of the onset of gross plastic deformation in a tensile specimen; final failure follows rather quickly. Fig. 40.3 shows the nature of the emission response from

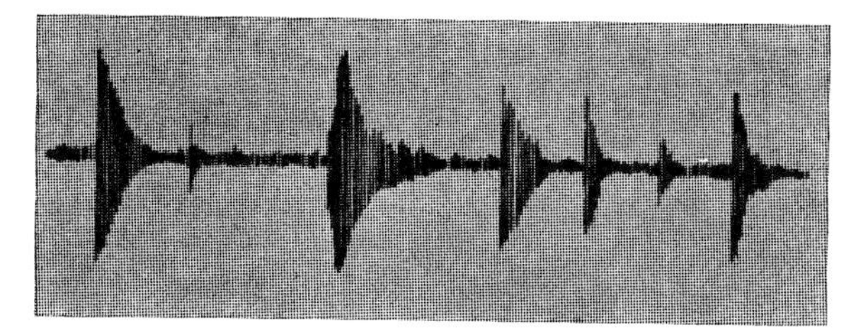

Fig. 40.1. Burst type acoustic emission

a specimen subjected to a fatigue environment in which a crack is growing in the specimen. A reduction in the oscilloscope sweep rate tends to obscure the discrete 'burst' activity.

These two forms of emission then, although not genetically different, are indicative of the specific time–energy behaviour of the deformation process. Generally speaking, the 'continuous' form of emission is observed when a relatively large portion of material is undergoing gross plastic deformation. In this instance the entire portion of the material is under a high state of stress, resulting in a multitude of activated sites of energy release. At this high level of stress the deformation process tends to progress in smaller but more frequent energy releases which tend to accelerate in occurrence frequency under a constant or increasing driving force.

Conversely, the 'burst' type of emission is typical of a very local region in the material under relatively high stress surrounded by material at significantly lower stress. In this instance the deformation takes place in a discontinuous fashion with sudden releases of r energy. Twinning deformation has been note of deformation which induces this 'burst' typ Another mechanism also producing signi emissions, and the mechanism of paramou this paper, is the initiation and propagation of a crack or crack-like defect.

It is the above two salient emission characteristics that find unique application in the utilization of acoustic emission in evaluation of pressure vessel integrity. Before leaving the subject of emission behaviour, however, it is worth noting one further emission anomaly; one which at present finds limited application to pressure vessel surveillance, but one which may become valuable in special circumstances. The latter anomaly is commonly referred to as the Kaiser effect, named after its discoverer Dr Joseph Kaiser, late of Germany.

The Kaiser effect refers to the characteristic that once a material is deformed to a given degree, during which emission is induced, a restressing of the material over the same stress range will not induce comparable emission activity. Hence, it is possible to determine the previous stress level to which a material has been subjected by establishing the stress at which the emission occurs. This characteristic may find useful application in assessing the maximum level of pressure transients over a period of time. It must be noted, however, that thermal ageing and other forms of metallurgical and physical alteration will, of course, affect the emission–stress level response such that emission activity may be restored in time, the latter ranging from minutes to months, depending on the

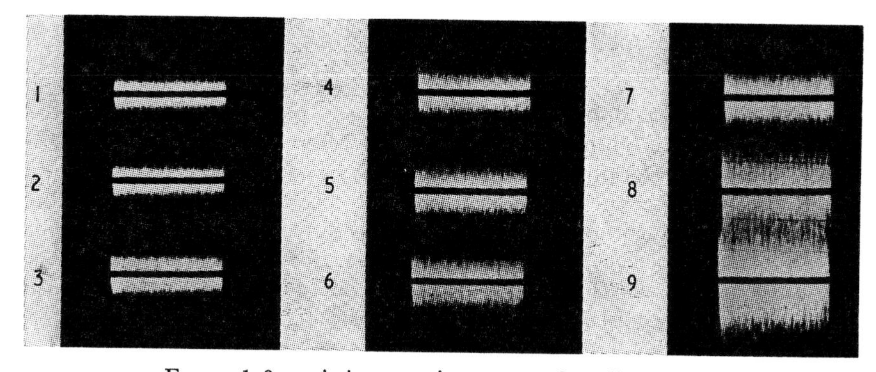

Frames 1–9: emission at various stages of tensile loadings.
1—General background noise at initiation of test. 9—Emission at onset of gross yielding.

Fig. 40.2. Photographs of the increasing level of the higher frequency spectrum for pure aluminium single crystal

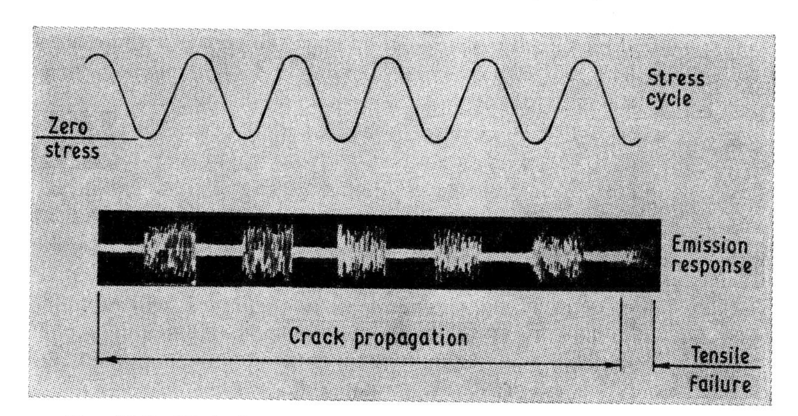

Fig. 40.3. High-frequency acoustic emission behaviour during fatigue stressing of magnesium

material environment. Radiation effects would be included in the latter.

For readers interested in exploring the fundamental aspects and research efforts on acoustic emission, a list of primary references is given in Appendix 40.1.

SCOPE OF ACOUSTIC EMISSION EVALUATION

Observance and analysis of the acoustic emission response from a pressure vessel, both the so-called 'burst' and 'continuous' emission, provides a three-fold assessment of the vessel integrity. First, emission behaviour provides a precursor response such that risk of sudden, catastrophic failure of the vessel can be significantly reduced if not eliminated. Second, emission 'bursts' signal the presence of defects in the structure and, third, such bursts can be used to physically locate the defect in the vessel.

In a number of historical instances it would undoubtedly have been tremendously advantageous to have been aware that failure was impending. Substantial projectiles have been unwittingly produced even under the safer liquid hydrostatic test.

By monitoring the total emission energy developed in a vessel in the course of a hydrostatic test, the onset of gross plastic deformation and/or the development of crack instability can be discerned. Fig. 40.4 is an ideal representation of the precursor to impending failure. Stage I is typical of the emission behaviour from a vessel containing non-critical defects. Emission is being produced, but at a relatively stable rate as pressure is increased. In Stage II the emission energy is seen to be increasing rather rapidly, but again in a relatively stable fashion. During this stage either gross plastic deformation is developing or a crack defect propagation rate is accelerating, or finally, there has been a substantial increase in the number of active defect sites. This is a critical state of test and would necessitate a decrease in loading of the vessel to prevent failure. Immediate analysis of the data would be undertaken to assess the specific nature of the anomaly and its location or locations on the vessel. Stage III depicted on the diagram represents the region wherein the failure mechanism is proceeding at such a rate that failure will ensue.

In order to assess and analyse impending failure, it is necessary to establish the impending failure curve, such as Fig. 40.4, for the same material as the vessel and for specimens bridging weld regions. These tests would comprise both tensile and fracture-type studies. The established emission response then becomes an integral parameter of the system as reference data to be discussed in a later section.

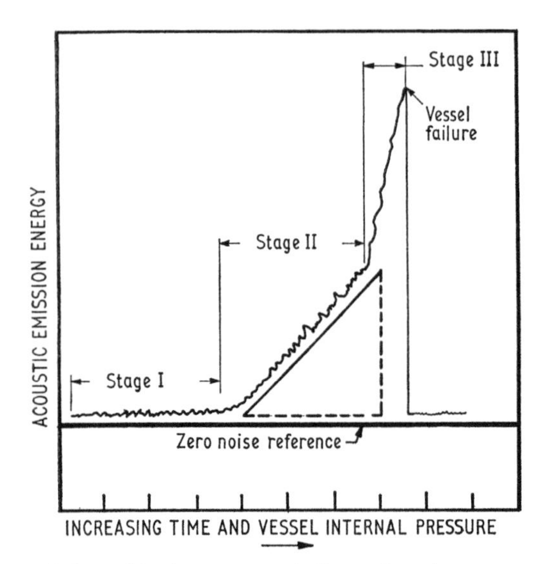

Fig. 40.4. Ideal representation showing precursor acoustic emission for impending failure of pressure vessel in hydrostatic test

The elapsed time period of Stage II is variable depending on the structure geometry, material, pressurization rate, and local stress states. The exact time cannot be stated *a priori*; nevertheless, it is relatively short, as little as 15 s in specific tests on small, thin-walled vessels. It is axiomatic, therefore, that automatic pressure release systems be incorporated in any test procedure requiring significant pressure levels.

The presence of the 'burst' type of emission, because of its discrete, sudden occurrence, provides information not only about the existence of the defect, but also about its location in the structure. As previously noted, the 'burst' emission is typical of crack or crack-like propagation—i.e. discontinuous growth—hence, regions of local stress concentrations, whether actual cracks or conglomerations of foreign material, are likely to produce local areas of acoustic emission which can be detected.

The location of the defect is accomplished through the judicial placement of several transducers on the structure under test. The 'burst' pulse propagates from the defect, as from a point source, and since signal attenuation is extremely low, the pulse is detected at each of several of the sensors or transducers. By measuring the time at which the sharply fronted pulse arrived at each sensor, and with experimental data on propagation velocities for the wave, the actual physical location of the emission source, and hence the defect, can be determined within fairly accurate limits. Current data indicate accuracies within a radius comparable to the wall thickness for thick-walled vessels, i.e. about 6 in.

Two aspects relating to the scope of the emission evaluation should be noted. First, the acoustic technique as described above monitors 100 per cent of the vessel simultaneously, with fixed position transducers which need engage only a fraction of the structure surface area. In contrast to other non-destructive techniques the sensor need not physically engage the precise area from which data are obtained, since the emission data are 'transmitted' to the sensor and only distressed regions produce desired data.

The second item relates to a shortcoming of the acoustic method which is the inability to *quantitatively* assess the severity of the emission-producing defect. The energy of a 'burst' pulse is related, primarily, to the rate of deformation and the volume of material undergoing the deformation. Consequently, it is not presently possible to precisely predict whether the crack is small and progressing in large discontinuities or vice versa. Qualitative information such as the repetition rate of the pulses can be utilized, especially a change in repetition rate, but here again some ambiguity may be extant, such as the simultaneous growth of two defects proximately located. Triangulation can be helpful in this last instance; however, the possibility of ambiguity is not entirely eliminated. Work now in progress by a number of investigators is expected to provide significant advances in assessing defect severity with greater precision.

The energies associated with the acoustic emission signals, either 'burst' or 'continuous', are extremely low relative to most environmental noise associated with mechanical/electrical components. The noise generated by high flow rates of fluid in pipes, for example, generates considerable noise of rather broad frequency band. Studies made on operating nuclear reactors show that the fluid noise encompasses the same range of frequencies as the acoustic emission with only limited mutually exclusive frequency windows.

Advances in signal detection within a general noise field have undoubtedly enhanced the emission technique; however, the procedures are rather costly and at best represent a statistical assessment of the desired information. Further, because of the somewhat limited experience in utilizing the emission technique on complex installations such as nuclear reactors, it is the writer's opinion that emission surveillance should be accomplished with a quiet system, i.e. under hydrotest-type conditions. Consequently, in-service monitoring would be undertaken on a periodic basis as opposed to continuous monitoring. The more often these inspections can be done—e.g. during yearly refuellings—the better will be the confidence in the structural assessment. Many refuellings and so-called shutdown periods on current reactor systems still involve rather substantial fluid flow rates and operation of numerous electromechanical components. Until techniques of sufficient confidence are fully developed to detect small signals within considerable environmental noise, the inspection procedure should be confined to the relatively quiet hydro-test conditions. It would appear that present operational conditions of the units will have to be modified by the system designer to accommodate the acoustic method.

SYSTEM CONSIDERATIONS

The development of multisensor acoustic emission systems for surveillance of pressure vessels has been under way for about five or six years by as many investigators. A review of the available units shows rather wide variety in the methods of signal conditioning and data display, but the basic concepts are generally identical. Fig. 40.5 is a schematic showing the basic elements common to these systems.

System operational frequencies range from as low as 1000 Hz up to 3·0 MHz, the upper frequency range generally being used where operational environmental noise is significant. Apart from the latter, there is little benefit in detecting emission at the higher frequency levels, and with the in-service inspection limited to quiescent conditions considerable advantage is gained by employing the lower frequency regions. Inasmuch as the high-frequency components of the acoustic signal attenuate significantly during propagation through the material, the high-frequency system necessitates use of considerably more transducers. For the larger vessels the signal conditioning, data display, and interpretation increase in complexity as the number of sensors increase. The latter, of course, increases system investment costs, not to mention the additional cost of installation of the transducers on the complex reactor systems.

Transducers are readily available in a variety of piezoelectric materials and frequency response. Probably the current most popular sensor is the PZT-5 ceramic (lead–zirconium–titanate). This material is quite stable and can be used at moderate temperatures, e.g. 400°F. The effect of neutron radiation on transducer response has been studied to a limited extent. The studies indicate no drastic influences for the modest nuclear exposures conducted to date; however, because of the rather severe temperature environment of reactor systems, techniques are necessary to remove the transducer from the immediate region of high temperature. By so doing the nuclear exposure is also significantly reduced and satisfactory sensor lives appear to be realizable.

Removal of the transducer from the high-temperature zone is accomplished by the use of extension rods, sometimes referred to as microwave extenders. This is simply the coupling of a 6–10 ft metallic rod or tube about $\frac{1}{8}$–$\frac{1}{4}$ in in diameter to the vessel surface, the rod extending radially outward through the insulation and shielding material. There has not been extensive experience in the use of such

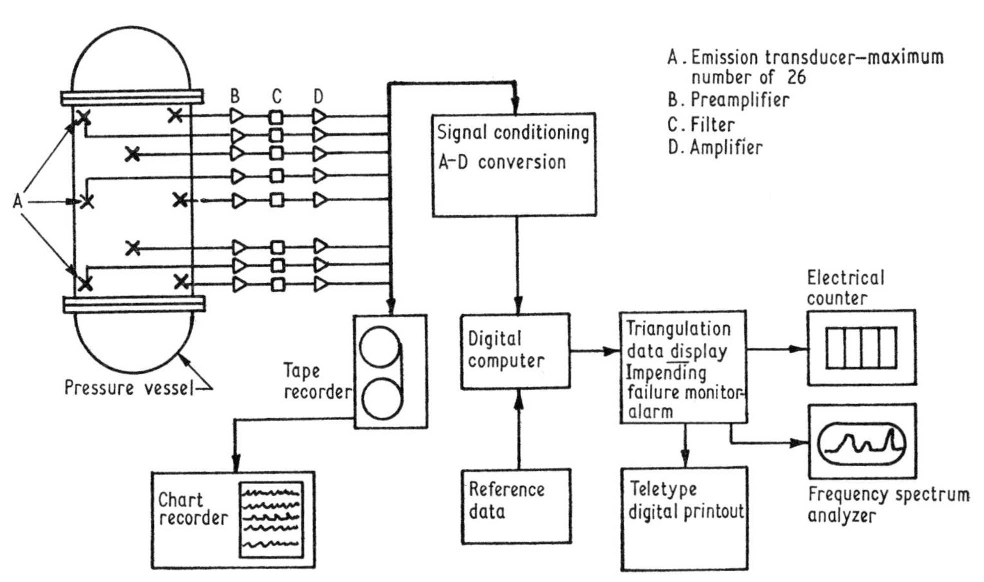

Fig. 40.5. Block diagram of acoustic emission in-service surveillance system

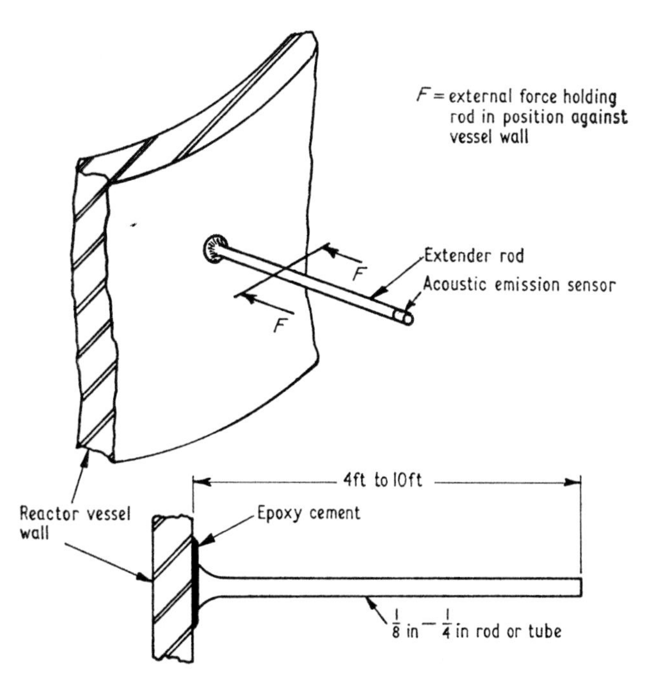

Fig. 40.6. Sketch showing acoustic emission extender rod and positioning of rod on reactor vessel

rods, but there do not appear to be any technical shortcomings which will affect the desired results. Fig. 40.6 shows a schematic representation of an extension rod in position.

The instrumentation from the transducer through the low noise preamplifier, filter, and amplifier is quite uniform in most systems except for frequency response and frequency band pass. As noted earlier, this will depend on the particular choice of the investigator depending on environmental circumstances. Total maximum gain of the systems ranges between 10^4 and 10^6 for most systems and is stepwise variable throughout a range of 1 to the maximum value.

The output from the preamplifier is customarily split for specialized signal conditioning while retaining the unmodified emission for other evaluations and data processing. The special signal conditioning involves discriminator gate circuits where conversion is made to a shaped pulse, a form convenient to digital manipulation of the data. The form of the shaped pulse may be uniform and identical or variable to reflect quantitative aspects of the data. Pulse count rate and count summation are most conveniently handled with the constant form shaped pulse. In some instances the determination of delay times for triangulation computations is performed on converted emission pulses, whereas in other systems the preference is to detect the time delta from the unprocessed signal from the amplifier. A wide variety of processing procedures may be used in the systems, some of which have been retained as proprietary information by several investigators.

Although there are variations, normally the raw acoustic emission data can be monitored from each of the individual sensors separately or some or all summed on a monitor. Three forms of the data are usually evaluated: (1) instantaneous total emission energy (from one or several transducers combined); (2) pulse repetition or count rate for a given transducer; and (3) integrated discrete pulse energy.

Almost all systems incorporate magnetic tape recorders to obtain a permanent record of the raw emission data and/or processed data. The necessity to operate at the higher frequency levels is again a factor here because of the necessary high tape speeds or incorporation of a heterodyne to permit tape recording at moderate speeds.

The other forms of data analysis and display are only limited by the imagination of the investigator. Oscillograph chart recorders for emission count rate, summation count, and burst emission energy are very common. These are generally complemented by oscilloscope displays (including memo-scopes), spectrum analysers, and electronic digital counter displays supplemented by digital paper tape printout.

Triangulation data are obtained from specific arrays of the transducers on the vessel under test. To provide real time analysis and defect location, the time analysis data are fed to a digital computer. Locations of the physical sites of emission source are then displayed visually or in coded digital form printed out on teletype or a comparable system. Visual display has taken several different forms. In one, an optical system displays light images on a screen containing a geometrical outline of the exact structure. The computed data points are also digitally tabulated for permanent record.

An alternative method displays the emission location on a scaled oscilloscope screen. Each computed location appears as a light dot on the scaled screen with each successive computation at the same location being positioned at successively higher amplitudes on the screen. This format provides a statistical assessment of the significant emission source. The latter is quite valuable because of the possible complexities arising from multiple emission sources which could result in erroneously computed locations.

A final but invaluable element of the inspection system is the reference data. The latter consists of information developed from experimental and analytical efforts. This information is stored in the computer memory and is continually compared as a criteria to those emission patterns being observed in real time. Impending failure data and fracture mechanics studies provide the bulk of reference criteria. Decisions concerning the instantaneous integrity of the structure under test can consequently be made rapidly and, possibly in the future, automatically.

STATE OF THE ART

At the present time there are about six firms or organizations in the U.S.A. at varying stages of acoustic emission system development for pressure vessel testing and in-service inspection. The systems vary considerably in sophistication, i.e. with or without computers for real time multiple sensor triangulation and the storage of reference data libraries. The maximum number of advertised channels is of the order of 32, while the simplest systems consist of four channels of sensor data. All are capable of detecting emission corresponding to impending failure.

The above systems can efficiently detect the emission response of pressure vessels under the quiescent conditions generally associated with a hydro-test in the hold or static condition, i.e. with pumps and other equipment momentarily stopped. To date none of the systems is capable of detecting significant emission within the environment of an operating nuclear system, although some recent advancements have been made towards improving this status. Recent reports suggest laboratory success has been attained at noise to signal ratios of about 2; however, this is yet far from field application at the much higher ratios of noise to signal that are extant. In the author's view success in this area is several years away and becoming even farther from reach as a consequence of reductions in sponsored research in recent times. The solution will undoubtedly require the utilization of data analysis methods developed in the sophisticated space communications technology, coupled with additional basic research on the emission source signal and its propagation in materials.

The unique feature of acoustic emission to signal the onset of impending failure may be a most valuable asset in certain instances, a feature not proffered by any other test method. The ability to stop a vessel from continuing to ultimate failure has been demonstrated time and again on small pressure vessels, a considerable number having been so tested by the author. The tests have been successful on moderately thin-walled vessels fabricated from materials exhibiting reasonable ductility. Total elastic energy in the vessel and in the fluid medium was relatively low at the time of developing failure. Although there is no apparent technical reason why the acoustic method would not be effective on large reactor vessels, no tests have been made on large, thick-walled vessels at relatively high fluid pressures where a defect or other condition produced a condition of impending failure. Consequently, it cannot be conclusively stated that failure can absolutely be precluded. The relatively larger amounts of stored elastic energy in both the vessel and fluid medium may pose a time problem in depressurization to a satisfactory level, rapid enough to arrest failure. On the other hand, the larger vessel may enhance the time available in the Stage II slope region. Clearly, full-scale tests on critical defect vessels would be most enlightening in this area.

Another aspect of which the acoustic emission investigator must be aware is the presence of a thermally induced defect in the vessel. To this point the acoustic surveillance has been premised on hydro-test loading and stressing of the vessel under study. This would appear to be completely adequate in the fabricator's shop test, prior to operation of the vessel. However, it is well known that thermal transients must be considered in the design of nuclear reactor systems. If the latter are sufficiently severe, it is conceivable that a thermally induced crack can be initiated and propagated in the vessel structure. It is also conceivable that the thermal stress state producing the defect is substantially different in the defect region from the pressure-induced stress state. In such circumstances a pressure test may not provide sufficient driving force to cause propagation of the existing defect and, consequently, no emission would be generated.

The above situation emphasizes two obvious points. First, the acoustic emission investigator and his colleagues must be fully competent in the theoretical stress and deformation aspects of pressure vessels for both the pressure and thermally induced states. Second, a comprehensive surveillance may further necessitate inspection of local discontinuity regions under thermal transient simulation in the vessel. This is considerably more difficult than the pressure test and will require developmental effort on the part of the acoustic investigator, as well as the system designer responsible for in-service monitoring. Notwithstanding that thermal problems may exist in special circumstances, the pressure test alone is indispensable in the structure evaluation inasmuch as a catastrophic-type failure would necessarily involve pressure-induced defect propagation.

Previously, mention was made concerning the difficulty in assessing the *severity* of a defect by acoustic emission. Clearly, defect severity can be discerned in a qualitative and indirect fashion. For example, a sudden change in the repetition rate of burst signals coupled with a significant increase in average 'burst' signal energy is indicative of increased crack 'severity'. A similar response could, however, be produced by the onset of a second propagating defect containing higher burst energy than the first. If the two are sufficiently separated to permit location resolution by triangulation, the status can be resolved. However, if the defects are within the circle of uncertainty of location resolution, an ambiguous result would be obtained.

Laboratory research efforts in progress have established preliminary and tentative correlations between emission and crack severity. Whether or not these correlative parameters are absolute, remains to be seen. Without question, resolution of this aspect bears heavily on the total success of the emission method, but even more important, bears on the general acceptance and utilization of the technique. Burying one's head in the sand to avoid the problem will obviously be neither a help nor a solution to the problems. Experience through utilization of acoustic emission will provide the answers for effective results.

The final item of discussion concerns the reactor system designer and his impact on the state of the art.

Reactor systems already in service and many under present construction were not designed with acoustic in-service inspection in mind. Provisions were not, of course, made for installation of acoustic transducers and generally it would be most difficult, even if economically justified, to instal effectively at this date. Many critical regions of the vessels are virtually inaccessible to any form of inspection. A specific and recent instance emphasizing the necessary involvement of the designer is seen in the simple hydro-test. In the past, nozzles were closed by integral weldments for the hydro-test. To effect reductions in costs, nozzle configurations were standardized where possible, and reusable plug inserts designed for use in the hydro-test. These plugs physically move under the influence of increasing pressure and consequently generate substantial friction noises adjacent to the very regions of interest, i.e. nozzle discontinuity areas. Fig. 40.7 indicates, by the circled regions, some of the more critical regions of a reactor vessel. The generated noise was of sufficient energy in the tests to significantly mask the acoustic emission data.

Recent discussions with systems designers evidences an awareness on their part to the desirability and need to accommodate acoustic emission surveillance. Such accommodation is required in the physical arrangement of the structure and also in the operational conditions during the in-service inspection periods. Inasmuch as this effort entails minor alternatives and variations in existing design concepts as opposed to design addition and innovation necessitated by other inspection methods, the cost advantage of acoustic emission over other less comprehensive methods is sustained.

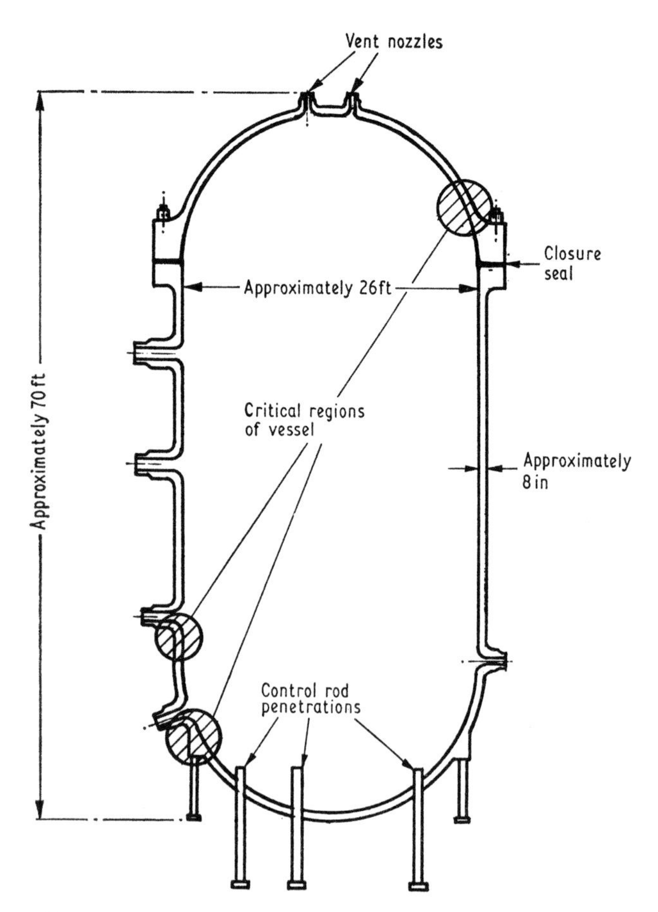

Fig. 40.7. Geometry of typical boiling water reactor vessel

SUMMARY

Acoustic emission can be utilized effectively for the surveillance of reactor pressure vessels under hydrostatic or similarly quiescent test conditions. The technique provides 100 per cent inspection of the vessel. Data are obtained on the presence and location of defects and conditions of impending failure. Developmental efforts and field tests need to be undertaken to improve the degree of reliability in specific areas and to verify overall capability of the systems to meet desired results under a variety of conditions and environments. Realization of the full potential of the acoustic method will require the co-operation of reactor systems designers in the accommodation of the emission method, which still appears to be the most economical and also comprehensive non-destructive method available for in-service inspections.

APPENDIX 40.1

BIBLIOGRAPHY

BAREISS, R. A., KYRALA, A. A. and SCHOFIELD, B. H. 'Acoustic emission under applied stress', ASTIA Doc. No. AD155674, WADC Tech. Rept 58-194, 1958 (Wright-Patterson Air Force Base, Ohio).

DUNEGAN, H. L. 'Acoustic emission—a new nondestructive testing tool', *Symp. Nondestructive Testing of Welds and Materials Joining*, Los Angeles, 1968 (11th–13th March).

DUNEGAN, H. L., HARRIS, D. O. and TATRO, C. A. 'Fracture analysis by use of acoustic emission', USAEC Rept UCRL-70323 (CONF-670604-1), 1967 (February).

GREEN, A. T. 'Stress-wave detection, Saturn S-II', Rept NASA-CR-61161, 1966 (Aerojet-General Corp., Sacramento, California).

HUTTON, P. H. 'Acoustic emission in metals as an NDT tool', paper presented at Society for Nondestructive Testing, 27th National Fall Conf., Cleveland, Ohio, 1967; abstracts of papers in *Mater. Eval.* 1967 **25** (No. 9, September), 37A.

KAISER, J. 'Untersuchungen über das auftreten Gerauschen beim Zugversuch', Ph.D. thesis, Technische Hochschule, Munich, 1950.

PARRY, D. L. 'Nondestructive flaw detection in nuclear power installations, in incipient failure diagnosis for assuring safety and availability of nuclear power plants', USAEC Rept CONF-671011, 1968 (January).

SCHOFIELD, B. H. 'Acoustic emission under applied stress', ARL 150, 1961 (Wright-Patterson Air Force Base, Ohio).

SCHOFIELD, B. H. 'Acoustic emission from metals—its detection, characteristics and source', *Proc. Symp. Physics and Nondestructive Testing*, San Antonio, Texas, 1963.

SCHOFIELD, B. H. 'Acoustic emission under applied stress', AST-TDR-63-509, Part I, 1963 (Wright-Patterson Air Force Base, Ohio).

SCHOFIELD, B. H. 'Acoustic emission under applied stress', ASD-TDR-63-509, Part II, 1964 (Wright-Patterson Air Force Base, Ohio).

SCHOFIELD, B. H. 'Investigation of applicability of acoustic emission', AFML-TR-65-106, 1965 (Wright-Patterson Air Force Base, Ohio).

SCHOFIELD, B. H. 'A study of the applicability of acoustic emission to pressure vessel testing', Rept AFML-TR-66-92, 1966 (Lessells and Associates, Inc., Waltham, Mass.).

C41/72 THE PRESENT STATUS OF IN-SERVICE INSPECTION FOR NUCLEAR POWER REACTORS IN JAPAN

Y. ANDO*

In order to ensure the operational integrity of nuclear power reactors in Japan, there has arisen the need for in-service inspection. In this connection, studies have been made concerning the necessary regulations by the Japan Electric Association, and jointly by the Government, electric power companies, manufacturing companies, and academic circles. The direction of in-service inspections in Japan, based on the A.S.M.E. Code Section XI, is described. Details are then given on in-service inspections carried out on the current power reactors; future trends; inspections made for certain difficulties occurring in the existing power reactors; and development efforts relating to techniques of in-service inspection.

INTRODUCTION

IN JAPAN, construction of nuclear power plants is proceeding at a high pace in an effort to satisfy the rapid increase in power demand. It is expected that by the end of 1975 about 8600 MWe will be produced by atomic energy. The power reactors in operation or under construction in Japan are mostly light water reactors, about half and half for B.W.R. (boiling water reactor) and P.W.R. (pressurized water reactor) respectively. The first power reactor to be introduced in Japan, however, was a gas-cooled reactor, which already has been operating for several years.

In order to ensure the operational integrity of these power reactors, there has arisen the need for in-service inspection. In this connection, studies have been made concerning the necessary regulations by the Japan Electric Association, and jointly by the Government, electric power companies, manufacturing companies, and academic circles.

In the present report, the direction of in-service inspections in Japan, based on the A.S.M.E. Code Section XI, which was deliberated by the Society, is first described. Details are then given on the following: in-service inspections carried out on the current power reactors, and future trends; inspections made for certain difficulties occurring in the existing power reactors; and development efforts for techniques of in-service inspection.

IN-SERVICE INSPECTION CODE

Background to the draft of a code

The code data available in Japan from overseas countries are the A.S.M.E. Code Section XI of the United States and the Oskarshamnsverket Programme of Sweden; of the latter, only part is available.

The MS. of this paper was received at the Institution on 13th October 1971 and accepted for publication on 12th January 1972. 23

* *Chairman, Committee on In-service Inspection, Japan Electric Association; Professor, University of Tokyo.*

Therefore, it was considered that the A.S.M.E. Code Section XI was the most suitable for use in preparing a code of regulations in Japan. The draft of the Japanese code for in-service inspections was made by referring to the A.S.M.E. Code after due consideration had been taken of the peculiar situation in Japan.

Draft of in-service inspection code in Japan

The A.S.M.E. Code Section XI has been studied, and it was concluded that its contents were generally appropriate for applications in Japan. The fundamentals were therefore adopted that in-service inspections following the practice of the A.S.M.E. Code Section XI would be carried out for power reactors to be built in Japan. Therefore the draft of the Japan in-service inspection code (Japan Electric Association Code) has been prepared accordingly.

The draft of the code is composed of six chapters, each chapter with its respective sections. Where necessary, figures, charts, etc., are given, to explain clearly the desired requirements.

The main differences between the present draft code and the A.S.M.E. Code are:

(1) In the A.S.M.E. Code, the reactor-coolant pressure boundary is defined in IS-120 of Section XI. In the draft of the Japanese code, the definition is according to the existing 'Guides for electric technology (JEAG-4602)', thereby aiming at uniformity in Japan. There is not much difference between the A.S.M.E. Code and the Japanese code, except for minor points.

(2) The conditions for application are given in the 'Foreword' to the A.S.M.E. Code. However, in the Japanese code they are stipulated in the text, providing for the date of enforcement, and other similar items.

(3) Concerning the qualifications required of inspecting personnel and of inspecting agencies, there is a great difference regarding inspecting personnel.

In Japan there exist no inspecting firms such as those essential for conformity with the A.S.M.E. Code Section XI. In addition, the system of examinations for inspecting personnel has only recently been started; therefore the number of such examinations so far held is correspondingly small. The recruitment of personnel of the kinds specified in the A.S.M.E. Code Section XI in large numbers is impossible. In-service inspection of reactor components is therefore made currently in each company by those of its own employees who have passed the qualification test by fabrication.

(4) Since the Japan in-service inspection code is, in fact, a series of 'self-imposed' regulations, and not the law, the liability of reporting to the competent authorities the results of inspection and evaluation is not stipulated. However, as with the A.S.M.E. Code Section XI, provision is made for keeping records, and for their inspection by the authorities when necessary.

IN-SERVICE INSPECTIONS CARRIED OUT IN JAPAN

Boiling water reactor, case 1

The Tsuruga power station of J.A.P.C. (Japan Atomic Power Company) went into commercial power generation in March 1969, and its first regular inspection was carried out in 1970.

In the Tsuruga power station it is very difficult to perform any inspection of the type stipulated in the A.S.M.E. Code. Any such inspection must be restricted, especially for the reactor pressure vessel. The reasons for this are:

(1) Inspection from the outside is impossible, except for the nozzle portions and top head, because the insulating material cannot be removed.

(2) Structures cannot be removed from the pressure vessel, excepting the steam separator and drier, thus making difficult the approach to the inside.

(3) There is no finish on the inner overlay, except at the nozzle portions and the weld portion between body shell and flanges. Therefore ultrasonic testing is difficult.

(4) The space between the external shielding and the pressure vessel is narrow because of the existence of heat-insulated pipes, supports, etc. Thus, access to this space is not easy, and any work to be done here is difficult to perform.

(5) It is laborious work to remove and reinstall heat-insulation plugs at the nozzle portions of the biological shields.

(6) There are many areas of high-level radiation. In addition, air contamination would result from the removal of the insulating materials and surface brushing. It is therefore necessary to take measures for the protection of the inspecting personnel. Due to the restrictions on space and structure, it is difficult to install the protective shields. Also, the necessity for ventilation equipment and protective clothing hampers any inspection.

(7) The procedure for cleaning the interior surface, essential for remote-controlled inspections (ultrasonic test, etc.) and periscope observation, is not yet established.

(8) It is difficult to establish any variation in radiation level within the time available.

As detailed above, the in-service inspection for such nuclear power plants is fairly difficult due to the prevailing circumstances. It is therefore desirable to accumulate the necessary data and thereafter develop new techniques of inspection.

Inspections of the reactor pressure vessel in the Tsuruga power station are divided into two operations: (1) 'inspection' centred on the top head, where the inspection results can be clearly grasped; and (2) 'preparatory investigation' to acquire the data necessary for the establishment of a long-range inspection programme.

In-service inspection of the Tsuruga power station was carried out during its regular inspection in October and November 1970. Since the radiation levels were unexpectedly low, inspection could also be made of the recirculation nozzles.

No defects were discovered in those portions inspected. Inspection was also made partially (the welds in piping, and bolts of less than 2 in diameter) during the reactor shut-down for poison-curtain removal in the spring of 1971.

Boiling water reactor, case 2

Construction was started on No. 1 plant of the Fukushima (atomic) power station, Tokyo Electric Power Company, in December 1966. It went into commercial power generation (460 MWe) on 26th March 1971. The main features of the plant are as follows.

No. 1 plant is the so-called 1965 type B.W.R. In its design, no consideration was made for the in-service inspection as prescribed in the A.S.M.E. Code Section XI. There are many positions in this type where in-service inspection as specified in the A.S.M.E. Code Section XI is impossible, due to the inaccessibility, and other restrictions in structure; this is the greatest difficulty experienced at present. The plan presently considered is that, wherever possible, in-service inspections in accordance with the A.S.M.E. Code Section XI should be made to the fullest extent. In the initial stage of in-service inspections, therefore, inspection will be made of the reactor pressure vessel and associated pipings at the time of a regular inspection and a fuel exchange, when disassembly, check, etc., are made. At this time, the following items are also carried out: measurement of the radiation levels at various positions; and investigations regarding any plant alteration that might give an increase in accessibility and permit development of inspection techniques and equipment. Three or four years after the start of the initial stage, the second stage of in-service inspections, extending the initial stage, is formulated on the basis of the results gained from the initial stage. Subsequently, these inspections are put into practice one after the other.

In the meantime, information gained from similar-type reactors and the developments in oversea countries will be introduced and utilized, where possible, in the in-service inspection of Fukushima No. 1 plant.

Pressurized water reactor

In the Mihama power station of Kansai Electric Power Company, No. 1 plant went into commercial power generation (340 MWe) in November 1970. Its first regular inspection is planned to take place during the period October 1971 to December 1971.

Locations for inspection

(*a*) *Pressure-resisting welds and weld overlay in the reactor pressure vessel.*—Since in the first regular inspection, reactor-core components are not removed and inspection from inside is thus rendered impossible, the following 'representative' locations are inspected. For the pressure-constraining welds, the longitudinal weld in the lower shell panels and the weld between the top-head panel and flange are inspected by ultrasonic testing; in this case, the adjoining base metal is also included. For the inner weld overlay, one patch at least in the area that is accessible through the top head is inspected by liquid penetration or visual tests. For the pressure-constraining welds between different metals, a testing tool is now being made that permits inspection of the nozzle from inside. In the first regular inspection, however, the nozzle part is inspected from the outside, where possible, by both ultrasonic and visual tests. The weld between safe-end and piping is also inspected similarly.

(*b*) *Bolting.*—For the fastening bolts of the (pressure vessel) top head, representative ones are removed, and these are inspected by ultrasonic, liquid-penetrant, and visual tests.

(*c*) *Support members.*—External welds in the steam generator and pressurizer are inspected by ultrasonic test and, if necessary, by liquid penetration.

Then for the support members or structures, representative locations are inspected by ultrasonic or magnetic-particle tests, and in addition by visual test.

(*d*) *Pressure-constraining welds in the piping.*—In addition to the weld between the safe-end and piping already mentioned, other representative welds in the pipings are inspected over the entire circumference, including the base metal, by ultrasonic or liquid-penetrant tests.

Hydraulic test

After completion of the tests on various positions, all systems of the reactor coolant are subjected to hydraulic test by a specified method. Then the boltings and pressure-resisting welds concerned are tested visually.

Measures taken before inspection

In order to increase the areas of accessibility and to reduce inspection time, alterations in the reactor plant to facilitate the removal and reinstallation of insulating materials, and also installation of the inspection footings, are made.

EXPERIENCES OF IN-SERVICE INSPECTION IN THE INITIAL POWER REACTORS IN JAPAN

Tokai power station in J.A.P.C.

In-service inspection

Commercial power generation in the Tokai power station began in July 1967. For the in-service inspection of the pressure boundary in a gas-cooled power reactor such as the Tokai power station, there is no code comparable with the A.S.M.E. Code Section XI which is being employed in power reactors.

In the Tokai power station, the inspection so far carried out has been a visual test of the welds in the S.R.U. (steam raising unit), mainly of the positions which have caused troubles. For the reactor itself, only test samples of steel (pressure vessel) and graphite (moderator) have been extracted because of the restrictions resulting from radiation and structure. These samples are taken out and tested at specified intervals for monitoring purposes. In these tests, however, the welds are not inspected specifically. In each regular inspection, the damage, etc., to internal structures are observed partially by photography.

In the in-service inspection carried out early in 1971, due to lack of adequate preparation only the following positions were inspected, it being considered that these would represent the overall construction: one position in the higher temperature region in the S.R.U., and one position on the high-temperature side in the gas duct.

In future, liquid penetration will be made once every four years (inspection of the S.R.U. and duct once each year). In addition, studies will be made on the positions of sampling, its frequency, and the methods of inspection.

Steel oxidation

In the Bradwell nuclear power station, damage to nuts and bolts due to oxidation was discovered in May 1968. Since that time, extensive studies in this respect have been made in the U.K. Therefore, information concerning these studies was acquired from the U.K. for use at the Tokai power station. The co-operation of educational institutions, research institutes, and private enterprises made possible the investigation of oxidation, by means of specimens, of components of the S.R.U. and the reactor pressure vessel. Laboratory studies were then made on the oxidation phenomena and mechanisms, using an autoclave.

In the U.K. the temperature of the gas coolant was lowered to 360°C at the reactor outlet. Following the U.K. example, it was also reduced in the Tokai power station, starting in the autumn of 1969. During the ensuing period, the condition and quality of structure of the primary coolant system were examined, and the life expectancy was evaluated for critical portions, and also the consequences of their rupture.

From the results of investigations, it was found to be extremely difficult to clarify the mechanisms of oxidation, especially as the phenomena of oxidation are very much in fluctuation. It is therefore not easy to reach any definite conclusions, though the accuracy of measurements may be raised somewhat. At each regular endurance check (i.e. life expectancy), the operation of the reactor must be continued.

The difficulties arising from oxidation may be reduced by lowering the coolant temperature. If operation at 360°C is continued, there is a real possibility that the tubes in the S.R.U. may be damaged due to the unavoidable load. When oxidation in the current range of gas temperatures is considered, there is no doubt that the temperature must not be raised too far in the Tokai power station. During 1970, however, the amount of oxidation was relatively small. It is therefore planned to raise the temperature by about 10 degC and to operate for one year at this temperature, and from the results revealed in 1972, the pattern of future operation will be decided.

Japan power demonstration reactor in J.A.E.R.I.

This reactor gave the nation's first electric power generation (B.W.R. with 12·5 MWe) in 1963. In the 1966 regular inspection, many fine cracks were detected in the internal overlay of the pressure-vessel top head. The cracks were concentrated on the hand-welded overlay. They were mostly from 5 to 10 mm in length, 25 mm at most; the cracks all disappeared at the interface between overlay and base metal, none penetrating into the base metal. Consequently, it was decided to carry out inspection on the hand-welded overlay of the pressure vessel and at the high-stress-concentrated nozzle parts.

Methods of inspection

It was decided to inspect the cladding of the pressure-vessel bottom head, which was produced similarly to the top head, and the base of large-diameter circulation nozzles where high stress concentration was expected. Since the usual methods of inspection could not be applied because of intense radiation, investigation was made as to the methods of inspection to be used.

External observation by a borescope was first considered. Therefore an order was made for a borescope with a maximum amplification of about five diameters, and with a resolution of 25 μm under optimal conditions.

In 1967, when the plant was shut down for inspection, in addition to the investigation on the interior condition of the pressure vessel, preliminary studies were made concerning the methods of inspection, mainly with a borescope. As a result, the following items were revealed:

(1) An improvement in illumination is necessary when inspecting by borescope.

(2) Coloration of the borescope lenses is considerable due to the intense radiation from irradiated fuel. Therefore any observation by borescope must be performed fairly quickly.

(3) It is necessary to have an accurate positioning of the borescope to attain high reproducibility.

(4) The oxide film adhering to the inspection surface must be removed.

Furthermore, it is difficult to judge whether or not the indications by a borescope are really cracks. Even if confirmed, it is essential to know the depth of such cracks.

The following preparatory procedures were carried out before any inspection:

(*a*) For borescope illumination, iodine lamps were used of a large wattage. In addition, underwater lighting was employed.

(*b*) For the positioning of a borescope, two separate supports were used for the forced-circulation nozzle and the pressure-vessel bottom head, respectively.

(*c*) In order to reduce the coloration of borescope lenses and to minimize the radiation exposure of inspecting personnel, prior to the inspection irradiated fuel and control rods were removed from the reactor core and stored in spent-fuel ponds.

(*d*) A sufficient number of stand-by borescope lenses were provided to replace the 'coloured' ones.

(*e*) For brush-cleaning of the inspection surfaces, various preliminary tests were made out-of-pile. The brushing agent 'MMM Scotchbrite' was chosen because it removed effectively the oxide film from the cladding without affecting the cracks.

To cope with crack depth, various available methods were tested. Though not standard nor established yet, the electric-resistance probe method was employed (which was developed jointly by J.A.E.R.I. and Mitsubishi Heavy Industries Ltd).

Radioactivity outside the pressure vessel was investigated. It was revealed that the radiation level at the position of the circulation nozzle, far away from the reactor core, was several roentgen/h. It was thus possible for personnel to inspect the nozzles by placing temporary shields within the pressure vessel and then removing the existing concrete shield outside the pressure vessel. In this manner, the nozzle and its area were inspected directly from outside the pressure vessel by ultrasonic testing; thus, any penetration of the cracks into the base metal could be seen. For such usually accessible positions as the flange of the pressure vessel, observation by the ultrasonic method was readily possible. The dye-penetration method was similarly used wherever access was possible.

The first inspection of the reactor pressure vessel was thus carried out in 1968, and the second one in 1969–70. The locations of inspections and the methods employed are shown in Table 41.1, for both the inspections.

Table 41.1. Inspection methods applied to the in-service inspection of J.P.D.R. vessel

Components and parts examined	Inspection technique	
	1st inspection	2nd inspection
Bottom head	B.S. R.P.	B.S. R.P.
Forced circulation outlet nozzles (nozzle to vessel welds)	B.S. R.P. U.S. (A.B. and S.B.)	B.S. R.P. —
Forced circulation inlet nozzles (nozzle to vessel welds)	U.S. (A.B. and S.B.)	—
Vessel shell (upper portion)	L.P. R.P.	L.P. —
Vessel flange to shell weldment	U.S. (A.B.)	—
Vessel flange seal surface	U.S. (S.W.)	—
Head flange seal surface	U.S. (S.W.)	—
Top head	L.P. U.S. (A.B. and S.B.)	L.P.

Note: B.S., borescope; R.P., electric resistance probe; U.S., ultrasonic testing; L.P., liquid penetrant; A.B., angle beam; S.B., straight beam; S.W., surface wave.

Results of inspections

In the first inspection, crack-like defects were detected at several locations in the pressure vessel.

Fig. 14.1 shows the borescope photograph of a crack-like defect detected at about the 90° position. The results of smeck-gauge measurements on the west outlet nozzle are shown in Fig. 41.2.

In the bottom head of the pressure vessel, indications were given by a borescope and smeck-gauge at several positions corresponding to the openings of the bottom grid plate. These were, however, much finer in size than those detected at the nozzle; it was not easy to determine whether or not they were really cracks.

A liquid-penetration test was made on the cladding in the upper portion of the pressure vessel. Several pointer line-like indications were given. The lengths of line indications were about 5 mm on average (maximum 12 mm); the depths were estimated to be about 2 mm by smeck-gauge measurement.

For the flange seal of the pressure vessel, very shallow flaws were indicated on the surface by ultrasonic testing.

As noted above, in the first inspection, cracks or crack-

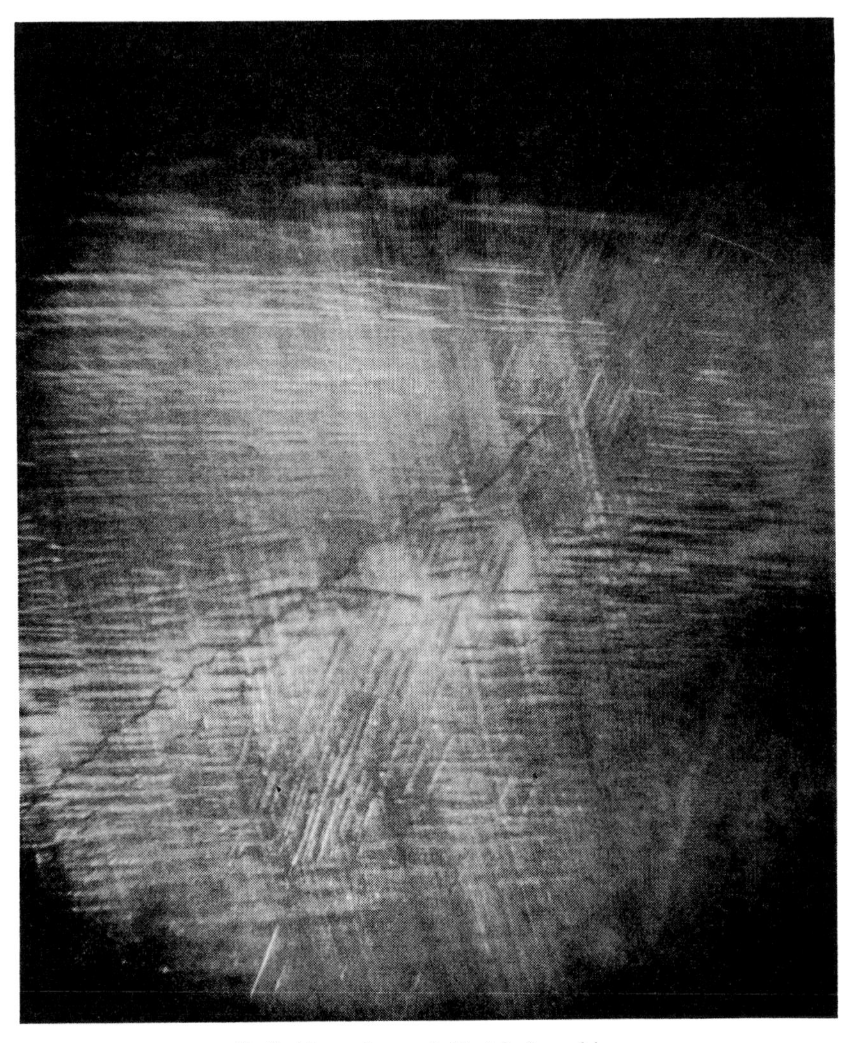

Shell side surface at 2.40 o'clock position.

Fig. 41.1. Crack-like flaws observed on the west forced circulation outlet nozzle

like defects were detected at a number of locations in the pressure vessel. These, however, were all up to several millimetres in depth, and were confined within the cladding. It was therefore concluded that they would not affect significantly the integrity of the reactor pressure vessel.

The second inspection was carried out over a period 1969–70. In 1969, inspection was made on the west forced-circulation outlet nozzle; and in 1970 on the east outlet nozzle, the bottom head, the interior surface of vessel shell, and the top head. The methods employed are shown in Table 41.1. The borescope position-determining equipment was improved upon from that used in the first inspection; strobe illumination was also utilized. Observations similar to those in the first inspection were obtained, though somewhat different in minor points from the first inspection. It was shown that no cracks other than those in the first inspection had occurred, nor any growth of the cracks revealed by the first inspection. The results of the observations noted in the first and the second inspections are shown in Fig. 41.3, especially with regard to 're-producibility'.

The in-service inspection specified in the A.S.M.E. Code Section XI could not be performed, since the reactor pressure vessel was not specifically designed with this in view. However, it must be considered as a success that the integrity of the pressure vessel could be confirmed by means of the data obtained. The borescope had a central role in the inspections carried out; it could detect even very fine flaws if the inspection surface was sufficiently clean. The smeck-gauge measurement, on the other hand, is very effective in the determination of the depth of surface flaws, and is also capable of remote operation; it is therefore a promising tool of in-service inspection. The ultrasonic and liquid-penetration methods were found to be sufficiently useful, wherever access is possible.

Some problems or difficulties were indicated as a result of the inspections. Excepting those areas where direct access by inspecting personnel is possible, the in-service inspection is restricted to a part only of the pressure vessel. This includes the circumference of the forced-circulation inlet and outlet nozzles and a portion of the vessel bottom head; other locations defy inspection due to the reactor-core structure, etc.

In the inspections carried out, fuel and control rods were removed to lower the radiation levels. This, however, is not practical in commercial power plants, for the time of inspection is prolonged. In the first inspection, the ultrasonic method was used to inspect the nozzle position. This

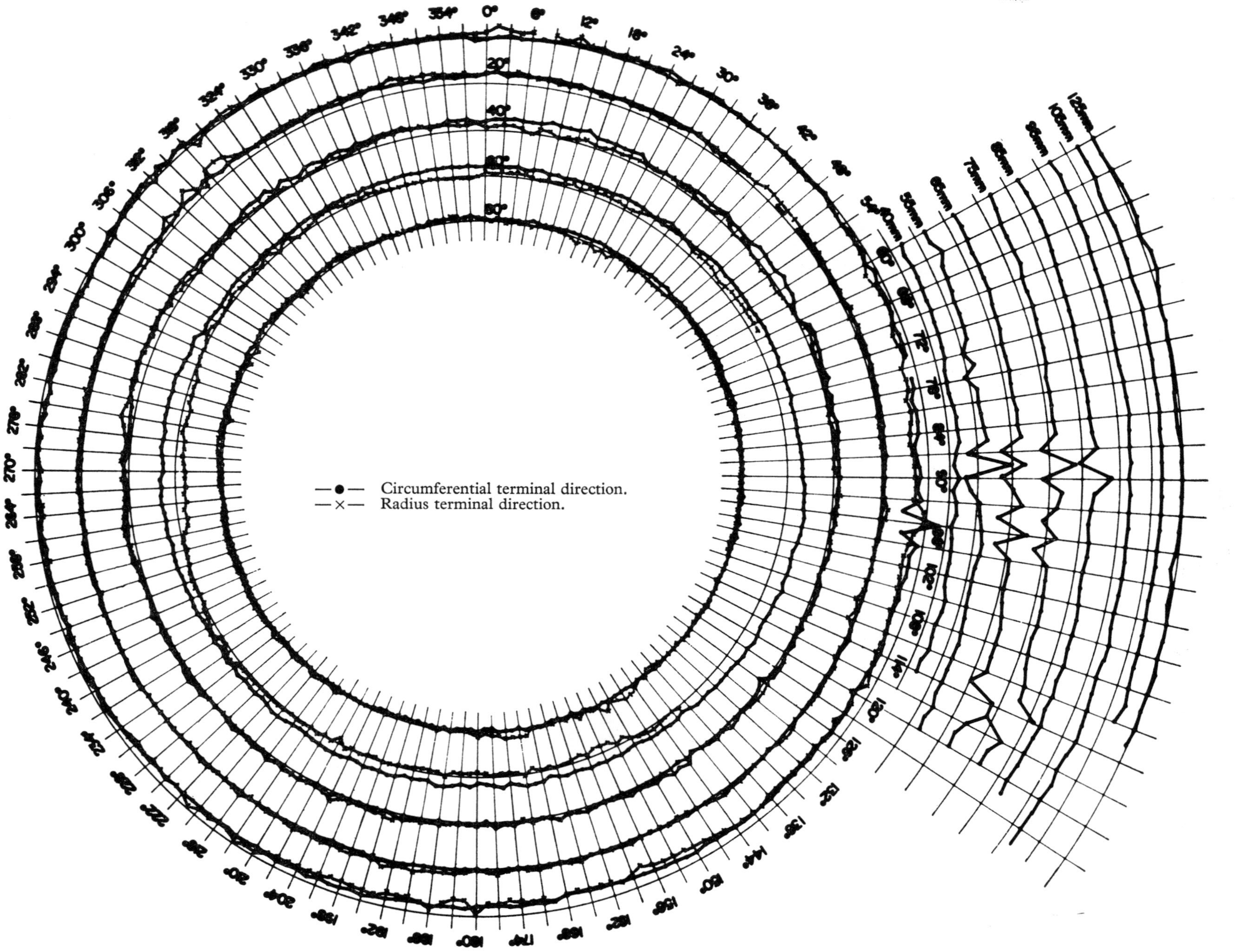

Fig. 41.2. Data measured by electric resistance probe method, west forced circulation outlet nozzle

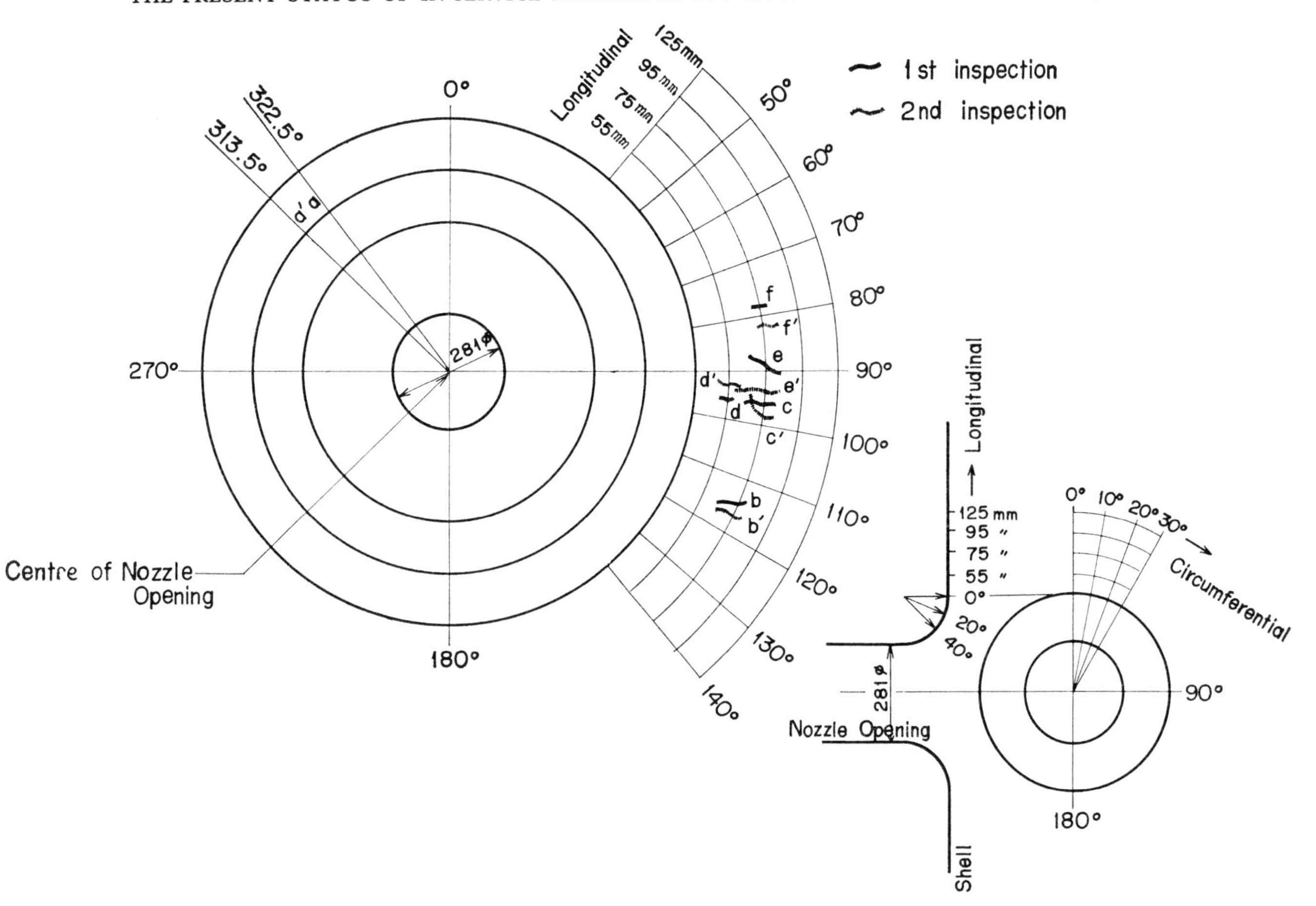

Fig. 41.3. Inspection reproducibility of the outlet nozzle of J.P.D.R. pressure vessel

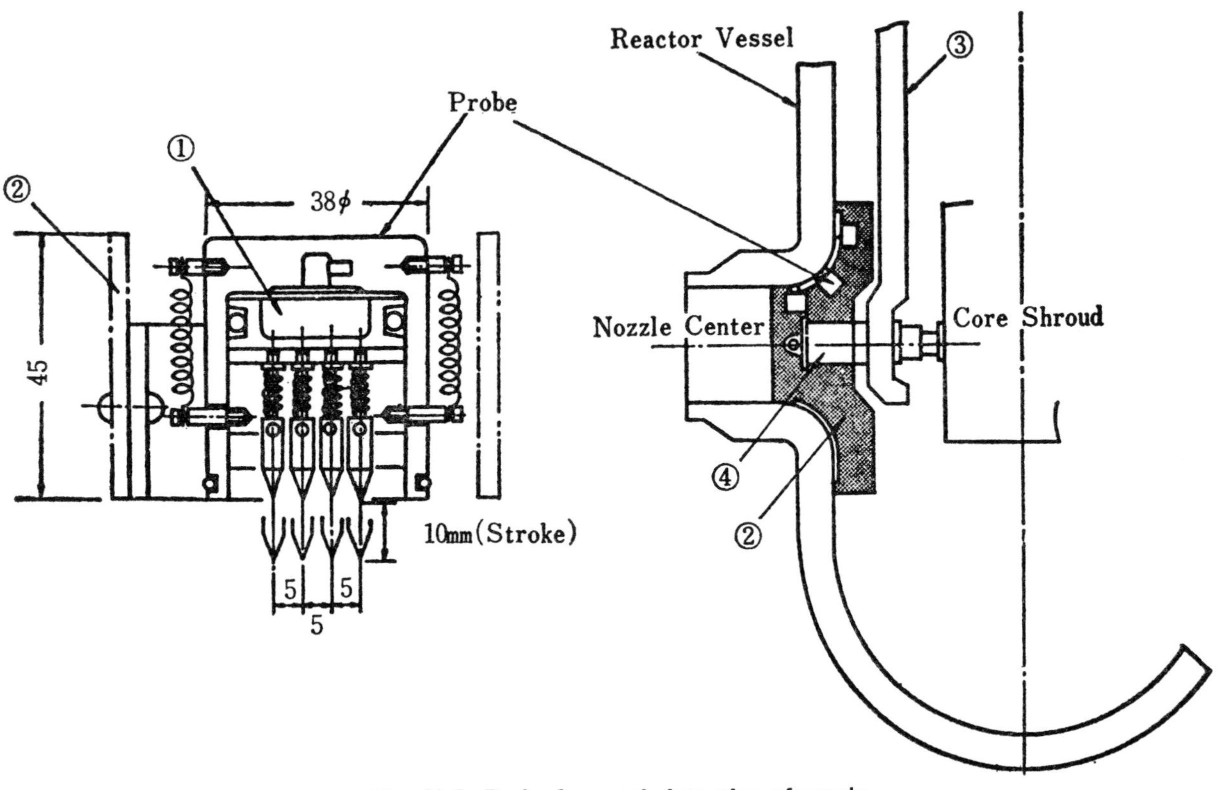

Fig. 41.4. Probe for crack detection of nozzle

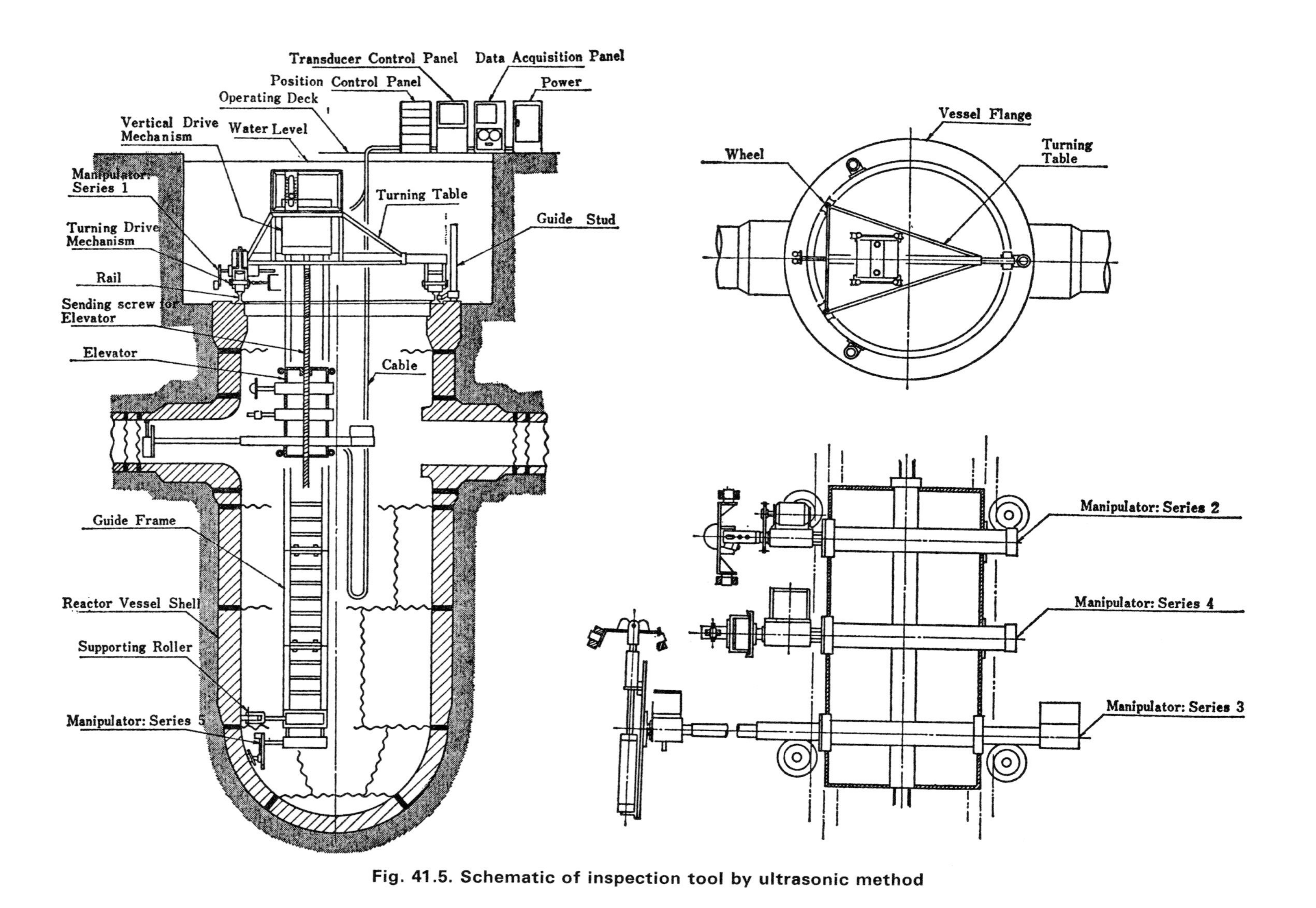

Fig. 41.5. Schematic of inspection tool by ultrasonic method

entailed the approach of personnel towards high-radiation areas; as a result, a personal exposure of maximum 950 mrem was recorded.

Each method of inspection utilized must be further improved; but, generally, it is satisfactory. By a suitable combination of the methods employed, highly reliable data may be obtained in in-service inspection of the reactor pressure vessel.

DEVELOPMENT OF IN-SERVICE INSPECTION INSTRUMENTS

Electric-resistance probing instrument

A probing instrument based on electric resistance was developed by the co-operation of J.A.E.R.I. and Mitsubishi Heavy Industries Ltd, and the first inspection with it was made in 1968. Details of the probing instrument were published at the First Conference on Pressure Vessel Inspection Techniques, held in Delft, the Netherlands, in 1969.

If a crack exists on the surface of a metal, the resistance of an electric current flowing transverse to the crack increases with increase in crack depth. The electric-resistance probing instrument is based on this principle; its schematic drawing is given in Fig. 41.4.

The probe itself consists of four contacting needles; the two outside are for current supply, and the remaining two, inside, are for measurement of the potential difference. The probing instrument can be remotely operated by means of a scanning mechanism and an air cylinder to press the needles. With this instrument, the metal surface can be scanned for any flaws.

By this inspection, cracks were detected about the circumference of the nozzles. These cracks were all less than 3·5 mm in depth and did not reach to the base metal; in this manner, the integrity of the pressure vessel was reconfirmed. Subsequently, inspection has been carried out once yearly with the electric-probing instrument. It has been found that the reproducibility of measurements is fairly high; no further growth of the cracks has been observed.

The probing instrument developed is very easy to handle, and the magnitude of cracks and their positions can be determined quantitatively. It is expected that the instrument will provide a highly effective tool for in-service inspection.

Remotely operated ultrasonic instrument

Development of the remotely operated ultrasonic instrument is proceeding in Mitsubishi Heavy Industries Ltd. It is specially designed so that inspection of the P.W.R. (pressurized water reactor) pressure vessel positions specified in the A.S.M.E. Code Section XI can be observed extensively.

The ultrasonic instrument being developed is shown schematically in Fig. 41.5. For the ultrasonic instrument, normal, shear, and pitch–catch techniques are utilized. The search units are fitted to the respective manipulators; and the inspection is made by an immersion method. The

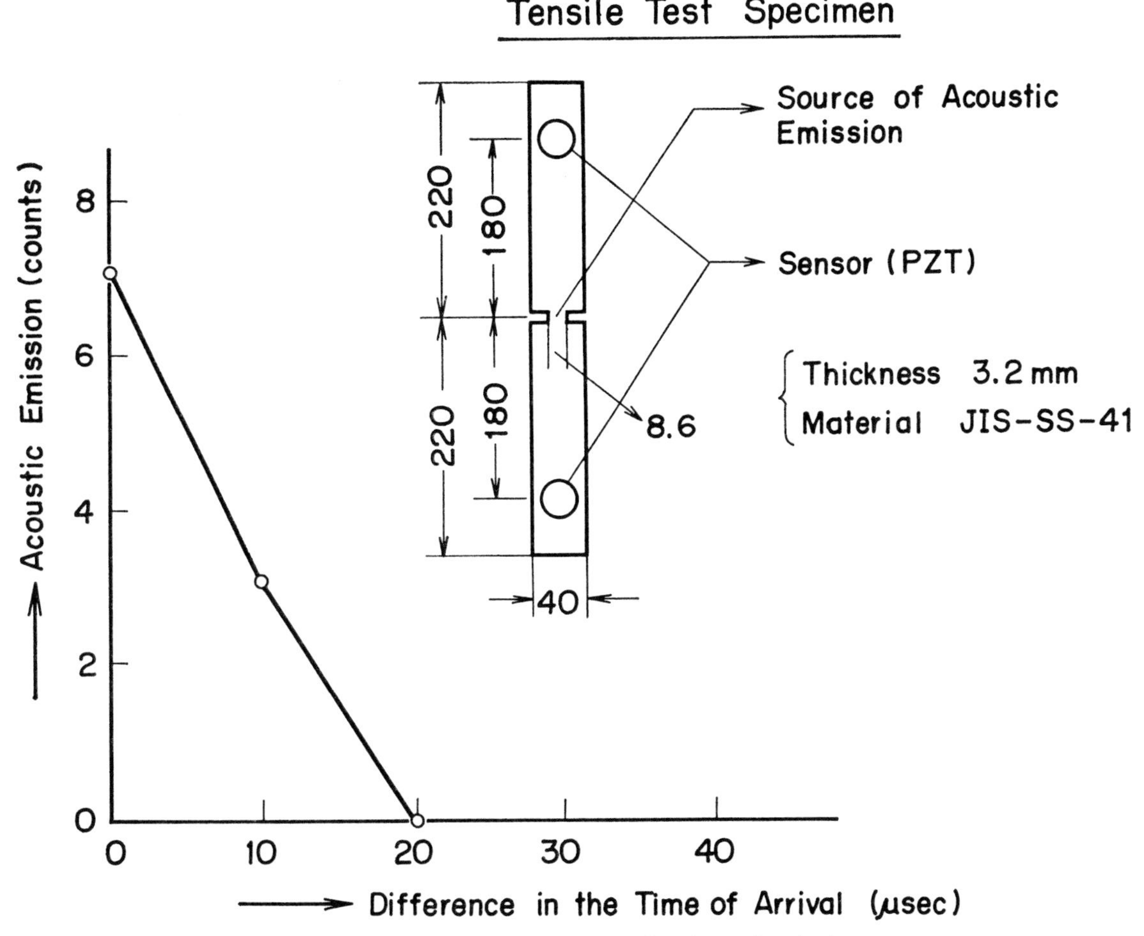

Fig. 41.6. Flaw location by time of arrival

over-all ultrasonic instrument consists of elevators to lift or lower manipulators, a rotary table to move the probe suitably, an operation console for remote operation, and a data acquisition unit. The dimensions of the respective sections of the instrument can be adjusted according to the size of the pressure vessel being inspected; such variations are 3300–4420 mm for the radial direction in the vessel, 8400 mm for the maximum depth, and 690–900 mm radially in a nozzle. Details of the instrument will be presented in a paper entitled 'The nuclear reactor vessel inspection tool using an ultrasonic method' (Paper C25).

The greatest problem encountered in the inspection of the reactor pressure vessel from inside by the ultrasonic method was the wave of the interior (stainless steel) overlay cladding. Studies in this connection are also in progress at Mitsubishi Heavy Industries Ltd.

Acoustic emission method

An acoustic emission technique for the inspection of a reactor pressure vessel is being studied extensively in the J.A.E.R.I. The method has the outstanding advantages of high efficiency in inspection and the possibility of 'overall' inspection. Moreover, information can also be obtained on the danger degree of defects, which is impossible with the conventional non-destructive methods. By use of the acoustic emission technique it is expected that the efficiency of inspection of a reactor pressure vessel can be improved and the safety standards of reactor operation raised in the future.

The basic experiments were made in J.A.E.R.I. by applying pressure to a sample plate in an Amsler tensile tester. As a result, the following data were determined:

(1) Burst-type acoustic emission is caused by the existence of a plastic deformation in a material.

(2) The pressure required for acoustic testing is achieved with difficulty, due to the Kaiser effect.

(3) By using two detectors, the detection of a defect position is possible via the difference in arrival time of signals reaching the detectors (Fig. 41.6).

In other experiments to be performed (starting at the end of 1971), defects will be made artificially in the experimental model of a pressure vessel either by fatigue or stress corrosion. In addition, studies will be made on the methods of determining the size of defects and their position; and on the relationship between the magnitude of a defect and the acoustic emission.

Experiments will further be made (starting at the beginning of 1972) on the noise in a reactor plant, in order to determine the actual situation of measurement and the feasibility of continuously monitoring for flaws in an operating reactor.

The problems arising in the actual application of acoustic emission techniques for pressure vessel inspection are the degree of pressure necessary, the accuracy required in the detection of the flaw position, and the method to be used in evaluating defect size.

The acoustic emission technique may be expected in the future to prove an effective means of monitoring continuously the integrity of the pressure vessel of an operating reactor.

CONCLUSIONS

Extreme emphasis is placed on the integrity of reactor plants in Japan, and therefore keen interest is shown in their in-service inspection. To raise operational efficiency of reactor plants, however, it is necessary to have effective and accurate means of inspection. For this purpose, strenuous efforts will be continued on the development of in-service inspection techniques in the future.

ACKNOWLEDGEMENTS

The author is deeply indebted to the following for their aid in the preparation of the present report: Tadato Fujimura, Japan Atomic Energy Research Institute; Tsutomu Fujimura, Japan Atomic Energy Research Institute; Ryoichi Nakamura, Tokyo Electric Power Co.; Toshiyuki Harada, Kansai Electric Power Co.; Nobuo Maekawa, Japan Atomic Power Co.; and Tomio Yamaguchi, Mitsubishi Heavy Industries Ltd.

C42/72 LEGAL REQUIREMENTS FOR PERIODIC EXAMINATION OF PRESSURE VESSELS UNDER THE FACTORIES ACT 1961

J. T. TOOGOOD*

The paper traces the history leading up to the present legal requirements relating to periodic examination of pressure vessels under the Factories Act. Its requirements are discussed and some questions are posed as to periodic examination in the future.

INTRODUCTION

THE TERM 'pressure vessel' covers a wide range of vessels and apparatus ranging from aerosol containers, gas bottles, and steam boilers to nuclear reactor containment vessels. In this paper only the legal requirements for those vessels covered by the Factories Act 1961 will be discussed.

Historically, the need for periodic examination of pressure plant was felt in the early part of the nineteenth century when the numbers of boiler explosions had risen, in rough proportion to the increase in boiler pressures, to an unacceptable standard even for those days. It was in the middle of that century that a public-spirited body of gentlemen formed a company, the Manchester Steam Users Association, whose object was to reduce the number of accidents at steam boilers by a voluntary system of independent examination.

However, it was not until 1901 that the first legislation requiring periodic examination was made. This was the Factories and Workshops Act 1901, which contained provisions as to the use, maintenance, and periodic examination of boilers used in factories and workshops. This was followed by the Factories Act 1937, which extended the requirements of periodic examination to other plant; with slight modifications, these are retained in the Factories Act 1961.

The Factories Act 1961 contains statutory requirements for periodic examination of pressure vessels of three types. They are steam boilers, steam receivers, and air receivers. Dealing firstly with steam boilers, the requirements for periodic examination are contained in section 33 of the Act and in the Examination of Steam Boilers Regulations (1964). Section 33, in dealing with the examination of a steam boiler, requires any steam boiler to be examined before it is used in any factory. The section also gives to the Secretary of State power to make regulations as to the manner of the examination and the period between subsequent examinations; this he has done by virtue of the regulations mentioned above. The regulations set out the manner in which a steam boiler and its fittings and attachments should be prepared for examination, and how they should be examined. The period between thorough examinations of a steam boiler must not exceed 14 months, except in the case of certain classes of steam boiler where the period must not exceed 26 months.

The MS. of this paper was received at the Institution on 10th November 1971 and accepted for publication on 17th December 1971. 33

* *H.M. Engineering Inspector of Factories, H.M. Factory Inspectorate, Baynards House, 1–13 Chepstow Place, Westbourne Grove, London, W.2.*

THE HONEYMAN COMMITTEE

The relaxation of the maximum period between examinations for certain classes of steam boiler from 14 to 26 months followed a report by a committee of enquiry set up by the Minister of Labour under the chairmanship of George Gordon Honeyman in 1958. The committee in their report said that they were impressed by the evidence given in support of the view that it was no longer necessary to require a thorough examination of all types of boiler at intervals not exceeding 14 months. They considered extending this period to 18 or 20 months, but thought this to be of limited value to industry. The preference was to confine any relaxation to boilers that could with safety be allowed an interval between thorough examinations not exceeding 26 months.

The committee came to the conclusion that water-tube boilers with an evaporation of over 50 000 lb/h could be included in the category mentioned above, not because such boilers had any special characteristics that other boilers lacked. Essentially the safe record of operation was a consequence of their intrinsic value and the standards of operation and maintenance that for economic reasons were observed in their use. The committee also applied similar considerations to certain groups of smaller water-tube boilers, and to waste-heat boilers that were an integral part of certain processing plant.

The main recommendations of the committee were enacted when the Examination of Steam Boiler Regulations were made in 1964. This was the first differentiation between boilers that were likely to be well looked after and those not so likely to be so.

The periodic examination requirements for steam receivers and air receivers are given in sections 35 and 36 of the Factories Act. Unlike the case of steam boilers, there are no regulations dealing with the manner of the examination. The subsections in both cases require the vessels to be properly maintained and thoroughly examined by a competent person at least once in every period of 26 months.

A requirement common to all these three types of pressure vessels is that they shall be thoroughly examined by a competent person at least once in the laid-down period. There is no definition given in the Factories Act relating to the competent person, and questions are often asked about the qualifications necessary for competency.

The Honeyman Committee considered evidence on the standards of boiler examination and on the suggestions that the qualifications to be required of persons carrying out these examinations should be defined in regulations. The committee were of the opinion that there would be serious practical difficulties in the way of producing a fair and workable definition, and that in any case the effectiveness of the existing arrangements made it unnecessary to attempt the task. Nevertheless, they stressed that it was desirable to safeguard the independent status of the competent person. The committee went on to recommend that the competent person should be an independent person who has no personal interest in and who is not employed in or about or in the management of the factory in which the boiler he examines is used.

In practice there is need for the application of common sense. In some cases the examinations are of a comparatively simple kind, requiring little or no theoretical knowledge and could be carried out by a reasonably intelligent and responsible person familiar with the workings of the plant; in most cases, however, examinations are more complex. There is considerable variation in technical knowledge, training, and experience required as the range of pressure vessels and plant to be examined covers a wide field.

Some guidance was given officially on this question a number of years ago, and it may be useful to quote it:

> The question whether a person who makes an examination under the Act is competent for that purpose depends on the circumstances. What appears to be contemplated is that the person should have such practical and theoretical knowledge and actual experience of the type of machinery or plant which he has to examine as will enable him to detect defects or weaknesses which it is the purpose of the examination to discover and to assess their importance in relation to the strength and functions of the apparatus.

Most of this work is, in fact, done by companies or associations specializing in the examination of pressure vessels and other plant.

EXAMINATIONS

The term 'thoroughly examined' has been construed in the past to mean visual examination internal and external, together with any other method of examination that the competent person considers necessary. The Examination of Steam Boilers in regulation 3 (3) specifically mentions particular methods of examination that the competent person may at his discretion use, and they include: '. . . examination, testing, or measuring by means of ultrasonic, radiographic, magnetic, or electronic devices or of tube calibration gauges, steam trial, and hydraulic testing'. He is also authorized to require brickwork, baffles, and coverings to be removed to a sufficient extent to make his thorough examination. Although it is not specifically required in sections 35 and 36, the competent person may require similar conditions at his discretion in order to complete his thorough examination of a steam or air receiver.

In the case of steam receivers, the Act requires them to be thoroughly examined only so far as the construction of the receiver permits. This relaxation was probably made because there are many steam receivers that are difficult, if not impossible, to completely examine internally and externally. Examples are steam-heated cast-iron rollers, platens of presses, moulds, and ironing beds. In such cases competent persons usually require periodic hydraulic tests to supplement their examinations.

In the case of the three types of pressure vessels mentioned above, the statutory period between examinations required by the Act are maximum periods; the wording used in each case is: '. . . examined at least once in every period of . . .'. With certain boilers or vessels known to suffer from severe operating conditions, the maximum statutory period may be excessive to ensure safety of the plant.

In the above circumstances the occupier of a factory may not fulfil all his statutory duties by complying with the statutory period because the Act also requires that a steam boiler or steam or air receivers shall be 'properly maintained'. This term is defined in section 176 of the Act as: '. . . maintained in an efficient state, in efficient working order, and in good repair'; there is therefore an absolute obligation to maintain the vessel and all its fittings in an efficient state, in efficient working order, and in good repair.

One may conjecture, with such a stringent requirement in the Act, why the legislators should feel it necessary to lay down statutory periods of examination at all, as it would seem that to ensure those pressure vessels covered by the Act meet this requirement, thorough examinations at appropriate intervals are essential. It appears that the legislators considered it necessary to state some minimum statutory requirements for periodic examination in order to ensure that the factory occupier is compelled to go at least some of the way to meet the general requirement of 'properly maintained'. As the last report of the examination must be kept available for scrutiny, it affords some sort of check as to whether a steam boiler, or steam receiver or air receiver, is being properly maintained.

The provision of a requirement for a definite period between examinations that must not be exceeded can perhaps be criticized on the grounds that it does not take into account the service conditions of any particular vessel, the materials of construction, the factor of safety, the corrosion allowance, or anything else. It says, in effect, that, come what may, it must be examined at least once in every period of 14 to 26 months, depending on the type of vessel.

In the chemical and petro-chemical industry, for instance, the statutory vessels, such as boilers and steam receivers, which are integral with a continuous process plant have to be shut down at specified periods for examination. However, it may be felt that technological improvements in materials of manufacture, improvements in corrosion inhibition, and the like, should and are being made that may make it possible to safely exceed the statutory period for the pressure vessels in the plant. Indeed, there are those who would say that in certain cases the statutory requirements are already unnecessarily restrictive, particularly in view of on-stream monitoring that has become possible in certain applications by the use of ultrasonic and other inspection techniques.

The above argument would certainly appear to have some validity. The difficulty is that the Act legislates a minimum requirement to cover the broad spectrum of users of pressure vessels. These vary from small factories, where there is little knowledge of the safety aspects of pressure vessels, to the larger plants, where a very high standard in design and use of pressure vessels may be practised. It would seem to some extent that the best users of pressure vessels may be handicapped by the bad ones as far as legislation is concerned.

In the case of steam boilers the committee of enquiry made recommendations that led to the Examination of Steam Boiler Regulations 1964; in these regulations the period of examination is almost doubled in the case of particular classes of boilers. Whilst the committee of enquiry no doubt considered the economics of the case, they certainly would have held safety of the plant to be of paramount importance. If one considers the type of boilers that benefited, it is likely that they will be operated by large companies having a high degree of technological know-how in the running of such plant. It seems likely that the committee took this factor into account in arriving at their conclusions.

EXCEPTIONS TO SECTIONS 32–36

No similar committee has ever been set up to consider pressure vessels other than steam boilers. The Chief Inspector of Factories is, however, empowered by section 37 (2) of the Factories Act to except from any of the provisions in sections 32–36 of the Act any class or type of steam boiler, steam receiver, or air receiver to which he is satisfied that the provision cannot be reasonably applied. In addition, he can subject any certificate of exception as he may care to make to any conditions he may wish to impose to ensure safety.

The Chief Inspector has used his powers to make exceptions from, or vary the requirements relating to, periodic examination. Examples of some such certificates that have been issued in recent years are:

Certificate of exception 43

This certificate applies to steam boilers used in connection with nuclear reactors using natural uranium with a graphite moderator in a steel pressure vessel. The certificate has the effect of extending the period of examination to 50 months for those parts of the boiler that are inside the containment vessel.

Certain conditions that relate to monitoring devices are made in the certificate.

Certificate of exception 48

This certificate applies again to steam boilers in nuclear reactors, but in this case is confined to those with concrete containment vessels and excepts those parts of the boiler inside the vessel from a statutory requirement for periodic examination, subject to certain conditions.

It should be noted that the plant covered by the above two certificates is subject to the Nuclear Installations Acts of 1965 and 1969, and the terms of the licence issued include arrangements for the approval of maintenance and periodic examination.

Certificate of exception 49

The 26 months' period between examination permitted in the case of certain waste-heat boilers by the Examination of Steam Boilers Regulations reverts after the boiler is 21 years old to 14 months. This certificate applies to particular waste-heat boilers connected with a particular type of oil-refining plant and permits the competent person after the boiler is 21 years old to extend the period to 26 months again. The certificate is subject to conditions which stipulate, in effect, that the boiler must be restored to an 'as new' condition.

Certificates of exception 53, 55, and 56

These certificates apply to air receivers, steam receivers, and certain steam boilers, respectively, which are used in connection with the generation of electricity for distribution throughout the national electricity supply system. The certificates have the effect of extending the period of examination of such plant to 30 months. This is of benefit to the Central Electricity Generating Board, who, it should be emphasized, only need to take advantage in certain comparatively rare circumstances.

OTHER PRESSURE VESSELS

What of pressure vessels not included in the definitions previously mentioned? There are probably more pressure vessels in use that are not directly subject to the Act than those that are. This does not mean that the factory occupier has no duty under the Act to maintain such vessels in a safe condition. Section 29 of the Act, for instance, requires that every place at which any person has at any time to work shall be, so far as is reasonably practicable, made and kept safe for any person working there.

Section 54 of the Act would allow an inspector of factories to make a complaint to a magistrate if a vessel was unsafe; and a magistrate can prohibit the use of a dangerous vessel either completely or until it is satisfactorily repaired.

NEW SAFETY, HEALTH, AND WELFARE LEGISLATION

In December 1967 the Ministry of Labour circulated some thoughts on the subject of new safety, health, and welfare legislation in the form of a consultative document upon which various organizations were invited to comment. Propositions in this document were that a wider field of pressure vessels should be subject to the Act, in that they should be of 'good construction, sound material, and adequate strength'; and that the Minister should be empowered to make regulations relating to examinations during manufacture and afterwards. Since then the Government has set up a committee to examine and report on the safety and health of employed persons under the chairmanship of Lord Robens.

No reasonable person would doubt that periodic examination of pressure vessels is essential to ensure the safety of vessels. The more difficult question is: what is a suitable period between examinations? The shut-down of pressure plant for examination is an extremely expensive business, and the achievement of longer runs with continuous-flow installations is not only desirable for economic reasons but may be essential in the face of world competition. However, safety considerations apart, it would be bad economically to stretch periods between examinations too far, because although examination of plant may be costly, failure of a pressure vessel in service is likely to be even more so. Therefore, any case for the extension of periods between examination would have to be very convincing before any responsible person would consider increasing the present periods. However, there are factors which should be closely considered in the future.

Two difficulties with a statutory period fixed by Act of Parliament are: there is no flexibility; and the periods between examinations cannot be adjusted in the light of experience, except by changing the Act itself. Even this latter may not be the answer, as flexibility is needed with the same type of plant. For example, should a steam boiler with good feed water be examined as often as one using a poor water supply, or a steam receiver subject to an innocuous process as often as one subject to a corrosive or erosive process? It would seem that there should ideally be a distinction between such cases, for which a statutory period applying to a particular class of vessel cannot cater. In fact, under such circumstances, the period between examinations must cater for the worst rather than the best case if overall safety is to be observed.

How, then, could a statutory requirement be made flexible enough to give a bonus where the previous history of the examination showing the rate of corrosion, etc., justifies it? It would seem that the person most qualified to make a proper assessment of the appropriate period to elapse before the vessel is examined again is the competent person who made the last examination. This might seem particularly so when he is a member of an inspecting organization, which will have experience over a wide field of vessels subject to particular processes or applications upon which he can draw.

Supposing the competent person making the examination under the Act was empowered to vary the period, perhaps between certain wide statutory limits, how may this work in practice? Firstly, it obviously places a great deal of responsibility on the competent person himself. Probably there is no difficulty here as competent persons have now to exercise their judgement at every examination of a vessel, so there may be little difference in their position.

Secondly, the relationship of the inspecting organizations, such as insurance companies, and their clients would have to be considered. It may be that inspecting organizations would be subjected to heavy commercial pressures to extend periods between examinations to the maximum. On the other hand, such inspecting organizations now carry out a tremendous number of periodic examinations of non-statutory plant, and in this area they are quite free to make their own arrangements. They do have, however, statutory requirements for other vessels, and these must tend to serve as a guide.

A competent person who has carried out an examination of a statutory vessel can now state on his report of the examination that he wishes to re-examine the vessel at some time which is less than that allowed by the Act. It seems, therefore, that if statutory periods were extended to become maximum periods, and it was made quite plain that within these limits a competent person should set the date of the next examination, then this could, indeed, provide a much needed flexibility that may benefit British industry, without, if it were properly done, decreasing the safety of pressure vessels.

Thorough examination of a pressure vessel in the past has always meant visual examination both externally and internally, together with the use of whatever inspection aids the person making the examination considered necessary.

It is interesting to conjecture whether internal examination will always be necessary with all vessels to establish their integrity in view of the progress of non-destructive forms of testing. On-stream examination of vessels is practised, for example in the oil-refinery industry, to a considerable extent in order to establish as far as possible the internal condition of vessels; and for this, various techniques are used. The information obtained, however, is used more to establish the extent of repairs that might be necessary on shut-down rather than to determine the date of shut-down. This may be to some extent because the date of shut-down is usually dictated by the statutory vessels in the complex. It could be that the results shown by on-stream monitoring of pressure plant, where this is possible, can be used together with the history of the vessel in order to assess the date of the next examination.

Another possibility would be not to state specific periods between examinations in an Act of Parliament at all, but to state a general requirement that pressure vessels should be thoroughly examined at suitable periods. Guidance on suitable periods for particular vessels in particular industries could be given in codes of practice prepared jointly by the industry, inspecting organizations, and other interested parties. These codes of practice could deal also with the manner and methods of examination. The periods between examinations would, of course, be subject to the overriding decision of the competent person in each particular case, but he would be guided by a code of practice.

The advantage of this would be that codes of practice could be easily and quickly changed, which is certainly not the case where an Act of Parliament is concerned. Incidentally, there would have been considerable advantage if the period of examination had been specified in regulations rather than the Factories Act. Regulations can be remade much more quickly and easily than altering an Act.

CONCLUSIONS

It can be said with confidence that the most effective single factor in reducing accidents due to explosions in pressure vessels is periodic thorough examination by competent inspectors, and the value of the legal requirements has been amply demonstrated in the past. Nevertheless, the legal requirements should be reviewed from time to time, perhaps extended in some directions and relaxed in others, to meet technological changes. Should the report of the committee chaired by Lord Robens lead to any revision of the Factories Act, the opportunity should be taken to examine closely the best method of achieving safety in pressure vessels, whilst giving the maximum benefit to industry of longer runs between shut-downs where appropriate.

The views expressed in this paper are those of the author and may not necessarily be those of the Department he serves.

C43/72

IN-SERVICE INSPECTION SYSTEM FOR THE FIRST ARGENTINE ATOMIC POWER PLANT

J. N. BÁEZ* C. VENTURINO* O. WORTMAN*

This work describes the in-service inspection system to be used at the Atucha nuclear power plant. The scanning device, the transducers, the inspection technique, and data acquisition system are outlined and general characteristics discussed.

INTRODUCTION

THE ATUCHA REACTOR, the first nuclear power plant in Argentina, is scheduled to produce 319 MWe by the end of 1973. This reactor, designed and built by Siemens AG of Germany, is a P.H.W.R. (pressurized heavy water reactor), utilizing heavy water moderation and heavy water cooling.

The original design of this reactor plant did not provide for in-service inspection of the pressure vessel. On the basis of the power reactor inspection codes in use in the U.S.A. and Europe at the time the reactor pit design was completed, C.N.E.A. (Argentine Atomic Energy Commission) considered it essential that provision should be made for in-service inspection.

The usual procedure for internal inspection could not be followed because of the nature of the reactor design, which precluded access to the internal pressure vessel wall; consequently, inspections would have to be performed externally. After considering several proposals, submitted by different organizations, C.N.E.A. awarded a contract to the S.W.R.I. (Southwest Research Institute) of the U.S.A. to design and fabricate an automatic scanning system for the inspection of welds in the lower part of the reactor pressure vessel, the welds between nozzles and piping, and the studs of the reactor pressure vessel. The equipment was to include positioning devices, drive units, inspection modules, transducers, and the control panel to operate the equipment. In order to allow this equipment to be installed, the thermal insulation had to be modified. The data acquisition system and the inspection programme were to be designed by C.N.E.A.

SPECIFICATIONS OF THE REACTOR PRESSURE VESSEL

The material of the reactor pressure vessel is specified as 22 NiMoCr 37 (type 508, class II, A.S.T.M.), clad with 5·5 mm of 4551 stainless steel, deposited by the strip-cladding process. The working pressure is 115 atm and the working temperature 271·1°C. The head is held by 60 studs. Other dimensions are given in Table 43.1; and the principal ones are shown in Fig. 43.1.

INSPECTION AREAS

Fig. 43.2 illustrates construction details of the lower part of the reactor pressure vessel. The vessel consists of a flange containing the nozzles, two cylindrical sections, and a bottom. The flange and the two cylindrical sections each consist of three forgings welded together. Excluding the bottom welds, which were not included in the in-service inspection programme, the welds can best be described as three circumferential welds. The first joins the flange to the first cylindrical section, the second joins the two cylindrical sections, and the third joins the bottom to the second cylindrical section. These welds are shown in Fig. 43.2 as C designations. Six vertical welds joining the forgings of the cylindrical sections are denoted in Fig. 43.2 by the nomenclature V. There are eight welds joining the nozzles to the flange, and eight welds joining the piping to the nozzles. Except for the three vertical welds in the flange, all these welds are included in the in-service inspection programme.

As noted above, the welds in the spherical bottom and the vertical welds in the flange will not be inspected. This decision resulted from the difficulty experienced in designing a system to inspect them, together with the fact that these welds, when considerations of neutron

Table 43.1. Dimensions of reactor pressure vessel

Item	Dimensions, mm
Lower part	
Internal diameter	5360
Bottom radius	2730
Internal height	9730
Height of the cylindrical part	5695
Wall thickness of the cylindrical part	220
Wall thickness of the bottom part	135
Flange height	1850
Flange thickness	455
Reactor head	
Internal diameter	4990
Radius	3452
Flange height	800
Wall thickness	340

The MS. of this paper was received at the Institution on 6th December 1971 and accepted for publication on 14th January 1972. 34

* *Comision Nacional de Energia Atomica, Gerencia de Tecnologia, Avda del Libertador 8250, Buenos Aires, Argentina.*

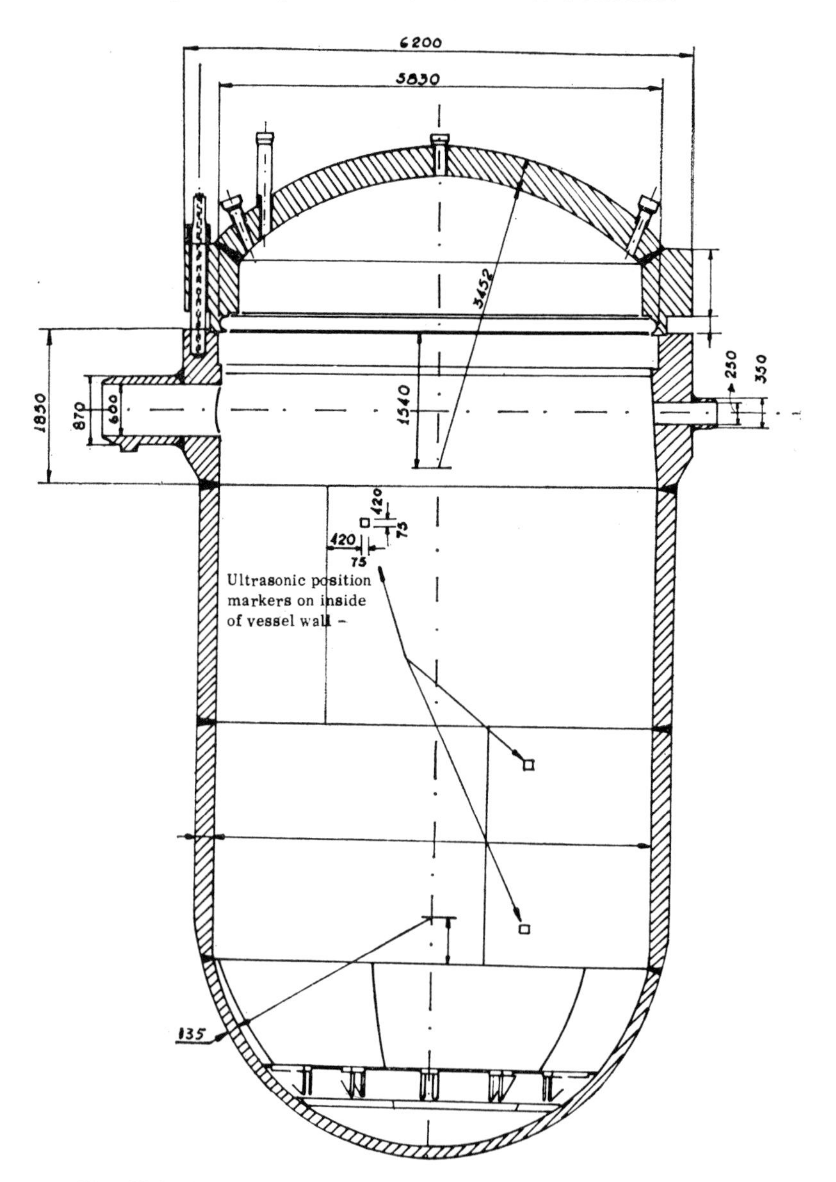

Fig. 43.1. Principal dimensions of the Atucha pressure vessel

damage and stresses are taken into account, are the least likely to fail. The in-service inspection programme also includes the ultrasonic inspection of the sixty studs that hold the head of the vessel.

DESCRIPTION OF THE AUTOMATIC SCANNING SYSTEM

The automatic scanning system provided by the S.W.R.I. consists of three separate devices for the inspection of the vessel, the nozzles, and the studs respectively; there is also a control panel.

Vessel inspection device

This device is designed to allow for the ultrasonic inspection of the circumferential and vertical welds of the pressure vessel. It consists of the following parts:

(*a*) A permanent structure located in the annulus between the vessel and the reactor pit.

(*b*) An inspection module with ultrasonic transducers.

(*c*) A drive unit for the inspection module.

(*d*) A drive unit to rotate the permanent structure.

The permanent structure, installed before the vessel was placed on site, consists of a rotary frame, which supports a vertical track. This structure can be rotated around the vessel through 360°. Fig. 43.3 shows schematically the vessel inspection device and its position inside the reactor pit.

Each time an inspection is to be made, the inspection module and its drive unit will be installed in the vertical track of the permanent structure. A temporary track is to be used to insert the inspection module, and a cable tensioning device from the top of the biological shield.

The drive unit for the permanent structure is engaged during each inspection through an access tunnel located in the bottom of the reactor pit. The scanning of the area to be covered by the inspection is obtained through the combined movement of the module along the vertical track and the rotation of the permanent structure. In this manner the transducers can be positioned in any area of the cylindrical part of the vessel, below the line of the nozzles. The characteristics of the Atucha nuclear reactor pressure vessel permit inspection from the outer surface.

In accordance with the S.W.R.I. conception for the

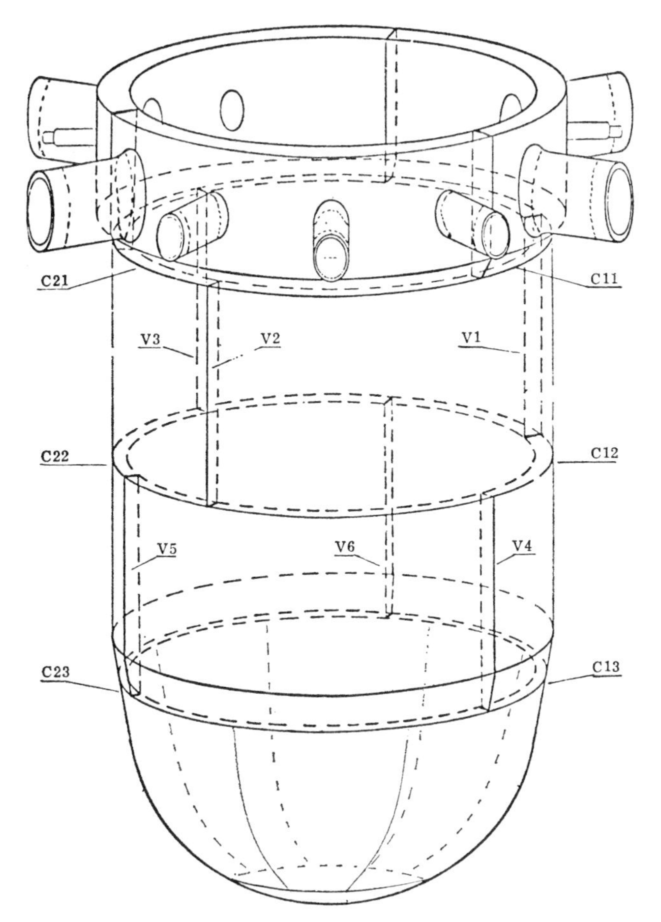

Fig. 43.2. Schematic drawing of the Atucha pressure vessel showing the inspection areas

ultrasonic in-service inspection system, an inspection module was designed that had three transducers, as shown in Fig. 43.4. The transducers incidence angle (T1 and T3, Fig. 43.5) can be remotely adjusted from 24° away from the centre of the assembly to 0–24° towards the centre of the assembly. This movement generates a longitudinal wave through all shear wave angles up to 60° in each direction. The remote angle and positioning capability of the two transducers allows the module to perform an analysis function, if so required.

The transducers are contained in captive-water chambers to ensure proper coupling with the surface. To

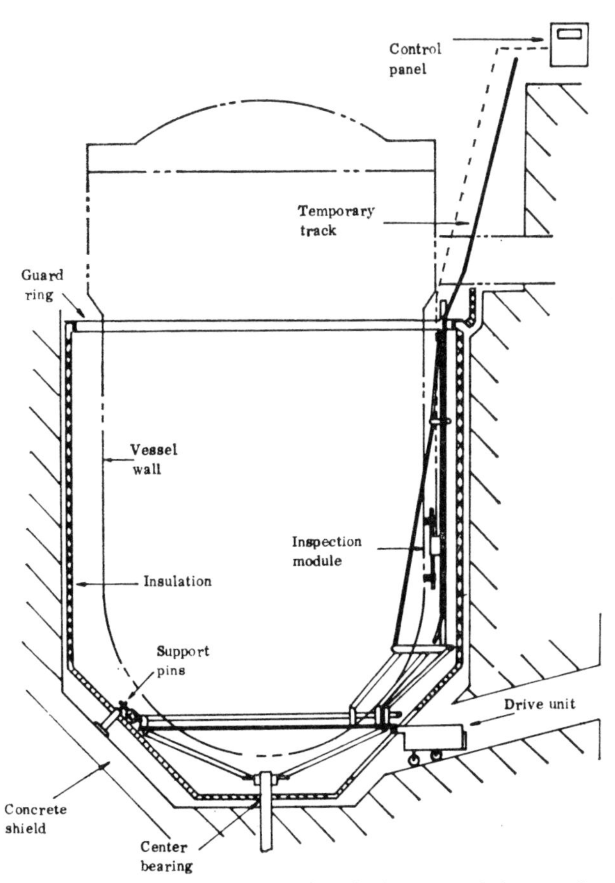

Fig. 43.3. Schematic drawing of the vessel inspection device

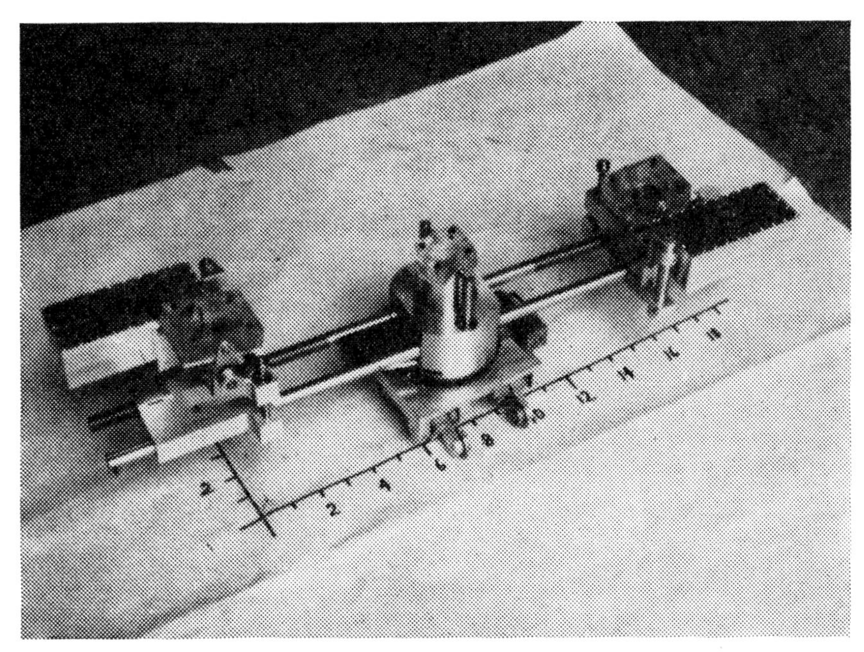

Fig. 43.4. Ultrasonic inspection module

accommodate the expected distance variation between the track and the wall surface a spring-loaded action is incorporated in each transducer housing.

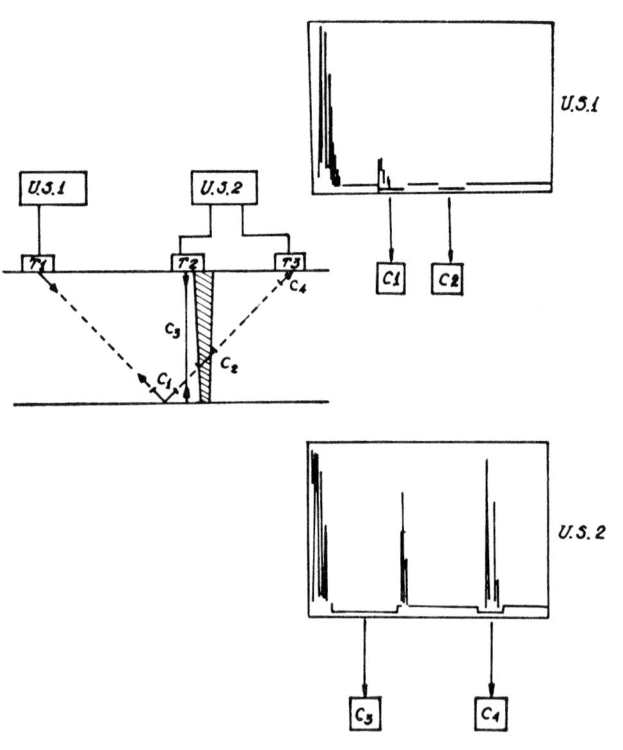

U.S.1 and U.S.2, ultrasonic instruments.
T1, shear wave transducer (emitter–receiver, 45° and 60°).
T2, longitudinal transducer (emitter–receiver).
T3, receiver–transducer (pitch and catch with T1).
C1, C2, C3, C4, gated areas.

Fig. 43.5. Drawing illustrating the inspection technique

Nozzle inspection device

To provide for inspection of the nozzle welds, circular tracks have been permanently installed at the correct distance from the areas to be examined. The ultrasonic inspection module is then mounted on these tracks whenever an inspection is to be made.

Studs inspection device

The studs have an axial orifice throughout their length that allows ultrasonic examination from inside by means of an auto-centring module guided by a bar fixed to the stud.

INSPECTION TECHNIQUE

The transducer array already described is basically conceived to detect any defect oriented perpendicularly to the surface and above a predetermined size.

The primary inspection technique will be single pass and pulse echo; the pitch-and-catch technique, the secondary method. This arrangement is illustrated in Fig. 43.5, where T1 and T3 are shown in the pitch-and-catch mode. The other transducer, with fixed longitudinal inspection capability, will be used mainly for the inspection of the base material adjacent to the weld.

The flaws of primary concern invariably have rough surfaces, so they will reflect part of the energy transmitted by T1. Such flaws will block the energy normally received by T3 (pitch-and-catch mode). Therefore they will be shown on the cathode ray tube, and consequently in the corresponding information channel. Fig. 43.5 also shows the selected or gated areas from which the ultrasonic information will be gathered through individual channels.

In normal conditions, incidence angles of 45° and 60° will be used; the latter only in the transmitter–receiver

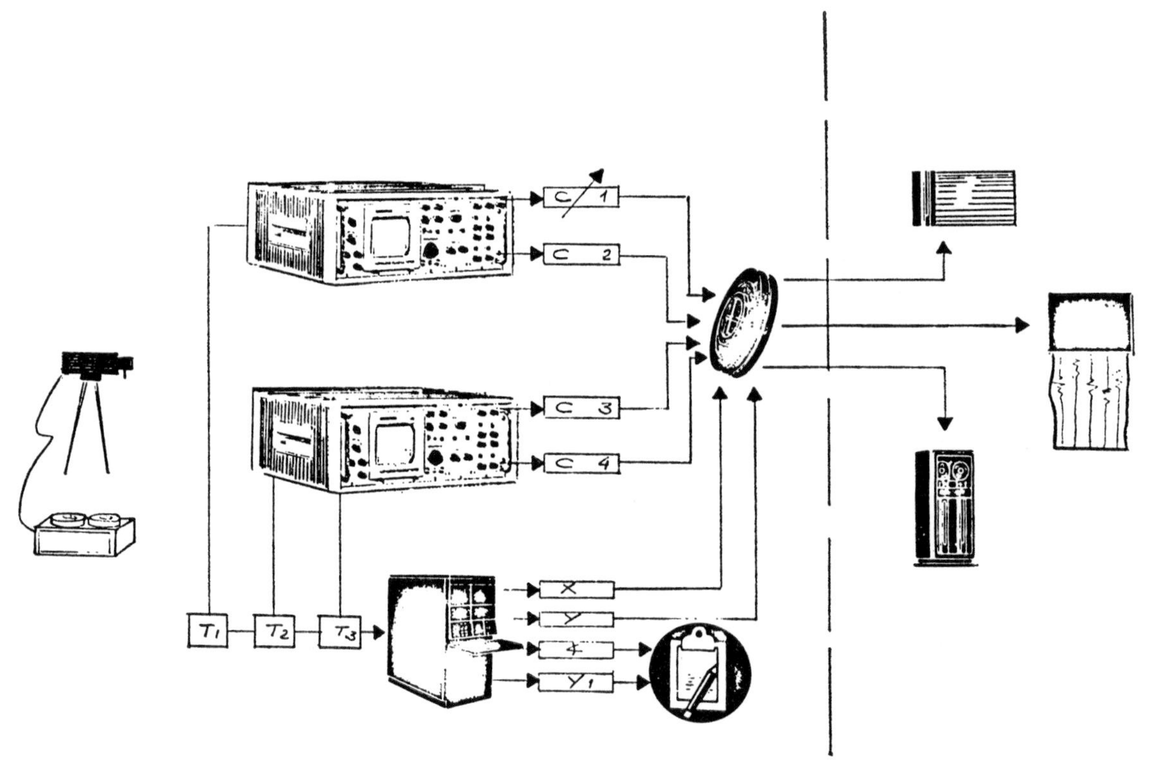

Fig. 43.6. Instrumentation and data acquisition system

mode (pulse echo). Normally an inspection will be performed from both sides of the centre line of the weld seams. Access problems limit this operation to one side only in the inspection of the upper and lower circumferential welds. To provide full coverage of the wall cross-section, several scans will be necessary.

Instrumentation and data acquisition system

Fig. 43.6 shows the instrumentation required to perform the pre-service inspection, according to the conception described above. Two ultrasonic instruments will be used, with a total of four ultrasonic information channels.

Linear progress, lateral transducer location, and transducer angles are controlled from the panel; their position being indicated on that panel. Positioning readout devices supply binary coded values of the X and Y location of the inspection module. All the data will be recorded magnetically in a conventional seven-channel instrumentation magnetic-tape recorder.

A special interface device will be employed to convert to serial the 16-bit B.C.D. parallel output from the position panel meters. The magnetic tape will be the master tape during the subsequent data handling and processing operations. Simultaneously with the magnetic recording, the data will be fed to a conventional multichannel, strip-chart recorder. Although such a system might be considered redundant, it has proved advantageous as a first estimation of the inspection results.

In the future it is intended to process by computer the large amount of data gathered during inspection. The data acquisition system as described is compatible with a computer.

A closed-circuit TV system, including camera, monitor, and video-tape recorder, will be used during inspection. The TV system will record cathode-ray tube images, thus allowing a complete or partial revision of the scanning after the inspection. As a permanent record, 35-mm pictures will be taken of visual signals appearing on the TV monitor.

C44/72 EXPERIENCE WITH THE RECURRING INSPECTION OF NUCLEAR REACTOR STEEL PRESSURE VESSELS

H. MAZUR* H. G. RUMPF† A. TIETZE‡ W. WERBER§ L. ZUTZ||

In the Federal Republic of Germany steam boilers and pressure vessels of power plants are among the facilities requiring inspection during operation. They must be subjected to recurring inspections (examinations by T.Ü.V. (the technical control authorities). The Gundremmingen, Lingen, and Obrigheim demonstration power plants were planned ten years ago when the problems of power generation were paramount, and little attention was then being given as to the feasibility of recurring inspections of parts subjected to pressure influence. In the Federal Republic, for the past four years, the aim has been to incorporate recurring inspections in the planning of modern power reactors, and to apply the new techniques to the demonstration power plants. Data is now available relating to the experience gained by T.Ü.V. with recurring inspections of the reactor pressure vessels of the three demonstration power plants named above. The essential experience gained during these inspections is summarized and conclusions given regarding further developments in this field. The importance of technical safety is stressed.

LEGAL BASIS OF RECURRING INSPECTIONS OF REACTOR PRESSURE VESSELS

IN THE Federal Republic of Germany, nuclear reactor pressure vessels are considered to be 'steam boilers'. They are therefore subject to the 'Dampfkesselverordnung' ('Regulations on steam boilers') (1)¶ and must comply with the requirements of the 'Technische Regeln für Dampfkessel (T.R.D.)' ('Technical requirements for steam boilers') (2). In references (1) and (2) it is stipulated that steam boiler plants must be subjected to recurring inspections. However, experience has shown that it is best not to put too narrow a limitation on the formal legal basis when carrying out the recurring inspections of pressure components in nuclear power plants. The approval documents relating to the construction and operation of nuclear power plants therefore contain only a requirement to carry out recurring inspections of reactor pressure vessels.

The technical institutions charged with the performance of these inspections [T.Ü.V. (technical control authorities)], however, allow a relatively wide range of discretion regarding the extent of the examinations and the methods of inspection to be used by the operators. There is, in addition, a legal liability to take into account newly developed techniques and past experience when carrying out each recurring inspection.

The MS. of this paper was received at the Institution on 6th December 1971 and accepted for publication on 14th January 1972. 33

* *Technischer Überwachungs-Verein Hannover, D-3000 Hannover-Wülfel, Loccumer Str. 63.*
† *Technischer Überwachungs-Verein Bayern, D-8000 München, Eichstätterstr. 5.*
‡ *Technischer Überwachungs-Verein Rheinland, D-5000 Köln, Lukasstr. 90.*
§ *Technischer Überwachungs-Verein Baden, D-6800 Mannheim, Richard-Wagner-Str. 2.*
|| *Technischer Überwachungs-Verein Rheinland, D-5000 Köln, Lukasstr. 90.*
¶ *References are given in Appendix 44.1.*

METHODS OF INSPECTION

The technical aspects of the recurring inspections of nuclear reactor components, especially of reactor pressure vessels, were matters of discussion at the two I.A.E.A. meetings in 1966 (3) and 1968. Both general and special recommendations resulted from these meetings. Special emphasis was placed on inspection techniques to be developed. By employing new inspection techniques an attempt was made to compensate for the limited accessibility, direct or indirect, of pressurized nuclear reactor components through radioactive radiation.

Meyer recently reported (4) on progress made in the development of special ultrasonic inspection methods for nuclear reactor pressure vessels in the Federal Republic. This work is being sponsored partly by the Federal Government. The work started in 1967 and resulted in the development of mainly automated inspection methods. The inspection principle is a fixed parallel arrangement of U.S. probes in a 'tandem technique' in order to detect flaws in various wall regions, and an arrangement of probes working in accordance with the normal echo technique for adjustment control. The limited accessibility of reactor pressure vessels also necessitated the development of new manipulators. To keep the inspection time as short as possible, for plant availability reasons, the evaluation of tests was automated to a large extent.

In the meantime, results are available of automatic ultrasonic inspections carried out on the reactor pressure vessel K.W.O. (Obrigheim). On newer plants, e.g. K.W.W. (Würgassen) and K.K.S. (Stade), finger prints were carried out. In addition, manual and semi-automatic ultrasonic inspection methods have been used, e.g. equipment developed by M.A.N. for the inspection of bolts, and a Sperry wheel (5) developed by Sperry, U.S.A., for the inspection of welding seams on thick sheets.

Irradiation methods with γ-rays and X-rays have not yet been used for recurring inspections of pressurized components in nuclear power plants. Lack of accessibility to the vessel, high levels of radiation of the nuclear pressure vessel, and the relatively limited possibility of flaw detection in heavy wall components are the reasons why, at present, there are no plans in the Federal Republic to develop known irradiation techniques, or to examine their applicability to recurring inspections. In general, ultrasonic inspections are preferred to irradiation techniques.

Extensive experience has been gained with optical inspection and dye penetration methods for the inspection of lead-in components. The limited accessibility of pressurized components in nuclear power plants led to the development of remotely controlled equipment. At present, practical experience has been gained with endoscopes, periscopes, telescopes, and remotely controlled TV equipment.

In the Federal Republic, special importance is being attached to recurring overpressure tests of those components in nuclear power plants whose inspection is deemed necessary. The necessity for a pressure test results from the above-mentioned definition of a nuclear reactor pressure vessel as a 'steam boiler'. Test requirements are determined for each individual case. Generally, the test pressure is taken as 1·3 times the design pressure if water is used as a pressure medium.

For some years, the results and applicability of fracture mechanics, in connection with the importance of recurring inspection, have been receiving increasing attention. At present, efforts are being made to determine critical flaw dimensions for operation parameters of interest with reactor pressure vessels. The recurring pressure test is increasingly considered to be a means whereby the reactor pressure vessel (or any other pressurized component) can be shown to be free of flaws exceeding a certain dimension. To determine absolutely that a pressurized component can be deemed safe for a certain period of time, additional information is necessary concerning the under-critical crack growth under operating conditions. Corresponding measurements are at present being made for the operating conditions of reactor pressure vessels of light water moderated reactors in the Federal Republic.

The Federal Government is now sponsoring research and development studies the object of which is to determine the applicability of new inspection techniques. Besides the above-mentioned work of M.A.N.–Krautkrämer, only the research work of the Battelle Institute on acoustic emission is of importance for the recurring inspection of steam boiler and pressure vessels in nuclear power plants. However, the real applicability of this method, which is predominantly being developed in the U.S.A., is not proved.

EXPERIENCE WITH THE RECURRING INSPECTION OF THE REACTOR PRESSURE VESSEL K.W.O.

The nuclear power plant K.W.O. has a pressurized water reactor with an output of 340 MWe. In the design and construction of this reactor pressure vessel, the special requirements which would now be incorporated as a result of recurring inspections were not taken into account. A special defect is that the thermal shield cannot be removed. This greatly limits accessibility to the vessel wall from the interior. From the outside, the reactor pressure vessel K.W.O. is practically non-accessible.

On the occasion of the first regular refuelling of K.W.O., the first inspection of the nuclear pressure vessel was carried out during August and September 1970. In order to apply the Tazzelwurm method (newly developed inspection techniques), it was essential to construct manipulating equipment suitable for the specific conditions of K.W.O.

The principal measuring arrangement can be seen from the schematic cross-section of the K.W.O. pressure water

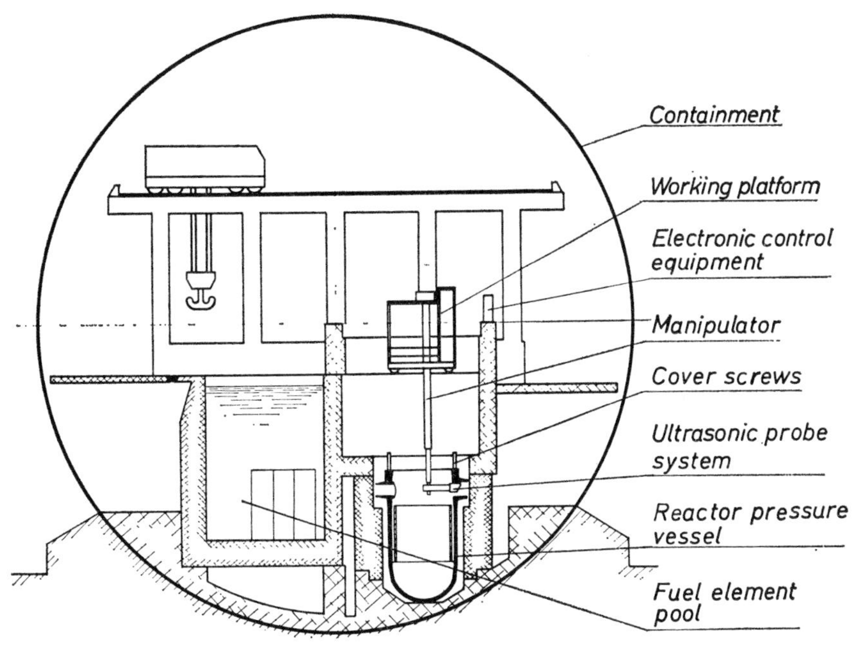

Fig. 44.1. Schematic cross-section of K.W.O. pressurized water reactor

reactor in Fig. 44.1. The manipulator is attached to the refuelling machine. The manipulating elements are on the working platform. Control equipment has been placed at the side of the platform (Fig. 44.2). Fig. 44.3 shows details of the measurement of a circumferential nozzle welding seam using the Tazzelwurm arrangement. In Fig. 44.4 this equipment, fixed at the lower end of the manipulator, is shown before introduction into the reactor pressure vessel. The individual test procedures were examined on a nozzle model (Fig. 44.5) in the fuel element pool.

The upper vessel circumferential seam and the four coolant nozzles were inspected with the Tazzelwurm (Fig. 44.6). For the ultrasonic inspection of the cover the Sperry wheel was used, the nozzles and the circumferential seam being additionally inspected through remotely controlled TV equipment. This optical inspection also served to determine the conditions for the later ultrasonic inspection. As the platform crane could be moved only in two directions, i.e. vertical to each other, it was only possible to introduce the Tazzelwurm into the nozzle on a saw-tooth curve path, which, however, did not present any special

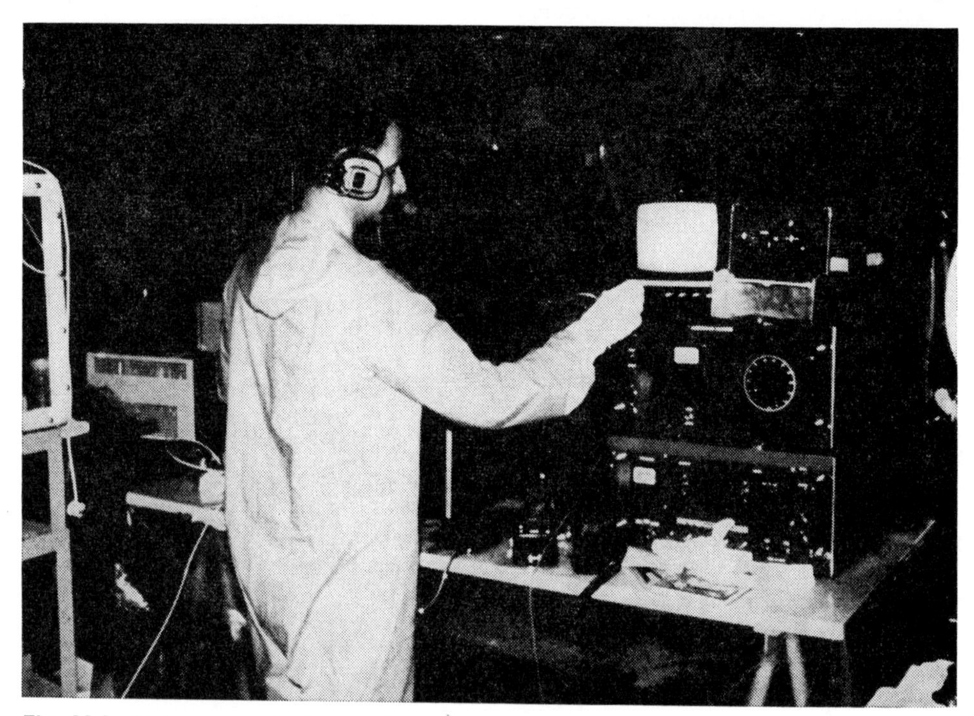

Fig. 44.2. Detail view of electronic control equipment for the K.W.O. pressure vessel inspection (after Allianz)

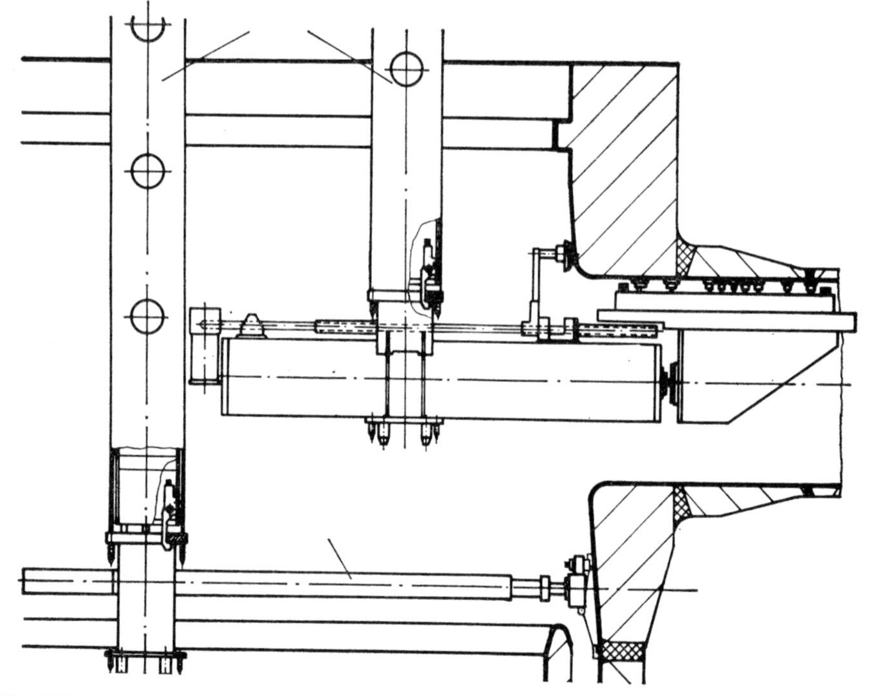

Fig. 44.3. Scheme of the automatic ultrasonic equipment 'Tazzelwurm' and cross-section of a nozzle area (after K.W.O.)

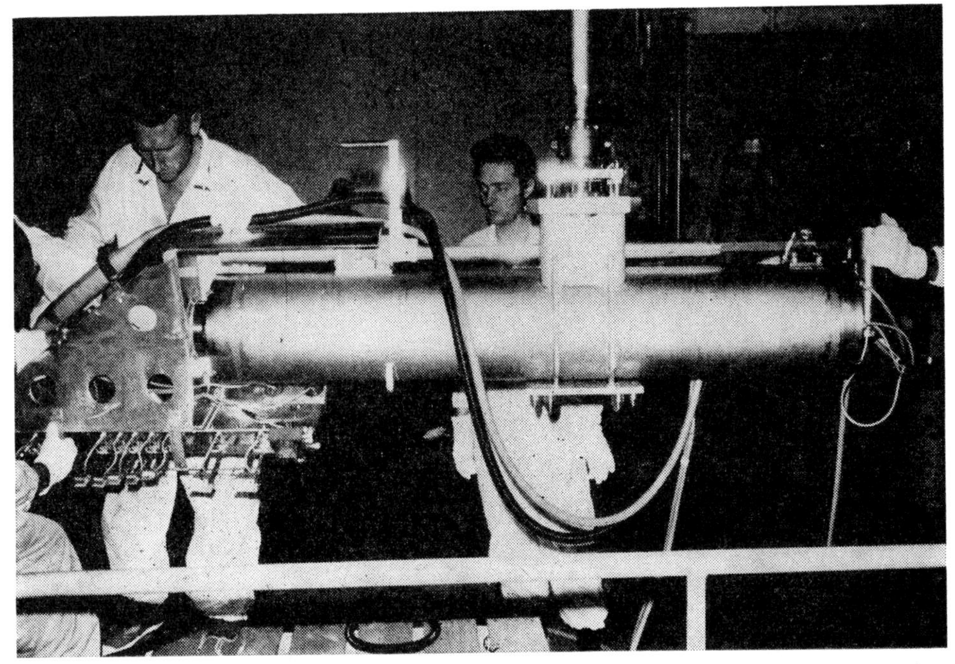

Fig. 44.4. View of the Tazzelwurm arrangement before inserting into pressure vessel (after K.W.O.)

difficulties. Contact between the Tazzelwurm and the vessel wall was prevented with the aid of sensors.

The inspections were completed in 50 h, and it was determined (**6**) that the circumferential seam and the segments of the nozzle seams, which could be inspected by use of the U.S. procedure, were free of detectable flaws (equivalent flaw: cylindrical hole, 10 mm diameter). In some cases there were flaw indications that have not yet been successfully interpreted. This applies, in particular, to one larger indication outside the circumferential seam of the coolant outlet nozzle. The interpretation of this test result has proved to be difficult, because the geometry of the welding seam is rather complicated and earlier examinations during the manufacture of the vessel did not show any indication of a flaw. This indication will be given special attention during the next inspection.

Company staff and external staff were employed on the inspection of the reactor pressure vessel K.W.O. According to the data given in Table 44.1, the total 'person' dose

Fig. 44.5. View of the K.W.O. coolant nozzle model (after K.W.O.)

Table 44.1. Irradiation doses during the first recurring inspection of the reactor pressure vessel K.W.O., autumn 1970

(A) *Ultrasonic testing of bolts*

Average dose	270 mrem
Maximum dose	600 mrem
Number of persons	10
Total dose per person	2·7 man rem

(B) *Vessel inspection (optical and ultrasonic)*

Dose, mrem	Number of persons	Total dose per person, mrem
1– 100	1	70
101– 200	1	130
201– 300	3	750
301– 400	4	1530
401– 500	—	—
501– 600	6	3000
601– 700	—	—
701– 800	—	—
801– 900	1	840
901–1000	1	950
1001–1100	—	—
1101–1200	1	1180
		Total 8450

(C) *Ultrasonic inspection of the cover (Sperry wheel)*

Dose at the cover	300 mr/h
Maximum dose	300 mrem
Number of persons	6
Total dose per person	approximately 6 man rem

Fig. 44.6. Testing of the Tazzelwurm equipment at a nozzle model (after K.W.O.)

was about 17 man rem. About 2·7 man rem of this figure can be allotted to the bolt inspection, about 8·4 man rem to the vessel inspection, and about 6 man rem to the inspection of the welding seams of the cover with the Sperry wheel.

For the optical inspection, remotely controlled TV equipment was used. This equipment was made by the Allianz Versicherung A.G. for special use in nuclear power plants. From the optical measurements no test results could be derived other than those compatible with safety.

EXPERIENCE WITH INSPECTION OF REACTOR PRESSURE VESSEL K.R.B.

The nuclear power plant K.R.B. (Gundremmingen) is equipped with a single-circuit B.W.R. (boiling water reactor) and has an output of about 280 MWe. The reactor pressure vessel (Fig. 44.7) is not accessible from the outside beneath the feed-water nozzles. The steam nozzles are accessible through press stones. To gain access to the inner vessel surface beneath the steam outlet nozzles it is necessary to remove the steam separators. At the time of the first start-up of the nuclear power plant, the licensing authority stipulated that the operator and the inspection organization must, from time to time, work out and agree on inspection programmes.

The 1971 inspection programme (7), for instance, provides for the inspection of 10 bolts, 10 nuts and washers, the cladding, the circumferential seam, four nozzles with nozzle seams, and the reactor cover. Moreover, the programme covers the inspection of the pressure vessel, from the outside, of four steam nozzles, two feed-water nozzles, two core spray nozzles, and the pressure vessel circumferential seam above the feed-water nozzle.

Inspection methods include manual ultrasonic inspection with angle probes and portable monitor, dye penetration, and TV techniques. Fig. 44.8 shows a section of the cladding in the region of a coolant nozzle. From the representation, the limited resolving power of the TV equipment can be seen.

Preparatory measurements showed a radiation dose rate of about 500 mr/h for the nozzles to be inspected. To permit access to the places to be inspected from the outside, press stone walls in the steam line openings had to be removed. The ultrasonic inspections were carried out by hand. The tests did not show any flaw indications that would affect further normal reactor operation. To keep the radiation doses as small as possible, the originally intended scope of the inspection was not fully carried out,

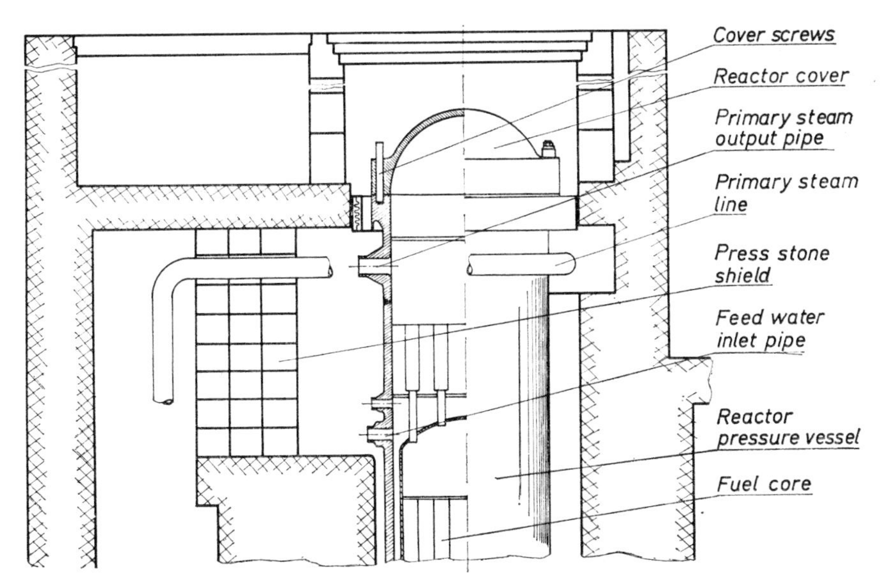

Fig. 44.7. Schematic cross-section of the K.R.B. reactor pressure vessel and its surrounding construction parts

especially since there were no objections concerning further normal reactor operation. Therefore it seemed advisable to keep the radiation doses for the staff as small as possible. This is why only two steam nozzles, parts of the steam line welding seams, and a length of 1·5 m of the pressure vessel circumferential seam above the feed-water nozzle were inspected for longitudinal flaws.

Optical measurements were carried through with a remotely controlled TV camera; there were no difficulties. The TV camera had been fixed to the refuelling equipment for these inspections.

EXPERIENCE WITH RECURRING INSPECTIONS OF REACTOR PRESSURE VESSEL K.W.L.

The nuclear power plant K.W.L. (Lingen) was designed as a two-circuit plant and is equipped with a B.W.R. The primary steam generated in the nuclear pressure vessel is condensed in two steam transformers and undercooled in heat exchangers before re-entry into the reactor pressure vessel. In these exchangers the preheating of the secondary feed water takes place, after which it is evaporated in the steam transformers. Before entry into the turbine this secondary steam is heated to 530°C in an oil-fired superheater.

The total electric output of the nuclear power plant is approximately 250 MWe. The thermal power of the reactor is 520 MWt. Of the 250-MW electric output, about 160 MW can be allotted to nuclear generation and about 90 MW to heat introduced into the superheater according to traditional methods.

Recurring inspection of the reactor pressure vessel is based on service instructions (**8**). Among others, these service instructions list in detail all tests to be carried out on the pressure vessel. For instance, the pressure vessel cover must be inspected visually, both externally and internally, and it must be examined by dye penetration and ultrasonic inspection methods. The bolts between the vessel flange and the cover are to be subjected to a random test (about 15 per cent at each inspection) with dye-penetration methods. An optical inspection of the total inner surface of the pressure vessel is required. The safety valves must be tested regarding their operation pressures. In addition, the reactor pressure vessel, with the total primary circuit, must be subjected to a recurring pressure test.

During the 1970 and 1971 inspections, the visual examination of the inner surface of the pressure vessel cover was widely disturbed by γ-radiation. Measured at the surface the dose rate went up to 10 r/h. The total inner surface was covered with a grey layer, which could not be removed by washing and staining. This layer made it impossible to detect flaws during the surface magnaflux inspection. No flaws were found during a visual inspection, which had to be performed quickly owing to the limited available time.

The outer surface of the reactor pressure vessel cover is much more accessible. Here the dose rate (measured at the surface) was only up to 50 mr/h. In view of this irradiation dose rate a complete inspection of this surface, with surface magnaflux testing and ultrasonic testing, was possible. In these tests the welding seams of the cover (two circumferential seams and four meridian seams) were inspected. For the ultrasonic inspection, commercial miniature angle crystals were employed. Transmission of sound was effected from one or two sides. Before the inspection, the U.S. equipment was adjusted with the aid of test flaws (cylindrical hole with 6 mm diameter). During the actual test, indications were recorded up to equivalent flaw diameters of 2–4 mm. They were not thought to be important enough to constitute safety hazards.

For the magnaflux surface testing of the welding seams a magnet yoke equipment of Tokushu, Tokyo, was used, and for the visual inspection an ultraviolet lamp was available. As a contrast medium, fluorescent magnetic powder in a hydrous solution was used. During this examination, no significant flaws were detected.

Inspection of the bolts did not reveal any flaw indications. The inspection was carried out from the inner hole with a semi-automatic test apparatus of M.A.N. equipped with commercial miniature ultrasonic angle crystals

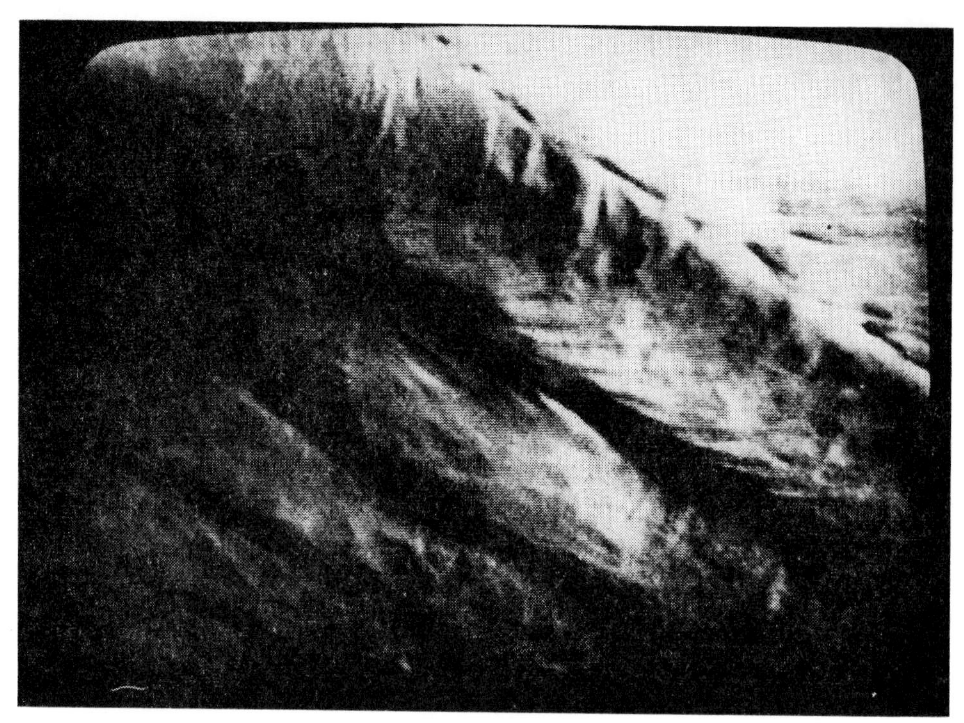

Fig. 44.8. Television record of the K.R.B. reactor pressure vessel cladding (after Allianz)

(70°C). A bolt with a saw cut of 2-mm depth was used as a calibration specimen.

The visual inspection of the pressure vessel had been impeded in 1970 because the steam separator above the core could not be removed. The gap between the steam separator and the vessel wall was too small to permit the TV camera used for the inspection (Tele-Thomson make) to be introduced with an accuracy sufficient to obtain a clear impression of the surface of the vessel.

Although in 1971 the steam separator could be removed, the gap between the core shroud and the vessel wall was too small for inspection purposes. For this reason, no better results were achieved than in the previous year. Attempts to inspect a circumferential seam of the reactor pressure vessel above the core shroud with the TV camera were not successful either, although the camera could be firmly guided in the manipulator of the refuelling platform. As the inner cladding had been applied by manual electric welding in the region of the circumferential seams, and been ground, there were no calibration points by which a well-focused picture could be obtained.

For this reason only a limited visual inspection of the pressure vessel could be carried out, and this was performed with the aid of a commercial telescope. At the time of inspection the filling height had been reduced to the flange height. During this test—by which only relatively obvious flaws can be detected—no flaws were found.

It was not possible to inspect the bottom of the reactor pressure vessel with the camera, as all attempts to remove a coolant inlet orifice in the below-core grid were unsuccessful. No suitable tools were available for this work.

At the conclusion of the inspections of the pressure vessel and pressure vessel cover, only the examination of bolts and pressure vessel cover proved to be satisfactory. However, the examination of the cover may be considered to be representative of the total pressure vessel, due to the relatively high stresses influencing the cover.

Two of the four anticipatory control valves for the safety valves of the reactor pressure vessel were also examined to determine their opening pressure. This examination revealed that the opening pressure of one of these valves was considerably above the originally adjusted pressure. A more detailed inspection showed that the cause of this defect was an error in the design of these valves, which had led to increased friction between lifting jack and casing. Therefore the two remaining anticipatory control valves were removed; one of these showed an excessive opening pressure. Subsequently, the design of all four anticipatory valves was modified, and the valves were readjusted.

To complete the inspection a pressure test was performed with 1·1 times the operating pressure. During this test the essential components were visually inspected. Again, there were no difficulties.

CONCLUSIONS

Results of recurring inspections of reactor vessels carried out in the Obrigheim, Gundremmingen, and Lingen demonstration power plants allow the following general conclusions to be drawn.

Inspections of the pressure vessel, especially of the inner pressure vessel wall, are considerably less effective than those for conventional pressure vessels and steam boilers. The detection of flaws, which is rather limited for various reasons, most probably will not be improved by chemically or mechanically removing the disturbing layer found on the vessel surface without risking later disturbances due to loose particles, e.g. the catching of control rods. Even the employment of auxiliary means (dye-penetration methods, remotely controlled TV equipment, telescopes) will probably not improve the results of optical inspection methods.

Ultrasonic inspection methods may be more effective when careful finger prints are made before the start-up of reactor pressure vessels, and when accessibility to areas of interest has been previously planned. This planning includes the establishment of a flaw catalogue to contain all essential indications of non-destructive testing methods after completion of the vessel.

As stated above, the recurring pressure test is deemed to be the only integral strength test of the reactor pressure vessel. The test conditions for the recurring inspection of the pressure vessels are not uniform. A development can be observed in the Federal Republic, whose object is to promote agreement on the test conditions for recurring inspections of reactor pressure vessels individually between the operator, the authorities, and the technical control organizations. Whilst there are no effective test methods applicable to the non-destructive testing of the whole vessel material, it will not be possible to renounce recurring pressure tests.

The irradiation doses sustained by members of the staff employed in the recurring inspections are, compared to the other irradiation doses during normal operation, encouragingly small. It may be expected, therefore, that any considerable extension of the scope of inspections by the use of automatic and remotely controlled inspection equipment will not involve any radiological problems.

APPENDIX 44.1

REFERENCES

(1) Verordnung über die Errichtung und den Betrieb von Dampfkesselanlagen (Dampfkesselverordnung—DampfkV) vom 8.9.1965, GB Bl. S. 1300 (Decree on the construction and operation of steam boilers).

(2) Deutscher Dampfkessel- und Druckgefäßausschuß (D.D.A.), 'Prüfung der Hochdruckdampfkesselanlagen', TRD 501 (Prüfung), Ausgabe Juli 1968 (German committee for boilers and pressure vessels, 'Inspection of high-pressure steam boilers', issued July 1968).

(3) I.A.E.A., 'Recurring inspections of nuclear reactor steel pressure vessels' (Report of a panel on recurring inspections of nuclear reactor steel pressure vessels, held in Pilsen (C.S.S.R.), 3rd to 7th October 1966), Tech. Repts Ser. No. 81, 1968.

(4) MEYER, H. J. 'Ultraschall-Wiederholungsprüfungen an Reaktordruckbehältern', *Kerntechnik* 1971 **13** (No. 2), 56 ('Ultrasonic recurring inspections of nuclear reactor pressure vessels').

(5) Sperry Bulletin 50-125, Sperry Products Division of Automation Industries Inc. (Danbury, Connecticut).

(6) T.Ü.V. Baden, Technischer Bericht Nr. II 71 2589, 15.9.1971 (TÜV Baden, Tech. Rept No. II 71 2589, 15th September 1971).

(7) K.R.B. Besprechungsergebnis vom 21.1.1971 über die amtlichen Untersuchungen während des Brennstoffwechsels 1971, Aktennotiz 5/71 vom 2.2.1971 (K.R.B. memorandum of 21st January 1971 on the official inspection during the refuelling process 1971, memorandum 5/71 of 2nd February 1971).

(8) K.W.L. Betriebsanweisung, Teil C, 'Wiederholungsprüfungen, 1.3.1970 (K.W.L. Service Instructions, Part C, 'Recurring inspections', 1st March 1970).

C45/72

AUTOMATED, MOBILE, X-RAY DIFFRACTOMETER FOR MEASURING RESIDUAL STRESSES IN OPERATING EQUIPMENT

J. R. CRISCI*

Residual stresses develop in structures during fabrication. Their presence may be detrimental to the life and performance of a structure during service. In recent years the need for stress analysis by non-destructive means has become apparent. X-ray methods are widely used for the non-destructive determination of surface stresses. The Annapolis Laboratory of the Naval Ship Research and Development Center has automated a mobile X-ray stress diffractometer that is capable of determining non-destructively the surface stresses in large immovable structures.

INTRODUCTION

ENGINEERS HAVE been concerned for many years with problems of weld failure, dimensional instability, unsatisfactory fatigue life, and stress corrosion. A contributing factor to these problems is the surface stress in a material. Large residual tensile stresses on the surface provide favourable conditions for initiating stress corrosion and fatigue cracking, and consequently may impair the reliability of a structure. The demands from industry for higher performance structural materials have increased the need for determining structural integrity. X-ray methods are widely used for non-destructive determination of surface stresses. This paper summarizes the techniques and instrumentation under development at the Naval Ship Research and Development Center, Annapolis, Maryland; techniques that provide more accurate determination of structural integrity using automated X-ray diffraction instrumentation.

OBJECTIVE

The objective of the work is to develop automated, mobile, X-ray diffraction equipment and techniques for non-destructive determination and surveillance of surface stresses in important service structures. The automation is required in order to reduce human error, increase the speed, and reduce the cost of the analysis.

THEORETICAL STRESS DETERMINATION

The theoretical basis for the stress determination by use of X-rays in polycrystalline materials has been well established. For completeness of the present discussion, the highlights of the theoretical background will be presented here. A detailed discussion may be found in reference (1)†.

The MS. of this paper was received at the Institution on 10th November 1971 and accepted for publication on 14th January 1972. 20

* *Materials Engineer, Annapolis Laboratory, Naval Ship Research and Development Center, Annapolis, Maryland 21402.*

† *References are given in Appendix 45.1.*

A mono-energetic and collimated X-ray beam that is directed on to an unstressed polycrystalline material, Fig. 45.1, will exhibit diffraction maxima whenever the Bragg condition is satisfied, i.e.

$$\lambda = 2d_{hkl} \sin \theta \quad . \quad . \quad . \quad (45.1)$$

where λ is the X-ray wavelength, d_{hkl} is the distance between two adjacent diffracting planes of Miller indices (hkl), and θ is the half-angle between the extended incident and reflected beams. The above are necessary, but not sufficient, conditions to describe the diffractometer geometry. Further conditions are:

(1) The plane formed by the incident, extended incident, and reflected beams must contain the normal to the diffracting crystallographic plane; and the plane must be perpendicular to and bisect both the X-ray tube slit and the detector slits.

(2) The normal to the diffracting crystallographic plane must bisect the angle formed by the incident and reflected beams.

If the 2θ angle of the scintillation counter is preset to a specific value [e.g. for martensite (211) in martensitic steels], then some of the crystallites in a polycrystalline unstressed material will always be properly oriented to produce a diffraction peak.

When a stress is applied to the material, the interplanar spacing d_{hkl} changes by an amount Δd, and θ changes by a small amount (1° or less) $\Delta\theta$, to restore the diffraction maxima. The resulting lattice strain, $\epsilon_{\phi\psi}$, is defined by

$$\epsilon_{\phi\psi} = \frac{\Delta d}{d_{hkl}} \quad . \quad . \quad . \quad (45.2)$$

Differentiation of equation (45.1) and rearrangement of terms yields

$$\epsilon_{\phi\psi} = \frac{\Delta d}{d_{hkl}} = -[\text{ctn}\ \theta]\,\Delta\theta \quad . \quad (45.3)$$

It is readily noted in equation (45.3) that the lattice strain

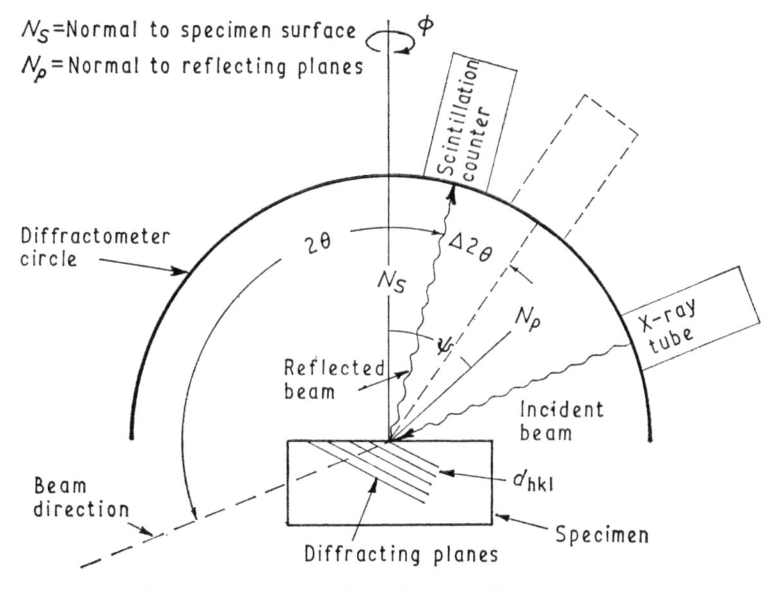

Fig. 45.1. Schematic of X-ray diffraction method

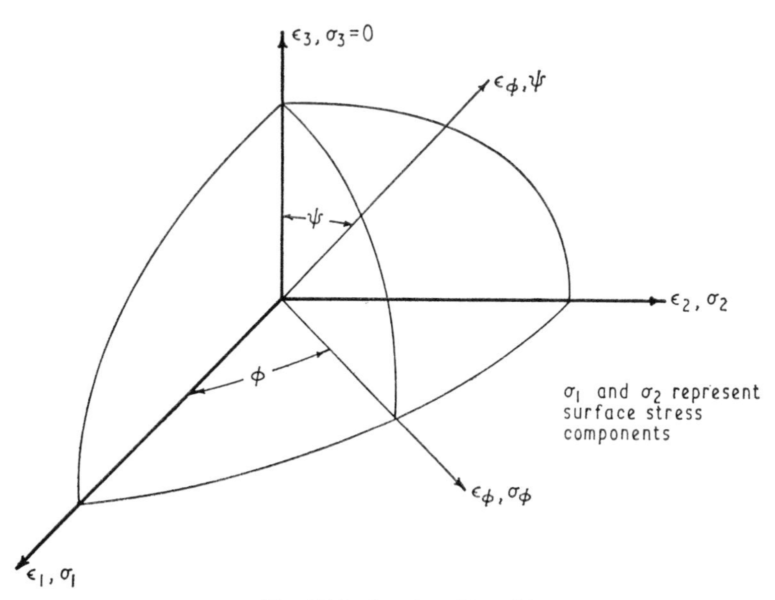

Fig. 45.2. Strain ellipsoid

can be found if θ and $\Delta\theta$ are known. Various methods to determine $\Delta\theta$ will be discussed subsequently. The subscripts, $\phi\psi$, define the strain component $\epsilon_{\phi\psi}$ with respect to both the strain ellipsoid orientation angles ϕ and ψ, Fig. 45.2, and the X-ray beam orientation. The angle ϕ indicates the projection of the X-ray beam on to the plane of the specimen surface, and ψ is the angle between the surface normal and the normal to the diffracting planes and lies in the plane of the ϕ-to-surface projection. Referring again to Fig. 45.2, the strains ϵ_1, ϵ_2, and ϵ_3 are the principal strains, with ϵ_3 normal to the surface.

The stresses in the surface, σ_ϕ, σ_1, and σ_2 ($\sigma_3 = 0$), can be related to the lattice strain by an application of the elasticity equation (**2**):

$$\epsilon_{\phi\psi} = \left(\frac{1+\nu}{E}\right)\sigma_\phi \sin^2\psi - \frac{\nu}{E}(\sigma_1+\sigma_2) \quad . \quad (45.4)$$

where ν is Poisson's ratio, E is the Young's modulus for the (hkl) plane, and σ denotes a stress on the (hkl) plane. Equation (45.4) enables the determination of the surface stress σ_ϕ in any given direction ϕ when a set of readings for $\epsilon_{\phi\psi}$ has been determined.

A detailed examination of equation (45.4) leads to a straightforward determination of σ_ϕ from experimental measurements. Let ϕ be constant and let the angle ψ be varied. A plot of $\epsilon_{\phi\psi}$ computed from equation (45.3) versus $\sin^2\psi$ for constant ϕ will result in a straight line, as shown in Fig. 45.3. The accuracy of this procedure is dependent on the resulting least square line fit and not on a single point. Consequently, scatter of data points can be tolerated while still achieving high accuracy. The intercept at $\psi = 0$ ($\sin^2\psi = 0$) is given by $-(\nu/E)(\sigma_1+\sigma_2)$. More important, however, is the slope of the resulting line. From equation (45.4) it is noted that the slope is equal to $[(1+\nu)/E]\sigma_\phi$ from which σ_ϕ is determined for the (hkl) plane. If the slope of the line is positive, then the measured surface stress is tensile; and if the slope is negative, then the surface stress is compressive.

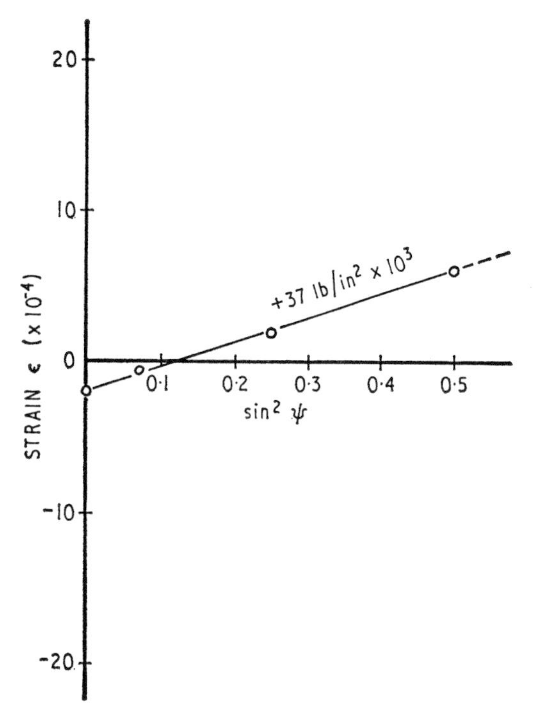

Fig. 45.3. Strain versus $\sin^2 \psi$

It is of interest to observe that the calculation of σ_ϕ does not require a knowledge of the sum of the principal stresses $(\sigma_1+\sigma_2)$. The accuracy of the determination of σ_ϕ can be improved by increasing the number of ψ angles used in the measurement. Naturally, the procedure can be duplicated for several values of ϕ, so that the magnitudes and direction of the principal stresses can be found. For the determination of the principal stresses, standard methods of elasticity are applied (**2**).

The direct quantitative determination of σ_ϕ requires a knowledge of the (hkl) elastic constants, ν and E. These values are not directly available in the literature. Determining a micro- rather than the more commonly measured and reported bulk macroscopic strain is accomplished by applying a known macroscopic surface strain to a calibration specimen, measuring the microscopic strain $\epsilon_{\phi\psi}$ using X-ray diffraction and determining the proportionality constant k by

$$k\sigma_{hkl} = \sigma_{\text{bulk}} \quad . \quad . \quad . \quad (45.5)$$

X-RAY DIFFRACTOMETER METHOD

Experimental aspects

In the previous section dealing with the theory it was indicated that θ and $\Delta\theta$ in equation (45.3) must be determined. Two methods are currently used to accomplish the measurement of θ and $\Delta\theta$, the camera and the diffractometer techniques (**1**). For the purpose of this discussion, only the diffractometer technique will be described since this technique is more accurate. The diffractometer technique has been illustrated schematically in Fig. 45.1. The incident X-ray beam is reflected from the material at the diffraction angle 2θ. The intensity of the X-ray beam and the diffraction peak maxima are displayed on a recorder as the scintillation counter scans 2θ. The X-ray tube and specimen remain stationary during the 2θ scanning operation. Stressing the material produces a strain and changes the spacing between the diffracting planes. The strain, or change in strain, is observed by a shift in the diffracted angle by an amount $\Delta 2\theta$.

Except for the mobile stress diffractometer, which will be discussed subsequently, laboratory diffractometers have several significant disadvantages. The laboratory diffractometers have large physical dimensions such that the centre of X-ray focus is too short to extend beyond the unit, thus precluding their use on immovable massive structures; and they lack the mobility and ruggedness for field applications.

Mobile stress diffractometer

At the present time unautomated mobile stress diffractometers are commercially available. These units require extensive operator involvement. The operator manually adjusts the diffractometer to all desired ϕ and ψ settings. For the work reported here, a mobile stress diffractometer has been selected and partly automated in order to minimize required operator involvement. Specifically, a Siemens stress goniometer was selected to form the basic component of the mobile automated X-ray unit. The instrument, as purchased, is shown in Fig. 45.4 and was initially assembled on a movable table. The complete unit is mounted on a vertical column in order to obtain 360° rotation in the horizontal plane. The X-ray tube and the scintillation counter are held in a mount that can be manually rotated through the ϕ and ψ angles, where ϕ is defined as rotation of the X-ray tube and all attached components around the emergent X-ray beam, and ψ is a sliding motion of the X-ray tube along the circular rail. A 2 kVA generator supplies power to a 480-W chromium X-ray tube. The scintillation counter position on the circular rail is changed by a 2θ motor drive after ϕ and ψ are set to predetermined positions. The intensity output from the scintillation counter is automatically recorded as a function of the 2θ position on the chart recorder. The goniometer 2θ range is between 113° and 163°. Usable back-reflection lines of most materials lie within that angular range. The unit shown in Fig. 45.4 forms the basic hardware used in the current automation development. This stress diffractometer was chosen because of its ability to determine the stress on a large structural member without the requirement of motion by the structural element. In the next section the automation that has been accomplished to date will be discussed.

AUTOMATION OF THE MOBILE STRESS DIFFRACTOMETER

Data requirements

To determine a stress component, the parameters in equations (45.3) and (45.4), namely ϕ, 2θ, $\Delta 2\theta$, and ψ, must be determined. Three ϕ, twelve ψ and twenty-four 2θ motions are required to collect the necessary data for a complete stress determination. (More or fewer values can be used, depending on the accuracy desired.) It is obvious that a cyclic procedure can be used and automated to obtain data needed for the complete surface stress determination.

Automation instrumentation

The present partly automated diffractometer, with automated motors, a teletypewriter, and the automation elec-

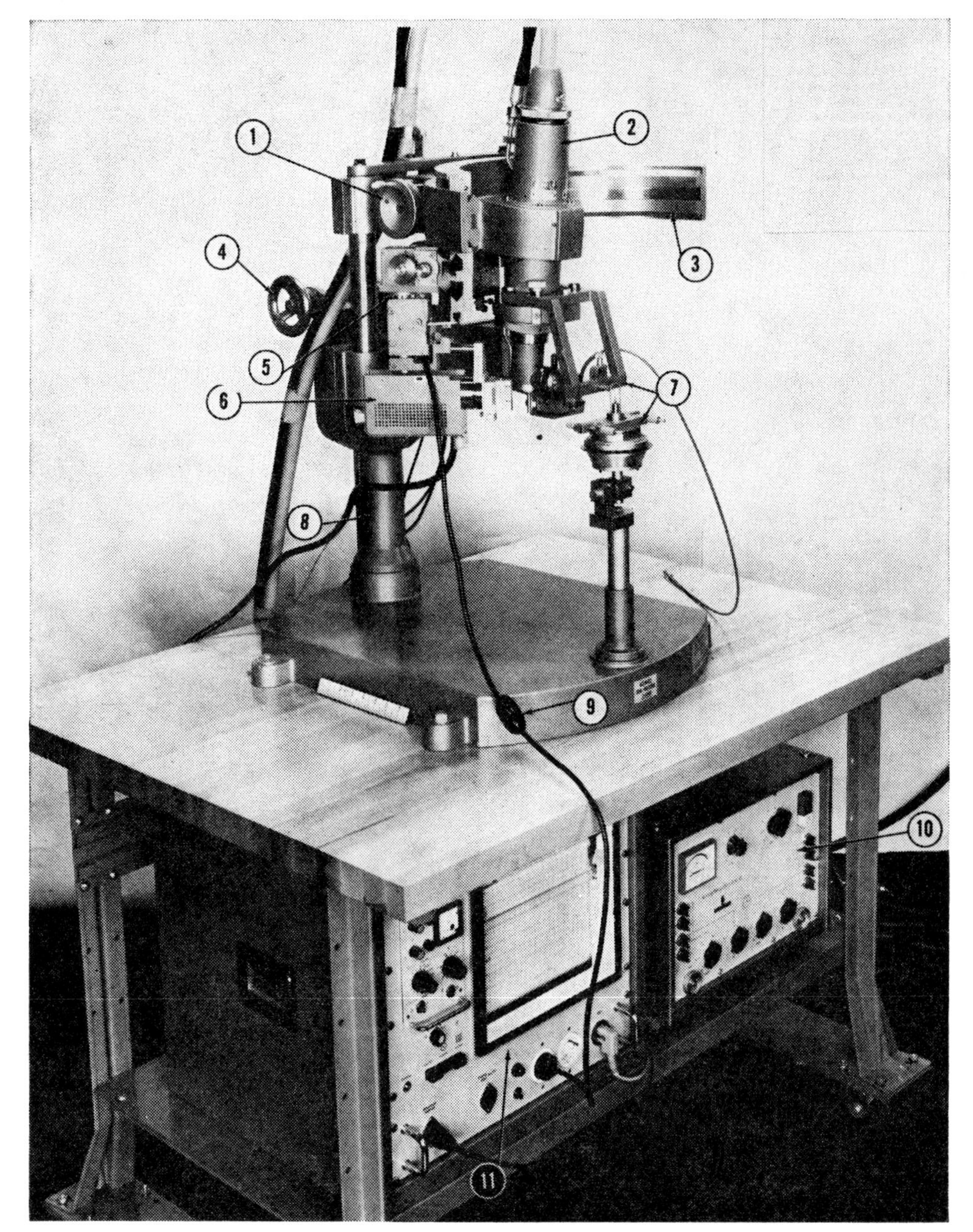

1 ψ motion control.
2 X-ray tube.
3 Circular rail.
4 ϕ motion control.
5 2θ control.
6 Scintillation counter.
7 Alignment fixtures.
8 Vertical column.
9 2θ cable switch.
10 Power supply.
11 Recorder.

Fig. 45.4. Original mobile stress diffractometer

tronics, is shown in Fig. 45.5. A block diagram relating all automation components is shown in Fig. 45.6. The essential functions and performance characteristics will be discussed in the following sections.

Motors

The ψ motion handwheel has been replaced by a precision stepping motor coupled directly to the ψ gear train drive. The stepping error of this motor is non-cumulative and is equal to ± 3 per cent/step. Each step is equal to 1/200 of a revolution of the armature. The motor can step with equal accuracy in either direction and is controlled by automated electronics. The electronic control results in a repeatability of $\pm 0{\cdot}005°$ for the ψ settings. The ϕ motion handwheel has been replaced by a heavy-duty precision stepping motor coupled to the ϕ drive by a sprocket gear arrangement with a 1:5 gear ratio. The stepping motor has the same general characteristics as the ψ stepping motor except that the repeatability of electronically positioning the ϕ axis is $\pm 0{\cdot}5°$. The 2θ manual 'on–off' motor switch has been replaced by relay circuits, which have been specifically designed to start and stop the 2θ motor upon receipt of commands from the automation electronics. The 2θ motor scans the diffraction peak twice by using a motor-reversing switch and two mechanical stops. A relay operating at the end of the second scan stops the 2θ motor and directs the teletypewriter to read the next programmed instruction.

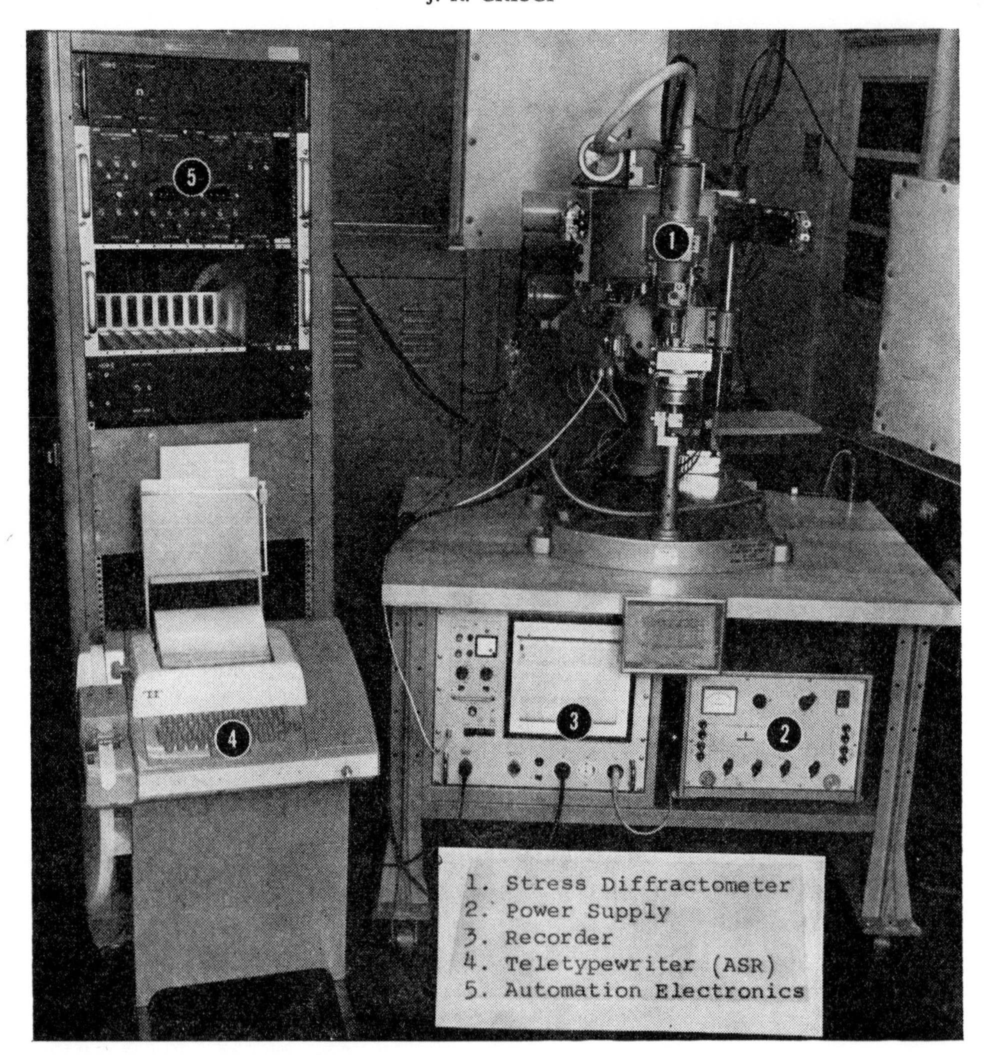

Fig. 45.5. Present partly automated diffractometer

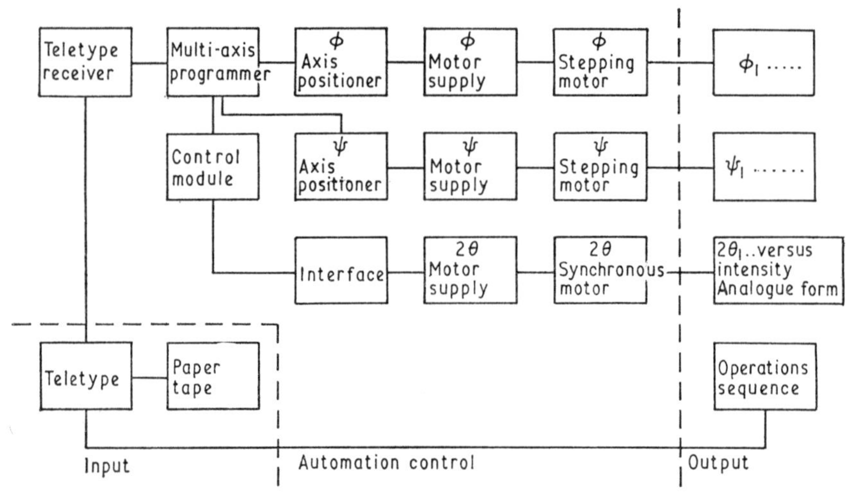

Fig. 45.6. Automation control system for the present partly automated diffractometer

Teletypewriter

The operator prepunches the complete automation instructions on a paper tape for later readout on the paper tape reading head of the teletypewriter. Briefly, a single paper tape is used and a single action (pressing the start button) makes all instrumental adjustments and prints out all the data required for complete stress determination. The punched tape can be used repeatedly with little wear.

Automation electronics

All the electronic controls have been specifically designed or modified from standard electronic components for the current automation of the diffractometer.

The teletypewriter receiver obtains information directly from the punched tape and converts this information into electrical signals suitable for processing by a multi-axis programmer. The teletypewriter receiver also turns the paper tape reader on when the multi-axis programmer has completed a specific task and turns the reader off after a set of instructions has been received. Now the programmer is ready to execute these instructions.

The multi-axis programmer accepts, recognizes, and stores the input instructions and organizes itself to control, set, and/or readout the system in response to the programme. The programmer also responds to components within the system itself, such as relays, special control modules, and scalers. The multi-axis programmer exercises direct control over the ϕ and ψ axis positioners and notifies each axis positioner how many revolutions, in what direction, and at what speed to run the stepping motor.

A control module, interface, and relay circuits were developed that utilize the original Siemens reversible 2θ motor switch to provide a single forward and reverse 2θ scan after each position change in the ψ motor.

The axis positioner transmits logic level pulses to a motor supply to determine the speed, direction, continuous or intermittent motion (step scan), and final position of the stepping motor. The axis positioner drives the motor at speeds ranging from 0·125 to 60·000 rev/min, monitors the exact motor position at all times, and feeds the position information back to the multi-axis programmer.

The output from the automated mobile stress diffractometer is presented on two records. First, a print-out of the command sequence appears on the teletypewriter. Second, a chart record of 2θ versus intensity is produced. Both outputs must be used in conjunction with each other to determine the stress at a point in a specific direction.

The partly automated diffractometer requires 6 h of instrument time versus 8 h for the unautomated unit. The major saving with the partly automated unit is that one push of the start button executes the entire instrument sequence, saving 6 h of operator time.

TYPICAL EXPERIMENTAL RESULTS OBTAINED USING X-RAY TECHNIQUES

In this section the accuracy of the data will be discussed, together with the mechanical alignment requirements of the diffractometer and finally some typical surface stress determinations using X-rays will be presented.

Data accuracy

A brief discussion is in order on the method employed to measure θ and $\Delta 2\theta$ from the chart recordings produced by the automated mobile stress diffractometer. This section will conclude with a discussion on the effect of errors in the measurements and the overall accuracy that may be achieved.

A typical 2θ diffraction scan is shown in Fig. 45.7. Every vertical chart line is equal to 0·6° using the current 2θ gears and chart speed. The height of the curves indicates the intensity of the diffracted X-ray beam. A strain in the surface causes a shift in θ producing a value for $\Delta\theta$. In Fig. 45.7, $\Delta\theta$ can be measured by determining the change in L. Once the values for 2θ and $\Delta 2\theta$ are determined at known values of ϕ and ψ, the desired stresses can be calculated.

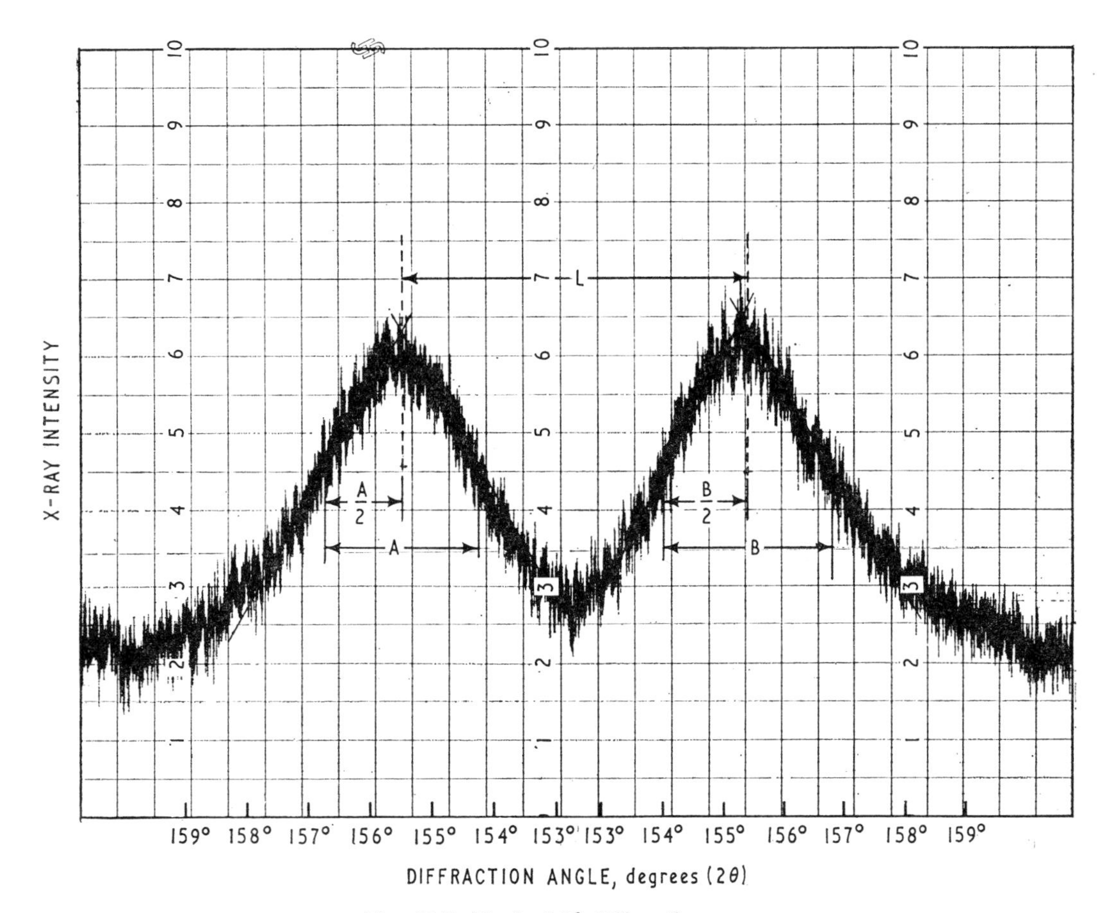

Fig. 45.7. Typical 2θ diffraction scan

X-ray diffraction stress analysis has limited precision depending on such parameters as X-ray intensity and scanning rate. In addition, when manual control is exercised over the motions of the goniometer, human errors are introduced. Some control can be exercised over the tube intensity and scanning rate but not over the operator. In the partly automated, mobile diffractometer, the unpredictable human element is virtually eliminated. Consequently, with the automated stress determination using X-rays, the attainable accuracy approaches the precision that is characteristic of this method.

The influence of measurement errors in ϕ, ψ, and θ is best examined by considering each of the parameters independently. Specifically, consider the errors in the surface stresses if the actual surface stress is known to be 100 $\text{lb/in}^2 \times 10^3$ in high-strength material.

First, consider an error in ϕ. Since ϕ is a measure of the direction of the surface stress σ_ϕ, ϕ does not influence the amplitude of σ_ϕ but has an effect only on the resulting principal surface stress direction and magnitude. Second, at constant ϕ and $\psi = 45°$, consider that an error in 2θ is made that equals 0·03°. It should be noted that such accuracy can be reliably achieved. The resulting error in the 100 $\text{lb/in}^2 \times 10^3$ stress level will be about 2·5 $\text{lb/in}^2 \times 10^3$. Finally, an error in the ψ measurement that equals 0·1° at a stress level of 100 $\text{lb/in}^2 \times 10^3$ will be less than 0·7 $\text{lb/in}^2 \times 10^3$. The instrument is capable of positioning the ψ setting with an accuracy of 0·1°. Thus, it is seen that in stress determinations that can be made using X-ray techniques, the obtainable accuracy is well within the experimental accuracy of any other available method, e.g. the destructive hole-drilling technique.

Mechanical alignment

An accurate determination of the state of stress at a point on the surface of a structure by X-ray techniques requires a precise alignment of the diffractometer with the point of interest. In the succeeding paragraphs, the procedures used for the alignment of the mobile diffractometer with the point on the structure will be discussed.

The most critical aspect of the mobile stress diffractometer is the mechanical alignment. The X-ray beam must strike the specimen surface at the same point while the X-ray beam inclination changes. The diffractometer components are mechanically aligned using a simulated surface (vertical needle tip) located in the stationary reference post. A flexible needle-pointed alignment jig with calibrated scales to permit needle motion is permanently attached to the X-ray tube. When aligned, the horizontal needle coincides with the X-ray beam, both needle tips virtually touch, and their junction represents the point in space around which the ϕ, ψ, and 2θ motions occur. The mobile diffractometer can be rotated up to 360° around its support base without disturbing the point in space located at the tip of the horizontal needle.

Alignment of the diffractometer with a real structure utilizes the following procedure. A rough positioning is made of the diffractometer table with respect to the structure. Then, the diffractometer is rotated around its support base to 'near alignment' with respect to the point of interest on the structure under study. The ϕ motion control may be used to orient the diffractometer with respect to the structure. Finally, the entire circular track is moved towards the structure using a fine adjusting knob. This phase completes the required preliminary alignment, and the actual stress determination at a point commences.

Experimental results

A number of experimental results have been obtained in the course of this investigation. The results to date are primarily qualitative, but more detailed analyses are under way and will be reported at a later date. In this section, the determination of the material constant will be discussed as well as analysis of the effect of stress relieving and electropolishing weldments.

The calibration data for a specific material is obtained by the use of a bending specimen with a known set of calibration constants, e.g. Poisson's ratio and the elastic modulus. The specimen is mounted in a four-point loading fixture and a known stress level is applied and monitored using a strain gauge. The stress is determined using the X-ray diffractometer. The measured stress has been plotted versus the applied stress, as shown in Fig. 45.8. Acceptable agreement exists.

Generally, the four-point bending fixture is used only to determine the material constant, k (equation 45.5), which relates an X-ray stress value on the crystallographic plane to bulk strain gauge determinations of stress. However, the above application does show the validity of using the mobile diffractometer for surface stress determinations. In connection with this, it is useful to note that no special surface preparations are required. The bending specimen has a smooth surface, but an as-welded condition is quite acceptable as a surface finish. Grease, paint, loose rust, or light mill scale are removed with a rag soaked in organic solvents.

Flat, machined tensile specimens were also mounted in a load frame and a known stress level was applied and monitored with a load cell. The X-ray measured stress has been plotted versus the applied stress in Fig. 45.9. Agreement is better than 5 $\text{lb/in}^2 \times 10^3$ at the 94 $\text{lb/in}^2 \times 10^3$ stress level, an agreement which again demonstrates the validity of surface stress measurements obtained with the mobile diffractometer.

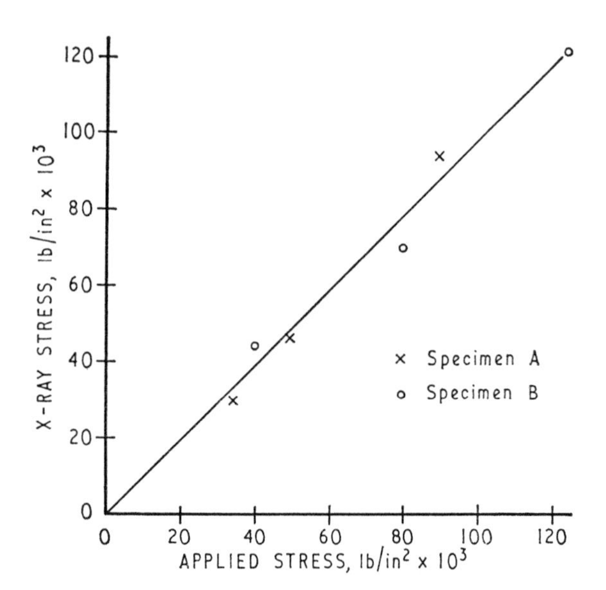

Fig. 45.8. Stress determined by X-ray versus actual stress for calibration of X-ray stress diffractometer (bending fixture)

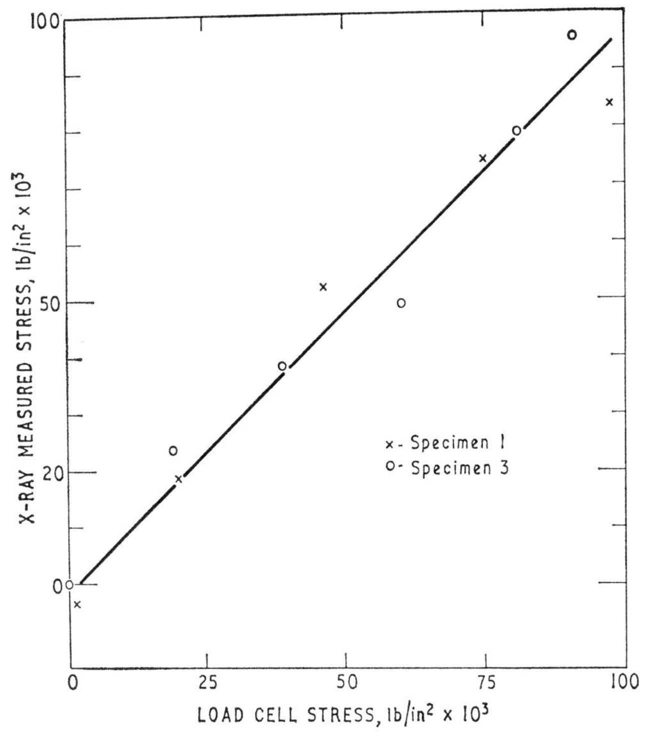

Fig. 45.9. Stress determined by X-ray versus actual stress for calibration of X-ray stress diffractometer (load frame)

A stress determination was made on the weldments shown in Fig. 45.10. Three steel, tee-welded specimens were cut from a single plate. The surface stresses at the toe of the welds were measured using the X-ray technique. The specimens were examined in the as-welded, stress-relieved and water-quenched, or stress-relieved and furnace-cooled condition. The results indicated that the stress relieving caused a reduction in the stress level of about one-half, in addition to showing that furnace cooling was better for stress-relieving purposes. The same specimens were then electropolished. The electropolishing removed about 0·003 in from the surface. For the as-welded condition, stress analysis showed that the electropolishing resulted in a reduction of the X-ray measured surface stresses by a factor of approximately one-half. The stress-relieved surfaces were not significantly affected by the electropolishing, a fact indicating the marked decrease in stress gradient resulting from stress-relief treatment.

Although the above results are not in themselves startling, they do show the flexibility of the X-ray stress analysis. Furthermore, at the present time, there is no other non-destructive method available that will provide an exact determination of the surface stresses.

MOBILITY, DURABILITY, AND WORKING SPACE

The diffractometer is marketed by the manufacturer as the 'Siemens Mobile Diffractometer' and the word 'mobile' has been retained in the text. In practice, the degree of mobility depends on the operator's ingenuity and proposed application.

Inherent mobility

Rotation of the diffractometer through 180° with respect to the baseplate provides the spatial clearance required to examine large convex or flat surfaces. The diffractometer, however, becomes unstable and topples unless its baseplate is bolted to a counterbalance of reasonable mass. Bolting the baseplate to an oak table equipped with castors was adequate for laboratory development work. The high-voltage power supply and the ratemeter–recorder chassis occupy a lower shelf. The loaded table, a total weight of 1000 lb, can be moved by one man on a smooth laboratory floor. A hydraulic unit such as commercially available welding, television, or camera booms

Fig. 45.10. Tee-welded, high-strength steel specimens

can be used to elevate and tilt the diffractometer for increased access to very large or irregular surfaces.

Separation of components and mobility

The diffractometer is currently separated from the teletypewriter and automation electronics by 30-ft cables. A separation of several hundred feet is feasible using a repeater station to amplify the small signals employed.

Separation of the diffractometer from the high-voltage supply would require lengthening the massive high-voltage cable and high-pressure water cooling lines and has not been considered. Separation of the ratemeter–recorder from the diffractometer is feasible but offers no real advantage.

Durability

A careful inspection of all components indicates that the teletypewriter, the high-voltage power supply, and the ratemeter–recorder should currently withstand moderately rough treatment. The diffractometer alignment is considered sensitive to rough handling and careful shock mounting is required.

Working space

The diffractometer, when centred on the baseplate and with the X-ray tube in a vertical position, occupies a box 24 in wide, 32 in deep, and 50 in high. Rotating the diffractometer by 180° with respect to the baseplate and rotating the X-ray tube to a horizontal position changes the dimensions to $42 \times 32 \times 30$ in. The dimensions of other components are not considered significant in normal stress analyses.

FUTURE WORK

The major future efforts in developing a fully automated mobile X-ray stress diffractometer will be concerned with applications, instrumentation, and the design of increased mobility. A brief discussion on each of these areas follows.

To date, the number of applications has not been large because of the emphasis on completion of the instrumentation. However, at the present time, the instrument is being used to determine surface stresses on simple structures. The mobile X-ray diffractometer will also be used to obtain stress analyses on a complex structure such as a large pressure cylinder with welded curved ends to simulate conditions in the field.

The instrumentation will be altered to provide a completely automated diffractometer. A 1000-W chromium X-ray tube has been purchased to increase the X-ray intensity, and the scanning rate will be increased by high-speed step scanning. The output will be in digital format and the resulting fully automated X-ray diffractometer for stress analysis will have increased speed and accuracy. The output will be in a form suitable for insertion into a digital computer.

Finally, effort will be expended to provide a detailed design which will allow rapid, precise positioning of the diffractometer with respect to a large in-place structure.

CONCLUSION

A mobile X-ray stress diffractometer has been automated and is capable of non-destructive determination of surface stresses on large immovable structures. The automation has produced major advantages, including the elimination of manual controls and a reduction in instrumental and analytical time for a complete analysis by a factor of two. Future work will include computer interfacing that will further reduce the time required for analysis.

AUTHOR'S NOTE

The opinion or assertions contained in this paper are the private ones of the author and are not to be interpreted as official or reflecting the views of the U.S. Navy.

APPENDIX 45.1

REFERENCES

(1) TAYLOR, A. *X-ray metallography* 1961 (John Wiley and Sons, New York).

(2) TIMOSHENKO, S. and GOODIER, N. *Theory of elasticity* 1951 (McGraw-Hill, New York).

C46/72

ULTRASONIC TECHNIQUES FOR REMOTE INSPECTION OF NUCLEAR REACTOR VESSELS

G. J. POSAKONY*

Remote ultrasonic inspection of nuclear reactor vessels is accomplished in a number of ways. The approach is dependent on access to the inspection surface. To inspect the vessel from the inside surface requires removal of the head and internals. To inspect the vessel from the outside surface requires a space between the vessel wall and thermal insulation for the test heads and the mechanical equipment. After establishing the vessel access there are various ultrasonic techniques, equipment, and instruments that can provide an accurate test. This paper describes techniques that are in current use.

INTRODUCTION

THE RADIATION ENVIRONMENT of an in-service nuclear reactor vessel precludes the use of non-destructive testing methods that require direct human access. For example, radiographic, magnetic particle, penetrant, and direct visual methods, used during the fabrication and construction phases, are not applicable to a vessel that has gone into service. For in-service conditions the present applicable non-destructive methods are limited to remote visual and remotely operated ultrasonic systems. This paper deals with the ultrasonic techniques in current use.

The selection of techniques for the ultrasonic non-destructive evaluation of nuclear reactor vessels is surprisingly complicated. The interrelationship of accessibility, inspection surfaces, mechanical scanning procedures, defect orientation, ultrasonic instrumentation, ultrasonic test heads, and the data acquisition system presents formidable engineering problems that must be satisfied. There is no single approach that is technologically superior. In fact, even the most accepted approaches require some compromise to fulfil the true inspection need.

INSPECTION PLAN

Irrespective of the approach used, the testing procedure must accurately identify the size, depth, and location of all ultrasonic reflectors that exceed predetermined levels. Further, it is most desirable to be able to characterize these ultrasonic reflections in terms of interface or defect orientation, type, shape, etc.

By its nature, ultrasonic energy will reflect from any impedance interface in its path. The source of reflections may be cracks, laminations, porosity, or other materials conditions at weld or clad boundaries. Since not every ultrasonic reflection represents a condition detrimental to the service or safety of the vessel, the interpretation of data in terms of defect character relative to the fracture analysis is mandatory.

At the present time the inspection plan followed is, first, to qualify the integrity of the reactor vessel through a careful non-destructive test programme (1)† during the fabrication and construction phases. Test programmes include the use of radiographic, ultrasonic, magnetic particle, penetrant, and visual methods. Once the vessel has passed the qualification tests and has been erected within the power station, the plan calls for a pre-service baseline inspection using equipment, techniques, and procedures that will have inspection continuity throughout the service life of the vessel. The baseline inspection provides a permanent record of the size, depth, and location of all ultrasonic information obtained from the structure. Since the vessel was previously qualified, the baseline information represents the reference base from which all subsequent test data will be compared. Volumetric tests of the reactor vessel are scheduled at periodic intervals throughout its life cycle. By comparing baseline data with data taken during service, a determination can be made concerning any defect growth or additional defects that may have developed.

The plan is fundamentally good, but it requires the development of basic mechanical equipment, electronic instrumentation, procedures, and data acquisition methods that can be accurately reproduced over an extended period of time.

REMOTE INSPECTION EQUIPMENT

A typical reactor vessel will range from 40 to 50 ft high, from 15 to 20 ft in diameter, and have an average wall thickness of 8–10 in (2). The weld areas that attach plates, sections, nozzles, piping, etc. together are the primary areas requiring inspection (3). Within the scope of this text it is not possible to cover the ultrasonic techniques for remotely inspecting each of the welds, but it is possible to select specific areas and amplify the procedures that could

The MS. of this paper was received at the Institution on 15th November 1971 and accepted for publication on 18th January 1972. 44

* *Automation Industries Inc., P.O. Box 950, Boulder, Colorado 80302, U.S.A.*

† *References are given in Appendix 46.1.*

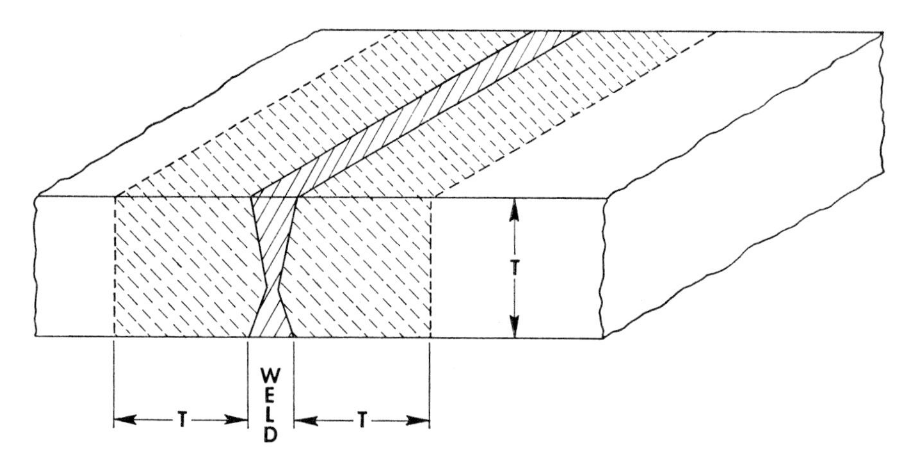

Fig. 46.1. Sketch of inspection zone

be used. Test techniques described are applicable to the primary welds in the wall of the vessel.

The key to remote ultrasonic inspection is the ultrasonic test head. Sound energy from a search unit or probe is highly directional, and to achieve adequate coverage it is common to use multiple search units in the test head. The remote handling and positioning equipment is designed to manoeuvre the test head in a programmed pattern that provides full coverage of the weld and the adjacent area. Each weld in the vessel is to be evaluated within the volume limits shown in Fig. 46.1. The sound beams must interrogate the weld from both sides as well as along the weld axis to ensure that potential defects lying in a variety of planes can be accurately described. The pattern of the sound beam is carefully selected to give full weld coverage from the multiple transducer test head.

Figs 46.2–46.4 are sketches of different types of remote positioning equipment being used for vessel inspections. Vessels can be successfully inspected from the inside or the outside surfaces, but the conditions of test are somewhat different. The technique has to be selected on some basic considerations such as access to outside or inside surfaces, location of interference supports which might limit the positioning of the test head, surface conditions, etc.

Fig. 46.2 illustrates a system used to inspect a vessel from the outside. This type of equipment relies on permanent tracks installed during construction. The test equipment is inserted at the time of inspection and positioned on the tracks. The scan programme is referenced to the track and the X–Y manoeuvrability of the test head. External scanning systems require between 9 and 18 in clearance between the vessel wall and the thermal barriers.

Fig. 46.3 shows a remotely propelled robot tractor vehicle which manoeuvres the test head over the inspection region. The tractor track is composed of permanent magnets which attach to the shell wall of the vessel. The vehicle is remotely controlled to give inspection coverage. About 9 in access space is required for this device.

The equipment shown in Fig. 46.4 is designed to test the vessel from the inside or clad surface. The internals must be removed to use the equipment. The equipment is a flange supported telescopic structure that positions and manoeuvres the test head in a prescribed pattern. Various fixtures are adapted to the horizontal or vertical booms to complete full surface and nozzle inspections.

The importance of the mechanical positioning equipment cannot be overstressed. Ultrasonic reflections from within the vessel must be located and mapped to accuracies of a fraction of an inch. The mechanical equipment must be capable of placing the test head in a precise location and must be sufficiently stable to obtain reproducible scans in both X and Y planes. Further, it must be capable of repeating the exact scans over both short- and long-term inspections. Position reproducibility within a fraction of an inch is required.

ULTRASONIC TECHNIQUES

The number of search units in the test head varies as a function of the type of test and the procedure used. In certain instances a single search unit is used while other test techniques require as many as 20 individual units.

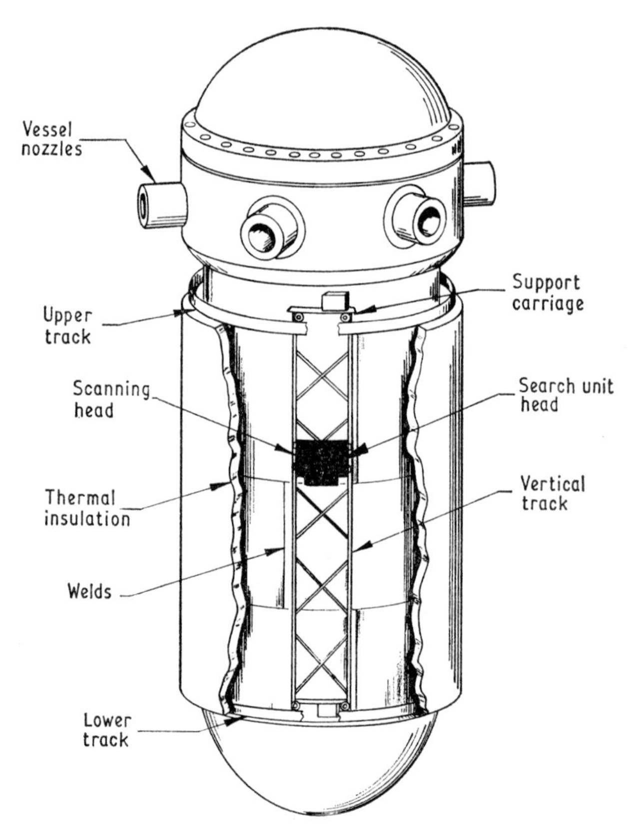

Fig. 46.2. External vessel inspection system

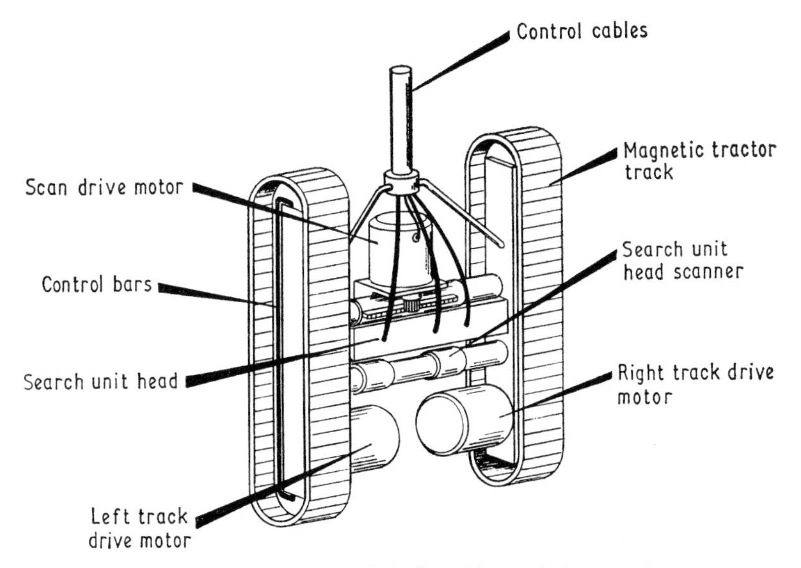

Fig. 46.3. Remote vehicle for ship weld inspection

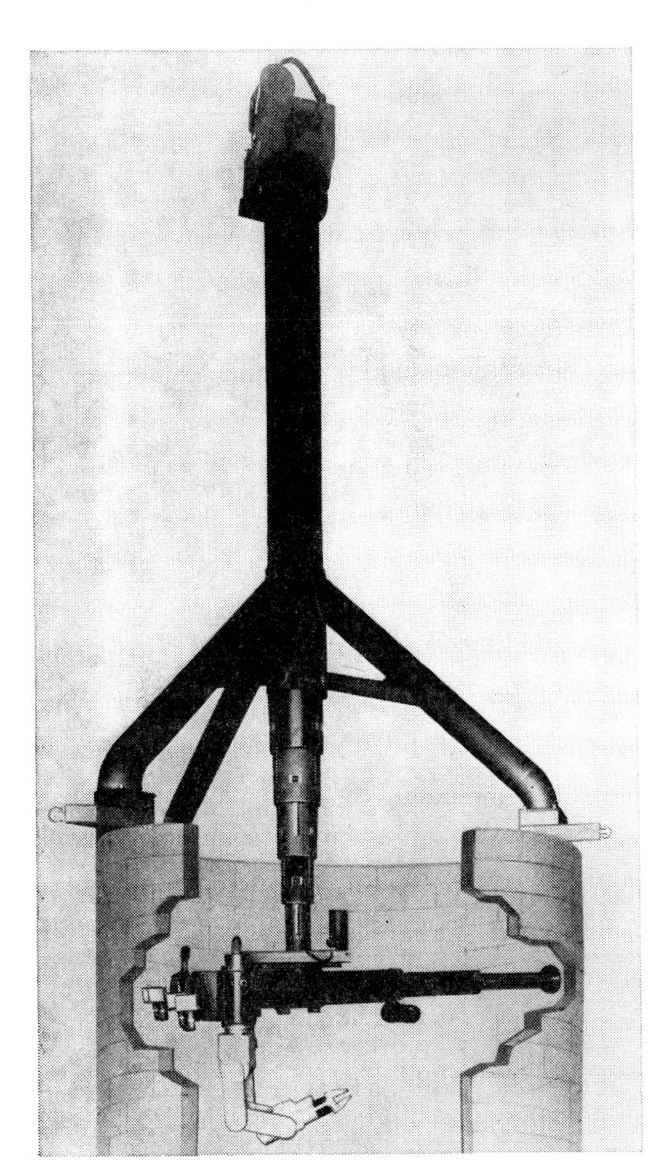

Fig. 46.4. Mechanical equipment for inside inspection

Each individual search unit directs a sound beam that is nominally 1 in (2·5 cm) in diameter. The size of the piezoelectric element determines the size and directivity of the sound beam. Elements larger than 1·5 in (3·75 cm) diameter have not proved practical. Broad coverage is achieved through multiple search units arranged in unique configurations.

There are a number of ultrasonic techniques employed and these may be integrated into one test head. These techniques include:

(1) Straight beam, pulse reflection (longitudinal wave).
(2) Angle beam, pulse reflection (longitudinal wave).
(3) Angle beam, pulse reflection (shear wave).
(4) Angle beam, pulse transmission (shear wave).
(5) Angle beam, pulse transmission, tandem technique (shear wave).
(6) Surface wave, pulse reflection.

Each of these test techniques has particular test capabilities that are well known.

Inspection from inside surface

Test techniques used for inspection from the internal surface of the reactor vessel are described in Figs 46.5–46.9.

Fig. 46.5 shows a three search unit test configuration.

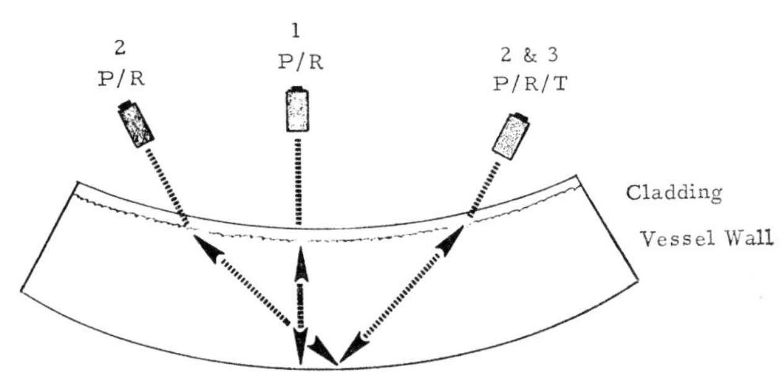

1. Straight beam, longitudinal wave, 0° pulse reflection.
2. Angle beam, shear wave, 45° pulse reflection.
3. Angle beam, shear wave, 45° pulse transmission.

Search units—Size: 1·0–1·5 in diameter.
Test frequency: 1·0 or 2·25 MHz.

Fig. 46.5. Test configuration No. 1

The three search units are a zero degree longitudinal wave used in pulse reflection, a 45° angle beam shear wave testing through the weld in pulse reflection from one direction, and a 45° angle beam shear wave testing the weld in the opposite direction. In addition the two angle beam units are used in pulse transmission in a full 'Vee' path. Defects are recorded from pulse reflection operation while loss of signal in the pulse transmission operation provides defect and couplant information. The search units in this head are fixed with the separation of the angle beam units dependent on the wall thickness. The test head is scanned to obtain full coverage of the weld zone. To scan along the axis of the weld the entire fixture is repositioned and the weld zone is rescanned. The test configuration shown in Fig. 46.6 represents an alternative type of three search unit test head (4). This head consists of a zero degree longitudinal wave and two 45° angle beam shear wave units. All units operate in the pulse reflection mode. The sound beams are designed to be incident over a small area, thus permitting the zero degree beam to be used for both flaw detection and couplant verification. The angle beam units are used in full 'Vee' to locate defects within the wall or on the o.d. or i.d. of the vessel. Weld coverage is a function of the scanning pattern but the data acquisition system is somewhat dependent on the pattern.

Fig. 46.7 shows a five search unit test configuration. This test head consists of the zero degree longitudinal wave unit and four angle beam shear wave units. All units are operated in pulse reflection. The zero degree longitudinal wave is used for defect detection and couplant verification. Two 45° angle beam units are used in half 'Vee' for internal defects as well as defects on the o.d. surface. Two units (typically 75°) are used for defect detection on the clad surface of the vessel.

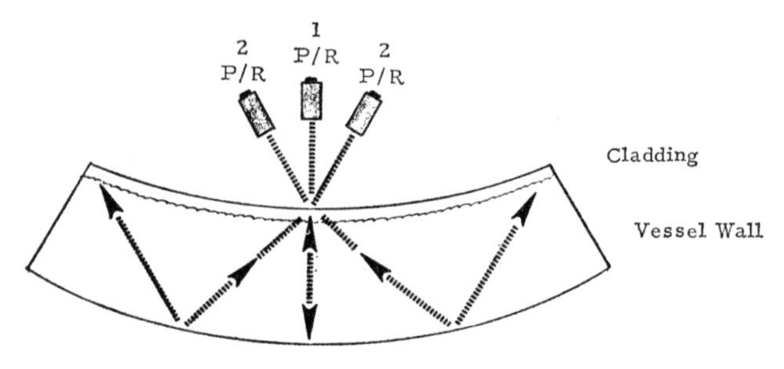

1. Straight beam, longitudinal wave, 0° pulse reflection.
2. Angle beam, shear wave, 45° pulse reflection.

Search units—Size: 1·0–1·5 in diameter.
Test frequency: 1·0 or 2·25 MHz.

Fig. 46.6. Test configuration No. 2

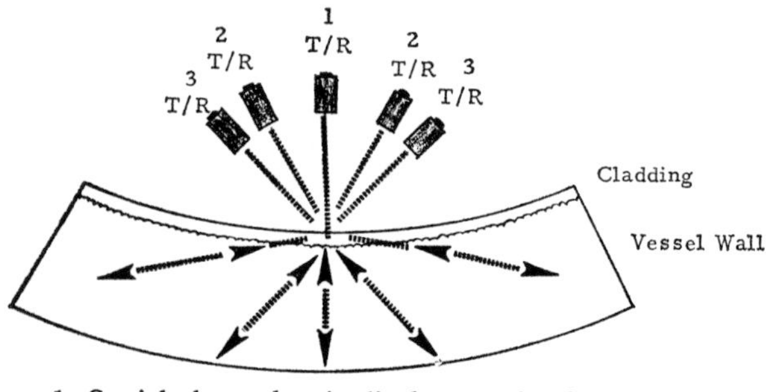

1. Straight beam, longitudinal wave, 0° pulse reflection.
2. Angle beam, shear wave, 45° pulse reflection.
3. High angle beam, shear wave, 70°–80° pulse reflection.

Search units—Size: 1·0–1·5 in diameter.
Test frequency: 1·0 or 2·25 MHz.

Fig. 46.7. Test configuration No. 3

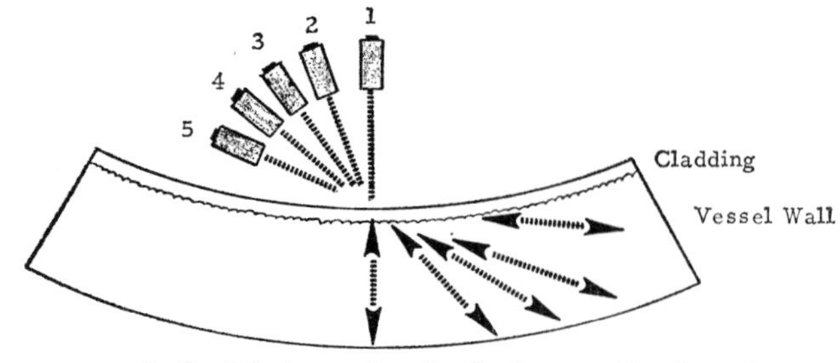

1. Straight beam, longitudinal wave, 0° pulse reflection.
2, 3, and 4. Angle beam, shear wave, 45°, 55°, and 65° pulse reflection, respectively.
5. Angle beam, surface wave, pulse reflection.

Search units—Size: 1·0–1·5 in diameter.
Test frequency: 1·0 or 2·25 MHz.

Fig. 46.8. Test configuration No. 4

Fig. 46.8 shows a five unit test configuration that generates a fan-shaped pattern. This is a modification of the test configuration described in Fig. 46.7. The test head consists of a zero degree longitudinal wave unit and four angle beam units operating at 45°, 55°, and 65° shear wave and one surface wave. All units are operated in pulse reflection. The advantage of this configuration is the ability to more accurately define defects which lie in random orientations within the wall of the vessel. The fan-shaped pattern is extended in only one direction. To achieve full coverage from both sides of the weld the fixture would have to be reversed or additional units would have to be added to the test head.

Fig. 46.9 shows a multiple search unit array used in one configuration of the tandem technique (5). This approach is unique in that it is a combination pulse reflection–pulse transmission test head. Defects within the vessel wall will reflect the sound energy in such a manner as to return energy to one or more search units. By using various combinations of units the depth position of the reflector can be determined. This array is applicable to inspection from either inside or outside surfaces.

Summary of inspection from inside surface

A summary of conditions that apply to inspection from the inside surface are listed below.

A. Surface flaws at or near the outside surface can be detected as follows:

1. Angle beam, shear wave, pulse reflection.
2. Angle beam, shear wave, pulse transmission, angle beam.
3. Shear wave, tandem technique, pulse transmission.
4. Straight beam, longitudinal wave, pulse reflection.

B. Base metal flaws in the wall thickness can be detected as follows:

1. Straight beam, longitudinal wave, pulse reflection.
2. Angle beam, shear wave, pulse reflection.
3. Angle beam, shear wave, pulse transmission.
4. Angle beam, shear wave, pulse transmission, tandem arrangement.

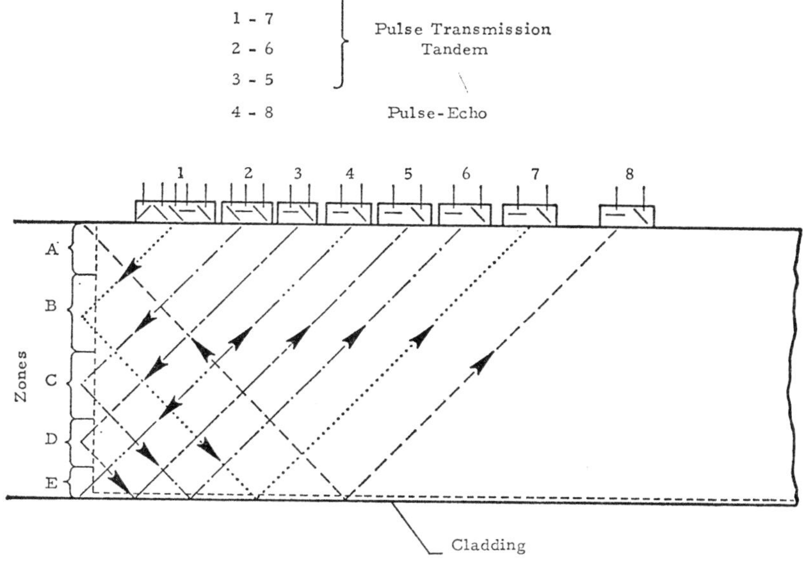

Fig. 46.9. Multisearch unit test configuration

C. Flaws at or near the clad base metal interface can be detected as follows:
   1. Straight beam, longitudinal wave, pulse reflection.
   2. High angle beam, shear wave, pulse reflection.
   3. Angle beam, shear wave, pulse reflection, full 'Vee' path.
   4. Angle beam, shear wave, pulse transmission.
   5. Angle beam, shear wave, pulse transmission, tandem arrangement.

D. Flaws in the clad metal on the inside surface, but not penetrating the base metal, can be detected as follows:
   1. Surface wave, pulse reflection.

E. Influence from the protective clad metal is chiefly loss by scatter of the ultrasonic energy caused by the irregular entry surface and the clad base metal interface. Surface grinding of the clad metal reduces the attenuation losses (as compared to non-clad surfaces) as noted from the following data at 1·0 MHz with a 1·5-in diameter search unit:

Clad condition	*Attenuation*
As deposited	8–10 dB
Ground flat	5–7 dB

The use of smaller size, higher frequency search units increases losses through the cladding. The following data at 1·0 MHz indicate the typical differences between a 1·5-in and a 0·75-in diameter search unit:

Diameter	*Attentuation*
1·5 in	8–10 dB
0·75 in	12–14 dB

Inspection from outside diameter

The test configurations that are applied to the outside inspection of the vessel are identical to those used for inside inspection. The techniques described in Figs 46.5–46.9 may be used for outside inspection. Depending on the tests, certain test angles may be modified slightly but the configurations are the same. The basic testing differences are caused by the location and influence of the cladding. On inside inspections, the cladding influences the sound beam as it enters the vessel. Once through the cladding the beam will follow the pattern prescribed by the search unit and the cladding. On outside inspection, the entry surface is smooth and the sound beam is influenced only by the cylindrical surface. However, if a full 'Vee' path is used for either pulse reflection or pulse transmission, the clad interface will affect the sound beam at the half 'Vee' position. This influence is a function of frequency. In certain instances the test frequency has been lowered to 0·4 MHz to reduce the cladding influence and maintain a uniform sound beam throughout the test distance. To compensate for cladding influence electronically, a step function electronic Distance Amplitude Correction (DAC) may be required. Whether a vessel is inspected from the inside or the outside, the problem of coupling is a primary consideration. Inspection from the inside surface when the vessel is filled with water reduces the coupling problem. However, the outside test or a dry inside test requires a positive couplant system with couplant verification to ensure the adequacy of the test.

Summary of inspection from outside surface

A summary of conditions that apply to inspection from the outside surface of the vessel is listed below.

A. Surface flaws at or near the outside surface can be detected as follows:
   1. High angle beam, shear wave, pulse reflection.
   2. Angle beam, shear wave, pulse transmission.
   3. Angle beam, shear wave, pulse transmission, tandem arrangement.
   4. Angle beam, shear wave, pulse reflection using a full 'Vee' path.

B. Base metal flaws in the wall thickness can be detected as follows:
   1. Straight beam, longitudinal wave, pulse reflection.

2. Angle beam, shear wave, pulse reflection.
3. Angle beam, shear wave, pulse transmission.
4. Angle beam shear, pulse transmission, tandem arrangement.

C. Flaws at or near clad base metal interface can be detected as follows:

1. Straight beam, pulse reflection (detection of clad unbonding).
2. Angle beam, shear wave, pulse reflection.
3. Angle beam, shear wave, pulse transmission.
4. Angle beam, shear wave, pulse transmission, tandem arrangement.

D. Flaws in the clad material or on the inside surface but not penetrating the base metal can be detected by the following technique:

1. Angle beam, shear wave, pulse reflection.

E. General effects of clad metal upon ultrasonic tests made from the outside surface involve undesirable ultrasonic noise and loss of sensitivity. Reflected signals from the clad metal interface during angle beam, shear wave, pulse reflection test at 1·0 and 2·25 MHz can exceed the primary reference sensitivity. Selective electronic gating is used to isolate the clad base metal interface noise at pulse frequencies of 1·0 and 2·25 MHz. When using the angle beam, shear wave, pulse reflector technique in the full 'Vee' path, sensitivity losses at the clad base metal interface have been significant. The following sensitivity losses have been typically observed in as-deposited cladding:

16–22 dB at 2·25 MHz

12–14 dB at 1·0 MHz

2–6 dB at 0·4 MHz

Test data recently obtained at 0·4 MHz for a 45° angle beam, shear wave, indicates that the attenuation effect of cladding can be minimized. Use of the lower frequency provides a usable ultrasonic response over the full 'Vee' path. Test data at 1·0 MHz and above indicate the need for step function electronic compensation circuitry to overcome the step attenuation losses at the clad interface.

Search unit selection

As previously noted, the selection of ultrasonic testing techniques is complicated. Five fundamental configurations have been listed as they apply to the inside and outside inspection procedures. There are others that could be used. Each configuration has some advantage in test procedure, ultrasonic instrumentation, data acquisition and mechanical equipment. With this complexity, the requirement for short- and long-term test continuity becomes paramount. Probably the most important component in the test equipment is the ultrasonic search unit. The initial selection and qualification of the search unit and procedures for requalifying these units involves exacting specifications. Search units can and do change with use, but by establishing fundamental performance requirements a unit can be requalified or its replacement can be qualified to duplicate the initial test. Test specifications for search units must involve frequency spectrum analysis, distance amplitude plots, and sound beam profile measurements.

Generally speaking, the ultrasonic search units should be as large as practical to provide a directional sound beam. As a minimum, the sound beam should be large enough to minimize the beam scatter resulting from the clad interfaces. At frequencies of 1·0 or 2·25 MHz, a 1-in (2·5-cm) diameter (or square) search unit provides a satisfactory beam.

CONCLUSIONS

Each nuclear reactor vessel design must be analysed to determine which of the applicable techniques will best satisfy the requirements for remote ultrasonic inspection. Access to the inspection surfaces is the first major consideration. A second consideration is the type of test head and the configuration of the search units within the test head. The mechanical scanning equipment, the ultrasonic and electronic instrumentation, and the approach to data acquisition are to a large degree dependent on these considerations.

The key to the inspection is the test head. The test head must scan in a programmed pattern to achieve full coverage of the inspection zone. The search units within the test head may be arranged in a number of configurations and still provide a good test. However, regardless of the pattern, the object is to provide test data that will describe the size, depth, and location of defects within the vessel.

The ultrasonic test head configuration and testing techniques described in the text are representative of current state of the art. Each configuration has certain advantages that involve the mechanical scanning equipment, test instrumentation, or data acquisition system. From the experience to date, no single test technique has been established as superior to others described. Any test approach must be examined from the overall system and not from individual components.

A major system consideration is the method used for data acquisition and data interpretation. Several major companies are using fast scanning mechanical systems (1–4 in/s) and computers to analyse the data. The more complicated the sound beam patterns, the higher the cost of the ultrasonic equipment and the computer software programme required to handle the data.

All systems are a compromise in the economics of testing. A rule of thumb in developing instrumentation for reactor testing is: analyse the data acquisition, storage, retrieval and interpretation, and build the testing system to suit.

APPENDIX 46.1

REFERENCES

(1) McMaster, R. C. (ed.). *Nondestructive testing handbook* 1959 (The Ronald Press Co., New York).

(2) *A.S.M.E. boiler and pressure vessel code*, Section III, 'Nuclear vessels', 1968.

(3) *Welding inspection* 1968 (American Welding Society Inc., New York).

(4) McGaughey, W. C. 'Ultrasonic investigation of heavy (6″ thick) carbon steel with stainless steel weld overlay', 1971 (March) (Automation Industries, Inc.).

(5) Meyer, H. J. 'Aspects of in-service inspections on reactor pressure vessels in Germany', *Materials Evaluation* 1971 (August).

BIBLIOGRAPHY

CROSS, B. T. 'Sound beam directivity: a frequency dependent variable', TR 70–23, 1970 (April) (Automation Industries, Inc.).

FILPCZYNSKI, L., PAWLOWSKI, Z. and WHER, J. *Ultrasonic methods of testing materials* 1966 (Butterworth).

GROSS, L. B. AND JOHNSON, C. R. 'In-service inspection of nuclear reactor vessels using an automated ultrasonic method', *Materials Evaluation* 1970 **28** (July).

HANNAH, K. J. 'Ultrasonic technique selection for in-service inspection of nuclear reactor pressure vessels', TR 71–3, EEI Project RP 79, 1971 (September)(Automation Industries Inc.).

HISLOP, J. D. 'Flaw size evaluation in immersed ultrasonic testing', *Nondestructive Testing* 1969 (August).

KRAUTKRAMER, J. H. *Ultrasonic testing of materials* 1969 (Springer-Verlag, New York).

DE STERKE, A. 'Automation in nondestructive testing of welds', Paper 15, *Sixth World Petroleum Congress*, The Netherlands, 1963.

C47/72 IN-SERVICE INSPECTION TOOL FOR NUCLEAR REACTOR VESSELS

H. W. KELLER* D. C. BURNS* T. R. MURRAY*

A complete description is offered of the remote vessel inspection tool developed by Westinghouse to examine the areas required to be inspected under Section XI of the A.S.M.E. Boiler and Pressure Vessel Code. Included in the paper are the design objectives, the selection of the tool concept, and also structural, mechanical, and electronic details. In addition, operational facets, including shipping, storage, and assembly, are discussed.

INTRODUCTION

SINCE THE reactor vessel is the most inaccessible and most radioactive area requiring periodic in-service volumetric inspection, the success of an in-service inspection programme depends to a great extent on the ability to perform this inspection remotely. For many years the problem of tooling to perform in-service inspection of reactor vessels has been under study. The objectives in this study, other than satisfying impending A.S.M.E. Boiler and Pressure Vessel Code inspection requirements, were to develop a tool that would allow the most meaningful non-destructive examination practicable, and to provide the tool with the flexibility to accommodate future technique developments in data acquisition and interpretation.

It was known that these goals would be difficult to achieve in view of the type of measurements to be performed, the adverse environment in which the tool must operate, and the remoteness of the operations. A tool has been designed to meet these needs and is presently in manufacture and test; it will be prepared for use in reactor vessel inspections starting in 1972. The tool was designed specifically for use in Westinghouse pressurized water reactors, but the concept can readily be adapted to accommodate other reactor designs.

This paper discusses the considerations regarding the selection of the tool concept, and the design and operation of the tool. The related operations of shipping, storage, assembly, and adaptability to particular plant and reactor conditions are also discussed.

CONSIDERATIONS IN SELECTING A TOOLING DESIGN CONCEPT

The primary consideration in selecting the tooling design concept was compliance with the A.S.M.E. Code requirements of volumetric examination categories, frequency, and characterization of data. Similar importance is assigned to the preference of performing the inspection from inside the reactor vessel rather than from the outside. This preference results from the greater accessibility of the inside once the reactor internals are removed.

The MS. of this paper was received at the Institution on 26th October 1971 and accepted for publication on 18th January 1972. 33

* *Westinghouse Electric Corporation, Nuclear Energy Systems, PWR-SD Nuclear Services Dept, P.O. Box 355, Pittsburgh, Pennsylvania 15230, U.S.A.*

However, the major concern in selecting a tool concept was compatibility of the tooling with the U.T. transducer configuration to ensure effective and efficient search, detection, and interpretation of signal. This compatibility must provide assurance of three functional requirements:

(1) The transducers must always be a uniform distance from the vessel surface.

(2) They must be arranged in an array to provide effective coverage.

(3) The position control must be accurate. The remote aspects of this operation make these considerations extremely critical.

The establishment of acceptance criteria for such design requirements was difficult if not impossible at the start due to the many uncertainties and inexperiences of performing remote ultrasonic inspection. Consequently, a programme of ultrasonic technique development was initiated in parallel with the tool development. Such problems as size of transducer, transducer arrays, effect of vessel cladding, water temperatures, etc., were included in that investigation.

To accommodate these problems it was decided to build into the tool:

(1) Maximum application flexibility in accommodating variations in transducer array and operation.

(2) Maximum positioning control within practical limits in accommodating reproducibility of measurements and studying flaw indications which may require continued surveillance. (In the event that a flaw is detected during the examination, it is very important to define the flaw's size, orientation, and location accurately. The tool must therefore be accurately located and structurally stable during operation.)

Along with these salient requirements the following are also important:

(1) Compatibility of tooling and equipment with present plant equipment, to provide minimum interference with normal refuelling and maintenance operations.

(2) Reliable remote operations in an underwater and radioactive environment.

(3) Built-in check points to verify proper orientation.

(4) Compatibility with the most sophisticated control and data acquisition and interpretation equipment. Obviously the most important item of the inspection is the handling of information gathered during the tool operation. Therefore it is beneficial to display information in an easily read manner, and in a form that requires little interpretation. The gathered information can then be recorded for use in interpretation for analysis. The system should be initially designed for manually recording with the capability of adapting to computer processing and automatic print-out in the future, if deemed necessary.

(5) Adaptability to various sizes of reactor vessels, and the capacity to accommodate specific differences in certain vessel and internals design features.

(6) Adaptability to existing operating plants (not specifically designed to accommodate A.S.M.E. Code, Section XI), as well as plants under construction.

(7) Mobility for ease in transfer from one plant to another and within a plant.

(8) Adaptability to decontamination and/or radioactive shipments.

SELECTION OF TOOL DESIGN CONCEPT

After evaluating the considerations described above, the concept illustrated in Fig. 47.1 was adopted. The figure shows the nozzle–flange scanning (mode 1) and vessel shell scanning (mode 2) operations.

It is important to note the portions of the tool that remain stationary and those that move during an inspection. These features are essentially the bases for satisfying positioning control and dynamic stability requirements with relatively simple structures. The primary structural components of the tool are the head assembly and column assembly. Both remain stationary. Attached to these are the various scanning assemblies for the mode of operation desired. In mode 1 the nozzle scanner and flange scanner assemblies are attached. In mode 2, the vessel shell scanners are attached. The flange scanner can also be used in this mode.

Description of basic structural concept

To describe the significance of the various features of the tool concept illustrated, it is appropriate to examine them in the light of the basic considerations discussed previously.

For positioning, the tool uses the reactor vessel internals support flange as a reference for all modes of operation. This reference is important for a number of reasons. It is a machined reference for all dimensions in the vessel and reactor internals. Consequently, all dimensions of interest in an inspection can be precisely related to it; it can be utilized during normal refuelling operations; and it is an integral part of the reactor vessel and not subjected to variations with time. Reference from one inspection period to the next is therefore constant.

Accuracy in position control is achieved by the structural stability of the tool. This stability is gained by minimizing the length of the structure and the number of sliding fits. A rigid head and column structure is employed rather than long and large telescoping columns, which could be sources of large errors. Dynamic stability is achieved by avoiding rotation of large masses over large distances. (For example, the long centre column is not rotated. If it were, the slightest dynamic instabilities would be amplified at the transducer mounting and result

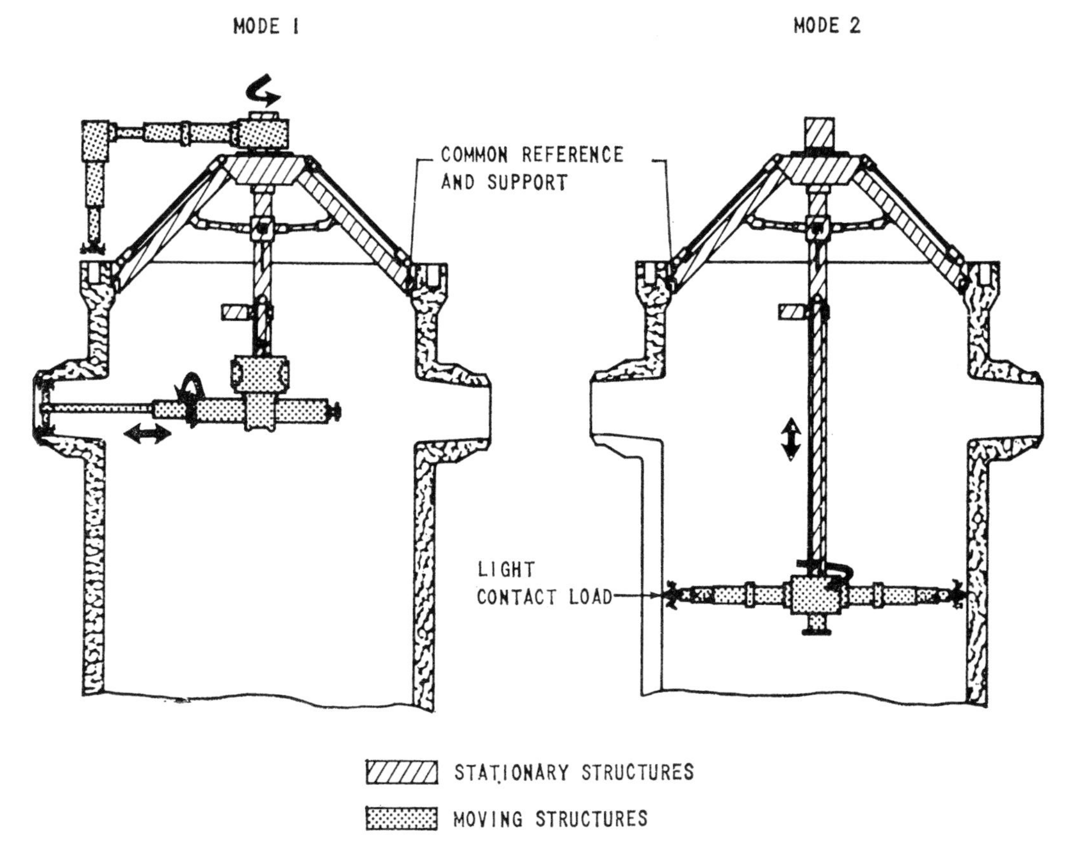

Fig. 47.1. Basic tool concept

in large errors in position indication.) In addition, structural stability does not depend on contact of the vessel scanner on the vessel wall; rather, the scanner loads against the vessel wall with a very light force to ensure contact for purposes of transducer position reference. Dynamic instabilities related to variations in cladding surface roughness and friction effects are therefore minimized.

Compatibility with normal refuelling operations is possible in mode 1 configuration examination of the outlet nozzles without removing the reactor lower internals. Both the nozzle and flange scanning operations can be performed in one equipment set-up, thereby minimizing interference with refuelling schedule.

Adaptability for various vessel sizes is possible because of the umbrella-like operation of the head assembly and varying the length of the column (Fig. 47.1). To adapt the head assembly to larger or smaller vessel sizes the umbrella-like section is opened or closed accordingly. To adapt to varying length, the column assembly is changed by adding or removing sections. To adapt the scanner assemblies, spacers are installed to accommodate ranges of sizes beyond the normal adjustments in the tool. These features make the tool nearly universally adaptable.

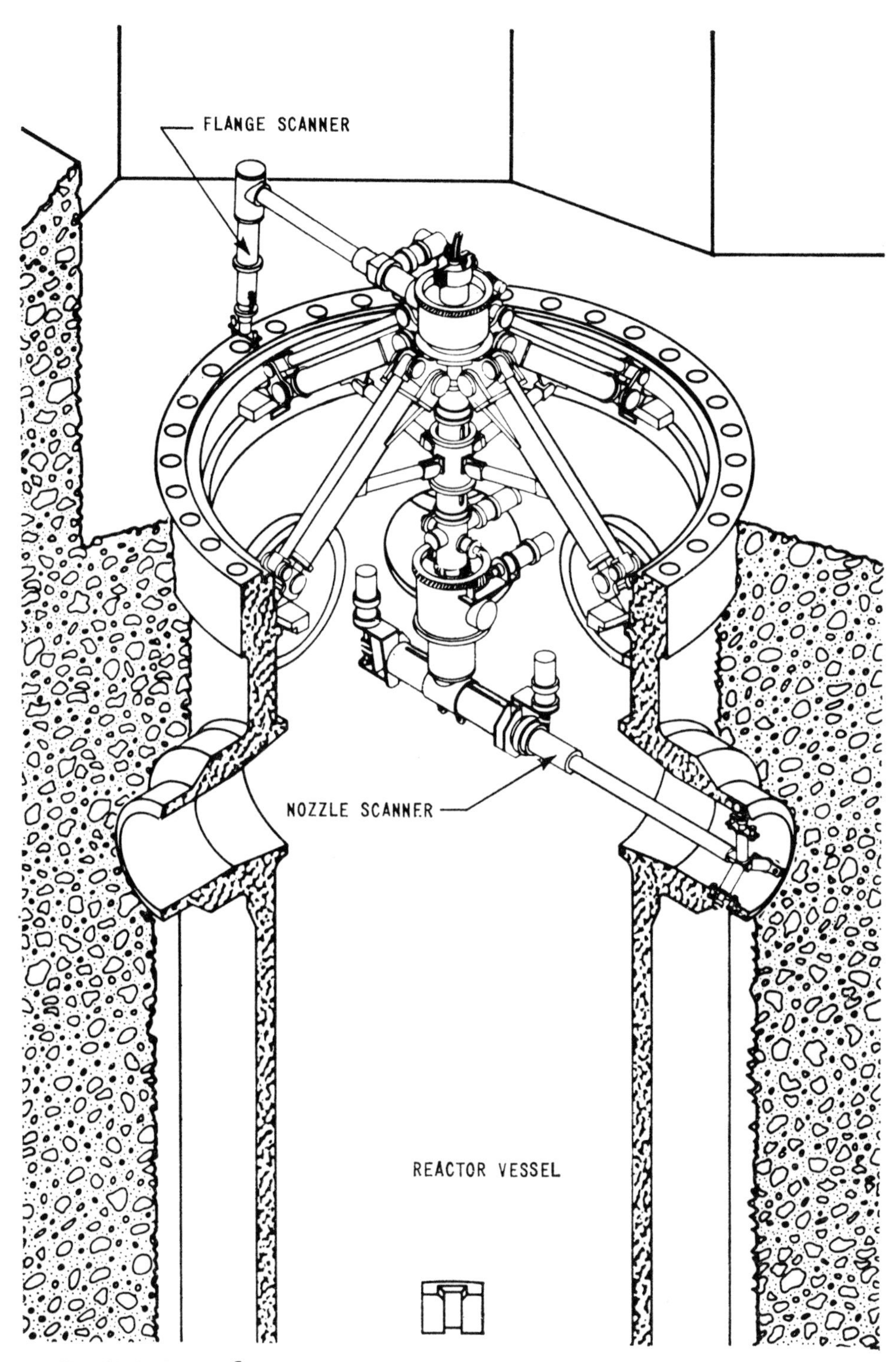

Fig. 47.2. General arrangement of tool in vessel (nozzle and flange scanner)

Drive mechanism

Drive mechanisms provide controlled motion to the scanner assemblies for the ultrasonic measurements. Pneumatic, hydraulic, and electric drives were considered for this operation. Electric drives, though the least attractive from the standpoint of operation under water, were selected for the drive mechanisms because (*a*) speed control is readily adjustable, (*b*) easy adaptability is afforded to the position indication and speed control system, and (*c*) the sizes of power lines and problems of running them into the vessel are minimized. The electric motor operation underwater is made reliable by the use of special seals.

Although rejected for control of the drive mechanisms, pneumatic and hydraulic drives are used where position control and speed accuracy are not essential. These mechanisms are used for actuating the head assembly arms and scanner arms during installation and removal of the tooling.

Position indication

The basic requirement of a position indication system is that it provides information as accurate as possible on the location of the transducer heads. Since an electrical drive system is used, encoders were an easy selection for the position indication. The encoder selected was an absolute optical type, which is coupled directly to the drive mechanisms. This combination ensures positive and absolute indication at all times. Such a system is readily adapted to automatic control of the electric motors.

Data acquisition, readout, and interpretation

The data acquisition system could vary in degree of sophistication depending on the amount of data to be processed and stored. Therefore our objective in selecting the tool concept was that it must be compatible with both the simplest and most complex plotting system. Hence, the integrated electric-motor drive and position indication system was selected.

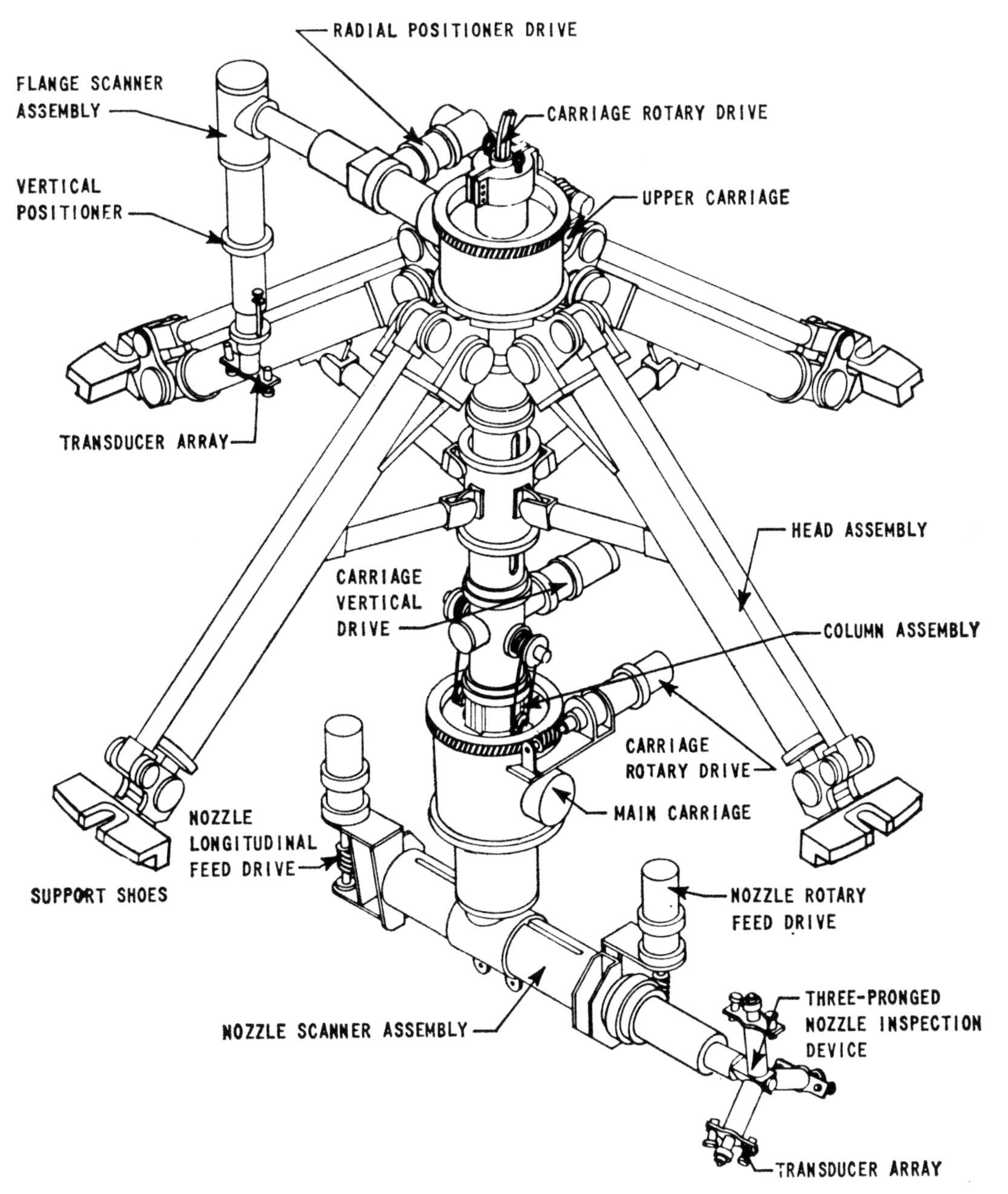

Fig. 47.3. Tool details (nozzle and flange scanner)

DESCRIPTION OF TOOL

Now that the basic tool concept has been discussed, more detail will be given on the design and operation features. The tool is designed to perform inspection in the categories of: primary nozzles to safe-end welds, primary nozzle-to-vessel welds and nozzle-to-vessel inside radiused section, ligaments between threaded stud holes, vessel-to-flange weld, and longitudinal and circumferential shell welds.

These categories cover all areas that are not readily performed manually. The first two categories are accomplished with a 'nozzle scanner', the second two categories with a 'flange scanner', and the last category with a 'vessel shell scanner'. Both the mode 1 (nozzle and flange examination) and the mode 2 (vessel shell examination) tool configurations consist of a head, a column, and the appropriate scanner assemblies.

Mode 1 configuration

Figs 47.2 and 47.3 show the mode 1 configuration with nozzle scanner and flange scanner. The head assembly and column assemblies are the primary structural components. These structures are the foundation for the various mechanisms and other related structures of the tool. The head assembly consists of a head weldment, arm assemblies, and an arm actuator assembly. The head weldment contains the pivot points for the four arms, which are attached to the weldment at 90° positions by a shaft that pivots on tapered roller bearings. The opposite ends of the arms contain the shoes that contact the vessel and engage the keyways for positive engagement. A rod attached to the shoe and head weldment ensures that the seating surface of each shoe is raised and lowered in unison and remain parallel to each other. In this manner, various vessel diameters can be accommodated. The shoes are designed specifically to fit the vessel being inspected and can be easily changed to fit other vessels if necessary.

The arm actuator mechanism is contained inside the head assembly weldment. The mechanism is a hydraulically operated system that performs the raising and lowering of the four arms by interconnection of the four short tie rods. Water is the hydraulic fluid. The entire system functions similarly to an ordinary umbrella—raising and lowering of the outer cylinder causes the four arms to raise and lower. When the head assembly is installed in the vessel the unit becomes self-locking and holds itself in place.

The basic design of this assembly will accommodate vessels ranging from 120 in to 180 in in diameter. Spacers are used to accommodate larger sizes of vessels.

Basically the column assembly is a centre-pole structure with an adjustable length. The columns contain three tracks, which are used to guide the lower carriage assembly, and a rack, which is used for vertical position indication of the carriage assembly. All the scanning drive systems attach to the column assembly either at the upper carriage assembly or main carriage assembly.

The main carriage assembly consists of an inner carriage and an outer carriage. The inner carriage is attached to the tracks on the column assembly and is raised and lowered by the wire rope system of the vertical drive assembly. The inner carriage contains six sets of linear ballways for accurate vertical tracking of the column, an encoder that engages with the rack of the column for vertical position indication, a worm wheel, and the mounts for attaching the outer carriage. The outer carriage is rotated about the inner carriage by means of an electric drive motor and worm gear.

The vertical drive assembly is attached to the head assembly. The drive is a worm-gear box with an integrally attached electric-drive motor that drives two drums. The drums wind and unwind dual wire ropes through pulleys to raise and lower the carriage assembly. The motor contains its own braking system to hold the gears in place, but an additional pneumatic brake is supplied for safety reasons. This motor is speed adjustable to obtain the necessary scan speeds. The main carriage assembly is raised or lowered to provide a vernier elevation adjustment for the nozzle scanner (mode 1), or to raise and lower the main carriage assembly throughout the column length for vessel scanning (mode 2).

The nozzle scanner is a telescoping-tube assembly capable of travelling 82 in. This assembly bolts to the main carriage assembly by means of a coupling and large clamp device. To accommodate various vessel diameters the carriage clamp is loosened and the scanner assembly is physically moved to a set-up dimension. The nozzle scanner has three primary mechanisms: a longitudinal feed mechanism, a rotary feed mechanism, and a three-pronged nozzle inspection device.

The longitudinal feed mechanism consists of a telescoping-tube assembly, which extends the nozzle inspection device into the reactor vessel nozzle. The motion is provided by an Acme screw-thread, which is driven by a worm-gear set. Coupled to the worm is an electric motor complete with an integrally mounted encoder and brake. This mechanism can position the nozzle inspection devices at any selected point inside the nozzle as far as the safe end welds. Once the devices are inside the nozzle, a rotary feed mechanism rotates the inspection device at the longitudinal setting. The rotation is again provided by a worm-gear set and an electric-drive motor with the integrally mounted encoder and brake.

The device that enters the nozzle consists of a three-pronged arm assembly, transducer mounting plate, and an arm-retracting system. Each individual arm is a spring-loaded telescoping assembly that contains a ball at one end for contacting the inside diameter of the nozzle. The arms are retracted to enter the nozzle by means of an actuator and cable drive. Once the longitudinal feed mechanism locates the nozzle-scanner prongs in the nozzle, the spring-loaded arms are extended out to touch the inside of the nozzle. Any ovality of the nozzle is readily accommodated by the compression springs. The transducers are held at a fixed distance from the nozzle surface by contact of the arms. The scan is accomplished by rotating the arm assembly one revolution, indexing with the longitudinal mechanism, and then rotating in reverse. The arm assemblies fit either straight or tapered nozzles varying in inside diameters from 27 in to 36 in.

It is possible to manually extract the nozzle scanner mechanism in the unlikely event that the longitudinal drive system malfunctions. The flange scanner assembly attaches to the column through the head assembly. This mechanism inspects the ligament area and the flange-to-vessel weld area. It is composed of an upper carriage, radial positioner, and vertical positioner. The upper

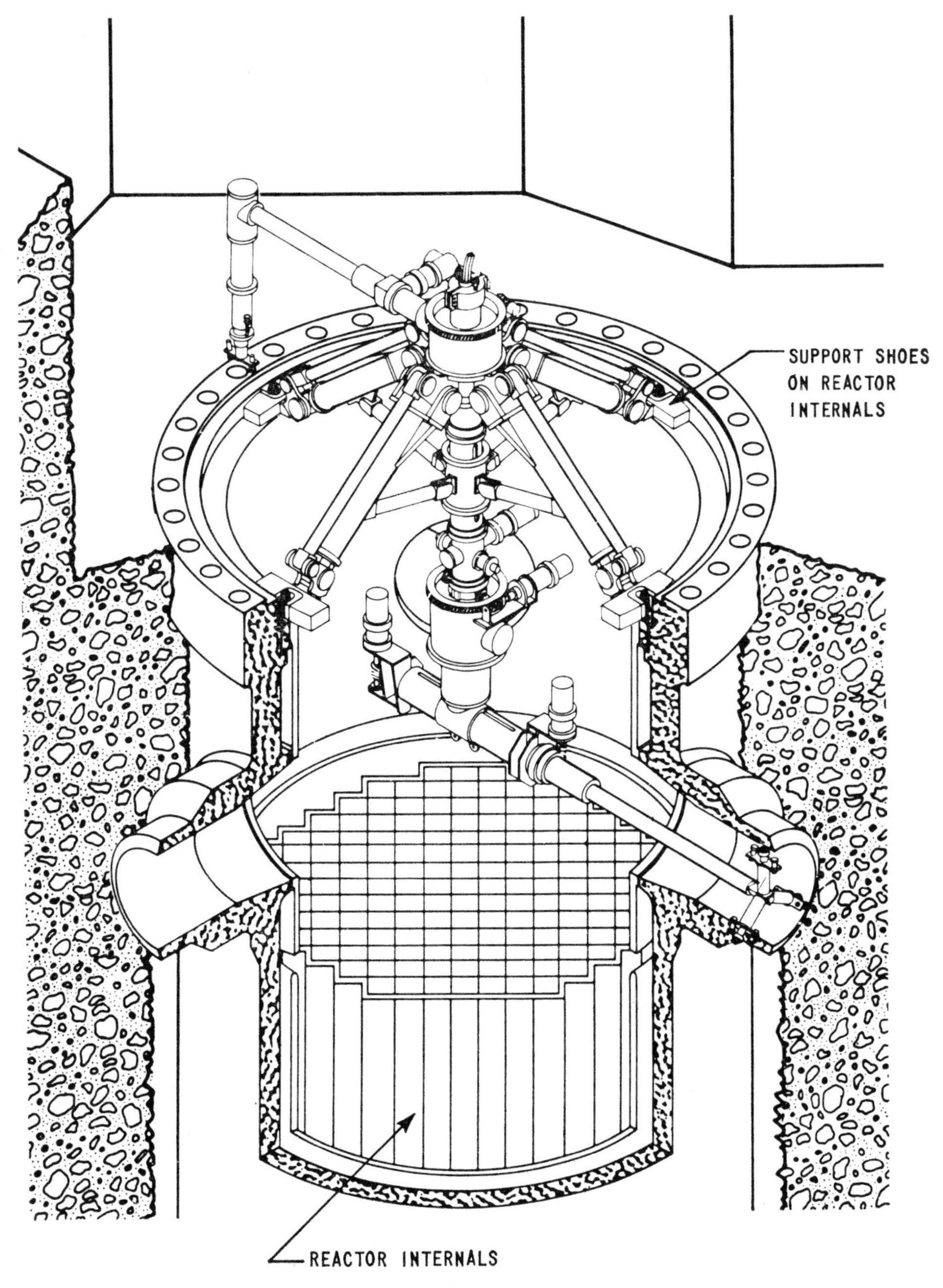

Fig. 47.4. General arrangement of tool in vessel (nozzle and flange scanner) (showing internals)

carriage is attached to the column above the head assembly and rotates about the column centre-line carrying with it the other equipment. The rotation is provided by a drive system identical to the main carriage drive system, i.e. worm-gear set, electric motor, encoder, and brake.

The radial positioner is a mechanism that moves the vertical positioner on a radial path in steps. Its construction is similar to that of the nozzle scanner but its operating distance is only 21 in. It is driven by a worm-gear set and electric motor with an encoder attached for position indication. A spacer is attached between the upper carriage and the radial positioner to accommodate various vessel sizes.

The vertical positioner is used to mount the transducers, and to move the transducers to a fixed distance from the vessel flange. Its construction utilizes a telescoping-tube concept with the telescoping motion supplied by pneumatically operated cylinders similar to the vessel scanner arms. Fixed stops are adjusted for various vessel sizes to maintain a fixed distance from the transducer to the flanged area.

Mode 1 configuration with reactor internals in place

Fig. 47.4 shows the arrangement for nozzle scanning and flange scanning with the reactor internals in place during a normal refuelling. In this configuration, special shoes are used on the arms of the head assembly to allow for tool positioning on the guide pins of the reactor internals rather than on the keyways of the vessel. Only the outlet nozzles can be reached through the nozzle openings in the internals.

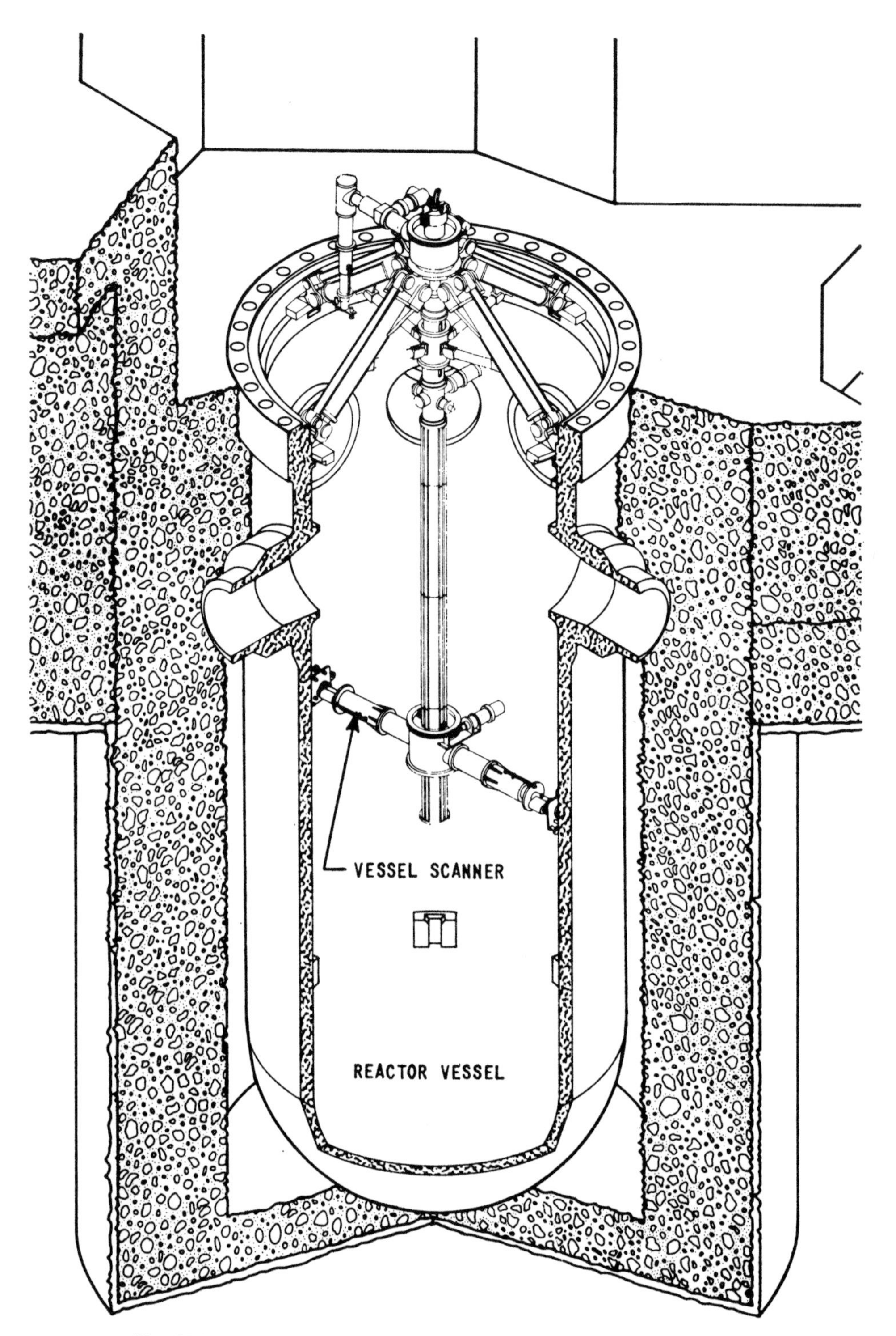

Fig. 47.5. General arrangement of tool in vessel (vessel scanner)

Mode 2 configuration

Fig. 47.5 shows the mode 2 assembly for scanning the inside diameter of the vessel. The head assembly is the same as for mode 1, but the column assembly is extended to a length suitable to the particular vessel being inspected. The column length is adjusted by adding sections of up to 10-ft increments, as required. The sections are internally piloted and bolted together. The main carriage assembly can traverse over the length of the column since sufficient cable is contained on the drums to accommodate vessel lengths up to 370 in. The vessel scanner assemblies attach to the main carriage. One of these assemblies is sufficient, but two can be used, as shown, for ultrasonic flexibility. More details of this assembly are illustrated in Fig. 47.6.

The vessel scanner assembly consists of a telescoping extension arm and a roller head assembly with a transducer mounting plate. After the tool is positioned, the extension arms are extended by means of pneumatic cylinders with fixed stops, adjusted to accommodate the vessel size being inspected. The end of the arm contains the spring-loaded roller head assembly, which contacts

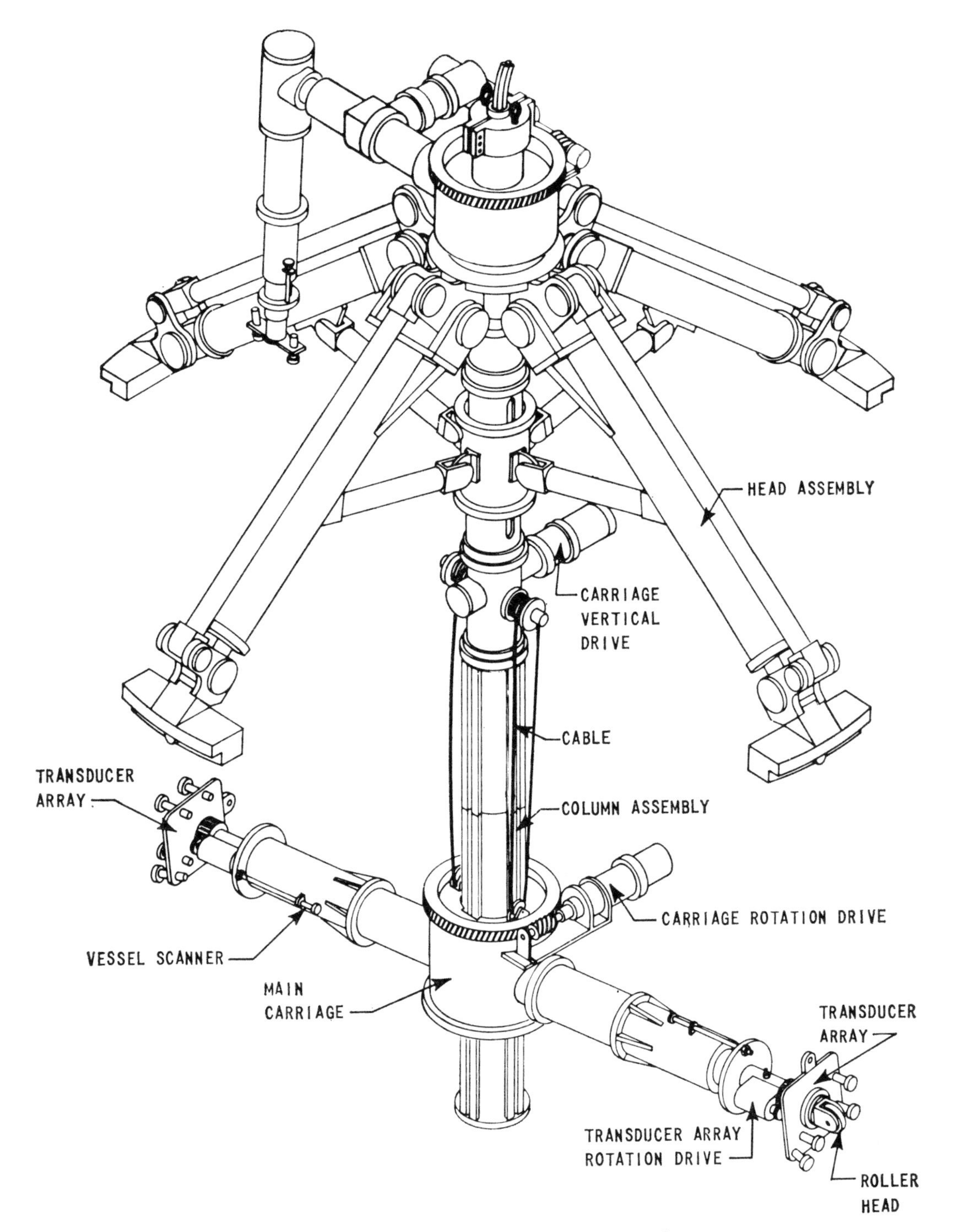

Fig. 47.6. Tool details (vessel scanner)

the vessel wall. Thus the only load applied against the vessel is the bias present in the spring, which is in the range of 5 lb. The roller contacting the vessel is coated with a non-marring material.

Attached to the roller assembly is the transducer mounting plate. The transducers are held at a fixed distance from the metal being inspected by virtue of the floating action supplied by the spring-loaded roller assembly. Therefore, any vessel inside diameter variations, such as ovality, do not affect the predetermined fixed water path for the ultrasonics transducers.

The transducer mounting plate is held in place by two pneumatically operated cylinders. It is possible to remove the mounting plate with the transducers and recalibrate or replace it without removing the tool. The head assembly with transducer mounting plate can be rotated in 90° increments by the operation of a small direct-coupled electric motor. This procedure permits the transducer search pattern to be varied.

Materials of construction

The tool is constructed mainly of stainless steel and structural aluminium. Stainless steel is used where close tolerances and potential wear conditions exist; aluminium for the basic structural components. The gears and high-duty bearings are stainless steel. Some bearings in locations of infrequent use, such as for the static structures, are carbon steel or bronze.

The lubricant is a silicon grease. Gearbox assemblies and certain bearing assemblies are sealed to contain the grease and keep out dirt.

Transducer heads

The tool employs a number of transducer head designs to accommodate varying reactor geometries and weld configurations. Mounts for transducer arrays are built to allow operation at various orientations. The transducer mounting plate is machined with transducer holes drilled at fixed angles. Up to five transducers are mounted in an array on each mounting plate. The major search units employ a $1\frac{1}{2}$-in diameter transducer driven at 2·25 mHz.

The transducer utilizes a high sensitivity, readily damped piezo-ceramic material, which has been tested for pressure sensitivity, thermal and gamma degradation, and is capable of operation in the hostile reactor environment.

Motor drive control

The electric-drive motors use a unique digital control scheme, which permits accurate stopping at a preselected

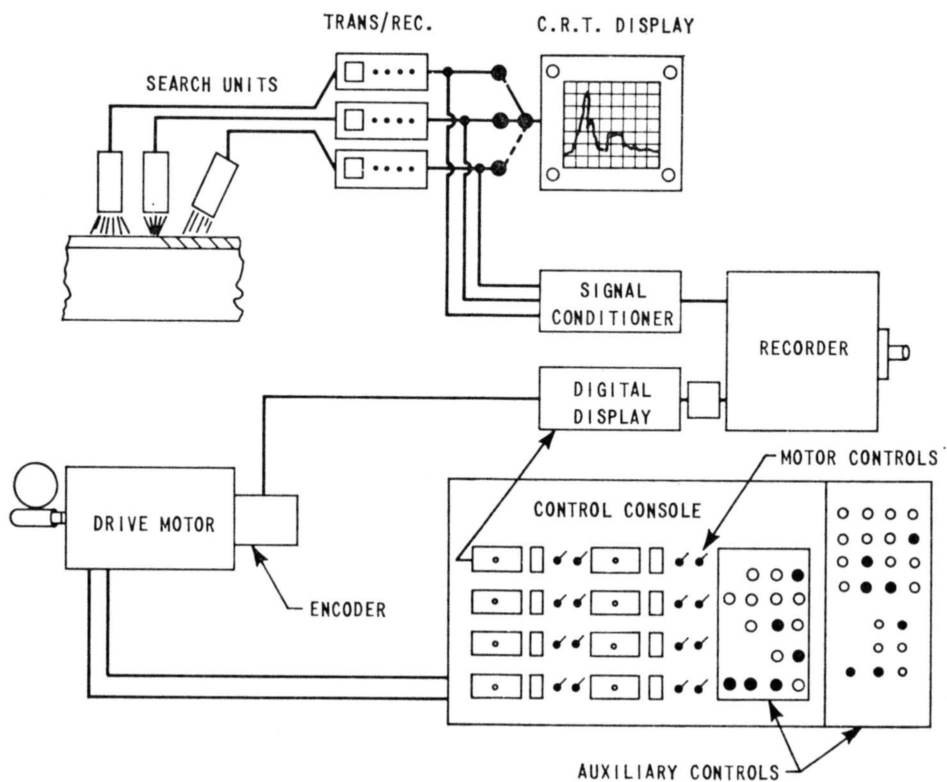

Fig. 47.7. Data acquisition equipment

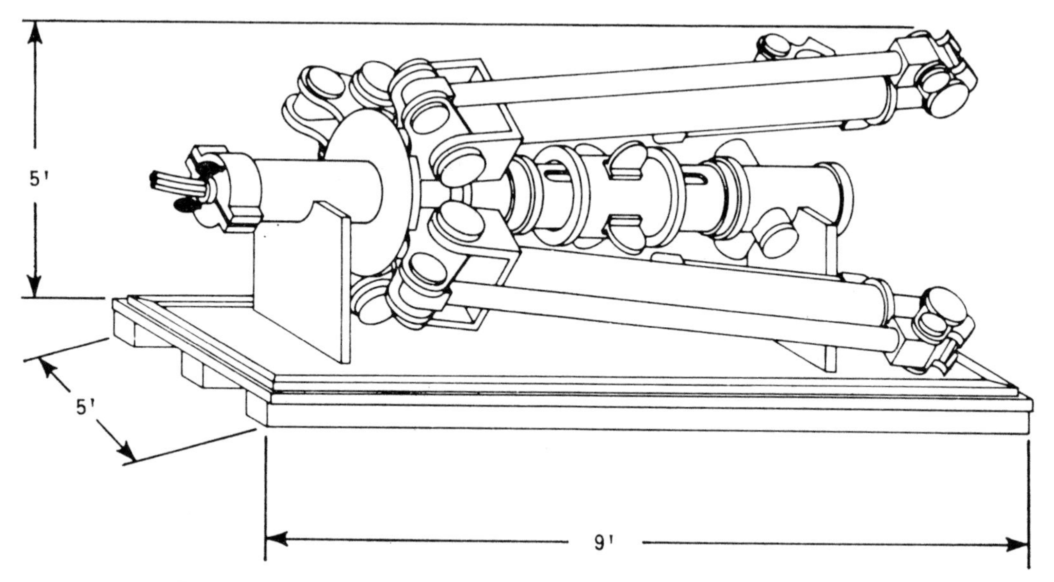

Fig. 47.8. Shipping and handling configuration for head assembly

spot or at the limits of a scanning range. Manual speed controls permit speed adjustment of each transducer head assembly.

Fail-safe (spring-operated) electric brakes hold motors when they are stopped. The electric system includes underwater cables and connectors, load current ammeters to warn of increased friction, and motor thermal protectors to avert possible burnouts.

Data acquisition system

The data acquisition system used for the vessel inspection tool combines simplicity with operational reliability and adaptability. As presently contemplated, it employs an electronic system with a receiver or data channel for each transducer unit. The signal received from the transducer is transmitted through an electronic distance amplitude correction device and is then gated to initiate digital position recording. A schematic of the data acquisition system is illustrated in Fig. 47.7.

TOOL SHIPMENT, ASSEMBLY, AND INSTALLATION

The tool and related components break down into assemblies that can readily be shipped and moved into the

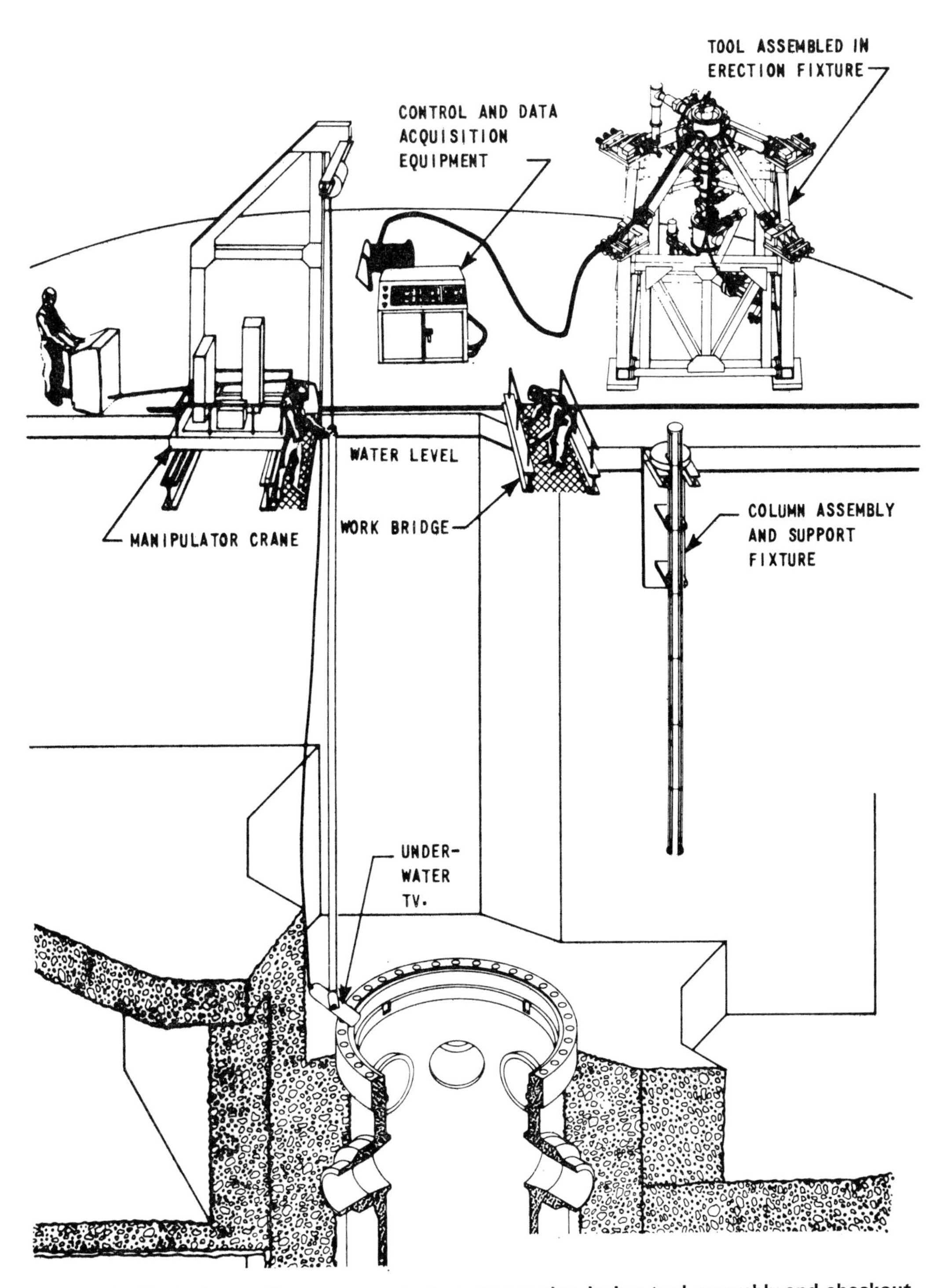

Fig. 47.9. Typical overall arrangement of reactor cavity during tool assembly and checkout

plant containment. Once inside the containment the components are assembled and operationally checked on an erection stand.

Shipment and handling

The equipment is shipped in about ten containers. All the equipment in the normal shipping mode will pass through personnel hatches, with the possible exception of one container, which houses the tool head assembly. In some cases this head assembly will need to be partially disassembled to gain entry through the personnel hatch if the equipment hatch is not opened. The largest container housing the retracted head assembly is illustrated in Fig. 47.8.

Assembly (mode 1)

The overall arrangement in the reactor cavity during tool assembly and checkout is illustrated in Fig. 47.9. The first thing assembled is the erection fixture. This is normally assembled in a work area adjacent to the reactor cavity. If floor space is not available, the erection fixture can be assembled on a platform or bridge installed over the reactor cavity. The tool assembled on the erection fixture is shown in Fig. 47.10 for the mode 1 configuration.

The procedure to assemble the tool is as follows. After the erection fixture is aligned and levelled, the tool head assembly is lifted from the folded shipping position and centred over the fixture (see Fig. 47.8). The arm actuator

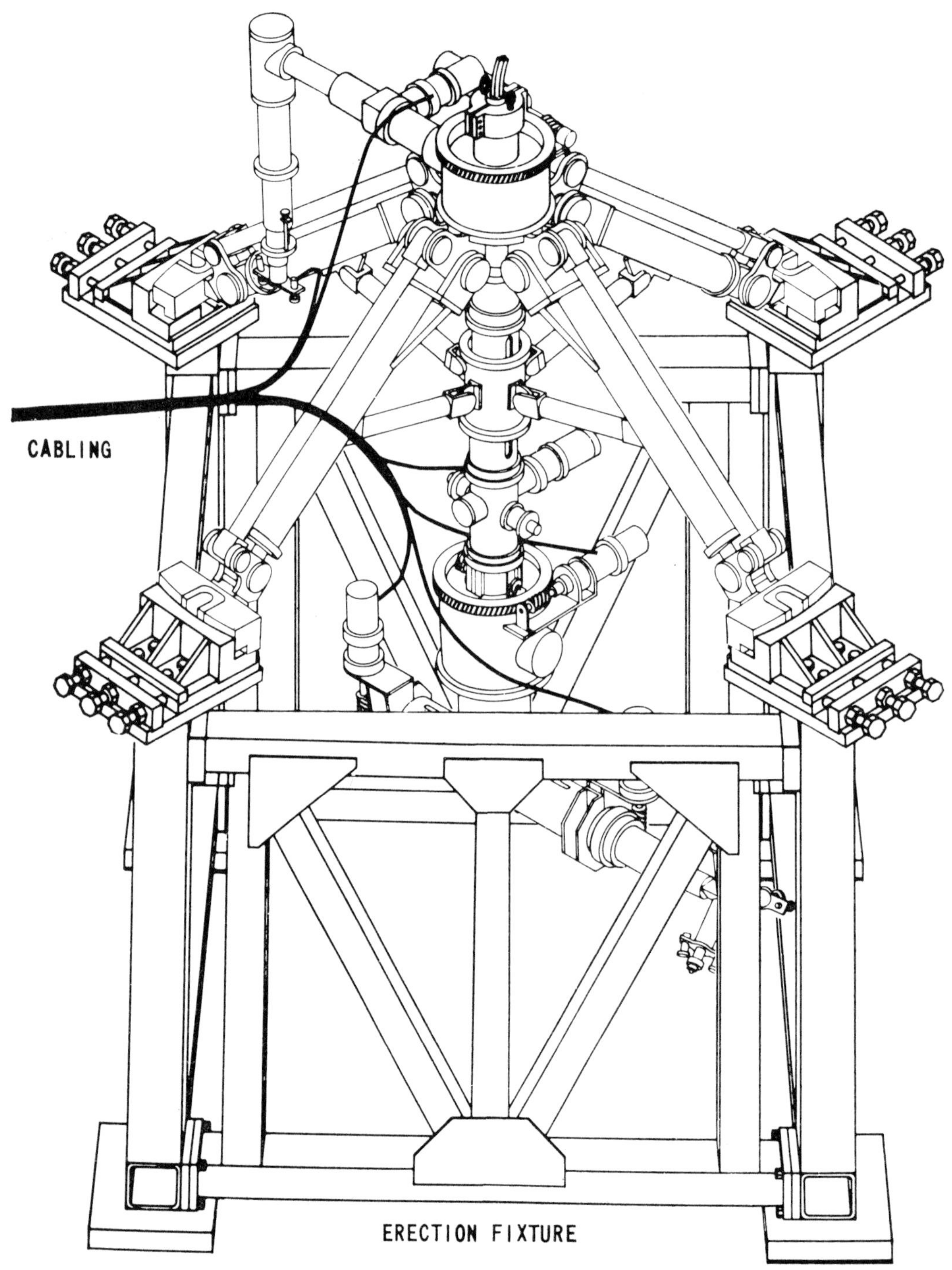

Fig. 47.10. Tool in erection fixture (nozzle and flange scanner)

is then activated, causing the arms to uniformly expand to an angle approaching 25°. The assembly is then slowly lowered and guided into position on to pads on the fixture. Once the tool is set down, the main crane can be released for other requirements until other assembly phases are completed. The additional components, such as the nozzle scanner and flange scanner, are then added along with the transducers and electrical cabling. In addition, all the necessary mechanical, electrical, and readout equipment checks are made. The equipment is then lifted out of the fixture, using the overhead crane, and positioned over the cavity above the reactor vessel, ready for insertion into the water.

Installation in reactor vessel

The overall arrangement in the reactor cavity during tool installation in the reactor vessel is illustrated in Fig. 47.11. The arms of the head assembly are extended to slightly less than the vessel flange diameter. Before lowering the tool, guide lines on the tool are in place and the underwater lighting and television camera are ready to monitor the operations. The tool is lowered to a point

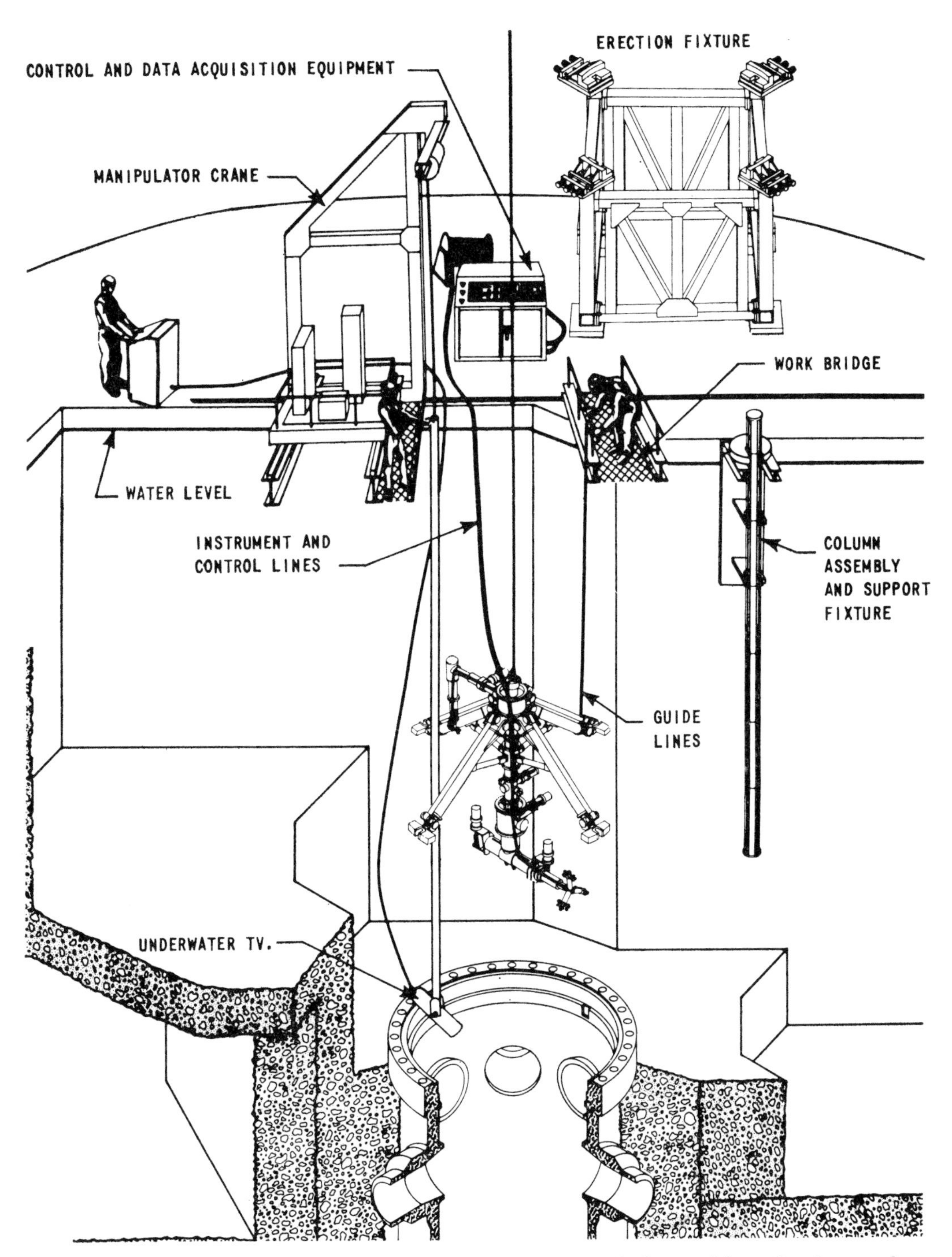

Fig. 47.11. Typical overall arrangement of reactor cavity during tool insertion in vessel (nozzle and flange scanner)

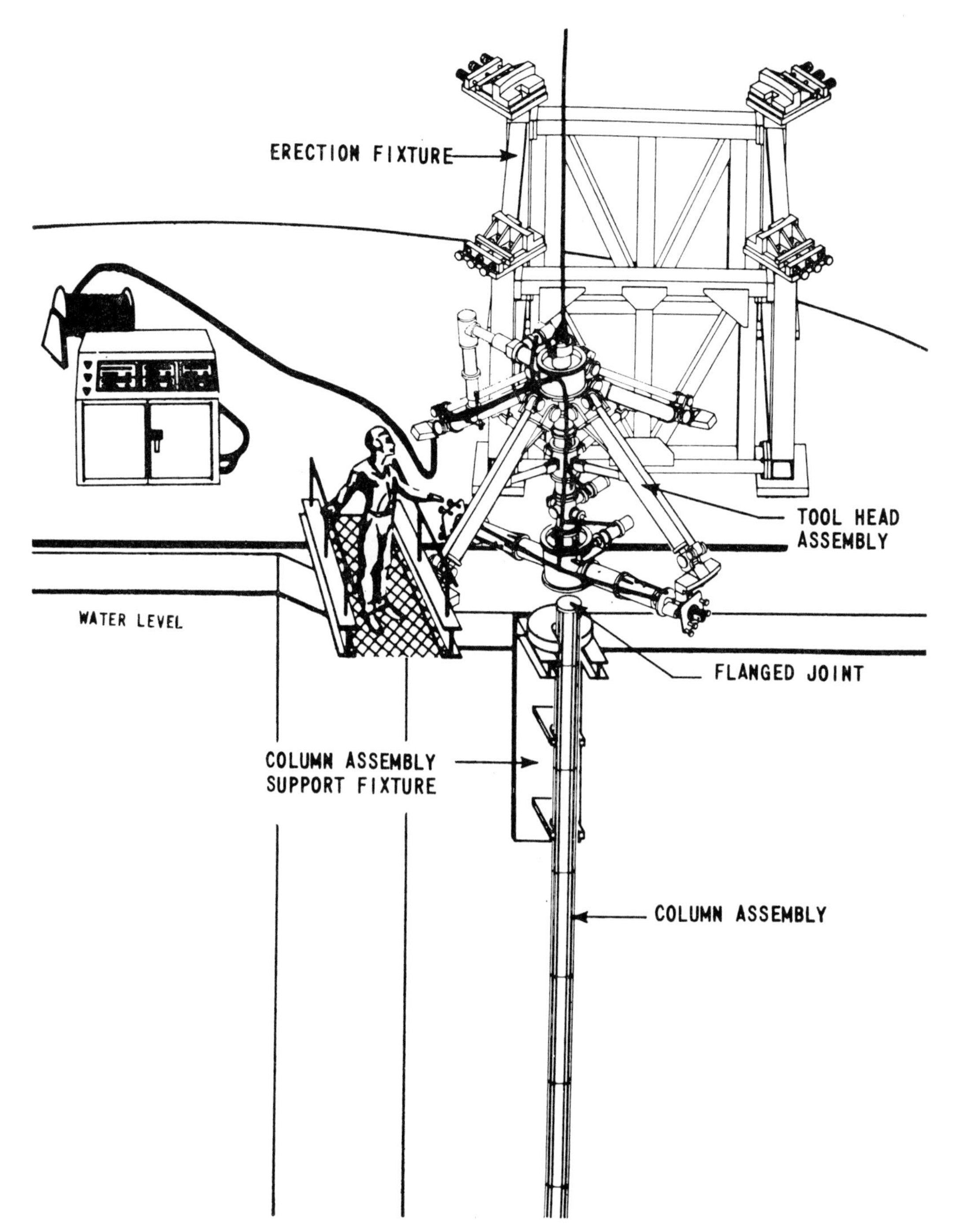

Fig. 47.12. Typical arrangement for assembly of column extension for vessel scanner

near the vessel flange. The underwater television is used to check alignment and orientation of the tool until it is finally seated on the vessel flange. Once at the final elevation, the arms are extended to a point where tight contact with the flange is obtained. Verification of proper seating is made with the underwater television camera. In-place operational checkout is made to verify proper tool functioning. Position indication is checked by comparing the built-in mechanical zero references with the previously verified digital readout zeros.

Assembly for mode 2 operation

This assembly procedure is the same as for mode 1 except that the final attachment of the column assembly is made at the edge of the reactor cavity, as illustrated in Fig. 47.12. The column sections are assembled on the floor while in a lay-down position. They are then lifted with the crane to a vertical position in the water and on to a support on the side of the reactor cavity. The head assembly, after being prepared for the vessel scanner configuration, is lifted from the erection fixture and suspended momentarily while the column assembly extension is attached. The tool is then lowered into the water over the reactor vessel.

CONCLUSIONS

A reactor vessel inspection tool has been designed to provide fine control on positioning ultrasonic transducer arrays, compatibility with existing plant equipment, and ease of transport, assembly, and operation. The tool can be adapted to fit a large range of sizes of reactor vessels. Since the tool inspects the vessel from the inside, scanning mechanisms for the vessel shell course nozzles, and flange are performed in one basic tool with a common reference point for vessel positioning. The tool can perform the required inspections of the reactor vessel without removing the reactor internals more than once every ten years.

C48/72

SAFETY AND RELIABILITY REQUIREMENTS FOR PERIODIC INSPECTION OF PRESSURE VESSELS IN THE NUCLEAR INDUSTRY

R. O'NEIL* G. M. JORDAN*

The use of pressure vessels to contain the primary coolant in nuclear reactors introduces the need for very stringent safety and reliability criteria. Expressed in terms of failure probability per year, these criteria imply conviction of failure rates considerably below that currently achieved in conventional pressure vessels, even if such a comparison was valid. This paper, after briefly deriving standards required, considers the factors used in building up the safety and reliability case, and their relative importance. These include design, manufacturing, construction, inspection, and testing standards, as well as the fracture mechanics methods involved in specifying tolerable defect sizes and inspection sensitivity. New inspection and testing techniques soon to become available are also outlined. Finally, the paper indicates, by way of example, the standard of safety and reliability likely to be achieved in the nuclear industry by pressure vessels which are built, inspected, tested, and repeat tested in accordance with the recommended procedures.

INTRODUCTION

THE REMARKS IN THIS PAPER are confined to steel pressure vessels, typical of those used in light water reactors. With the important exception of reinforced concrete pressure vessels, very little generality is lost by this restriction and the principles outlined are applicable to the majority of pressure vessels in the world's nuclear power plants.

Owing to the high potential risk to the public in the event of pressure vessel failure, such terms as 'highest possible standards', 'Grade 1 construction', etc. are used to define standards of design, manufacture, testing, and inspection. However, these terms represent an open-ended commitment and are meaningless to the engineers responsible. This paper examines the standards necessary and considers the possibility of linking standards of inspection and testing to the real requirements of safety and reliability.

RELATIONSHIP BETWEEN RISK AND CONSEQUENCE

The concept

In considering the safety of nuclear plant Farmer (1)† introduced the concept of a risk/consequence relationship. Farmer reasoned that the permissible probability of encountering a given hazard can and should be related to the consequences of that hazard.

The parameters governing this relationship should be chosen so that the likelihood of a given consequence arising from nuclear power should represent a specified small proportion of the likelihood of the same consequences arising from natural effects. Since the somatic effects of radiation exposure have their counterparts in natural arisings, and since the likelihood of each may be specified in probabilistic terms, such a relationship can be formed. The central problem of nuclear safety *policy* is to specify the relationship in quantitative terms. The central problem of nuclear safety *practice* is to ensure that the relationship is not violated.

The relationship in quantitative terms

The isotope of iodine, I-131, usually provides (2) the greatest external hazard following pressure vessel failure, and may conveniently be used as a quantitative measure of consequence. The amount of I-131 (in Curies) released to the atmosphere may thus be related to the maximum permissible target for frequency of the release.

Fig. 48.1, reproduced from reference (2), represents the proposed release/frequency limits for land-based power reactors. This will be used as a basis against which to evaluate the requirements for pressure vessel integrity in this paper.

REQUIREMENTS FOR NUCLEAR PRESSURE VESSEL INTEGRITY

Fig. 48.1 represents limits for atmospheric release from all causes. However, pressure vessel failure is likely to be the dominating mode, and for major accidents these limitations may be applied singly and directly for the purposes of this paper. Since the requirements relate to release to *atmosphere* the role of all 'post-failure' safeguards must be considered explicitly for any specific proposal. However, of these, the integrity of the containment structure is probably the most important at the large accident end of the spectrum.

The MS. of this paper was received at the Institution on 6th December 1971 and accepted for publication on 18th January 1972. 33

* *Safety and Reliability Directorate, U.K.A.E.A., Risley, nr Warrington, Lancs.*

† *References are given in Appendix 48.1.*

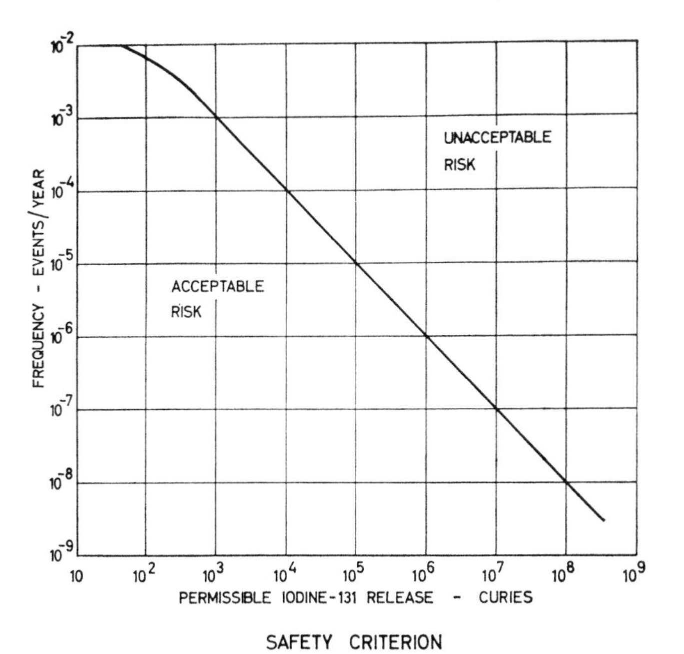

Fig. 48.1. Risk/consequence relationship

Again, for the purpose of this paper it is convenient to restrict attention to three broad pressure vessel failure categories which, for convenience, will be designated A, B, and C.

Category A. Catastrophic failure of the main pressure vessel with fragmentation, which is capable of compromising the structural integrity of the containment.

Category B. Gross ductile rupture of the main pressure vessel without fragmentation, or with fragmentation, but limited either by the nature of the defect or surrounding structures to the point where the containment structure is not at risk, although the leakage rate may be increased.

Category C. Small-scale rupture or limited leakage within the design capacity of the containment.

It is important to realize that the safety and reliability assessment of a complete reactor system embraces all possibilities of failure, not only of the main reactor pressure vessel but of a number of subsidiary vessels, pumps, valves, pipes, and other components. These are related to the reliability and capability of emergency cooling devices, make-up and flooding systems, containment sprays and filters, containment, melt-through protection, etc. While such an analysis is outside the scope of this paper, the categories proposed and supported by simplified arguments are typical of those met in practice.

Category A failures

The I-131 inventory of a land-based power reactor will be about 10^8 Ci. However, it is possible to reasonably postulate that not more than about 10^7 Ci will be released from the reactor installation to the containment area. Such a release to atmosphere could, from Fig. 48.1, only be acceptable with a frequency not greater than 10^{-7} per reactor year. For such a release to take place it is necessary to suppose that not only does the pressure vessel fail, but that it fails with such widespread rupture and fragmentation that all engineered safeguards, such as emergency core cooling, containment sprays, and clean-up systems, are rendered inoperative and that the missiles and/or pressure pulses destroy the containment structure to the extent that substantially all fission products released from the reactor become an airborne hazard to the public.

The frequency with which all engineered safeguards and containment might completely fail following a catastrophic failure requires the detailed evaluation of specific reactor designs, but for purposes of this paper it might reasonably be assumed to be in the region 10^{-1}–10^{-2} per demand. Thus the target catastrophic failure rate for reactor pressure vessels would be 10^{-5}–10^{-6} per vessel year.

Category B failures

In this situation the vessel is assumed to fail in a ductile mode, limited perhaps to a few square feet in equivalent leakage area, with any missiles producing only limited damage, such as a control rod penetrating the containment. In these circumstances the probability of the containment structure failing catastrophically and all engineered safeguards being compromised is significantly more remote than under Category A, and a further decontamination factor (d.f.) of 10^{-2} from these sources might be anticipated. This would reduce the release of I-131 to 10^5, acceptable from Fig. 48.1 with a frequency of 10^{-5} per reactor year. However, since every reactor pressure vessel failure in the circumstances outlined could lead to such a release, this vessel failure rate should be limited to about 10^{-5} per year.

Category C failures

Such minor failures, giving rise to only limited leakage, are not likely to compromise the containment structure and should be within the design capacity of the engineered safeguards. Depending on the efficacy of these measures, and on the size of the leak, the release of I-131 to atmosphere might lie between 1 and 10^3 and would be acceptable with a frequency of 1–10^{-3} per reactor year on the basis of Fig. 48.1. However, because (*a*) such small failures may become larger, (*b*) there are so many other ways in which a minor leakage from the primary circuit might develop, and (*c*) the economic consequences of such events are so severe, the acceptable frequency for Category C leaks is more likely to be assessed at about 10^{-3}–10^{-4} per vessel year.

The above requirements are summarized in Table 48.1. The arguments which support the table are somewhat simplified, but the numerical conclusions are typical of those that arise from detailed safety assessment.

STATISTICAL EVIDENCE OF NUCLEAR PRESSURE VESSEL INTEGRITY

In order to decide initially whether or not it is possible to design, construct, and operate reactor vessels of the standard required by Table 48.1 and to bring the strategy and tactics of periodic nuclear inspection into perspective, it is appropriate to examine the available statistics on plant that has been designed, operated, and inspected in accordance with the practices of the last decade. The major difficulty with such an approach is the lack of comprehensive statistics. Although a wealth of such data must exist, it is in a fragmentary and dispersed form and thus

Table 48.1. Summary of pressure vessel reliability requirements

Category	Vessel state	Reliability requirement
A	Gross failure of the main pressure vessel coupled with significant fragmentation. May compromise containment structure	10^{-5}–10^{-6} per vessel year
B	Gross failure of the main pressure vessel, without significant fragmentation. May compromise containment leakage rate	10^{-5} per vessel year
C	Small-scale rupture within the design capacity of the containment	10^{-3}–10^{-4} per vessel year

difficult, if not impossible, to collate. Since such data represent the final proof of the efficacy of procedures, it is disturbing to observe that no serious attempt has yet been made to collect the data on a national scale, despite the recommendation of such bodies as the Committee of Enquiry on Pressure Vessels (**3**).

Nuclear pressure vessels

From published world sources (including military reactors) it is possible to estimate that there is available about 2×10^3 reactor vessel years of experience. No category A or B failures have been reported, although the possibility that a failure has occurred cannot be rejected. However, no matter how these figures are viewed, they cannot be used to substantiate the requirements of Table 48.1.

Conventional pressure vessels

Here, meaningful statistical evidence may be obtained. Although not directly designed, constructed, or operated to nuclear standards, it is possible to obtain useful indications from conventional Class I vessels built and operated to standards considered comparable with nuclear plant. About 200 000 vessel years of recorded U.K. experience on the latter vessels have been reviewed by Phillips and Warwick (**4**). This review covered the period from 1962 to 1967, and involved 1352 reactor years of nuclear experience and 100 300 vessel years of Class I conventional vessel experience. The review continues and will be reported for the period 1967–72. The indications are that the current review will confirm the earlier findings (**5**). The predictions of Phillips and Warwick are also in reasonable agreement with those of Kellermann and Tietzle for German vessels (**6**).

Table 48.2, based on reference (**4**), indicates a catastrophic failure rate of about 7×10^{-5} per vessel year. The total number of catastrophic failures reported is seven, of which four were due to operational reasons, leaving only three instances where vessel quality led to catastrophic failure. However, even if the review figures are adjusted to 3×10^{-5} per vessel year, it would not be possible to justify the required failure rate for Category A or Category

Table 48.2. Detection of defects in service

Method of identification	No. of cases	Percentage
Visual examination	75	56·9
Leakage	38	28·8
Non-destructive testing	10	7·5
Overpressure test	2	1·5
Catastrophic failure	7	5·3
Total	132	100·0

B on the basis of what is currently being achieved in the conventional pressure vessel industry. It would appear that the Category C requirements are well within the demonstrable grasp of current technology.

The conclusions are, of necessity, generalizations in which mean failure rates have been assigned to different pressure vessel genera. These conclusions merely provide some indication of the norm that has been achieved with current standards of design, material control, construction, and inspection. The decision as to whether a specific vessel meets the norm requires considerable engineering skill and judgement. The process might be termed 'structural validation'—an essential element of which is inspection.

STANDARDS OF SAFETY AND RELIABILITY

The target failure rate for nuclear vessels, as summarized in Table 48.1, is 10^{-5}–10^{-6} per year. While it is possible that vessels currently being used in nuclear installations meet this target, this cannot be demonstrated by reference to experience in the nuclear field, or statistical evidence from the non-nuclear field summarized in Table 48.2. This suggests a gap of about 10^{-1}–10^{-2} and poses two questions:

(*a*) Can reactor pressure vessels generally be made to the required standard?

(*b*) Is the specific vessel up to the required standard?

Unless both these questions can be answered in the affirmative, there may be a serious limitation on the use of steel vessels in the nuclear industry.

In reliability engineering there are well-established techniques for synthesizing overall failure rates by consideration of the component parts which contribute to this overall failure rate. Such consideration may be based on direct statistical evidence or, where this is not available, on informed judgement. Such methods are invaluable in showing up the degree of reliance on people or processes, and assist in deciding whether or not such reliance is soundly placed. These techniques are more commonly applied in the field of electronics or in mechanical systems using mass production components, but in this section of the paper consideration is given to the use of such methods in the production of pressure vessels.

For a pressure vessel the probability of gross vessel rupture occurring between service inspections might be expressed as follows:

$$\begin{aligned} P_F = {} & P_D \times P_{PT} \times P_{AT} \times P_{US} \times P_L \times P_{VE} \\ & + P_M \times P_{PT} \times P_{AT} \times P_{US} \times P_L \times P_{VE} \\ & + P_C \times P_{PT} \times P_{AT} \times P_{US} \times P_L \times P_{VE} \\ & + P_X \qquad (48.1) \end{aligned}$$

where P_F is the probability of gross vessel rupture between inspection periods, P_D, P_M, and P_C the probabilities that failure in design, material, and construction will lead to failure in a vessel built and inspected to the Class I standards of reference (4), P_{PT} the probability of failure of pressure test to reveal potential failure, P_{AT} the probability that stress wave acoustic testing will fail to reveal potential failure, P_{US} the probability that ultrasonic testing will fail to reveal potential failure, P_L the probability that leakage will fail to reveal potential failure, P_{VE} the probability that visual examination will fail to reveal potential failure, and P_X the probability of some totally different cause of pressure vessel failure (e.g. aircraft crashing, sabotage, etc.).

From the foregoing it will be seen that the total target for pressure vessel failure might be built up from a series of dependent and independent factors. These are now discussed and tentative values ascribed to the various components of the equation. It is important to recognize that these factors are in addition to, and independent of, any process quality control applied during manufacture. Essentially, it is the application of an initial inspection and subsequent periodic inspection in accordance with the philosophy of Section XI of the A.S.M.E. Code. When a vessel is inspected or subjected to any process intended to detect catastrophic failure before it occurs, then if such failure does occur, not only has the vessel failed but so has the inspection process. A number of these factors are now examined using the five-year cycle covered by the survey as a convenient inspection cycle.

Failure in design, materials, and construction

The probability of a failure in design would require a gross error in estimated duties, e.g. stressing, fracture mechanics, critical crack size, crack growth, etc. There are no statistical data on the frequency of such errors, and there are considerable uncertainties in all of these headings. However, pressure vessel designs are not carried out *ab initio*, but rather as a series of limited extrapolations from previous platforms on the basis of established design codes where relevant. Where a number of vessels with novel features are built in the same class or series, it may be necessary to take special measures to clear the design for the first of class only. The reason for this need is that while some design deficiencies can be revealed by test and inspection procedures before service, others will not be revealed until the vessel has been in operation for a significant number of service cycles. This means that a novel vessel design is particularly at risk until it receives the first in-service inspection. Thereafter the risk is mainly due to increasing cycles and deteriorating properties, both of which should have been predicted and can be monitored. On the basis of the review by Phillips and Warwick (4), it is suggested that a competent design organization should produce vessels that are unlikely to fail due to design, with a frequency of not greater than 10^{-5} per vessel year or 5×10^{-5} over the first inspection period.

The probability that the vessel will not have material properties up to the standard specified by the designer is significant, and this is particularly true where alloy steels and other special materials are involved. Such mistakes, once made, may be difficult to detect in initial inspection unless they result in some identifiable defect, and may produce defects which grow at a completely unacceptable rate resulting in gross failure before the first in-service inspection. Defects which are identified after a period of service can be analysed before further service is permitted, but to ensure that the rate of gross vessel failure due to material deficiencies does not exceed 10^{-5} per vessel year or 5×10^{-5} over the first inspection period requires the most rigorous quality control and oversight. This will require material identification, bonding, and inspection to standards not commonly achieved in the pressure vessel industry.

The probability that there will be some errors in construction that could lead to gross failure due to, say, faults in heat treatment, fabrication, repair, or after a process inspection should be almost zero. Data are limited, but from Table 48.1 there is a probability of a failure rate of 3×10^{-5} per vessel year or $1{\cdot}5\times10^{-4}$ for the first inspection period.

The overall failure rate of a vessel suffering from some failure in design, materials, or construction might appear to be greatest during the period up to the first inspection and is assessed at 5×10^{-5} per vessel year or $2{\cdot}5\times10^{-4}$ over the first five-year cycle.

Failure of pressure test to reveal potential failure, P_{PT}

A successful overpressure test carried out on a pressure vessel will indicate freedom from a critical defect. If, subsequently, the vessel is operated at some lower pressure, it may be possible with knowledge of the crack growth mechanism to estimate the number of cycles that may be permitted until the next pressure test. It is important to recognize that the pressure test applies a membrane load, although in service the vessel may be subject to thermal fatigue, bending loads, etc. Furthermore, although a pressure test may, by causing failure, demonstrate gross deficiencies in design, material properties, or construction, these are more likely to require a number of cycles to propagate defects to a dangerous extent. Even in the well-documented Cockenzie vessel (7) failure did not occur until after four overpressure tests. The failure was due to a defect that could have been readily detected by recognized non-destructive testing methods.

The limited evidence reproduced in Table 48.2 shows that over a five-year period only two potentially serious defects (1·5 per cent of the total) were detected. If it is assumed that each vessel was tested once during the period, this represents a probable failure per demand of 0·985. Since some types of potential failure (e.g. certain design failures) may not be detected by pressure testing, the probability of failure per demand is assessed to be in the range 1–0·985. However, pressure testing is more useful than these figures would suggest, particularly when coupled with other forms of non-destructive examination, e.g. stress wave acoustic testing (SWAT) and ultrasonics. It is also useful for checking the strength and tightness of the overall system.

Failure of stress wave acoustic test to reveal potential failure, P_{AT}

When a vessel is loaded to the point where defects are propagating, even at the crystal level, bursts of elastic energy waves, known as stress wave emissions (SWE), take place. By the use of suitably placed sensors these bursts can be detected and located. Defects of low significance can be eliminated, and those which are or may be of safety significance can be explored by specific NDT methods. The system has potential for initial overpressure testing, periodic overpressure testing, or on-line monitoring.

If the initial pressure test provides a meaningful load on the structure, then any defects present can be detected and located and can then be subjected to more localized forms of NDT. The likelihood that suitable equipment, once commissioned, will fail to detect a meaningful signal is considered small, say not more than 10^{-2} per test, but whether the pressure test is completely meaningful can only be determined by detailed stress and fracture mechanics analysis and the physical presence of a defect. The overall probability that the technique will fail to reveal a potentially serious defect is therefore assessed at 1–10^{-2} failures per test.

On repeat pressure tests the above assessment is also likely to be valid, except that the emissions being analysed are likely to be smaller until the pressure of the initial test is approached, particularly when the tests are close together in time ('Kaiser effect'). However, the vessel will have been in operation for a number of service cycles, and the probability of failure of the technique is now assessed at 10^{-1}–10^{-2} per demand.

On-line monitoring is still at an early stage of development, but no insurmountable problem has yet been reported in the U.K. or the U.S.A. Since on-line monitoring will provide realistic cycling, the rate at which SWAT will fail to detect significantly growing defects is currently expected to be about 10^{-2} per vessel year.

Failure of ultrasonics to reveal potential failure, P_{US}

Ultrasonic examination has now been developed to the stage where defects likely to be of significance (say about 1 cm) can be detected and defined. A full 'fingerprinting' examination of the type defined in the A.S.M.E. Code requires detailed examination of the area considered critical on the basis of a typical stress analysis and fracture mechanics evaluation.

Subsequent examination is on a sample basis over a 10-year cycle. The process relies on the skill and responsibility of the operator, and in some circumstances may not detect defects in uniquely unpreferred orientations. It is also taking place on a vessel where process inspection might have been expected to locate any obvious defects. Furthermore, in the case of an initial inspection any defects that are initiated and propagated by fatigue will not yet be present. The probability that ultrasonic testing will fail to detect a potentially serious defect is assessed at 1–10^{-1} per test over the first inspection period and 10^{-1}–10^{-2} over subsequent inspection periods. For maximum effectiveness the ultrasonic test should be carried out before as well as after the final commissioning pressure test.

Failure of visual examination to reveal potential failure, P_{VE}

Table 48.2, prepared from the survey of reference (4), suggests that visual examination has been the most useful method of detecting pressure vessel failure, and A.S.M.E. XI attaches considerable importance to surface inspection. It may be, however, that the table only indicates the present strong bias to visual examination in routine in-service inspection of conventional vessels by insurance inspectors.

In a nuclear pressure vessel visual in-service inspection must be carried out remotely, either by optical means or by television. Enormous improvements have been made in the quality of television inspection, which can also detect under water virtually anything that can be detected by direct surface examination. However, nuclear vessels are likely to be stainless steel clad on the inside surface and it is possible that a potentially critical defect which propagates into the parent metal from the cladding interface would not be seen on examination of the exposed clad surface. In-service inspection from the outside of the vessel may be limited by design consideration.

Visual examination is an ideal method of detecting physical damage to the reactor structure and core internals but may not be highly reliable for detecting potentially dangerous pressure vessel defects, and the probability that it will fail to locate such flaws is tentatively assessed at 1–10^{-1} per inspection.

Failure of leakage to reveal potential failure, P_L

In conventional vessels, as indicated in Table 48.2, leakage provides a useful advance warning of a potentially dangerous situation. However, the conventional vessels on which the study was based tend to be comparatively thin walled (less than 2 in) and made of ductile materials. In nuclear vessels, wall thicknesses of 4–12 in are normal, and alloy steels are used. As a result, there are areas where the critical defect size is considerably less than the through thickness of the wall, and the classic 'leak before break' situation may not be applicable. Analyses based on tenuous calculations are made, but no direct experimental evidence is available, and the probability that leakage will fail to reveal a potentially catastrophic failure on the pressure vessel is tentatively assessed at 1–10^{-1} over an inspection cycle.

Other modes of failure, P_X

These, by definition, are not considered further in this paper, but in an overall assessment such external modes of failure as sabotage, aircraft crashing, etc., have to be examined to ensure that their probabilities are acceptably remote.

The values presented in Table 48.3 are now substituted in equation (48.1):

$$\begin{aligned}
P_F = {} & (5\times10^{-5})\times1\times0{\cdot}5\times0{\cdot}5\times0{\cdot}5\times0{\cdot}5 && (=31{\cdot}25\times10^{-7}) \\
& +(5\times10^{-5})\times1\times10^{-1}\times0{\cdot}5\times0{\cdot}5\times0{\cdot}5 && (=6{\cdot}25\times10^{-7}) \\
& +(1{\cdot}5\times10^{-4})\times1\times10^{-1}\times10^{-1}\times0{\cdot}5\times0{\cdot}5 && (=3{\cdot}75\times10^{-7}) \\
& && \overline{40\times10^{-7}} \\
= {} & 4\times10^{-6} \text{ over a five-year inspection cycle}
\end{aligned}$$

This is equivalent to a mean vessel failure rate of 8×10^{-7} per year.

Table 48.3. Summary of assessed individual failure probabilities

Item	Assessed probability of failure per five-year cycle	Selected values for equation (48.1)
P_D	5×10^{-5}	5×10^{-5}
P_M	5×10^{-5}	5×10^{-5}
P_C	$1{\cdot}5\times10^{-4}$	$1{\cdot}5\times10^{-4}$
P_{PT}	1–0·985	1
P_{AT}	1–10^{-2} (first inspection)	0·5 for design defects
	10^{-1}–10^{-2} (repeat inspection)	10^{-1} for material and construction defects
	10^{-2} (on-line inspection)	
P_{US}	1–10^{-1} (first inspection)	0·5 for design and material defects
	10^{-1}–10^{-2} (repeat inspection)	10^{-1} for construction defects
P_{VE}	1–10^{-1}	0·5
P_L	1–10^{-1}	0·5

DISCUSSION

In the example of equation (48.1) it is seen that the requirements of Table 48.1 are apparently satisfied with quite modest assumptions on the reliability of the various validation processes. Many inspecting organizations, using sophisticated techniques, might expect to fare better against defects likely to lead to vessel failure during the next inspection period. This may well be true for deficiencies in construction, and to some lesser extent for material deficiencies, but is a difficult argument to sustain in the case of design deficiencies which may not produce any physical manifestations until after a period of service. Indeed it is possible to postulate circumstances that would completely negate all the inspection processes currently available (on-line acoustic testing not yet being available). It is for this reason that a complete and independent assessment of the design is a necessary part of the overall quality control of a nuclear pressure vessel. Such an assessment would have to be capable of detecting design deficiencies likely to lead to vessel failure, and would require a success rate of 90 per cent to give an improvement of 10^{-1} in overall vessel reliability.

Failures in vessels of the type under discussion are likely to be few, and the likelihood of producing direct evidence on the reliability of in-service inspection techniques is correspondingly low. Nevertheless, standards of safety and reliability must be defined if pressure vessels are to be freely used in the nuclear industry. The figures used in the example involve an element of judgement and are presented as a basis for discussion.

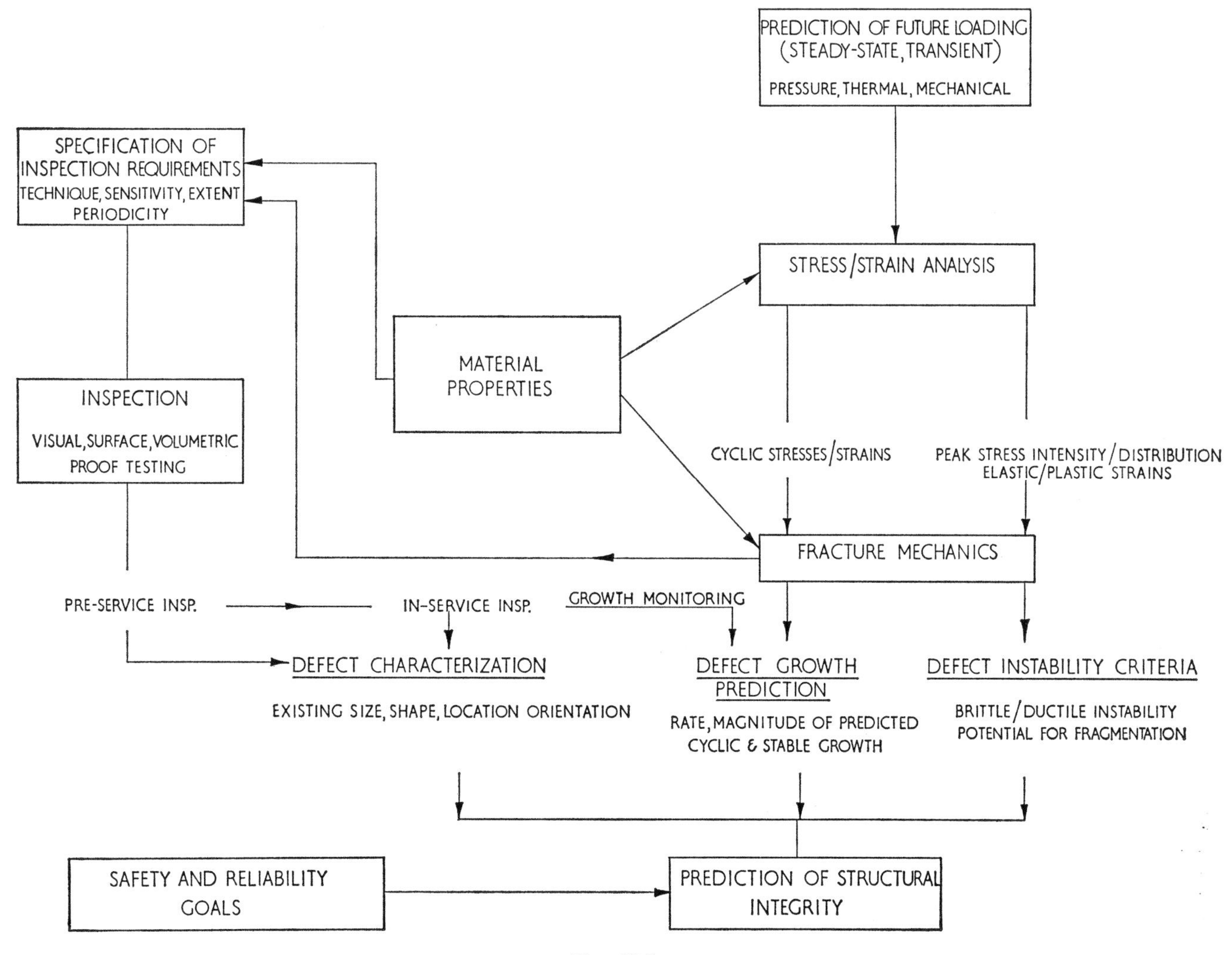

Fig. 48.2

STRUCTURAL VALIDATION—THE CONTRIBUTION OF PERIODIC INSPECTION

All too often in the past periodic safety inspection programmes have been aimed at 'searching for any defects' with little emphasis on the need to decide which defects are significant to the safety of the structure. 'Workmanship' acceptance standards more appropriate to manufacturing quality control are frequently invoked. Such practices can result in costly and time-consuming 'over-inspection' and have often been the cause of unnecessary repair—the 'repair' frequently leaving the structure in a worse state.

Structural validation is concerned primarily with answering four important questions:

(1) What are the characteristics of features having a real significance in the integrity of the structure?

(2) What is the current condition of the structure?

(3) Given the anticipated future usage of the plant until the next inspection, in what way and to what extent will the current significant defects change (in terms of number, size, orientation, etc.)?

(4) What constitutes a 'critical' condition for each significant defect?

These activities are illustrated in Fig. 48.2.

Section XI of the A.S.M.E. Boiler and Pressure Vessel Code (8) defines rules for in-service inspection of coolant systems in light water cooled-and-moderated reactors. Adoption of this code is virtually complete among U.S.A. utilities and is employed either directly or in modified form (9) by many European operators. The code recognizes that the choice of areas for inspection is largely governed by considerations of safety; areas of highest service factors (i.e. highest crack growth potential) are especially singled out and the majority of the inspections relate to welds and adjacent base metal.

While the code provides a useful basis for defining a programme of inspection, it is the opinion of the authors that the code does not give sufficient recognition to the important aspect of defect significance and thus, to some extent, perpetuates the philosophy of 'searching for any defects' which can be counter-productive.

CONCLUSIONS

(1) Steel pressure vessels for nuclear reactors must be designed and produced to the highest standards currently available, but even these cannot be demonstrated to be to the standard required. Further improvements in safety can only come from an independent detailed assessment of design and an independent regime of inspecting and testing.

(2) Quantitative safety and reliability goals for such inspection and testing can be specified and interpreted in terms of acceptable failure rates for pressure vessels.

(3) Periodic inspection is fundamental to the process of structural validation, and despite the difficulties associated with radioactive areas, inspection must be extended to such areas by the development of sophisticated equipment for remote handling and data presentation.

(4) More attention needs to be given to the avoidance of nugatory inspection. The latest code practices ensure that the areas selected for safety inspection are apposite. However, the effectiveness of inspection is also measured by the degree of its application to 'significant' defects.

(5) The requirements of reliability or availability are best satisfied by producing the best possible pressure vessel backed by a periodic inspection regime. However, the requirements of safety are best satisfied by a device which continuously monitors the vessel state, giving such sufficient advance warning of inspecting failure as to permit a safe shutdown and reduction of risk. SWAT, when developed for use on-line, appears to offer the best prospect for meeting such a requirement and present work on such a device is warmly endorsed.

APPENDIX 48.1

REFERENCES

(1) FARMER, F. R. 'Siting criteria—a new approach', SM-89/34.

(2) BEATTIE, J. R., BELL, G. D. and EDWARDS, J. E. 'Methods for the evaluation of risk', AHSB(S)R159, 1969.

(3) *Ministry of Technology Report of the Committee of Enquiry on Pressure Vessels*, vol. I, 1969.

(4) PHILLIPS, C. A. G. and WARWICK, R. G. 'A survey of defects in pressure vessels built to high standards of construction and its relevance to nuclear primary circuit envelopes', AHSB(S)R162, 1968.

(5) PHILLIPS, C. A. G. and JORDAN, G. M. Private communication.

(6) KELLERMANN, O. and TIETZLE, A. 'Proposal for recurring inspection of nuclear steel pressure vessels', *IAEA Techn. Rept, Series No. 99*, 1969.

(7) 'Report of the brittle fracture of a h.p. boiler drum at Cockenzie power station', 1966 (January) (S.S.E.B.).

(8) A.S.M.E. Boiler and Pressure Vessel Code, Section XI: 'In-service inspection of nuclear reactor coolant systems' (Am. Soc. Mech. Engrs, New York).

(9) KELLERMANN, O. and SEIPEL, H. G. 'Analysis of the improvement in safety obtained by a containment and by other safety devices for water cooled reactors', presented at IAEA Symp. on the Containment and Siting of Nuclear Powered Reactors, Vienna, 1967 (April).

C49/72 RESEARCH AND DEVELOPMENT PROGRAMMES ON IN-SERVICE INSPECTION IN THE UNITED STATES OF AMERICA

R. D. WYLIE*

This paper summarizes the extent of the research and development in progress in the U.S.A. on in-service inspection. The U.S. research programme has developed a unique systematic approach that has attacked the problem on a broad technological front. Since the start of the programme it is apparent that all the questions have not been answered, but many of the more important ones have. The availability of inspection techniques for in-service quality assurance should lead to reduced failure probability, but the paper emphasizes that inspection is only part of the system required to assure plant safety.

INTRODUCTION

RAPID ESCALATION of the number and capacity of nuclear power plants in the U.S. led to an industry–government reappraisal of the basic engineering knowledge of components which affect the safety of nuclear reactor systems. This reappraisal in the mid-1960s resulted in a number of actions by industrial organizations, codes and standards committees, the research and development community, the electric utilities, and government regulatory agencies. The main thrust of these actions was directed towards what has come to be known as 'quality assurance', in the broadest sense of the term.

A definition is in order at this point. 'Quality assurance' refers to actions by any party designing, constructing, or operating nuclear power plants which assure that the quality level of the installed systems is sufficient to permit them to be operated safely and reliably throughout their design life. The concept of 'cradle to grave' quality assurance was not new to utilities operating conventional power plants. The main difference in nuclear plants was the criteria of reliability which resulted from the projected consequences of failure of such a plant and the radiation environment.

In 1965–66 it became clear to the technical representatives of industry that there was insufficient technological information on procedures with which to determine the required quality level for safe and reliable operation. As a result, a 'system oriented research' approach was drafted by representatives of industry and government operating under the structure of the Pressure Vessel Research Committee of the Welding Research Council. This approach was designed to provide detailed information on design and analysis procedures, fracture processes and fracture mechanics, materials properties of heavy section steels and weldments, fabricating inspection processes, flaw identification and significance, and in-service inspection techniques. It was an ambitious programme in which every segment of the nuclear industry was asked to contribute. In the author's experience it was a unique, massive research and development and engineering programme which, although developed in a number of separate programmes, managed to achieve the integrated effect necessary to solve the systems problem.

The object of this paper is to indicate the scope of research and development programmes in the U.S.A. directed towards in-service inspection and to indicate the interdependence of these programmes on the other portions of the system.

THE SYSTEM APPROACH TO COMPONENT SAFETY

In order to describe the system of component safety, three generalized diagrams are presented in Figs 49.1, 49.2 and 49.3. Fig. 49.1 shows the various organizations that have direct input to assuring the safety and reliability of a plant. The ultimate responsibility belongs to the utility which delegates certain phases of the programme, particularly during construction, to other organizations. In the 'systems'

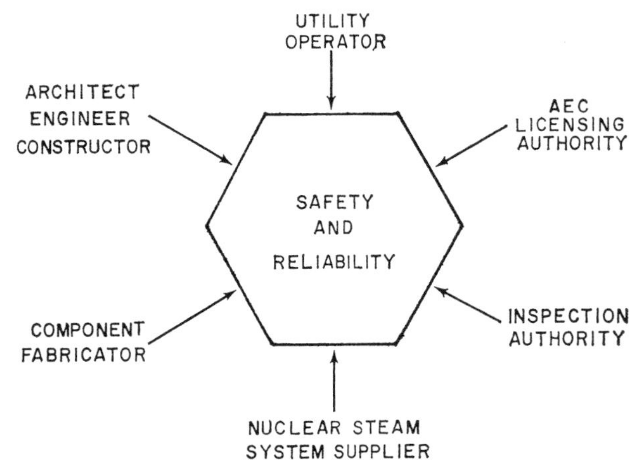

Fig. 49.1. Schematic diagram indicating major organizations which have input to nuclear plant safety

The MS. of this paper was received at the Institution on 18th January 1972 and accepted for publication on 25th January 1972. 33

* *Southwest Research Institute, 8500 Culebra Road, P.O. Drawer 28510, San Antonio, Texas 78284, U.S.A.*

research programme, therefore, the utilities assumed a very significant role.

The nuclear steam system supplier and the component supplier also assumed a necessary role in the development of information on component safety and reliability.

The inspection authority in the U.S.A. was the only group which remained outside the development effort. Individual organizations developing inspection techniques and equipment, however, did play an active role in the programme.

Finally, owing to the interest in public safety, the Atomic Energy Commission also contributed significantly to the research and development effort both in financial support and direct work performed by government laboratories.

These organizations undertook several major programmes of national scope. These were:

(1) The HSST (Heavy Section Steel Technology) programme funded by the Atomic Energy Commission and managed by Oak Ridge National Laboratory.

(2) The Industrial Cooperative programme, funded by materials suppliers, vessel fabricators, and nuclear steam system suppliers and managed by PVRC.

(3) The EEI–TVA Inservice Inspection programme, funded by utilities (Edison Electric Institute, Tennessee Valley Authority) and managed by Southwest Research Institute.

In addition to these major programmes, other significant work was carried out by individual companies, and the Atomic Energy Commission has contributed much additional information. Of particular interest and importance in this regard was the work done on radiation effects on structural materials and the resultant fracture technology performed by the U.S. Naval Research Laboratory.

Fig. 49.2 shows the input factors which determine component failure probability. The HSST programme was dedicated to developing a mathematical relationship between these various factors in order to permit an assessment of the significance of existing flaws, as well as the probability of initiation or growth of new flaws. Much work, of course, has been done and considerable progress made in achieving this objective. Model vessel confirmation tests are planned in the near future to give further confidence to that developed in specimen test programmes.

A large amount of data were required on materials properties, stress analysis, fabrication flaws, etc., of the construction materials used in nuclear reactor vessels. This information was developed, collated, and analysed in the Industrial Cooperative programme. Special testing

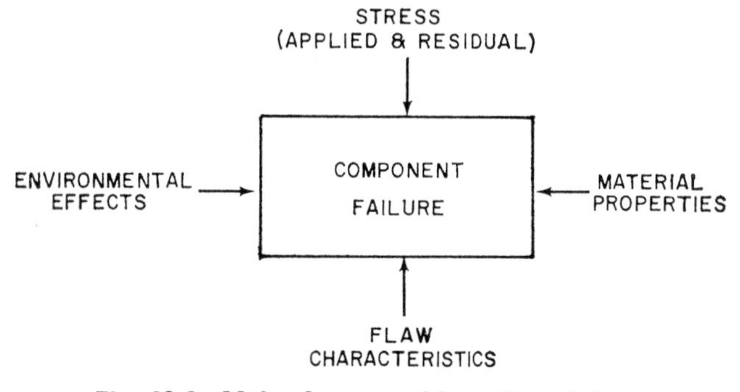

Fig. 49.2. Major factors which affect failure

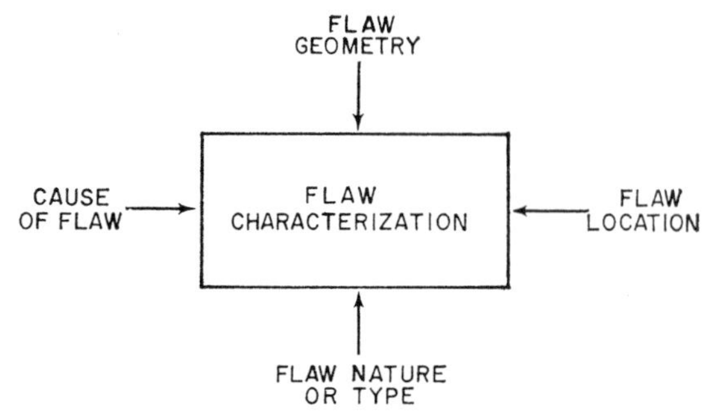

Fig. 49.3. Components of flaw characterization

programmes on reactor vessels provided more data than would normally be available on vessels constructed to Section III of the A.S.M.E. Boiler Code. These data are stored for reference and comparison in a computer program.

Another important phase of the co-operative effort was the evaluation of non-destructive testing techniques for fabrication inspection. For the most part, industry had depended on radiographic inspection for confirmation of weld quality. Since the application of ultrasonic testing had not reached the level of development that had been attained in Europe, there was a concerted effort by fabricators to compare the detection efficiency of radiography and ultrasonics and to improve the reproducibility of ultrasonic inspection. A number of 'round robin' type samples were prepared for ultrasonic testing. The test programme resulted in an ultrasonic procedure prepared by the PVRC with which reproducible results were normally obtained (**1**)*.

The EEI–TVA programme was charged with developing the information required for flaw characterization, as indicated by Fig. 49.3, with particular emphasis on volumetric procedures useful in in-service inspection programmes.

There has been considerable interprogramme activity and evaluation by the Pressure Vessel Research Committee which has brought about a fair, balanced result.

THE EEI–TVA IN-SERVICE INSPECTION PROGRAMME

The utility industry initiated its portion of the overall programme in 1967. The development of the programme was based on work that had been performed independently under sponsorship of various industries and government agencies. The object of the research and development was to promote the development and acceptance of new volumetric inspection procedures. The intention was to produce a system that could assure the quality of the reactor pressure vessel and components continuously throughout the service life of the plant.

Southwest Research Institute was selected to manage the programme under the direction of a utility Steering Committee appointed by the Edison Electric Institute and the Tennessee Valley Authority. As indicated above, close liaison was maintained with the Pressure Vessel Research Committee which set up an advisory panel of experts who periodically reviewed the progress of the programme.

The research programme was divided into a number of projects which were carried out at different laboratories

* *References are given in Appendix 49.1.*

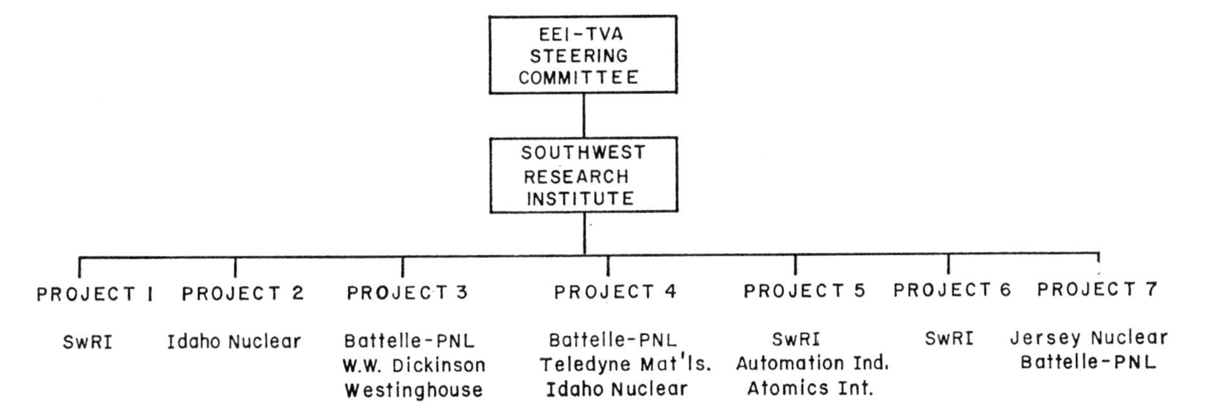

Fig. 49.4. Organization of the EEI–TVA in-service inspection research programme

throughout the U.S.A. A summary of the organization of the project has been reported previously (2)–(4). Fig. 49.4 shows the generalized project organization and lists the principal subcontractors who have participated. The nature of effort in each project is given below:

Project 1. Review of recent utility experience with power reactor coolant pressure boundary inspection regarding service conditions, defect detection capability, defect size, and defect orientation.

Project 2. Development of a non-destructive test method evaluation facility.

Project 3. Investigation of the acoustic spectrometer as a semi- or fully-continuous scanning method. (This project has been discontinued.)

Project 4. Investigation of acoustic emission techniques for continuous monitoring.

Project 5. Improvement in reliability and reproducibility of conventional ultrasonic systems for reactor inspection.

Project 6. Investigation of acoustic holography for visualization of flaws in thick-walled pressure vessels.

Project 7. Information collection, reporting, and dissemination.

It should be pointed out that, in general, the philosophy of this programme was to evaluate existing techniques and to encourage development of new techniques. Because of the state-of-the-art of ultrasonic inspection system development, it was decided early in the programme not to sponsor the development of mechanical or electronic hardware except as needed to evaluate the technique. Rather, organizations were encouraged to continue the parallel development of the inspection hardware.

The major effort was devoted to the development of acoustic emission technology for potential on line monitoring as well as encouraging the evaluation of pressure test systems. The paper by Vetrano *et al.* (5) describes some of the accomplishments of this programme.

A list of reports and papers that have been generated from this research programme is included in the bibliography of Appendix 49.1.

The final demonstration of the acoustic emission system will be performed on a pressure test facility at the National Reactor Test Site in Idaho Falls, Idaho. A pressure vessel called EBOR (Experimental Beryllium Oxide Reactor), which was originally constructed for the Atomic Energy Commission programme, was offered to the EEI–TVA programme as a test facility. This has been modified to permit water flow under pressure, as well as the addition of noise sources simulating both cracking and flow noises.

Many of the accomplishments of this programme are being described by the principal investigators in the seminar. The programme is nearing completion, and at least many of the objectives have been accomplished and further developments by inspection companies, utilities, and others have been encouraged.

SOUTHWEST RESEARCH INSTITUTE PROGRAMME

Over the last 10 years the Southwest Research Institute has invested a considerable amount of internal research funding and capital equipment money to the development of mechanized scanning equipment to accomplish reactor vessel inspections. It has also worked for utilities and government agencies in the U.S.A. and other countries to provide unique systems to accomplish the inspection objectives. Of particular importance to this effort were: the pilot programmes on the Elk River Reactor, sponsored by the Atomic Energy Commission; San Onofre Unit 1 Reactor, sponsored by Southern California Edison Company; the Oskarshamn Unit 1 Reactor, sponsored by the Oskarshamnsverkets Kraftgrupp AB; and the Atucha Reactor, sponsored by the Comision Nacional de Energia Atomica of Argentina. The equipment developed required unique engineering solutions that qualify them as research and development programmes. Each of these has been described in separate publications. Early inspection work on various power plants in the U.S.A. provided the necessary practical experience to enable our engineering staff to inspect reactor vessels and piping. This practical experience was supplemented by many in-house programmes which permitted the staff to solve problems identified in the field. Valuable experience was also gained by monitoring flaw growth in fatigue and fracture tests that were being undertaken at Southwest Research Institute in connection with other programmes.

Lautzenheiser (6) has described a summary of the equipment we have developed, and Whiting and Stolle (7) have presented our concept of data acquisition and storage.

In addition to the development of techniques and equipment, SwRI has been actively engaged in co-operative programmes to improve accessibility and design for inspectability in more recent plants in an effort to permit the utility to achieve the results desired by Section XI of the A.S.M.E. Boiler and Pressure Vessel Code. Changes have been made in insulation design, weld configuration, vessel land marks, biological shielding, piping layouts, etc., to facilitate inspection programmes. This is considered to be a most important part of our development programme.

MISCELLANEOUS RESEARCH PROGRAMMES

As might be expected in a rapidly expanding technological field, there are a number of organizations who are involved in developing techniques. This is especially true in the area of equipment and procedures for acoustic emission testing. On the hydrostatic test of the EBOR facility, there were five organizations with equipment to be evaluated. These were Jersey Nuclear Corporation, Battelle-Northwest, Southwest Research Institute, Westinghouse Research Laboratory and Teledyne Materials Research. In addition to these, Aerojet General Corporation and Dunegan Research Corporation have also been active in acoustic studies.

Two university research programmes are of interest. The University of Michigan has been working on acoustic spectroscopy, and the Johns Hopkins University has been evaluating the use of acoustic attenuation for detecting fatigue damage.

Another programme on acoustic holography by Holosonics, Inc. is extending the work done by Battelle-Northwest for the EEI–TVA research programme.

Many of the organizations who are planning to enter the in-service inspection field have development programmes in progress. General Electric Company, Westinghouse Electric Corporation, Combustion Engineering Corporation, Babcock and Wilcox Company, and more recently Gulf Energy Systems and Atomics International have all studied in-service inspection techniques and many have active contracts for inspection service.

APPENDIX 49.1

REFERENCES

(1) 'Nondestructive examination of PVRC plate-weld specimen 201', *Weld. J.* 1971 **50** (No. 12, December), 529.

(2) WYLIE, R. D. 'A general review of the status of the EEI–TVA research program on in-service inspection', Paper No. 31, 4th Annual Information Meeting of the Heavy Section Steel Technology (HSST) Program, Oak Ridge National Laboratory, 1970.

(3) REINHART, E. R. and WYLIE, R. D. 'Nondestructive testing techniques applied to in-service inspection', *Nuclear Metallurgy* 1970 **16**, 199.

(4) REINHART, E. R. 'Nondestructive testing techniques applied to in-service inspection', Paper No. 26, 5th Annual Information Meeting of the Heavy Section Steel Technology (HSST) Program, Oak Ridge National Laboratory, 1971.

(5) VETRANO, J. B., JOLLY, W. D. and HUTTON, P. H. 'Continuous monitoring of nuclear reactor pressure vessels by acoustic emission techniques', Paper C58 at this Conference.

(6) LAUTZENHEISER, C. E. 'Mechanized equipment for in-service inspection of nuclear reactors', Paper C57 at this Conference.

(7) WHITING, A. R. and STOLLE, D. E. 'Automated data handling systems for in-service inspection: ADDCOM system', Paper C60 at this Conference.

BIBLIOGRAPHY

'In-service inspection program for nuclear reactor vessels', Biannual Progress Rept No. 1, Southwest Research Institute and The National Technical Information Services, Accession No. PB 184 359, 1969 (2nd June).

'In-service inspection program for nuclear reactor vessels', Biannual Progress Rept No. 2, Southwest Research Institute and The National Technical Information Services, Accession No. PB 188 347, 1969 (2nd December).

EDISON ELECTRIC INSTITUTE. 'Progress report on reactor vessel testing', *EEI Bull.* 1970 (May), 135.

'In-service inspection program for nuclear reactor vessels', Biannual Progress Rept No. 3, Southwest Research Institute and The National Technical Information Services, Accession No. PB 193 433, 1970 (8th July).

'In-service inspection program for nuclear reactor vessels', Biannual Progress Rept No. 4, Southwest Research Institute and the National Technical Information Services, Accession No. PB 198 274, 1971 (7th January).

'In-service inspection program for nuclear reactor vessels', Biannual Progress Rept No. 5, Southwest Research Institute and The National Technical Information Services, Accession No. PB 200 518, 1971 (28th May).

LAUTZENHEISER, C. E. 'Automated techniques for ultrasonic scanning as applied to in-service inspection', Paper No. 34, 4th Annual Information Meeting of the Heavy Section Steel Technology (HSST) Program, Oak Ridge National Laboratory, 1970.

WORLTON, D. C. 'The potentials of acoustic holography for nondestructive testing', Paper No. 32, 4th Annual Information Meeting of the Heavy Section Steel Technology (HSST) Program, Oak Ridge National Laboratory, 1970.

SAWICKY, W. 'Recent activities in determining the effect of nuclear reactor environment on the ultrasonic characteristics of vessel materials', unpublished report presented at the Symposium on Nondestructive Testing in AEC Programs, Battelle Northwest Laboratories, Richland, Washington, 1970 (28th April).

JOHNSON, C. G. and ORTEGA, O. J. 'Review of RP 79 EEI–TVA in-service inspection project', American Society of Nondestructive Testing, Cleveland, Ohio, 1970 (19th–22nd October).

ABBOTT, D., GLYNN, W. and HANNAH, K. 'Ultrasonic technique selection for in-service inspection of nuclear reactor pressure vessels', Paper No. 27, 5th Annual Information Meeting of the Heavy Section Steel Technology (HSST) Program, Oak Ridge National Laboratory, 1971.

VETRANO, J. B. and JOLLY, W. D. 'In-service acoustic emission monitoring of reactor pressure vessels', Paper No. 28, 5th Annual Information Meeting of the Heavy Section Steel Technology (HSST) Program, Oak Ridge National Laboratory, 1971.

C50/72 EQUIPMENT DESIGN TO MEET IN-SERVICE INSPECTION REQUIREMENTS

L. R. KATZ*

The Westinghouse Pressurized Water Reactor complies with the requirements for accessibility and inspectability imposed by the A.S.M.E. Code for in-service inspection of nuclear reactor coolant systems. This compliance is generally based upon plant arrangement and component design. A general arrangement drawing and a layout indicating the in-service inspection plan for each primary loop component is presented to describe the specific examination requirements and the method to be used for each examination to meet the Section XI requirements for a typical pressurized water reactor. The specific design changes which were implemented on the primary loop components to permit code compliance are also outlined. The current development programmes and future plans for compliance with Section XI are also described.

INTRODUCTION

THE PRESSURIZED WATER REACTOR TYPE, because of its arrangement and inherent design, generally complies with the requirements for accessibility and inspectability imposed by the A.S.M.E. Code for in-service inspection (Section XI). The Westinghouse PWR design has many additional special features which permit complete compliance with the code. The general design features of a typical PWR which permit code compliance, as illustrated in Figs 50.1 and 50.2, are as follows:

(1) All reactor internals are completely removable from the reactor vessel, allowing access to the entire inside surface of the vessel for inspection. The tools and storage space required to permit removal of the internals are also provided.

(2) The reactor vessel closure head is stored in a dry condition on the operating deck during refuelling, allowing direct access for inspection.

(3) All reactor vessel studs, nuts, and washers are removed to dry storage during refuelling, allowing inspection in parallel with refuelling operations.

(4) Removable plugs are provided in the primary shield just above the reactor vessel primary coolant nozzles to permit access for inspection of the welds joining the nozzles to the safe-ends, and the welds joining the safe-ends to the primary coolant piping. Readily removable insulation is provided over these weld areas.

(5) Access holes are provided in the lower core structure barrel flange to permit inspection access to the inside surface of the reactor vessel without removal of the lower core structure.

(6) Manways are provided in the steam generator channel head to allow access for internal inspection.

(7) A manway is provided in the pressurizer top head to allow access for internal inspection.

The MS. of this paper was received at the Institution on 26th October 1971 and accepted for publication on 25th January 1972. 43

* *PWR Systems Division, Westinghouse Electric Corporation, P.O. Box 355, Pittsburgh, Pa. 15230, U.S.A.*

(8) The insulation covering all component and piping welds and adjacent base metal is designed for ease of removal and replacement in areas where external inspection is planned.

(9) Removable plugs are provided in the primary shield concrete above the main coolant pumps to permit removal of the pump motor to provide internal inspection access to the pumps.

(10) The primary loop compartments are designed to allow personnel entry during refuelling operations, and to permit direct inspection access to the external portion of piping and components.

(11) The reactor vessel shell in the core area is designed with a clean, uncluttered cylindrical surface to permit future positioning of test equipment without obstruction.

EXAMINATIONS

Reactor vessel examination

Figs 50.3 and 50.4 illustrate a typical design of a reactor vessel for a PWR plant. The examination category A welds in the core region shell courses and the examination category B welds in the nozzle shell course are accessible for volumetric inspection from the inside diameter of the vessel. This inspection will require removal of the fuel and upper and lower core support structures, and the use of a specially designed remotely operable ultrasonic test device that will be described in greater detail in another paper. The number of category A and B welds will vary, depending upon whether plate or forging construction is used in manufacture of the reactor vessel. Forging construction in the core and nozzle regions eliminates the longitudinal welds in categories A and B. Examination category C, the flange welds in both the reactor vessel and closure head, are accessible for volumetric inspection during refuelling shutdowns. The head-to-flange weld inspection will be performed by normal ultrasonic techniques by direct access to the outside of the closure head while in storage. The vessel-to-flange weld inspection will be performed from the flange surface

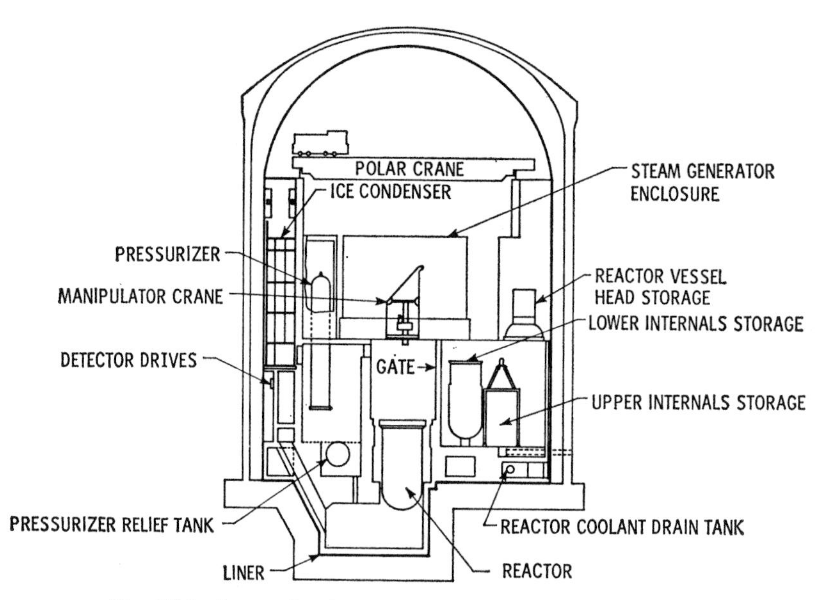

Fig. 50.1. In-service inspection plan, plant containment

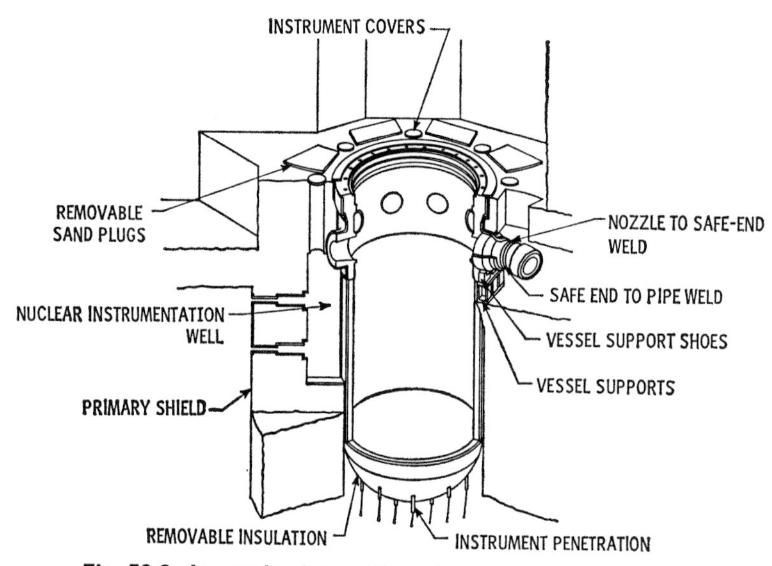

Fig. 50.2. In-service inspection plan, reactor arrangement

utilizing the special ultrasonic device. The examination category D and F welds at the nozzle-to-vessel and nozzle-to-safe-end welds will also be inspected volumetrically from the vessel inside diameter utilizing the special ultrasonic device. The partial penetration welds joining the control rod drive head adaptors in the closure head and the instrument penetrations in the bottom head can be classified examination category E-2, requiring only visual examination during periodic hydrostatic testing. The reactor vessel studs, nuts, and washers are inspected in accordance with examination category G-1. The studs are inspected by both surface and volumetric techniques during refuelling. The nuts and washers require only visual inspection while the ligaments between vessel tapped stud holes will receive volumetric examination from the flange surface by remotely utilizing the special ultrasonic device. The reactor vessel supports, consisting of integrally welded pads beneath each nozzle, are included under examination category H. Nozzle type supports are covered by the inspection requirements, under category D, for nozzle-to-vessel welds. The internal cladding of the vessel and closure head are included under examination category I-1, requiring visual examination in the vessel and a combination of visual and surface examinations in the closure head. The internal surfaces and the radial support block inside the vessel fall under examination category N, requiring remote visual inspection after removal of the fuel and upper and lower core support structures.

Pressurizer examination

Fig. 50.5 illustrates a typical design of a pressurizer for a PWR plant. All the longitudinal and circumferential welds joining the shell and head sections are included under examination category B, requiring volumetric inspection.

This inspection will be performed by normal ultrasonic techniques by direct access to the outside of the pressurizer after removal of the insulation. The welds attaching the surge, safety, and relief nozzles to the shell and heads are included in examination category D, requiring volumetric

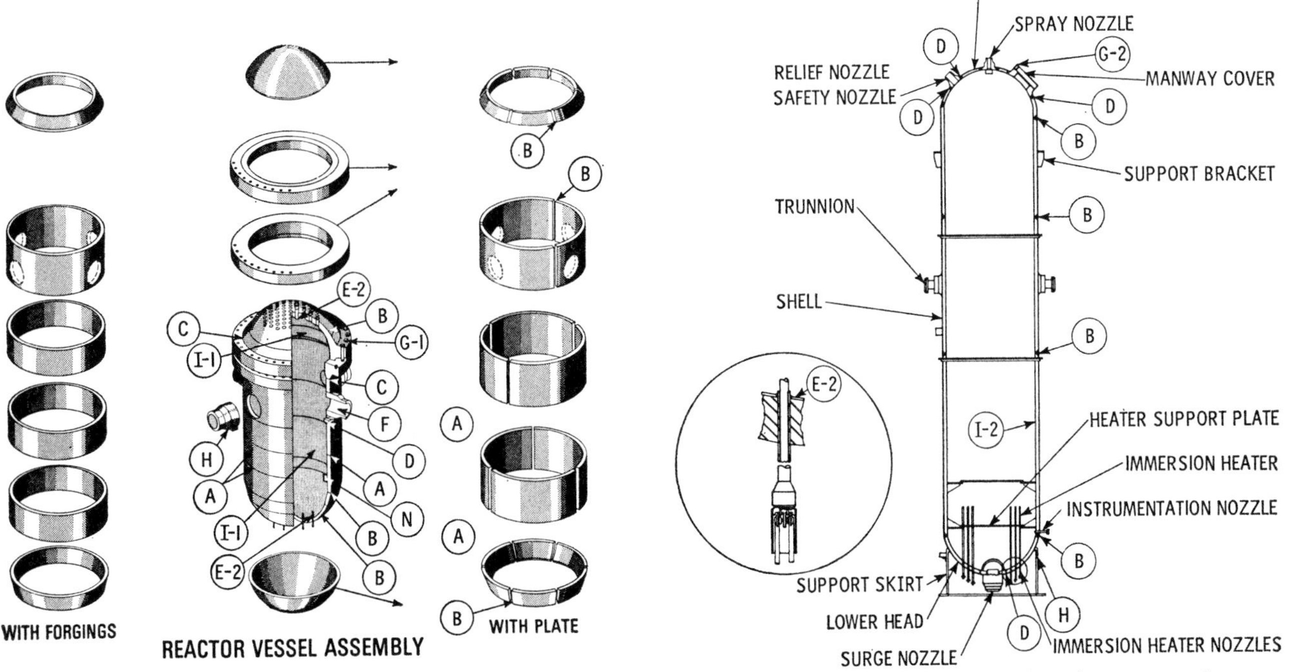

Fig. 50.3. Reactor vessel assembly

Fig. 50.5. In-service inspection plan, pressurizer

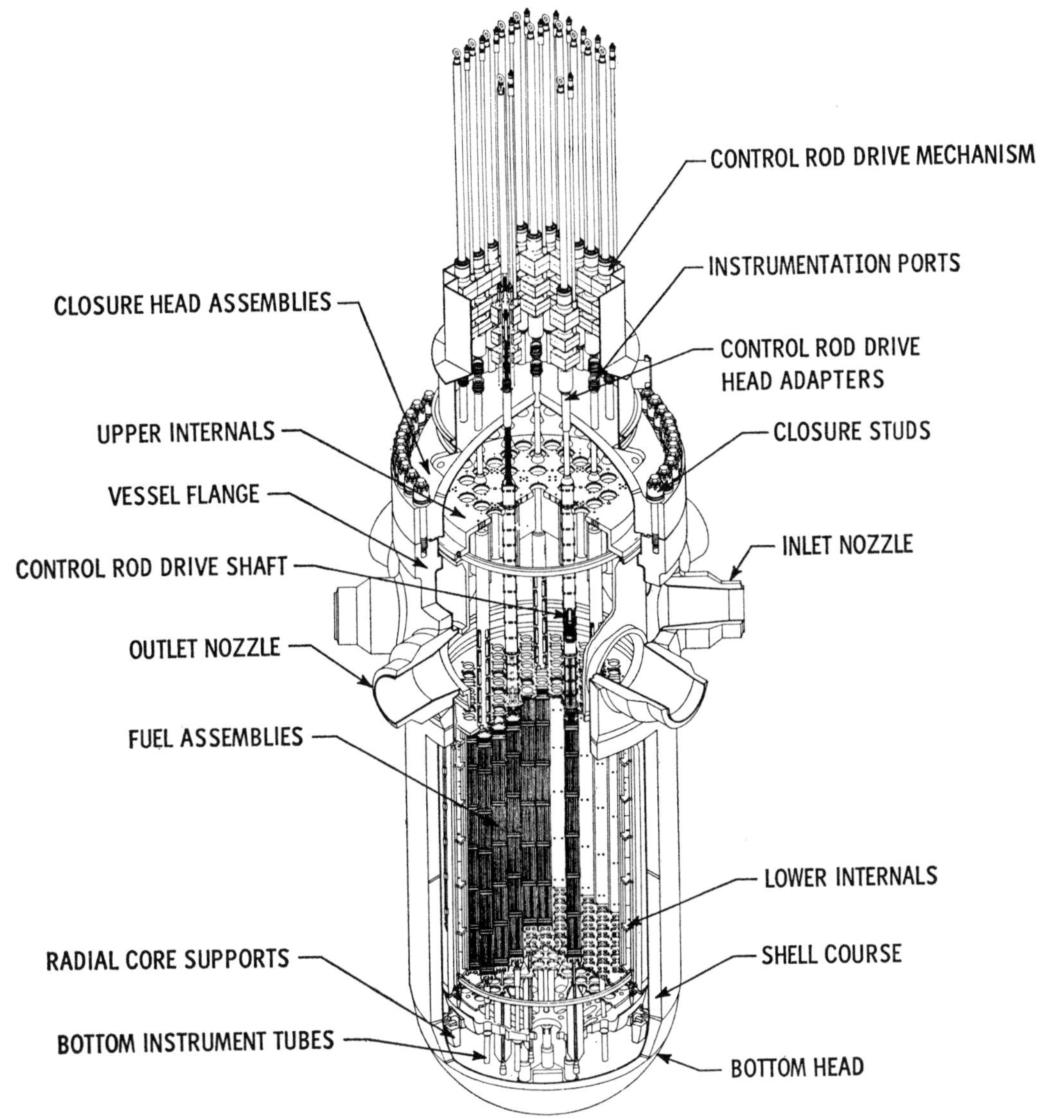

Fig. 50.4. In-service inspection plan, reactor vessel

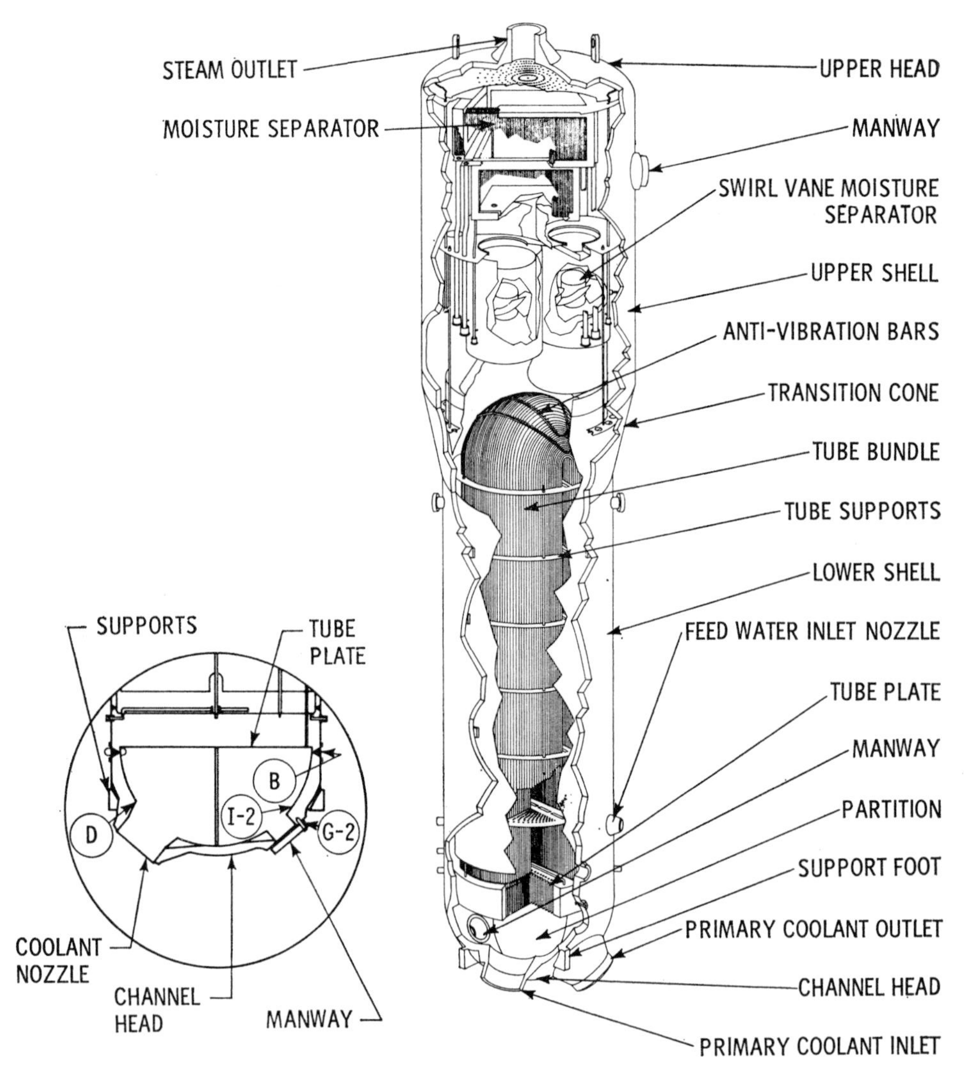

Fig. 50.6. In-service inspection plan, steam generator

inspection from the outside of the pressurizer using standard ultrasonic techniques. The heater nozzle and instrument penetration or partial penetration welds to the shell and bottom head are classified examination category E-2, requiring visual inspection during periodic system hydrostatic tests. The pressurizer manway bolts are category G-2, requiring visual examination only. The skirt type support is category H, requiring visual and volumetric inspection at the skirt to bottom head weld. Direct access for the use of standard ultrasonic techniques is afforded by removal of insulation. The pressurizer internal cladding is category I-2, requiring visual examination by either direct or remote means with access through the manway.

Steam generator examination

Fig. 50.6 illustrates the typical design of a steam generator for a PWR plant. The inspections on the steam generator are limited to the weld areas in the primary coolant portion of the unit. The weld between the tube sheet and channel head is classified as category B, which requires volumetric inspection from the external surface of the unit utilizing standard ultrasonic techniques. The channel head manway bolting is category G-2, requiring only visual examination. The channel head internal cladding is category I-2, requiring visual examination by either direct or remote means. Access for this examination is afforded through the manway. The steam generator support pads are exempt from inspection because they are integrally cast on the channel head. Although the primary inlet and outlet nozzles are integrally cast on the channel head, examination by volumetric methods applies to the inside radiused section at the head inside diameter under category B. External access for standard ultrasonic inspection is afforded by removal of insulation.

Pipe examination

Fig. 50.7 illustrates the typical design of primary coolant

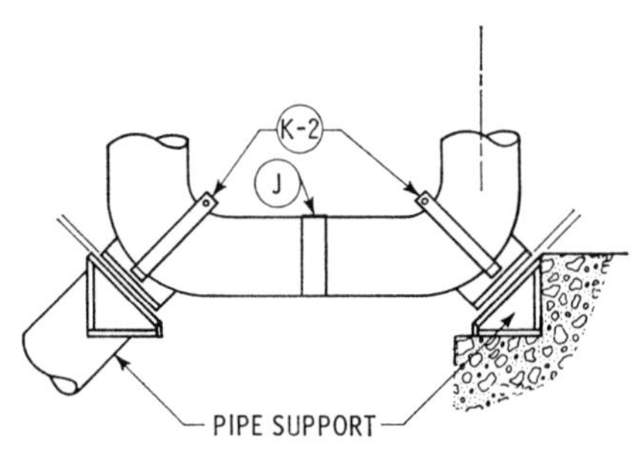

Fig. 50.7. In-service inspection plan, primary coolant pipe

pipe for a PWR plant. All circumferential joints between pipe sections are included in examination category J-1, requiring both visual and volumetric inspection. Removal of insulation at the weld joints affords external access for both direct visual and standard ultrasonics. No integrally welded pipe supports under category K-1 are employed. However, category K-2 supports, requiring direct visual inspections, are employed for all standard pipe hangers. No bolts requiring either category G-1 or G-2 inspections are employed in the piping design.

Primary coolant pump examination

Fig. 50.8 illustrates the typical design of primary coolant pumps for a PWR plant. The pump casing consists of two cast sections joined by a circumferential weld. The weld and adjacent base metal will be volumetrically inspected by radiography, as required in examination category L-1. In parallel, examination category L-2, requiring visual inspection of the internal portion of the pump casing, will be conducted. Both categories L-1 and L-2 require disassembly and removal of the pump motor and main flange. The pressure-containing bolts fall into examination categories G-1 and G-2, requiring volumetric and visual inspection respectively. The integrally welded pump supports, category K-1, will be volumetrically inspected by standard ultrasonics by direct access to the supports after removal of insulation. The support hangers for the pumps, category K-2, are available for direct visual inspection as required.

Primary coolant stop valve examination

Fig. 50.9 illustrates the typical design of a primary coolant stop valve for a PWR plant. The currently supplied valves consist of one-piece castings, eliminating the need for conducting examination category M-1, covering the inspection of welds in valve bodies. The internal visual inspection required under category M-2 will be conducted after valve disassembly. Category G-1 and G-2 valve bolting inspections will be conducted by visual and standard ultrasonic inspections with direct external access, or after removal of the bolts during category M-2 inspection. No external supports are employed on these valves,

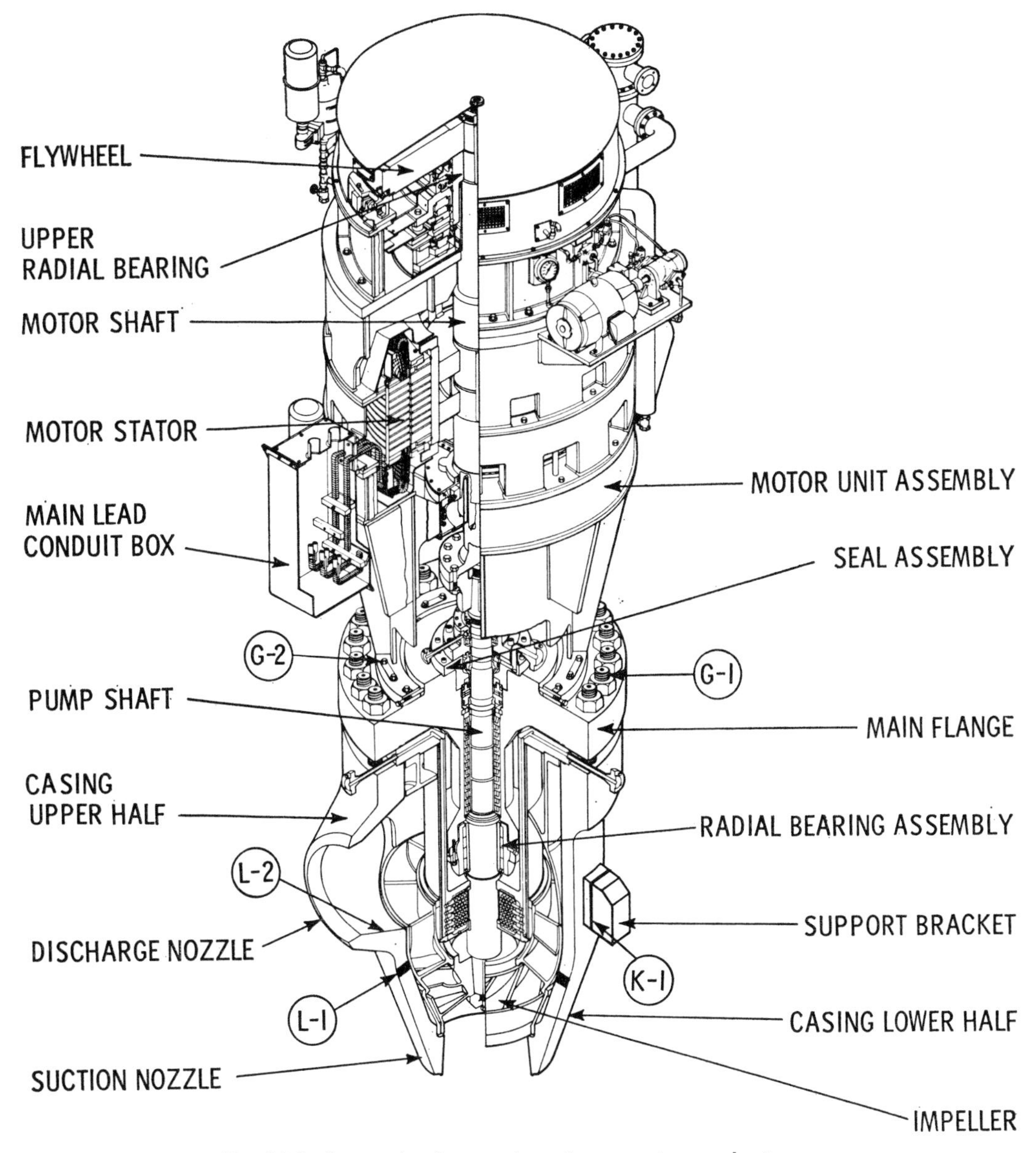

Fig. 50.8. In-service inspection plan, reactor coolant pump

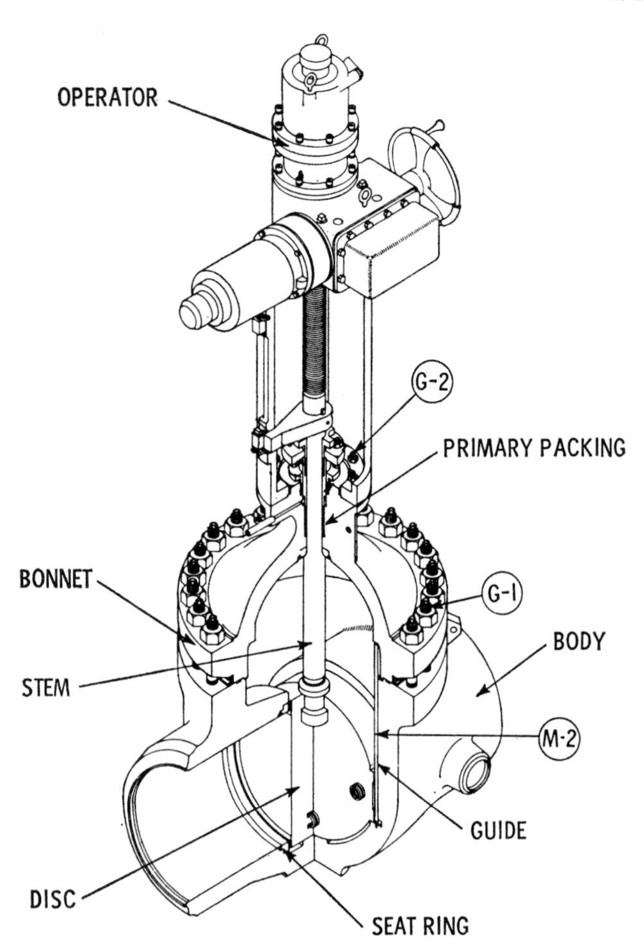

Fig. 50.9. In-service inspection plan, primary coolant stop valve

eliminating the need for examination categories K-1 and K-2.

DESIGN FOR INSPECTION

Although the Westinghouse PWR plant design complies with the requirements of Section XI of the A.S.M.E. code, many specific design changes were adopted to ensure strict compliance in all areas.

Reactor vessel

(1) The closure head and bottom head designs were modified to move all full penetration welds outside of the baffle support ring and instrument penetration pattern, respectively.

(2) The insulation on the closure head and bottom head of the reactor vessel were designed to be removable.

(3) The control rod drive mechanism head adaptors and instrumentation tube nozzles were modified to incorporate a collar on the portions within the vessel head to prevent the loss of the head adaptor or instrument tube in the event of a failure of the partial penetration welds. This change, in conjunction with the thermal sleeve redesign described in (4) below, places the head adaptor and instrument tubes in examination category E-2, requiring only visual inspection during periodic system hydrostatic tests.

(4) As illustrated by Fig. 50.10, the thermal sleeves were redesigned to serve as a flow restriction in the event that the head adaptor or mechanism fails in operation. Restricting flow to the equivalent of 1-in pipe size is required to preclude the necessity for volumetric inspection of the partial penetration weld joining the head adaptor to the head. This thermal sleeve redesign also places all welds in the control rod drive mechanism pressure housing into category E-2.

(5) The reactor vessel cladding is improved in finish by grinding to the extent necessary to permit meaningful ultrasonic examination of the vessel welds and adjacent base metal in accordance with the code.

(6) The cladding-to-base metal interface is ultrasonically examined to ensure satisfactory bonding to allow the volumetric inspection of the vessel welds and base metal from the vessel inside diameter.

Pressurizer

(1) Special tests were performed to ensure that the heater and instrument nozzle rolling operation, prior to partial penetration welding, results in sufficient interference between the nozzles and their openings to prevent the loss of the nozzles in the event of a failure of the partial penetration welds. These tests were performed on mock-ups

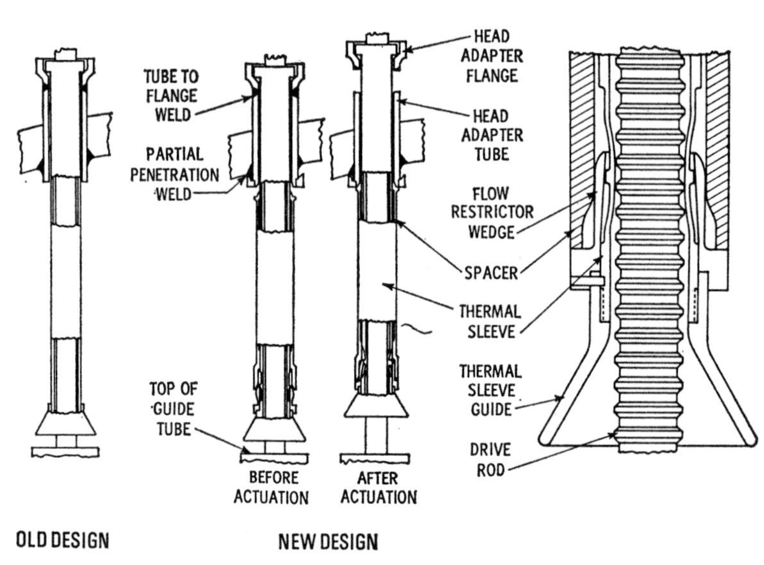

Fig. 50.10. Reactor vessel head adaptor flow restrictor actuation

simulating the actual conditions in the pressurizer. The tests were run at operating temperature utilizing forces simulating those resulting from the operating pressure. These test results place the instrument penetrations and heater penetrations in examination category E-2, requiring only visual inspection during periodic system hydrostatic tests.

(2) The heater spacing in the bottom head was rearranged to allow ample clearance around the surge nozzle to perform the required inspections on the surge nozzle to bottom head weld.

(3) The support skirt attachment to the bottom head was redesigned to allow better access for performing the required inspections on the support skirt to bottom head welds.

No design changes were required to accommodate Section XI on the steam generator, main coolant pumps or primary piping for the Westinghouse PWR plants.

DEVELOPMENT PROGRAMMES

Several programmes are currently under way to develop special non-destructive testing techniques for in-service inspection. These include:

(1) A volumetric inspection technique for use on the welds in cast austenitic pipe and fittings.

(2) A volumetric inspection technique for the electroslag welds joining the cast portions of the primary coolant pump casing.

(3) A remotely operated ultrasonic test device for the inspection of the reactor vessel welds. This device will be discussed in detail in another paper.

CONCLUSIONS

With the advent of the A.S.M.E. Code for in-service inspection of nuclear reactor coolant systems (Section XI), nuclear power plant designers have placed particular emphasis on those equipment design features and plant arrangements that must meet the requirements of that code. The typical Westinghouse Pressurized Water Reactor generally complies with the requirements for accessibility and inspectability, but does require some specific design changes for complete compliance. These changes include moving welds for accessibility, providing removable insulation to expedite inspections, and incorporating special features which limit the flow of primary fluid from pressure barrier penetrations in the event of structural failures. Several development programmes for special non-destructive testing techniques that are applicable to plants in-service are currently under way. Accessibility and inspectability have become two new basic design parameters for nuclear power plants.

C51/72 THE PERIODIC INSPECTION OF TRANSPORTABLE GAS CYLINDERS AND CONTAINERS

W. DEAN*

To ensure safety, the handling and the transportation of gas cylinders and containers are subject to control in one form or another by many countries. While the techniques of examination and test are common to all countries, the requirements of some vary regarding the intervals of examination and the type of hydraulic test to be employed. All these requirements are regularly modified and updated, therefore only those of the United Kingdom have been quoted. The method of control exercised in the U.K. is explained, the requirements for periodic inspection detailed, and the practices described.

INTRODUCTION

THE INSURANCE of safety in the handling and the transportation of gas cylinders and containers carrying at pressure permanent, liquefiable, or dissolved gases—many of which have toxic, corrosive, and inflammable characteristics, singly or in combination—has had considerable attention from governments and other bodies responsible for regulating the carriage of dangerous goods in their respective countries. The resulting legislation and regulations of the different countries vary in their detailed requirements, and it is essential that those responsible for the despatch of these cylinders and containers are knowledgeable not only as to the requirements of the country in which the despatch occurs but of those of the country of destination, and of all countries through which the items may pass *en route*. For many years, efforts have been made to bring order into this complex situation. In Europe, regulations to cover the carriage of dangerous goods by rail (the R.I.D. regulations) and by road (the A.D.R. regulations) have been made. The former are the agreed rules of the International Convention concerning the Carriage of Goods by Rail. The latter were prepared by a working party of the Economic Commission for Europe and are the substance of the European Agreement concerning the International Carriage of Dangerous Goods by Road, to which the U.K. is a signatory. In a wider field, a technical committee of the I.S.O. (International Organization for Standardization) has for some time been discussing a number of aspects relating to compressed gas cylinder design and to their conveyance.

In requirements laid down for periodic inspection, i.e. the actual routine practices for examination and hydraulic test, are those universally employed by competent inspection authorities for most types of pressure vessels. The differences in the requirements of the various countries are in the frequencies of inspections, the necessity or otherwise of carrying out a hydraulic test, and the type of hydraulic test to be employed. Apart from these differences, the requirements for periodic inspection are common to all countries.

Because the requirements of the various countries are frequently modified or updated, any list of differences would soon become obsolete; therefore no such list will be included in this paper. For guidance, however, it is necessary to refer to some requirements, and those referred to are of the U.K.

U.K. MECHANICS OF CONTROL

The safety of gas cylinders and containers has not been the subject of comprehensive legislation in the U.K.; nevertheless, government influence since 1895 has been considerable and effective. To understand this anomalous situation it is necessary to refer to the late nineteenth century, when the railways had a virtual monopoly of transport. By their general powers they could impose their own conditions for the acceptance and transportation of dangerous goods.

In 1894 the compressed gas trade and the railway companies agreed on the terms whereby compressed gases would be accepted for transportation by rail. At about the same time the Home Office, as guardians of public safety, set up a committee to advise on the manufacture of compressed gas cylinders. The report of this committee, published in 1895, endorsed and amplified the agreement reached in 1894. Presumably the report reassured the Home Office that the parties to the agreement had responsible attitudes to the potential hazards involved, because at that time none of the recommendations of the report became the subject of legislation.

These arrangements formed the established pattern in the years that followed. Since 1895 various reports (1)–(11)† covering different aspects of gas container design, filling,

The MS. of this paper was received at the Institution on 26th November 1971 and accepted for publication on 3rd February 1972. 30

* *Inspection Section Manager, I.C.I. Ltd, Mond Division, P.O. Box 14, The Heath, Runcorn, Cheshire.*

† *References are given in Appendix 51.1.*

and maintenance have been prepared by government department appointed committees. Following the 1895 report (1), another (2) was published in 1921, the first of a series made by the D.S.I.R., and it reported that no statutory regulations existed to deal with the manufacture and transport of cylinders for compressed gases. It implied that the railway companies had largely ensured the safety of gas cylinders by their general powers to restrict the transit by rail of dangerous goods, and by their ability to enforce the terms of the agreement reached in 1894 between the railways and the compressed gas trade. That agreement was endorsed and amplified by the committee responsible for the 1895 report (1). It was hinted that the introduction of mechanical road transport might have an adverse effect on the prevailing situation.

In the event, road transport became a reality, and matters came to a head in 1929 when a firm persisted in the conveyance by road of substandard imported cylinders. These had been rejected for rail carriage, and the Home Office had urged their withdrawal from use on the grounds of public safety.

The outcome was a decision to use the powers available under the Petroleum (Consolidation) Act 1928 to bring under control the conveyance by road of certain gases when compressed in metal cylinders. Effect was given to this decision by the Petroleum (Compressed Gases) Order 1930 (S.R. & O. 1930, No. 34) and by the Gas Cylinders (Conveyance) Regulation 1931 (S.R. & O. 1931, No. 679). The latter was subsequently amended in a number of detailed respects by the Gas Cylinders (Conveyance) Regulation 1947 (S.R. & O. 1947, No. 1594), and again by the Gas Cylinders (Conveyance) Regulations 1959 (S.R. & O. 1959, No. 1919). They were also qualified in their practical application by a number of exemption orders made by the Secretary of State. Although they were not published as statutory instruments, some of them still remain in force.

The above instruments were applied to a selection of the so-called 'permanent' gases. The gases covered are air, argon, carbon monoxide, coal gas, hydrogen, methane, neon, nitrogen, and oxygen. This list is not comprehensive even as a list of permanent gases, as opposed to liquefied or dissolved gases. Nevertheless, with the exception of dissolved acetylene, which is subject to the Explosives Act 1875 and to statutory instruments made thereunder, the gases listed are the only ones whose conveyance by road is at present regulated. The various reports of the committees appointed by government departments became, in effect, codes of practice for the design, utilization, and maintenance of gas cylinders and containers. The B.S.I. standards for various classes of cylinder are based on the recommendations of these reports.

The major authorities concerned with transport, i.e. the British Railways Board for transport by rail and the Department of Trade and Industry for transport by sea, have by their regulations for the acceptance of dangerous goods for transport imposed requirements for this particular class of dangerous good based on their shipment in containers to appropriate B.S.I. standards; or where such B.S.I. standards are not available, in containers approved by the authority. In the latter case, containers complying with the requirements of the appropriate Home Office report will usually be approved by the authority.

Transport by road imposes on the consignor the need to safeguard against actions at common law; therefore compliance with the recommendations of the appropriate Home Office report is an obvious necessity.

The latest report was published by the Home Office Fire Department in 1969 under the title 'Report of the Home Office gas cylinders and containers committee'. It ranges over a much wider field than any previous report, taking into account the unparalleled rapid development during the post-war years in the methods and techniques of conveying substances under pressure.

REQUIREMENTS OF U.K. FOR PERIODIC INSPECTION OF TRANSPORTABLE GAS CYLINDERS AND CONTAINERS

The following are based on Home Office recommendations (10) initially issued in January 1958 as part of a more complete report under preparation at that time and eventually issued in 1969 (11). The recommendations have been accepted in full by the Ministry of Transport and in part by the B.R. Clearing House; the latter retain their own rules for large, welded rail tanks.

Approved inspection authorities

It is recommended that the periodic inspections should be carried out by an inspecting authority approved by the Secretary of State. If the filler is not himself approved for this purpose, he is required to have the inspections carried out by an approved inspecting authority.

Responsibility of the filler

It is recommended that part of the filler's responsibilities should be:

(1) the subjection of the cylinders and containers to an external examination prior to filling, and the segregation for examination and assessment by a competent person of any having apparent defect; and

(2) to ensure that the stipulated period for the periodic examination has not elapsed at the time of filling or will not have elapsed unduly by the time it can be expected to be returned for refilling.

Requirements for hydraulic tests at periodic inspections

In the report (11), gases have been classified into one or other of the following categories according to their critical temperatures (T_c):

Permanent gases	$T_c < -10°C$
High-pressure liquefiable gases	T_c $-10°$ to $70°C$
Low-pressure liquefiable gases	$T_c > 70°C$

Also the toxic, corrosive, and inflammable characteristics of a large number of gases have been indicated.

The report further recommends that in the case of cylinders and containers intended to be used for the purpose of conveyance, the following shall apply.

Permanent and high-pressure liquefiable gases

The periodic inspection shall include a hydraulic test of the type and at the test pressure called for in the design specification.

Low-pressure liquefiable gases

The periodic inspection need not include a hydraulic test provided the container is of fusion-welded construction and is fitted with a manhole or hand-holes of such a size and in such positions as to permit the examination of the whole of the interior of the container either directly or by means of mirrors. Cylinders and containers not thus equipped shall at each alternative periodic inspection be subjected to a hydraulic proof test at the test pressure applicable to the particular design specification, except that, if they are intended for the conveyance of a low-pressure liquefied gas designated as being corrosive, they shall be subjected to the hydraulic proof test at each periodic inspection.

Containers of forge-welded construction

These shall be subjected to a hydraulic proof test at each periodic inspection.

Periodic inspection intervals

The report (11) recommends that the interval between the periodic inspections of each cylinder or container should not exceed five years, except that:

(*a*) Containers which form part of breathing apparatus shall be examined and tested three years after the initial test carried out by the manufacturer and thereafter at intervals of two years.

(*b*) Containers filled with non-corrosive fire-fighting products and used for fire-fighting need not be subjected to a first examination and test after the initial manufacturing test until a period of 10 years (20 years in cases where an external examination of the containers and examination for loss of content is carried out by a competent person at intervals not greater than one year) has elapsed from the time of installation. This relaxation would not apply to any container which had been discharged, showed loss of weight, or was excessively corroded externally. If after the first periodic examination and test the containers are found to be in a satisfactory condition, then the inspecting authority should certify that they may continue in service for a further period of 10 years before being again submitted to examination and test, subject to the exemptions mentioned above. Subsequently the containers should be examined and tested at intervals of five years.

Markings on containers

Without having certain basic information about the cylinders and containers it would be impracticable for the filler or the inspecting authority to fulfil their obligations. To meet this situation the various design specifications stipulate the permanent markings with which the cylinders and containers are required to be marked at the time of manufacture, to which must be added the dates of the periodic inspections as and when carried out. The different design specifications vary to some extent in the markings required, but in all cases there is sufficient information to identify the manufacturer and the manufacturer's serial number, and in most cases the specification and test pressure. From these sources it is possible to determine the suitability of the cylinders and containers for use with a particular product and other information that the inspecting authority may require.

Preparation for examination

The report (11) recommends that prior to an examination the cylinder or container should be thoroughly cleaned. External paint if in good condition need not be removed. In the case of a lagged, fusion-welded container provided with a manhole or hand-holes of such a size and in such positions as to permit the examination of the whole of the interior either directly or by means of mirrors, the competent person may dispense with the external examination if he is satisfied that external corrosion is likely to be negligible; alternatively a portion or portions of the lagging may be removed for examination of part of the external surface most likely to be affected, e.g. manhole and nozzle to shell areas; welds at the ends.

Where excessive rust or foreign matter is observed it shall be removed by methods such as wire brushing, shot blasting, flail or rumbling, scraping, boiling or steaming, or by heating to a temperature not exceeding 300°C. It is particularly important to observe this temperature limitation when heating containers that have been zinc sprayed for the purposes of external protection. At temperatures below 300°C the danger of the zinc diffusing into the iron is negligible.

EXAMINATION

The examination should be directed to observe the following defects: (*a*) damage by indentation; (*b*) damage by cuts, gouges, stab marks, etc.; (*c*) damage by corrosion and pitting; (*d*) bulging of the walls; (*e*) laminations and cracks; and (*f*) unathorized alterations to the container.

Particular attention should be given to the places where attachments to the cylinder or container shells occur. Cylinders having foot rings that meet the cylinder wall at an acute angle are liable to corrosion in the crevice between the foot ring and the cylinder wall. Detachable rolling bands and anchorage straps induce similar corrosion at the areas of contact or near contact with the shell. Undercuts at the toes of fillet welds attaching supporting brackets to large welded transport containers can be the starting place of cracks penetrating into the shell. Severe local pitting of the shell can occur as a result of moisture dripping on to a limited area from a projecting surface.

The taper-tapped holes of cylinders into which are screwed taper-threaded stem valves or plugs can be considerably distorted as a result of overtightening. The threads should be cleaned and examined to ensure they are complete and of good form. Gauges should be used to verify that the threads are of correct diameter and taper, and the walls of the tapered holes should be scrutinized for possible cracks.

The fabrication of seamless steel cylinders involves a forging operation to close the open end. This results in the wall thickness gradually thickening towards the centre, a difficult 'upset' operation which occasionally leaves the underside of the shoulder with a series of radial depressions. The depressions act as stress raisers to some degree, dependent on their form, and have been the source for the development of cracks. On the rare occasions that such defects have been detected, the indications were seen as cracks in the lower edge of the cylinder valve hole. The inspection of the taper-tapped valve hole should be extended to look for cracks developing from the above source.

Basically the examination is a thorough visual scrutiny assisted by such necessary aids as sources of suitable lighting, mirrors, introscopes, etc. The exploration of suspect findings may require the aid of crack-detection equipment, ultrasonic flaw-detection and thickness-measuring equipment, and radiography equipment. The weight of the cylinder or container alone may form a useful check on the extent of corrosion, and this figure should be preserved in the records for occasional comparison. Equal corrosion over all the surface of a cylinder or container seldom occurs, it tends to be localized.

The assessment of defects found requires knowledge of the cylinder or container design specification and the effects of stress-raising damage. When the periodic inspection has to include a hydraulic test it is preferable to make a visual examination before the test to avoid the possibility of unnecessary damage occurring whilst undergoing test. Following a hydraulic test, a further examination should be made to detect any defects that may have occurred as a result of the test.

HYDRAULIC TESTS

The periodic inspection includes a hydraulic test, except for the exemptions detailed earlier. There are two types of hydraulic test: the hydraulic stretch test and the hydraulic proof pressure test.

The hydraulic stretch test is a requirement for the testing of permanent gas and high-pressure liquefiable gas containers. Such containers are normally of the cylinder type; they are fairly highly stressed at ordinary temperatures, and have relatively thick walls not readily dented. The hydraulic proof pressure test is considered to be adequate for low-pressure liquefiable gas containers as the thinner walls are more readily dented, and the relief of such dents under test pressure would render the results of a stretch test unreliable.

The hydraulic stretch test

The hydraulic stretch test apparatus is fully described in Home Office reports (5) (11). The basic principle is that the container is subjected to a test pressure that induces stresses in the container walls approaching the yield stress for the material. Any thinning of the wall will allow a permanent strain to occur whereby the volume of the cylinder will have a permanent increase. The rejection limit imposed is that the permanent increase in volume should not exceed 10 per cent of the temporary increase in volume occurring under the test pressure. This test entails having a method of accurately measuring small changes in the volume of the cylinder, and the two practical means of accomplishing the measurement are known as the 'water jacket method' and the 'non-jacketed method'.

Both methods require the exercise of considerable care to ensure that the results are accurate, because if the test rejects the cylinder and doubts arise as to the validity of the test, the cylinder is required to be reheat-treated in accordance with its manufacturing specification before a retest can be carried out.

The 'water jacket method' is preferred as it is less liable to operational error. It requires the cylinder to be enclosed in a vessel that is filled with water and fitted with a gauge glass projecting from its upper cover. The expansion in volume of the cylinder under the test pressure is indicated by the rise of the water level in the gauge glass, and the permanent increase in the volume of the cylinder after the test pressure has been released is similarly indicated. The 10 per cent rejection limit is thus readily ascertained. The results can be rendered invalid if a leakage of the test water into the jacket should occur from the cylinder connection.

In the 'non-jacket method' the cylinder, connecting pipes, and pump are filled with air-free water. The additional water required to raise the cylinder to the test pressure is drawn from a suitable level gauge of known bore. Following the test the water is released back into the level gauge; if any permanent expansion of the cylinder has occurred, it will be indicated by the reduced level of water inside the gauge level.

Substantial corrections to the readings are necessary to allow for the increase in the density of water at the test pressure, and the volume of the cylinder has to be known before the ratio of permanent stretch to temporary stretch can be ascertained. Results invalidating the test will occur if the cylinder has a film of rust or other foreign matter that absorbs water under the test pressure and retains it when the pressure is released. Air entrainment in the water, and failure to locate the pump plunger in the same position at the important stages of the test, will also lead to unsatisfactory test readings. The test pressure to be applied must be that specified in the applicable specification for the cylinder or container, and it should be held for at least 30 seconds.

Hydraulic proof pressure test

For this type of test the cylinder or container is filled with water, care being taken to ensure that all the air is displaced at the filling and in the early subsequent stages of building up the test pressure. The test pressure to be applied shall be that specified in the applicable specification for the cylinder or container. The test pressure shall be held for at least 30 seconds and throughout the subsequent time it takes to carefully scrutinize all the external surface area for signs of leakage or of abnormal deformation.

CYLINDERS OR CONTAINERS REJECTED AS A RESULT OF EXAMINATION AND/OR HYDRAULIC TEST

Any cylinder or container found to be unsatisfactory after visual examination or which gives any sign of leakage or visible permanent deformation in the proof test, or which shows a permanent stretch exceeding 10 per cent of the total temporary stretch in the hydraulic stretch test should be rejected and made unserviceable for containing gas under pressure. The action taken should ensure that there is no possibility of repair by unscrupulous or ignorant persons.

The report (11) states that in certain circumstances it may be justifiable for a container that is not considered suitable for use at its designed duty to be used for the conveyance of gas at a lower pressure, provided that the container remains in the ownership of the firm concerned, who must be the filling firm, and that the container is always filled and serviced by that firm, and that the approval of an inspecting authority is obtained before the container is taken into use for a pressure duty less than that for which it has been designated.

APPENDIX 51.1

REFERENCES

(1) HOME OFFICE. *Home Office Committee of 1895 on the Manufacture of Compressed Gas Cylinders* 1896. (This dealt with lap-welded wrought iron and lap-welded or seamless low-carbon steel cylinders. These recommendations are incorporated in S.R. & O. 1931, No. 679.)

(2) D.S.I.R. 'Recommendations for medium-carbon steel cylinders for "permanent" gases', *First Report of the Gas Cylinders Research Committee* 1921. (These recommendations are incorporated in S.R. & O. 1931, No. 679.)

(3) D.S.I.R. 'Periodical heat treatment', *Second Report of the Gas Cylinders Research Committee* 1926. (This report recommended that the periodical reheat treatment of carbon steel gas cylinders served no useful purpose and should be discontinued.)

(4) D.S.I.R. 'Alloy steel light cylinders (nickel and nickel–chromium–molybdenum steels) and duralumin', *Third Report of the Gas Cylinders Research Committee* 1929. (This report recommended Ni–Cr–Mo steel cylinders for 'permanent' gases.)

(5) D.S.I.R. 'Cylinders for liquefiable gases', *Fourth Report of the Gas Cylinders Research Committee* 1929. (This report deals with low-carbon steel cylinders. British Standard 401 was prepared to meet those recommendations.)

(6) D.S.I.R. 'Recommendations for the design and construction of forge-welded containers for chlorine, phosgene, ammonia and sulphur dioxide in the liquefied state', *Report of the Committee on Welded Containers.*

(7) D.S.I.R. Paper S.13, 'Liquefied petroleum gases', 1940 (4th December). (Tentative recommendation for small, fusion-welded containers.)

(8) D.S.I.R. Paper S.12C, 'Revised recommendations for welded containers for the commercial transport of chlorine, phosgene, ammonia and sulphur dioxide in the liquefied state', 1942 (11th September).

(9) HOME OFFICE. 'Note on document S.12C (reference 24)', 1957 (July).

(10) HOME OFFICE. 'Periodic examination and test of seamless cylinders and welded containers used for the conveyance of compressed gases', Note 256–58, 1958 (31st January).

(11) HOME OFFICE. *Report of the Home Office Gas Cylinders and Containers Committee* 1969. (Recommendations for the design, construction and maintenance of seamless and fusion-welded cylinders and containers.)

C52/72 THE EXAMINATION BY NON-DESTRUCTIVE TESTING METHODS OF WATER-TUBE BOILERS IN H.M. SHIPS

A. E. LINES* C. J. HOLMAN*

Non-destructive testing techniques have been developed by the Ship Department of the Ministry of Defence for the examination of the pressure parts of water-tube boilers in Her Majesty's ships. The paper discusses the suitability of the equipment and methods employed for the pattern of deterioration of boiler parts normally encountered and observed over many years. The use of ultrasonic viewing, measuring, and photographic devices to permit realistic assessment of present condition, and how these methods lead to accurate prediction of remaining service life or repair requirements are considered. Details are given regarding the application of these methods in the fleet, including the training of operators and documentation; also of on-board trials, which resulted in the establishment of a central boiler inspection unit.

EXAMINATION OF WATER-TUBE BOILERS IN H.M. SHIPS

THE METHODS of examining and testing the boilers of those of H.M. ships that work in the pressure–temperature range 400 lb/in^2 at 700°F to 700 lb/in^2 at 950°F are designed to facilitate reasonably accurate forecasting of the probable remaining safe life of the pressure parts, taking into account the future programme for operating and refitting the ship. This paper is primarily concerned with a description of the periodic examinations carried out by a boiler inspection authority to determine what is called the 'durability' of the boiler installation, and to recommend the repairs or renewals required immediately and those to be expected at the next period in a dockyard.

There are several special features about boilers in H.M. ships which must be recognized, particularly when considering any maintenance and inspection routines. These features can be summarized as: (*a*) the size, (*b*) the flexibility of output and hence variation of load on pressure parts, (*c*) the fuel burnt in the furnaces and resultant fireside corrosion pattern, and (*d*) the corrosive environment.

Some idea of the influence of these special features was given by Macnair (1)†. Figs 52.1 and 52.2, reproduced from Macnair's paper, illustrate the compact nature of a naval boiler. The generator and superheater tube banks are very closely pitched around a furnace that is only about 6–7 ft long. The steam drum is relatively small and contains elaborate steam separation gear to ensure steam dryness over wide and rapid changes of steam flow.

Although the utilization of a boiler in a Royal Navy ship (8000 h steaming may take up to four years to accumulate) is lower than, for example, a land-based power station, the operating conditions are far more arduous due to frequent changes of power. The boilers will frequently be required to cycle between 10 and 90 per cent of full load in 20 s or so. Exceptionally, steam will sometimes be required to be raised from cold in 20–30 min; and the turn-down range of the combustion equipment may be about 16–20:1.

At present there is a necessary logistic requirement to use a common fuel in the mixed propulsion systems fitted in a number of warships. Most of H.M. ships now burn a distillate fuel (Dieso) in the boilers, but must remain capable of burning the F.F.O. (a residual blended with a distillate fuel), which all ships previously burnt, and which is still used in some older ships.

The boilers of H.M. ships are subjected to a severely corrosive environment because they operate in a salt-laden atmosphere, which affects external corrosion, and because internal corrosion can be initiated by chloride contamination of boiler feed water occurring through heat

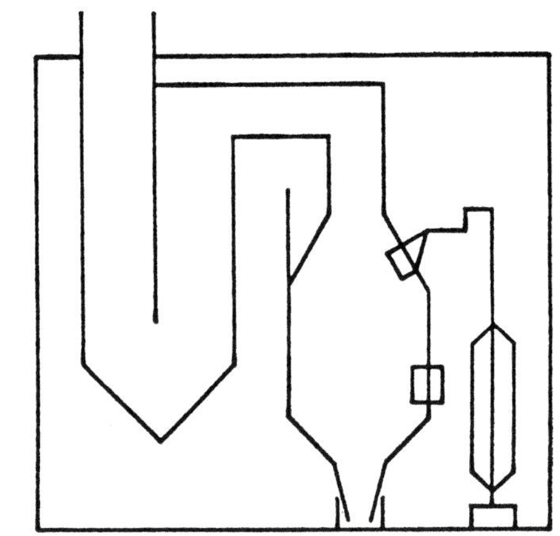

(*Left*) An oil-fired marine boiler. Evaporation 240 000 lb/h.
(*Right*) A pulverized-fuel-fired land boiler. Evaporation 240 000 lb/h (after Carlson).

Fig. 52.1. Comparison in size of marine and land-based boilers

The MS. of this paper was received at the Institution on 19th January 1971 and accepted for publication on 7th February 1972. 33

* *Ministry of Defence, Ship Dept, Foxhill, Bath BA1 5AE.*

† *References are given in Appendix 52.1.*

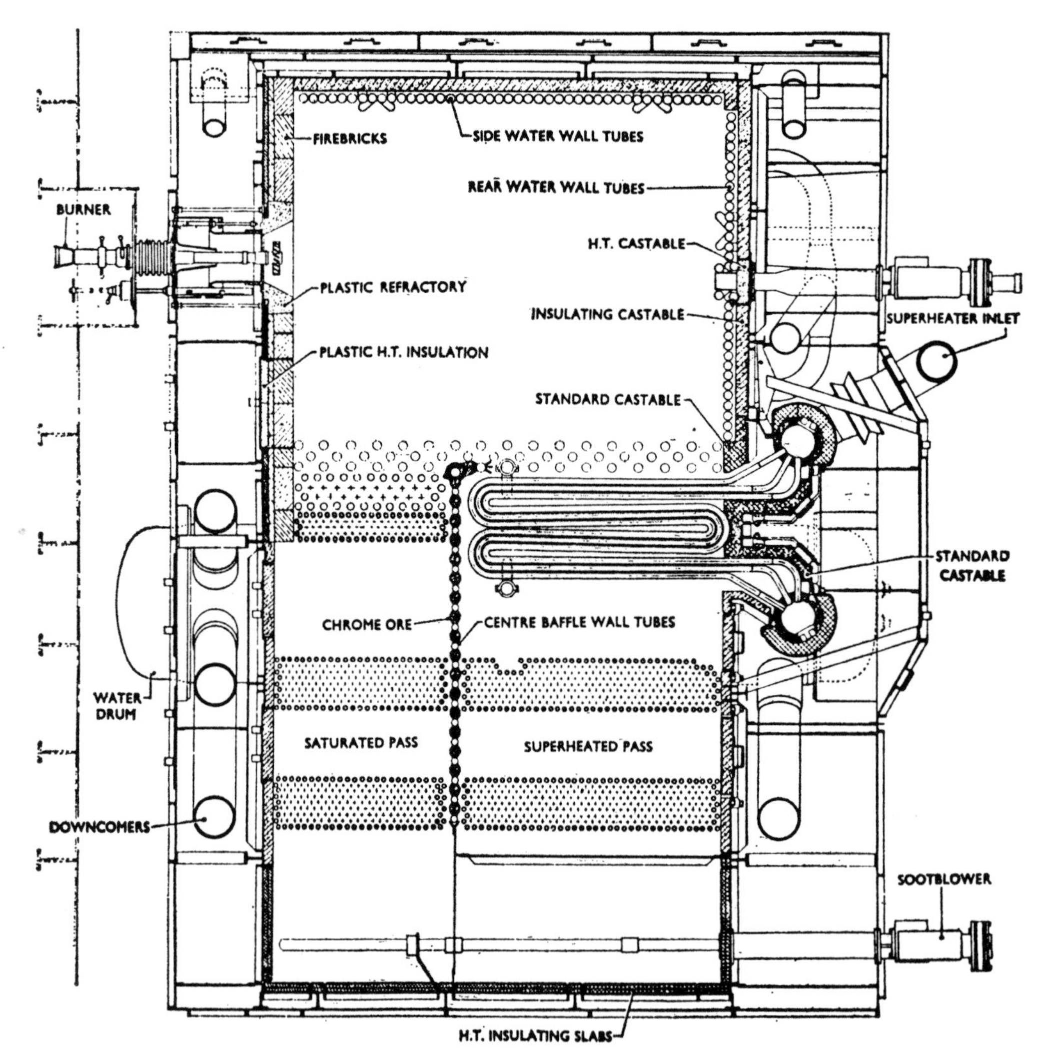

Fig. 52.2. Sectional plan of marine boilers

DURABILITY IN YEARS	TUBE ORIGINAL THICKNESS IN INCHES									
	▼ 0·092	0·104	0·116	0·128	0·144	0·156	0·160	0·176	0·192	0·250
	REMAINING MINIMUM THICKNESS BY GAUGE IN INCHES									
8	0·092	0·104	0·116	0·128	0·144	0·156	0·160	0·176	0·192	0·250
7	0·088	0·095	0·106	0·117	0·132	0·143	0·146	0·161	0·176	0·229
6	0·080	0·086	0·096	0·106	0·119	0·129	0·133	0·146	0·158	0·207
5	0·072	0·077	0·086	0·095	0·107	0·116	0·119	0·131	0·143	0·186
4	0·064	0·069	0·077	0·084	0·095	0·102	0·105	0·115	0·126	0·164
3	0·056	0·060	0·066	0·073	0·083	0·089	0·091	0·100	0·110	0·143
2	0·048	0·051	0·056	0·062	0·070	0·075	0·078	0·085	0·093	0·121
1	0·040 ▲	0·042	0·046	0·051	0·058	0·062	0·064	0·070	0·077	0·100

The above durabilities of tubes are in respect of pitting or local wastage and are subject to amendment at the discretion of the Examining Officer in cases of rapid deterioration, connected pitting, cracking or external wastage.

Fig. 52.3. Estimated durability of boiler tubes

exchanger–condenser tube end leakage. A disciplined inspection routine is therefore essential.

Until about two years ago it was the practice to carry out a 'wear and waste test' of boilers at about three-year intervals. Sample tubes, considered to be typical and representative of the worst tubes (in terms of material condition), in each tube bank were removed from the boilers. These were cut longitudinally, visually examined for scale, deposits, etc., before cleaning, and, after acid cleaning, the remaining wall thicknesses at pits and areas of external corrosion were measured by micrometers. In general, the estimated remaining safe life of the tubes left in the boiler bank under consideration was determined by reference to a table (Fig. 52.3). There were occasions when direct reference was misleading, and a knowledge of corrosion patterns to be expected was necessary for a reasonable interpretation of the figures available.

It will be evident that such a method of boiler examination had the following serious disadvantages:

(*a*) The destruction of nominally sound equipment in order to assess its likely future reliability. This is a most undesirable technique.

(*b*) The removal of representative tubes had to be carried out with considerable care to avoid damage to the drums or headers. But no matter how carefully carried out, mutilation of tube ends was inevitable—with the consequent loss of valuable evidence about the material condition of the tubes.

(*c*) Tubes selected for removal for wear-and-waste test purposes were expected to be those considered to be representative of the boiler as a whole. This was not always an accurate assessment. Frequently, it resulted in the removal of those tubes that were the easiest to get at. Sometimes those removed merely for access for repairs to brick work or baffle walls were used as the 'representative' tubes (Fig. 52.4).

Other pressure parts of boilers were carefully visually examined. In cases of excessive corrosion, test holes were drilled to permit determination of remaining thicknesses. In the type of boilers now in use it is generally not possible to remove individual superheater tubes or elements, or economizer elements, without disturbing others. For this reason it was ruled at the outset that a six-year limit was to be placed on the life of these tubes. When renewed, the

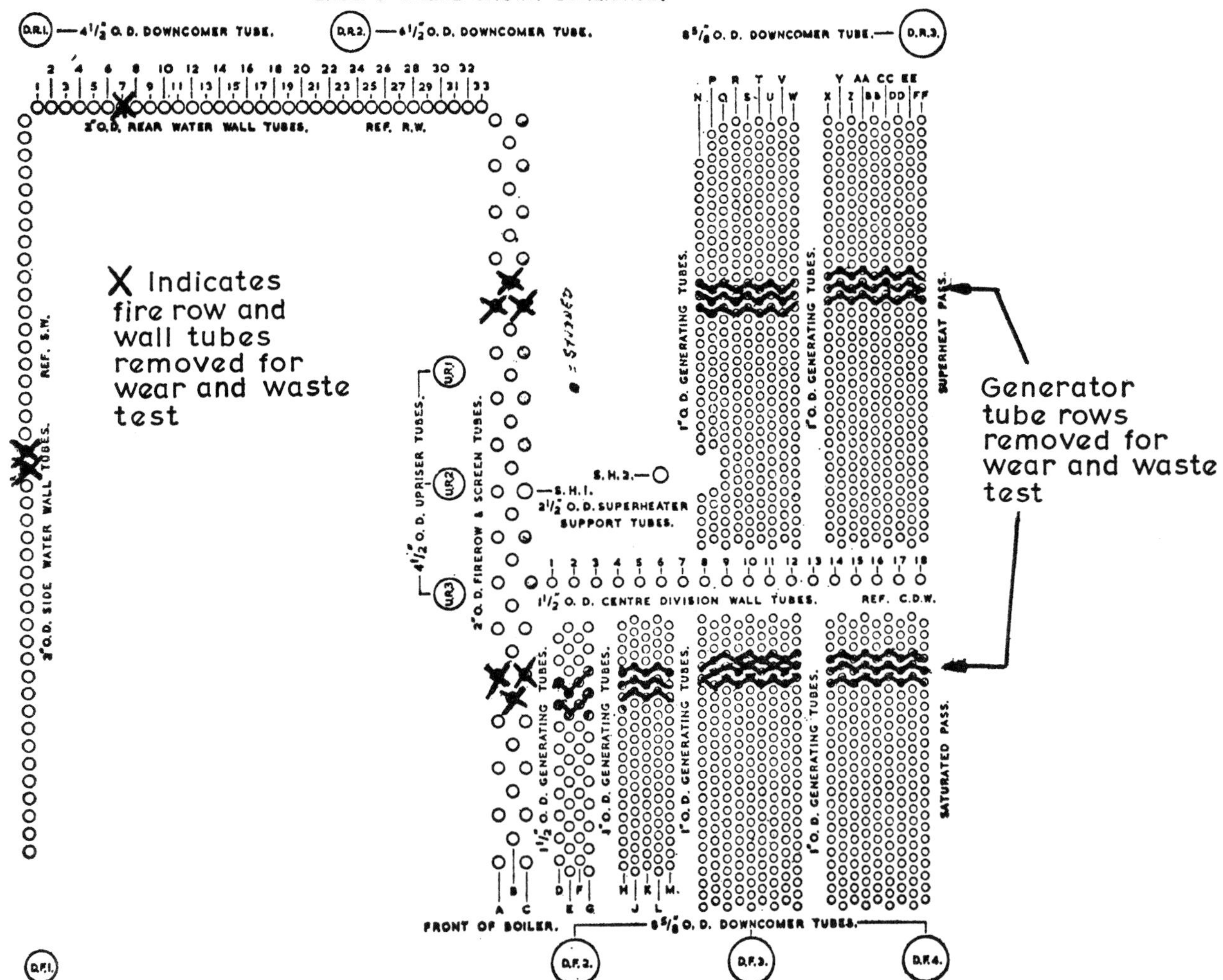

Fig. 52.4. Typical tube chart showing tubes removed for wear-and-waste test

old elements were cut up and gauged and the thicknesses recorded. An analysis of these records was carried out in 1969; as a result, superheaters are now considered not to require renewing until the mid-life refit of the ship, and may remain in service up to 10 years. Nevertheless, there will still be unavoidable isolated tube failures. Economizers still require renewal at six-year intervals.

The valuable experience gained from this practice of tube removal has been utilized in developing the techniques now used in assessing the durability of boilers. A pattern of corrosion had been identified, and provided there were no abnormal operating conditions, e.g. a failure of the boiler water treatment routine, it was reasonable to assume that this pattern would be repeated. In general, it had been observed that the air-bubble pitting was found mostly at the top ends of generator tubes and around the belling in the steam drum; that reduction of external diameter due to fireside corrosion occurred at the bottom ends of generator tubes, particularly at the tube roots; and that if hard-scab pitting were to be found at all, it would be in the hottest parts of the fire-row tubes.

A very full description of marine boiler deterioration is given in reference (2). B.R. 1335 is also considered to be a useful document of general interest (3), and is available to the public.

The development in the M.O.D. (Ministry of Defence), Ship Department, of N.D.T. (non-destructive testing) techniques to determine the material condition of a boiler has to some extent been governed by the identifiable pattern mentioned above.

N.D.T. EQUIPMENT

The equipment that is in current use on board H.M. ships comprises: a mirror probe set, endoscopes and pit-depth gauges, ultrasonic flaw detectors and specially designed transducers, a camera, and a portable tape recorder. A description of each set of equipment and how it is used now follows.

The mirror probe set

This set is a series of fixed, flexible, and telescopic rods (extended length, 45 in) fitted with a variety of mirrors, which can be readily attached to the viewing end. Illumination is provided by means of a torch housed in the handle. There is also a hook and magnetic end attachment, which has proved most useful for the retrieval of items in otherwise inaccessible positions.

The mirror probe set is relatively cheap, robust, packed in a wooden carrying case, and has withstood on-board service conditions. It is used for visually checking the external conditions of remote areas, between banks of tubes. The surface condition and the presence of corrosion and/or chemical deposits are noted; from these indications it can be seen where special attention should be paid during the internal examination.

Endoscopes and pit-depth gauges

The endoscopes, which are 8 mm diameter, are of the rigid, projected, optical, cold light type, 15 in and 30 in long, capable of forward or lateral viewing. The lens system is wide angle: 50° in the forward looking and 45° in the lateral probe (Fig. 52.5).

The power pack is fitted with a quartz iodine bulb, which provides the light transmitted to the endoscope by means of a flexible fibre, optic 'light pipe'. The amount of luminosity at the viewing end is controlled by the operator, who adjusts the selector switch fitted to the power pack; this can be operated from 110–240-V a.c. supply.

The normal subject distance, i.e. the distance between the object lens and the subject at which the latter will appear normal size, is approximately 2 in for endoscopes of this type with the angles of view quoted (Fig. 52.5). Inside this distance there will be magnification, the degree of which increases with the decrease of subject distance. It is important that these properties should be recognized during initial training, so that correct recognition of subject matter may be made and misjudgements avoided. Correct

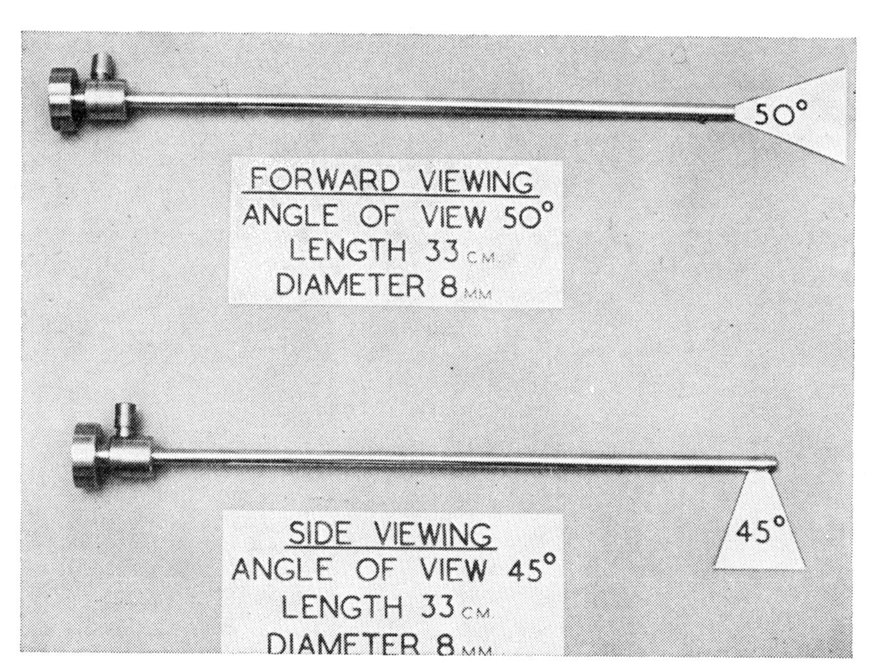

Fig. 52.5. 8-mm endoscopes, showing angle of view

assessment of defective areas is a matter of experience, plus readily available comparison between the view obtained on site and that given by specific defects–surface conditions when viewed at a similar subject distance.

The endoscopes are used for the internal examination of the top and bottom lengths of the generator tubes, the operator working from the steam and water drums of the boiler. By these means, defects such as pits, surface scars, scabbing, etc., can be detected.

Where a pit is detected by endoscope during the inspection, the actual depth is measured mechanically by the specially designed pit-depth gauge (Fig. 52.6). The pit depth can be measured to an accuracy of 0·025 mm (0·001 in) when used by an experienced operator. The pit-depth gauge was developed by the M.O.D., Ship Department, to be used in conjunction with the endoscopes. Basically, it consists of two tubes, one sliding over the other and actuating the moving part of a mechanical gauge. The instrument, which fits over the endoscope (the depth-gauge tube is cut away at the object lens, permitting the stylus to be observed), is located in the pit, the depth being measured by rotating the micrometer drum at the eye end.

The pit-depth gauge is manufactured in 15-in or 30-in lengths. During the inspection the locations of all pits in excess of 0·020-in depth are recorded. Should these coincide with areas of external corrosion, then this internal measurement is also taken into account in determining remaining wall thickness.

Ultrasonic flaw detector and probes

The ultrasonic flaw detector used is battery operated and of conventional design. Its overall dimensions must permit it to pass with ease through the manhole (16 in × 10 in) into the steam and water drum, and it must be capable of resolving thickness changes of a few thousandths of an inch. A suitable flaw detector, with built-in monitor and audible alarm, was commercially available. This was not the case with respect to a suitable ultrasonic probe.

Therefore special ultrasonic probes were designed by

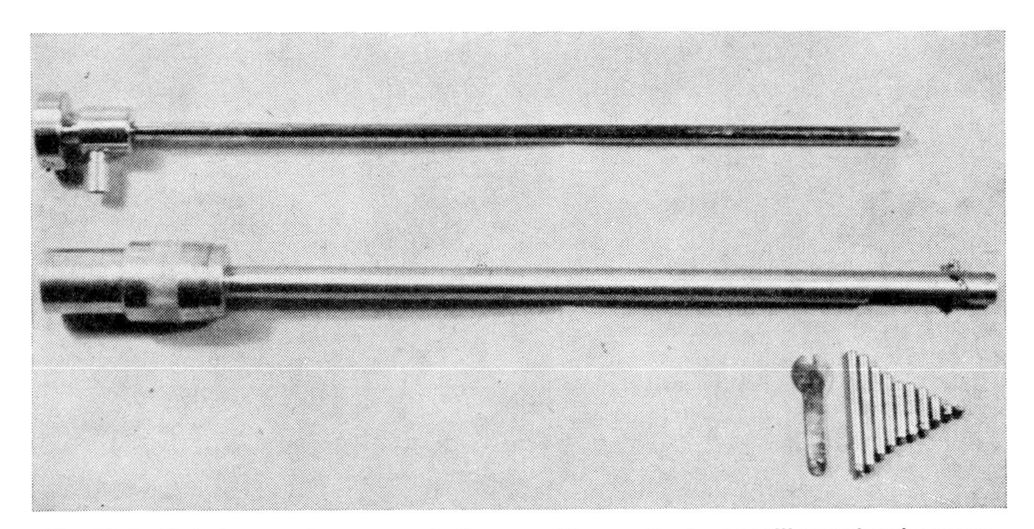

Fig. 52.6. 15-in-long endoscope and pit gauge. The anvils shown will permit tube up to 2-in bore to be measured

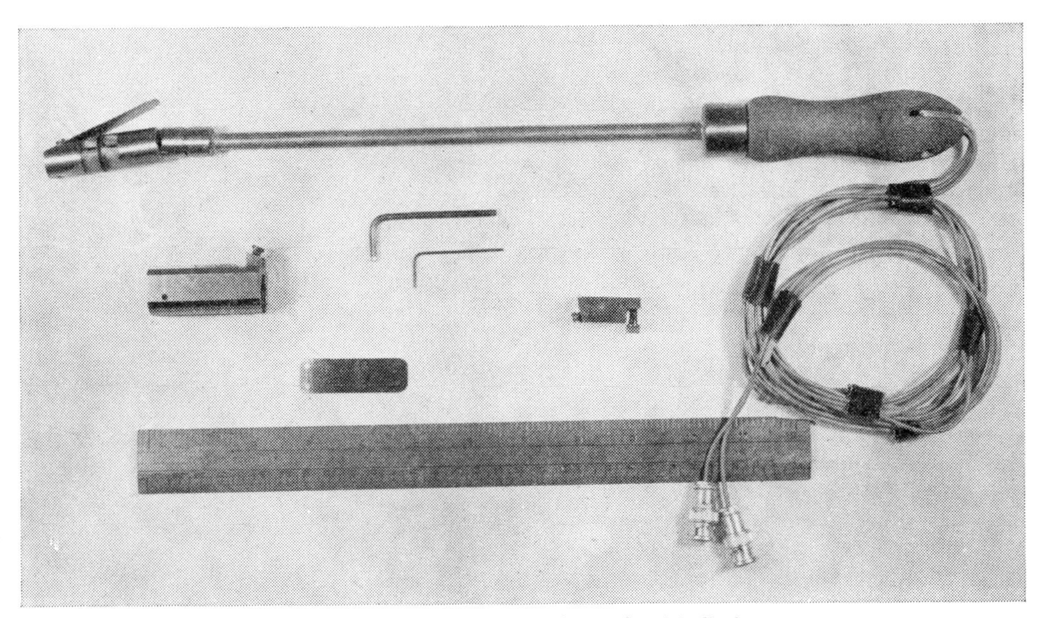

Fig. 52.7. Ultrasonic probe (short) with fittings

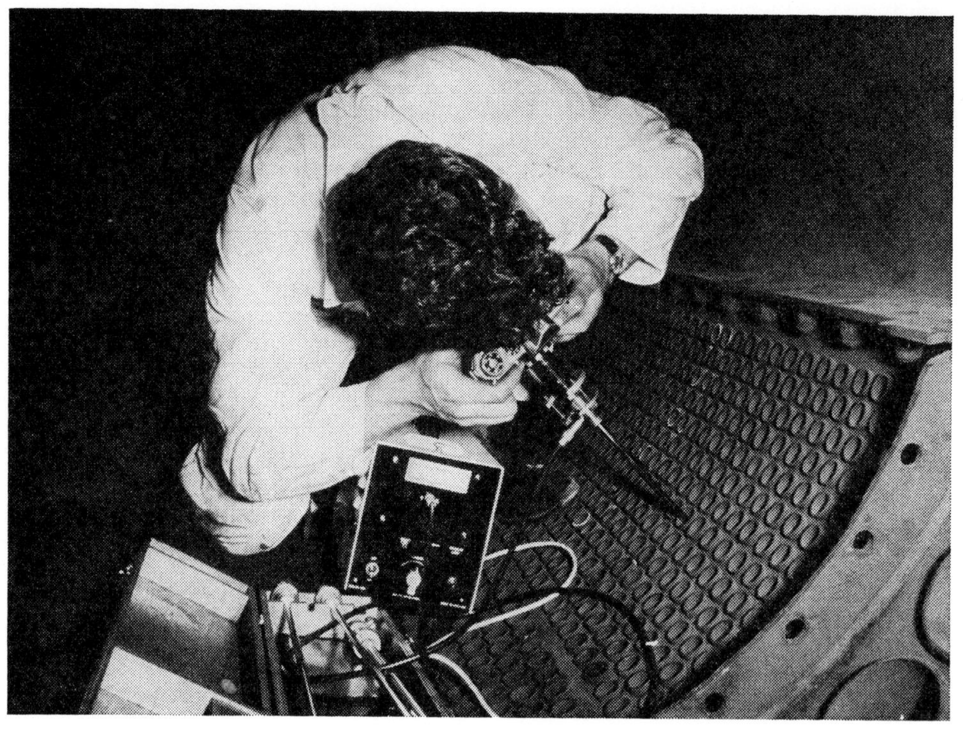

Fig. 52.8. Endoscope and camera attachment in the steam drum

the M.O.D., Ship Department, N.D.T. section (Fig. 52.7). Two types were manufactured; one was short-handled of approximately 16-in length, the other was 15 ft long to cover the total length of the generator tubes from steam drum to water drum. The ultrasonic examination is carried out from the base of the tubes, the smallest of which are of carbon steel 1 in o.d. with a wall thickness of 0·092 in or 1·104 in. The inspection requirement was to thickness gauge the tube to the nearest 0·25 mm (0·010 in) over the range 0·50–0·104 in. The estimated service life is then assessed by the thickness recorded.

The transducer head is standard for each probe and incorporates twin 5MHZS crystals 5×4 mm separated by an acoustic barrier. An irrigation tube is drilled through the brass head to the front of the crystal, which permits the couplant to be applied ahead of the probe. Whilst in the tube, the probe is held in the correct orientation by a flat plate spring. A series of springs and attachments is provided to cater for tubes from 0·8-in to 2-in bore, as shown in Fig. 52.7. The long probe is attached to a nylon tube, which is marked at 12-in intervals; the head is attached by a universal joint, which allows it to negotiate the bends in the tubes.

Photographic facilities and recording facilities

Where an interesting or unusual internal condition is detected by endoscope, e.g. internal pitting, chemical deposits, hard or soft scabbing, a camera is attached to record in colour the actual internal conditions of the tubes prevailing at the time (Fig. 52.8). The camera used is a single-lens reflex camera with clear-glass screen and a specially adapted lens. A high-speed colour film has proved to be most satisfactory; a typical example of an internal photograph is shown by Fig. 52.9, unfortunately not reproduced in colour. The M.O.D., Ship Department, has compiled a library of several hundred coloured slides, which register the internal conditions of the tubes throughout the durability inspections. A pocket-size, portable tape recorder is an asset under the cramped conditions, especially in the water drum where it was far easier to speak and record findings than to write notes.

The method of producing permanent records for the entire durability inspection was devised by the fleet work study team. All the equipment referred to is currently available from commercial sources. The approximate cost of a complete set is £2,500.

The organization and administration of the central boiler inspection unit will now be discussed.

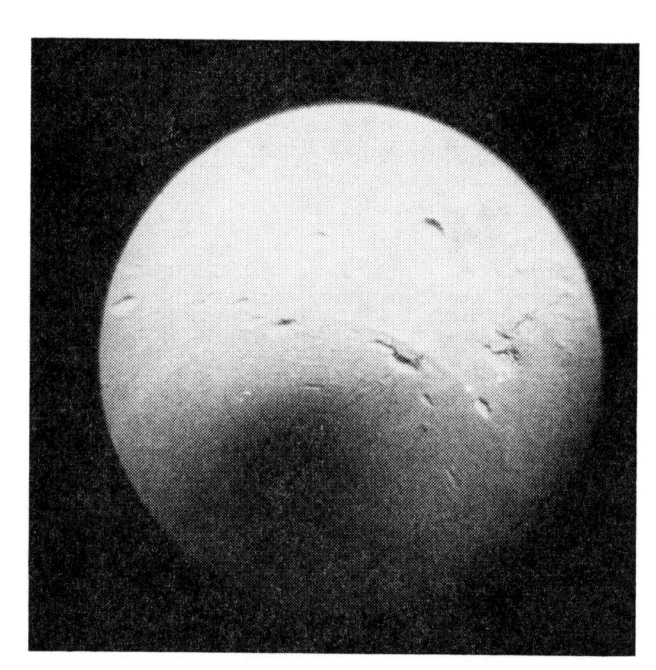

Fig. 52.9. Typical example of internal pitting

APPLICATION IN THE FLEET

Development of the techniques continued to the point where it was considered that pilot schemes could be implemented in the fleet and training of the operators begun. It was decided that a central boiler inspection unit would be established within the fleet technical staff under the Commander-in-Chief. The Fleet Marine Engineer Officer was designated as the examining officer, and a lieutenant-commander appointed as officer in charge of the unit. The unit consists of four C.M.E.As (P.) [chief marine engineering artificers (propulsion)], who operated as two teams, each team being capable of carrying out a full durability inspection of one boiler in five days. The capacity of the unit—bearing in mind that the work is extremely taxing, and allowing for leave, travelling, etc.—is about 60 boilers each year. It is not possible to maintain the very high standard for periods longer than 5 or 6 h each day.

Technical support for the unit

The unit is accommodated at the Admiralty Marine Engineering Establishment, Haslar, where it has access to stores, equipment, maintenance facilities, transport, and office services; here all the expertise is concentrated, this being the M.O.D. (N.) principal boiler research and development establishment. During the early days of the unit's existence the availability of this expertise was invaluable, and it supplemented that of the M.O.D., Ship Department, boiler specialist group. This department continues to monitor the performance of the teams and to investigate improvements in techniques.

TRAINING

The C.M.E.As (P.) were selected without reference to any special boiler experience they possessed. Indeed, the selection and subsequent training were aimed at avoiding preconceived conclusions on boiler condition that might have been derived from past experience. A two-week period at the ultrasonic equipment manufacturer's school was supplemented at the M.O.D., Ship Department, N.D.T. training school, until a thorough grounding in ultrasonic thickness measurement was achieved, and a measuring accuracy tolerance of 0·1 mm (±0·005 in) was consistently achieved. The method of interpreting corroded area, pit size, and pit depth through the monocular image presented by the endoscope was practised over a period of several months until, by using shadow effect and perspective, the error in estimating pit depth was seldom greater than 0·2 mm (0·010 in). The pit-depth gauge probably presented the greatest challenge, considerable dexterity being necessary to avoid straining the operating head or under-assessing pit depth. Quite remarkable standards were achieved after almost six months' training, and the M.O.D., Ship Department, constantly auditing progress, was satisfied that the accuracy tolerance was consistently within 0·1 mm (±0·005 in), which is the wear figure used in calculating one year's loss of durability of a boiler pressure part. The C.M.E.As (P.) were then given instruction in boiler design and maintenance by the M.O.D., Ship Department, boiler specialist group, and at H.M.S. *Sultan*, the Royal Naval Marine Engineering School. At this point it was necessary for the unit to accumulate knowledge of wear and corrosion patterns in modern boilers so that the areas of greatest wear could be anticipated and thus receive special attention. The environmental problem had not been neglected, and great determination was required to overcome the difficulties of operating sophisticated equipment in cramped, dirty, and wet furnaces, and in drums and headers. The equipment has proved to be remarkably durable, surviving not only use in boiler drums but also rough handling in transit.

REPORTS

The central boiler inspection unit produces an interim report on the final day of the inspection with the sole purpose of informing the operating and repair authorities of the condition of the boilers and the recommended repair work necessary to ensure the required durability, i.e. the interim report is a summary of action required and does not attempt to tabulate wear dimensions. The main report is produced in very comprehensive detail as a permanent record. The ship produces a report of water pressure test on completion of all boiler work, and this document details the work *actually* carried out and is distributed to the holders of the main report.

TEST PROCEDURE

The method by which the remaining tube wall thickness is measured has been defined in the M.O.D., Ship Department, 'Ultrasonic test procedure', No. 21. The ultrasonic measurement has been combined with visual and mechanical search to produce the following overall test procedure.

(*a*) Visually examine all generator tubes externally, using endoscopes and mirror probe set, with particular attention being paid to the water-drum tube ends, where the majority of deposits collect and corrosion occurs.

(*b*) Visually examine the internal surface of not less than 25 per cent of all the generator tubes, using endoscopes to examine up to 1 m of tube from both the steam and water drums. Particular attention is given to those tubes on which external corrosion was detected in (*a*). The location and depth of all significant pits, generally those in excess of 0·2 mm (0·010 in) deep, are recorded.

(*c*) The tubes examined in (*b*) are then searched ultrasonically at the top and bottom ends, 10 per cent are examined along their full length by an ultrasonic probe, and the remaining wall thickness measured and recorded (Fig. 52.10).

(*d*) The minimum remaining wall thickness is calculated by deducting the measured pit depth from the wall thickness, and this dimension is used in estimating the remaining tube durability.

(*e*) The significant features of the examined area are photographed for record purposes, deposits both internal and external are collected for analysis, and in cases where tube durability has been assessed as less than three years a few tubes are removed for destructive examination to confirm the non-destructive analysis.

The procedure is constantly under review as experience is gained of each type of boiler. The percentage of tubes examined is always in excess of the minimum required and the unit has complete discretion to extend its search in particular areas should the symptoms suggest that such a course is necessary. Tube remaining thickness is not the only factor considered in determining the probable serviceability of the boiler, and where there are other

Fig. 52.10. Ultrasonic thickness gauging in water drum

features considered to be significant, the examining officer may specify that the date of the next inspection should be advanced.

Non-pressure parts

This paper is not concerned with the non-pressure parts of the boiler, i.e. casings, structure, brickwork, combustion equipment, etc., but the central boiler inspection unit is required to examine and report on all these features concurrently with the examination of the pressure parts.

FUTURE DEVELOPMENTS

The area particularly vulnerable to corrosion from damp deposits is the external tube surface immediately above the water drum. This area gives erratic response to longitudinal ultrasonic waves due to the interface of tube and tube plate with the additional complication of tube expander marks. In order to detect this corrosion more readily, the M.O.D., Ship Department, N.D.T. section, is developing alternative probes. It is increasingly clear that viewing equipment capable of examining the whole tube length (approximately 3 m) is necessary. The device must be flexible, and the viewing head capable of circumferential scanning and indexing. Suitable fibrescopes are being evaluated for performance and to assess the vulnerability of optic fibres to damage in a boiler.

CONCLUSIONS

The assessment of the durability of a boiler and the basic repairs necessary are now determined without the need to remove tubes or to destroy its pressure integrity. This practice has been in use in H.M. ships for approximately two years, and the procedure and equipment are continually being improved to take account of problem areas, such as the generator tubes root external corrosion. Recently, rapid wastage has been experienced in a very narrow band above the tube plate. This is believed to be due to the gathering of cold end deposits, associated with Dieso burning, which have gravitated to this area. In our boilers some parts of this area are very difficult of access, and particular care is now being taken with this aspect of inspection.

It is considered that the inspection costs arising from the use of this N.D.T. technique are adequately compensated by the elimination of risk of damage to tube plate and tube hole tightness which has been associated with tube removals for destructive testing. It is estimated that a saving in dockyard direct labour costs, solely for the removal and replacement of wear and waste test tubes, of the order of £15,000 per year has been achieved.

It should be noted that this inspection procedure achieves a wider coverage of the boiler parts than the earlier destructive method, and hence a more reliable prediction of general condition is provided.

ACKNOWLEDGEMENT

This paper is published by permission of the Director-General Ships, Ministry of Defence, but any opinions expressed are those of the joint authors. Neither the Director-General Ships nor colleagues within the Department may necessarily agree with all that has been written.

APPENDIX 52.1

REFERENCES

(1) MACNAIR, E. J. 'Naval boilers', Paper 1, *Symp. Combustion in Marine Boilers* 1968 (10th January) (Inst. Mar. Engrs and Inst. Fuel).

(2) SLATER, I. G. and PARR, N. L. 'Marine boiler deterioration', 1949 (4th February) (Instn Mech. Engrs, London).

(3) B.R. 1335, 'Boiler corrosion and water treatment'.

C53/72 INSPECTION OF PRESSURE VESSELS IN A LARGE OIL REFINERY

J. E. MACADAM*

The refinery inspection department has specific objectives and is organized and staffed to meet them. Pressure vessel inspection is an important part of its work, beginning during the design stage, continuing with outside inspection at manufacturers' works, then with inspection during site installation and commissioning, when all initial information is recorded. The frequency of subsequent inspections is determined by the vessel's duty, what happens to it on stream, and its condition when shut down, as well as other factors. The interpretation of the measurements given by the tools, instruments, and techniques used for inspection is important, as is a good recording system and an understanding of the process and the factors influencing corrosion.

INTRODUCTION

PRESSURE VESSEL INSPECTION forms a large part of the work of a refinery inspection department and is necessary to ensure plant reliability and safety. The intervals at which inspection is carried out and the success in avoiding unplanned shutdowns are important factors in the achievement of profitability of the plants of which the vessels form a part.

REFINERY INSPECTION DEPARTMENT, ITS OBJECTIVES, ORGANIZATION, AND STAFFING

Objectives

The refinery inspection department is charged with the task of improving short-term and long-term plant availability. This it does by inspecting and reporting on present condition, recommending repairs, replacements and suitable materials of construction, advising on limiting operating conditions, and proposing the length of time that equipment can operate before further inspection is required. Inherent in the concept of improving plant availability is the need for safety. Plant is not available in the full sense of the word unless it can be operated and maintained without danger to equipment and people.

It is important that the long-term position, as well as the short-term, is considered when advice on improving availability is given and steps avoided that would give immediate relief to a problem, only to increase its size and cost in the future. Stop-gap solutions are, however, sometimes necessary, providing more permanent measures are planned for the near future. In addition to the safety aspect, the overall economics of a situation must be considered when advice is given; such factors as production losses by extending an unscheduled plant shutdown, availability of maintenance resources, future plant programming requirements, etc., have to be taken into account. The advice given must in itself be the least expensive from both the long- and short-term points of view; the use of expensive materials that are too good for the purpose, and repairs and replacements that are difficult and costly to implement, must be avoided.

Organization

Although the inspection department is concerned not only with unfired pressure vessels but also with piping, boilers, furnaces, tanks, etc., nevertheless pressure vessels form a large part of the work. The organization shown in Fig. 53.1 has been found to function satisfactorily and to provide the necessary flexibility. As can be seen, the inspectors are organized on a geographical as well as a specialist basis. The area inspectors in the sections of the refinery where the pressure vessels are situated, in addition to those in the offsites area and on new construction, are disposed geographically; but they have the support of specialist inspectors concerned with statutory inspection, welding, and non-destructive testing, and of the corrosion and metallurgy section. The numbers in the geographical areas vary depending on the workload, and when a plant in a particular area is shut down for inspection and maintenance the inspectors in the other areas are reduced and concentrated on the shutdown.

Staffing

The ability of an inspection department to attain its stated objectives is dependent on the quality of its inspectors and on the way they are led and organized. The following are the tasks of inspectors:

(*a*) Forecasting maintenance requirements to ensure that equipment is maintained in a safe, efficient, and economically reliable condition.

(*b*) Ensuring that acceptable standards and codes of engineering design and practice are adhered to.

(*c*) Devising suitable modifications to machines, structures, and materials to improve economic plant operation and reduce maintenance effort.

To carry out these tasks adequately they must be conversant with the plants in their areas and their past history.

The MS. of this paper was received at the Institution on 2nd November 1971 and accepted for publication on 7th February 1972. 23

* *Superintendent, Engineering Services Department, Shell U.K. Ltd, Stanlow Refinery, Ellesmere Port, Wirral, Cheshire L65 4HB.*

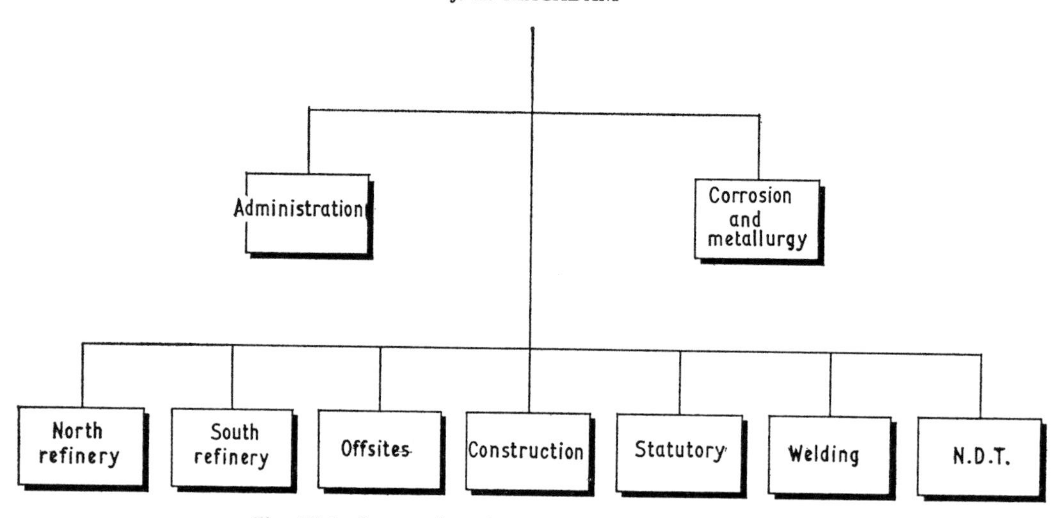

Fig. 53.1. Inspection department organization chart

They must have daily contact with operating and maintenance staff in order to know what is happening from a process point of view, what work the maintenance people have in hand or are planning, and they must put forward inspection priorities.

The sort of person who makes a good inspector is, obviously, one who is interested in the type of work that he must perform, who does not mind crawling into hot, dirty, inaccessible places, who is curious and sharp-eyed, and who knows what to look for. He must be able to persuade others to do as he recommends and earn their respect as someone who is knowledgeable and authoritative. He must also be able to express himself clearly and simply, both orally and in writing. Such people have a variety of backgrounds, but they have all had practical engineering experience and have attained the necessary technical qualifications. So well do those with engineering sea-going experience fit the technical and practical requirements that the minimum qualifications can be put at 1st Class M.O.T. Certificate, 2nd Class M.O.T. Certificate with Steam and Motor Endorsements or equivalent (e.g. H.N.C. and, say, 10 years' experience, half of which has been in a junior inspecting capacity).

CHECKS DURING PRESSURE VESSEL DESIGN STAGE TO ENSURE PAST EXPERIENCE IS INCORPORATED

The responsibility for obtaining vessels that are in every way suitable for the intended service rests with the project engineering department, it being the job of the inspection department to advise and carry out its inspection role, as opposed to a supervisory one.

Inspection interest in pressure vessels commences during the design stage. Data/requisition sheets (Figs 53.2*a*, *b*, and *c*) and preliminary drawings are spot-checked to ensure compliance with the appropriate design codes and with company standards, which may be more stringent than the codes. Suitability of the materials, the finishing heat treatment, and the type of construction for the intended service conditions are checked. In general, every endeavour is made to ensure that local experience and lessons learned from past problems are incorporated into the design.

Pressure vessel manufacturers' drawings are again spot-checked, and this may lead to further comment over their interpretation of requirement, although this is generally the responsibility of an independent inspection authority. It is generally possible at this stage to decide which vessels should have corrosion test spools and corrosion test probes installed, and arrangements can be made to have the appropriate brackets and nozzles fitted.

INSPECTION AT MANUFACTURERS' WORKS

Inspection of pressure vessels continues at manufacturers' works, though this is not normally the responsibility of the refinery inspection department. The company employs outside inspectors who, when purchase orders are placed, have an overall responsibility for this inspection, independent inspection authorities being assigned specifically for pressure vessel and 'statutory' orders, such as boilers and steam and air receivers, etc. The inspection authority takes responsibility for scrutinizing the manufacturers' drawings and designs, inspects the materials incorporated into the vessels, and ensures that fabrication is up to standard. The authority will ensure that specific requirements incorporated into the purchase order or data/requisition sheet are observed, e.g. for vessels that are to operate in hot hydrogen service, full ultrasonic flaw detection for laminar and other defects in the alloy material will be called for. A final certificate of acceptance is obtained by the purchaser from the inspection authority on satisfactory completion and testing of the vessel, and a manufacturer's data sheet prepared and sent to the refinery inspection department for incorporation into their records.

When orders are placed by design and engineering contractors on behalf of the company, the contractor will employ his own inspectors in addition to those of the independent inspection authority nominated by the company. But whether or not orders are placed by the company direct or by a contractor, good communication between the inspection bodies is essential, and intermediate reports are called for and circulated. By this means, early notification of intended repairs and agreement to deviate from specifications can be obtained and records amended as required.

The opportunity is taken during fabrication to obtain offcuts of construction materials, which, in certain cases,

Data/requisition sheet for PRESSURE VESSELS (Columns, reactors, accumulators, etc.) - cont. sheet	Design book No.: page
	Contr. Job No.:
	MESC No.:

MATERIAL SPECIFICATION

Parts	DIN & ASTM spec. No.	Parts	DIN & ASTM spec. No.
Shell	HII	Downcomers	
Liner of shell	–	Baffles	
Heads	HII	Internal pipe fittings	
Liner of heads	–	Stud bolts, external	A 193 - B7
Reinforcing rings	HII	Nuts, external	A 194 - Gr. 2H
Skirt, base plate, etc.	ST.37.2	Bolts, internal	
Saddles	–	Nuts, internal	
Jacket	–	Gaskets, external	G.A.F.
Shell flanges	–	Gaskets, internal	
Nozzles (line pipe/plate)	ST.35.8/HII	VORTEX BREAKER	ST.372
Liner of nozzles and manholes		SCHOEPENTOETER S.S.	
Flanges (USAS)			
Flanges (Non-USAS)			
Welding fittings			
Stiffening rings			
Insulation support rings	ST.37		
Cleats for platforms, etc.	ST.37		
Internal parts			

FABRICATION AND INSPECTION REQUIREMENTS

Construction according to: BS.1515 PART 1 1965

Inspection	full [illegible]
Inspection authority	NOMINATED BY SHELL PER EACH CONTRACT
Stress relieving	NO
Special heat treatment	NO
Radiography	FULL
Other non-destructive testing	
Chemical analysis	
Manufacturer's certificate - chemical analysis	YES
- mechanical data	YES

WEIGHTS

Erection weight (shipping weight):	60,600 kg	Weight of internals:	700 kg
Total weight, operating:	201,330 kg	Weight of insulation:	4470 kg
Total weight, full of water:	358,400 kg	Weight of fireproofing:	35300 kg

REFERENCE DRAWINGS/LISTS

Arrangement - construction [illegible]:	T1051606-A		
Standard vessel:			
Additional drawings:			
List of appr. welding electrodes, rods, etc.: DEP 40.10.64.10–Gen., LIST E1,2,3,4 and 6			
General remarks for vessels:	T.1051680	Anchor bolt ring and base plate:	S.20004
Flanged pipe nozzles:	S.10.001	Lifting lug:	
Thermowell nozzles:		Name plate:	S.10026
Mild steel flanges:		Support ring for insulation:	S.20003
Vortex breaker:	T.919910	Inspection hole/hand hole/	
Skirt/[illegible]:	S.20001	manhole/davits, etc.:	R61005

Fig. 53.2*a*. Typical pressure vessel data/requisition sheet, Part 1

can be used subsequently as service test pieces within the vessel. They can be used also for destructive mechanical tests or for chemical tests to determine the effect of operations on the vessel, thus helping to decide on inspection intervals and any modifications to the vessels or process that may become necessary.

INSPECTION AND CHECKS DURING SITE INSTALLATION AND COMMISSIONING

The next stage in the inspection of pressure vessels takes place when they are delivered to the refinery. Initially they are given a cursory inspection to check for external transit damage, e.g. bent nozzles, damaged joint faces,

EQUIPMENT No: V-201			Number required : 1	
		OPERATING/MECHANICAL DATA		
Description				Unit
Contents				
Working temperature, max./norm./min.	180	203		°C
Working pressure, max./norm./min.	5.5 4.3 3.0	3.3		bars
norm./min. vac. conditions				mb.
Design temperature, ~~upper/lower~~		310		°C
Design pressure, internal/~~external~~		8.2		bars
Test pressure, hydrostatic/~~pneumatic~~		18.978		bars
Liquid: quantity	1047.800	831.250		kg/h
specific gravity 15/4	0.881	0.861		
specific gravity at working temperature	0.775	0.737		
Vapour: quantity	57400	45500		kg/h
molecular weight	67.5	82		
density at working temperature	10.2	9.7		kg/m³
Heating/cooling medium				
max. quantity required				kg/h
Diameter of shell OD/~~ID~~		5200		mm
Length between tangent lines		10500		mm
Total packed height				mm
Height per bed				mm
Size and type of packing				
Number of packed sections				
Number of redistributors				
Height of skirt to bottom tangent line		6000		mm
Type of heads		"KORBBDGEN"		
Wall thickness - shell/head		24 & 26/27 & 29		mm
Corrosion allowance/~~lining~~		3		mm
Insulation thickness		50		mm
Trays: spacing/number required				mm
type				
lay-out acc. to sheet(s)				

Total volume	257.85	m³	Relief valve(s): Type/size :	
Normal liquid volume	:	m³	Set pressure :	bars
Volume range required for level control:		m³	Number required :	
Wind pressure	:	bars	Earth quake factor :	
			INFORMATION TO BE SUBMITTED WITH THE TENDER	

REMARKS AND/OR DESCRIPTION OF REVISIONS

Made by	Date	EQUIPMENT: PREFLASH VESSEL V201	Rev. letter			
Checked by	Date	PLANT: UNIT 200	Date			
Appr. by	Date	CONSIGNEE: STANLOW	Sign.			
			Sheet No. 1	cont'd on sheet No. 2		

Fig. 53.2*b*. Typical pressure vessel data/requisition sheet, Part 2

areas of flattening of shell plates, etc., and a quick inspection to ensure that nozzle sizes and orientation are correct. Once installation proceeds and piping is being connected, it is expensive in time and labour to make corrections.

During installation and commissioning, inspection proceeds in parallel and in accordance with a check list (Fig. 53.3). This list may, of course, be added to as necessary, most of the items on it being fairly obvious. However, the following are the points that experience has found require special attention.

NOZZLE DATA

Mark	Number	Service	Nom. dia. and flange rating	Remarks
N1	1	Feed inlet	24" - 150 RF	W/vaned inlet device
N2	1	Vapour outlet	24" - 150 RF	
N3	1	Liquid draw off	30" - 150 RF	Vortex breaker
N4	1	Vent	2" - 300 RF	
N5	1	Utility conn	2" - 300 RF	

INSTRUMENT CONNECTIONS

Mark	Number	Service	Nom. dia. and flange rating	Remarks
K-1	2	Level Controller	2"/¾"-150 RF	Range 5800 mm
K-2	2	Level gauge	2"-300 RF	On bridle
K-3	1	Pessure Ind	2"/¾"-150 RF	
K-4	2	Level alarm	2"/¾"-150 RF	Range 5800 mm

MANHOLES ETC.

Mark	Number	Service	Nom. dia. and flange rating	Remarks
A-1	1	Manhole	24"-150 RF	With davit

SKETCH

PREFLASH VESSEL V-201
UNIT 200
STANLOW

Sheet No. 3 cont'd on sheet No.

Eng. by :
Principal :

Req. No.

Fig. 53.2*c*. Typical pressure vessel data/requisition sheet, Part 3

Vessel supports

Most horizontal vessels are fitted with one fixed and one sliding support. The latter should be checked to ensure that it is, in fact, free to move, as lack of attention can result in damaged foundations or even stress in the vessel structure, which may promote failure.

Site changes

Alterations during plant construction usually involve welding and are checked to ensure they comply with the fabrication code. Retesting or restress relieving may be required, and the independent inspection authority is consulted. The alterations agreed are carefully recorded.

Equipment No. V-C-R-E*

GENERAL INFORMATION

Plant :
Duty :

Max.allow.working pressure/vacuum : shell: bundle:*
Max.allow.working temperature :
Hydrostatic test pressure :
Main dimensions :
Stress-relieved/X-rayed :

Purchase order No. :
Manufacturer :
Serial No.and type :

Erection contractor :
Contractor's inspector :
Company inspector :

Date of inspection :

* to be deleted where not applicable

CHECK LIST

Actions. Refer (5.1, 5.2, 5.4 and 5.6)	Check mark	Remarks
1. Check nameplate rating		
2. Check nameplate attachment		
3. Check and inspect foundation bolts and shims		
4. Check and inspect insulation and fireproofing		
5. Inspect wall for out-of-roundness, bulges and dents		
6. Check and inspect welding quality of "non-inspection vessels"		
7. Check and inspect alterations made during plant construction		
8. Check and inspect wall thickness of shell and nozzles		
9. Check and inspect internals		
10. Check, test and inspect lining of shell and nozzles		
11. Check, test and inspect reinforcement plates and test holes		
12. Check, test and inspect nozzle facings, gaskets and bolts		
13. Check insulation protection		
14. Check painting quality and specification		
15. Check outside bolting and stiffening rings		
16. Check and inspect correct material of plugs		
17. Check whether design of vessel and foundation allows vessel to be filled with water		
18. Test shell hydrostatically, if required		
19. Check and inspect sacrificial plates		
20. Check whether bundle jack screws have been retracted		
21. Check and inspect floating head clamp rings, gaskets and bolts (only when cover has been removed)		
22. Check for internal cleanliness before final boxing-up		
23. Check whether proper relief valve is installed		

Fig. 53.3. Pressure vessel inspection check list

Linings

Vessels may be lined internally to protect the walls against corrosive or erosive conditions; the linings take the form of clad plate, strip metal, weld deposit, paint or rubber or similar coatings, or of refractory. The metal linings are checked at the manufacturer's works for correct bonding or attachment and rarely present problems, though difficulties can arise when it is necessary to fit internal membranes on site. Paint and similar coatings and refractory linings, however, are easily damaged by rough handling, intense local heating, moisture, etc., and construction staff must be warned to take special care. Final inspection of the linings for damage should be left whenever possible until the time when all work on the vessel is complete.

Small-bore connections

So many leaks, resulting sometimes in fires and explosions, are caused by the failure of small-bore connections that special attention is given to them. Although at the design stage, minimum diameters are stipulated, screwed connections prohibited, forged material (rather than fittings made from bar stock) specified, and the geometry of fittings given attention, the inspectors would ensure that these requirements have been met. They would also spot-check the welding. It may be noted that forged material is preferred because its superior tensile, impact, fatigue, and welding properties combine to make a connection that could accept a maximum of arduous conditions without failing.

Excessive weld penetration sometimes causes partial nozzle blockage, and cases of incomplete weld penetration are found not only in small-bore connections but also in larger nozzles, particularly where the design does not call for 100 per cent radiography.

Pressure testing

If necessary, vessels are tested hydrostatically in the field, but this is not normal when the shop test is satisfactory and there have been no field modifications. It is sometimes convenient to include vessels in piping systems when they are tested, but the effect of the water on the materials of the vessels has to be considered. For example, to avoid stress corrosion cracking, water containing chloride is not allowed to contact austenitic stainless steel; and testing when the water temperature approaches a lower critical value, depending on the specification of the steel with respect to impact properties, is not carried out, to avoid the possibility of brittle fracture.

For safety reasons, pneumatic testing is not generally employed, but leak testing, using a soap solution, is sometimes necessary. In the case of tank bottoms, all welds are tested using a vacuum box. This has a glass window and an open bottom, around which is secured a continuous rubber seal. Soap solution is applied to the welds, a partial vacuum is created, and any leaks in the seams are positively located by direct visual examination. With large vessels that have been site erected, have one or more field-welded seams, and with foundations that would not support them full of water, field pressure and leak tests are omitted. The individual components are pressure tested at the manufacturer's works and field welding carried out under carefully controlled conditions. The weld procedure is agreed in advance and frequent inspection takes place during welding. Afterwards, welds are checked by 100 per cent radiography and ultrasonic flaw detection, as appropriate.

Throughout the construction period, spot checks are made of the materials used in pressure vessels and their components, to ensure compliance with the specifications.

OBTAINING INITIAL DATA AND INSTALLING TEST PIECES

The recording of initial data is vitally important, as information obtained from future inspections will be compared with it to determine corrosion rates, decide on subsequent inspection intervals, and determine the suitability of the vessel for continued service under the same or changed conditions. Not only are the details of the original requisition and manufacturer's test report recorded, but dimensional checks are made, particularly of wall thicknesses, using ultrasonic measuring equipment.

Corrosion test spools (Fig. 53.4) are installed in selected vessels and corrosion rates calculated when the spools are subsequently withdrawn and losses in weight measured. The corrosion spools contain metal from which the vessel has been constructed, as well as other metals and non-metallic materials. The sample of the metal of construction gives information about the rate and the type of corrosion, while samples of other metals and non-metallic materials give information about the most suitable protective linings should service conditions change or prove to be more arduous than expected.

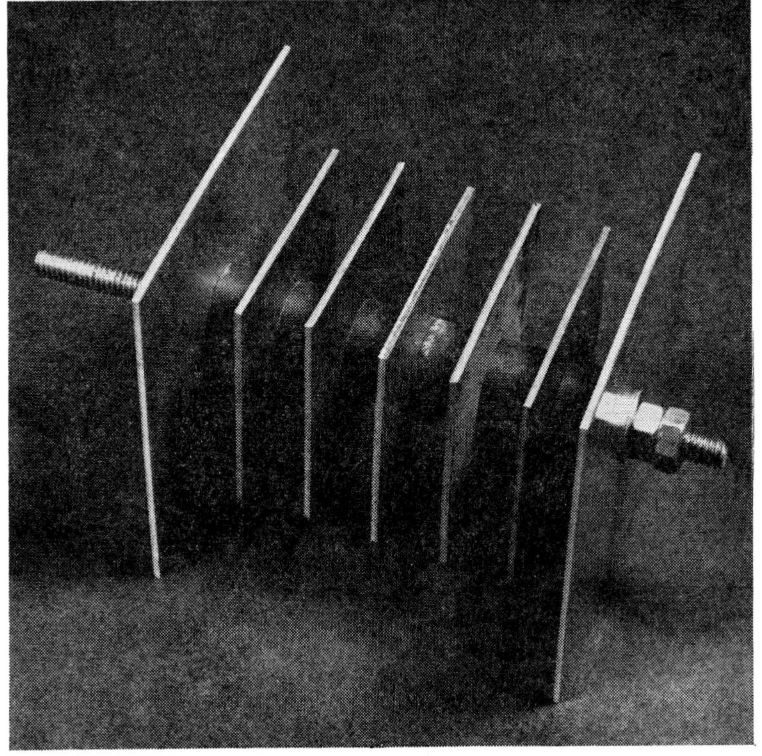

Fig. 53.4. Corrosion test spool

FACTORS DETERMINING FREQUENCY OF PRESSURE VESSEL INSPECTIONS AND PREPARATION OF PROCESS PLANT SHUTDOWN SCHEDULES

Statutory requirements sometimes determine the intervals at which pressure vessels are to be inspected. For example, steam drums associated with waste heat boilers have to be inspected at yearly or two-yearly intervals, depending on age and type of construction. If, as they frequently do, these inspections entail the shutting down of the whole process plant of which they form a part, it is often convenient to extend the scope of the inspection to non-statutory vessels.

In most cases there is no statutory inspection interval and decisions are made on the basis of the type of service in which the vessel is employed, its type and materials of construction, and its known condition. Usually, service conditions within a refinery have many parallels, and comparisons can be made with other vessels and with complete process units. These give a guide to inspection intervals, but the most important requirement is the gaining of knowledge about what is happening inside the vessels.

After first commissioning, a pressure vessel is not allowed to run for more than two years before inspection, so that the actual effects of service conditions can be verified. Although two years is an arbitrary length of time, experience has shown that this is about right; a shorter length would not always allow measurable differences to be detected, and a longer length might lead to excessive corrosion or even failure.

The condition of the vessel at the two-year and subsequent inspections is the main factor in determining when the next inspection will take place. Actual corrosion rates, the need to re-inspect repairs carried out, judgement as to when future repairs will be necessary, and, as already mentioned, experience gained from vessels in similar service set the service run to be allowed. However, no vessel is allowed to continue in service longer than six years without inspection.

Pressure-relieving devices associated with the vessels are allowed to continue in service for no longer than about two years before being inspected, overhauled, and tested. Their condition at the time of inspection is used as a guide to the inspection interval.

Changes in service condition and the effects of operations during the run may affect an inspection interval already decided upon. For instance, an increased pressure drop across the vessel caused by the build-up of corrosion products or the deposition of sand, scale, or foreign particles from the process stream may make a change necessary, as would a leak or other evidence of wall thinning.

In determining a pressure vessel inspection interval, not only is it important to ensure reliability and safety, but economic considerations must also be included. Too long an interval might result in a costly failure or in expensive and lengthy repairs being necessary when inspection takes place; and too short an interval means unnecessary loss of products and the cost of gas freeing, isolation from other equipment or units, opening up, scaffolding, etc., to say nothing of the employment of engineering resources that are invariably required on other work.

Operational and engineering considerations must always be taken into account when inspection intervals are being decided. The complexity of refinery process plants and their interdependence generally means that pressure vessels can only be depressurized, isolated, and gas-freed when the whole unit or even several units are completely shut down. Thus shutdowns have to be carefully planned well in advance, to ensure that seasonal market demands are met and that, at all other times, products can be obtained from stockpiles, other company refineries, or obtained from competitors at the most favourable terms.

Shutdowns for process reasons, such as catalyst regeneration, have to be included in the plans, and it sometimes happens that inspections can be carried out when acid, catalyst, or other inventories have to be changed. From an engineering point of view, it is important to reduce the need for peak labour forces, avoid more than one major shutdown at any given time, and have the necessary labour and equipment mobilized in advance. Materials, some of which take a long time to obtain, must be ordered in advance, and shutdowns are best conducted outside peak holiday periods.

In order to meet operational, engineering, and inspection needs, shutdown schedules are produced at regular intervals. Long-term schedules (Fig. 53.5) cover five-year periods, while short-term schedules (Fig. 53.6) cover four-monthly periods and are regularly updated.

DETERMINATION OF WHAT IS HAPPENING WHILE VESSELS ARE ON STREAM

Whilst the conditions of a pressure vessel as determined by its inspection when shut down are the most important factors in determining when it next requires inspection, the way it behaves during operation and knowledge of what is happening to it may lead to the inspection interval being either shortened or lengthened. As mentioned earlier, service conditions are monitored, such things as increasing pressure drop being noted. The following are other means of determining on-stream measurements and changes:

(*a*) Visual inspection for leaks.

(*b*) Checking of hot-spots on vessels with internal refractory linings. In some cases this can be done by feeling with the hand, and in others by the use of indicator crayons, paints, and infrared photography.

(*c*) Sampling process streams for corrosive components.

(*d*) Using corrosion probes (Fig. 53.7) and plotting corrosion rates.

(*e*) Monitoring injection rates of inhibitors and neutralizing agents.

(*f*) Checking for changes in dimensions of vessels with internal refractory linings by visual observations, sometimes aided by the use of oblique lighting, by looking for Lüder's lines, and by measurement with tapes, rules, and, on occasions, strain gauges.

(*g*) Using on-stream radiography of the vessels themselves and of adjoining pipework. Wall thickness measurements of the pipework—where flow velocities are higher than in the vessels and which, in any case, are easier to obtain than the vessels—give advance warning of possible changes in vessel walls. Radiography of the vessels is useful to check displacement of internals and locating blockages.

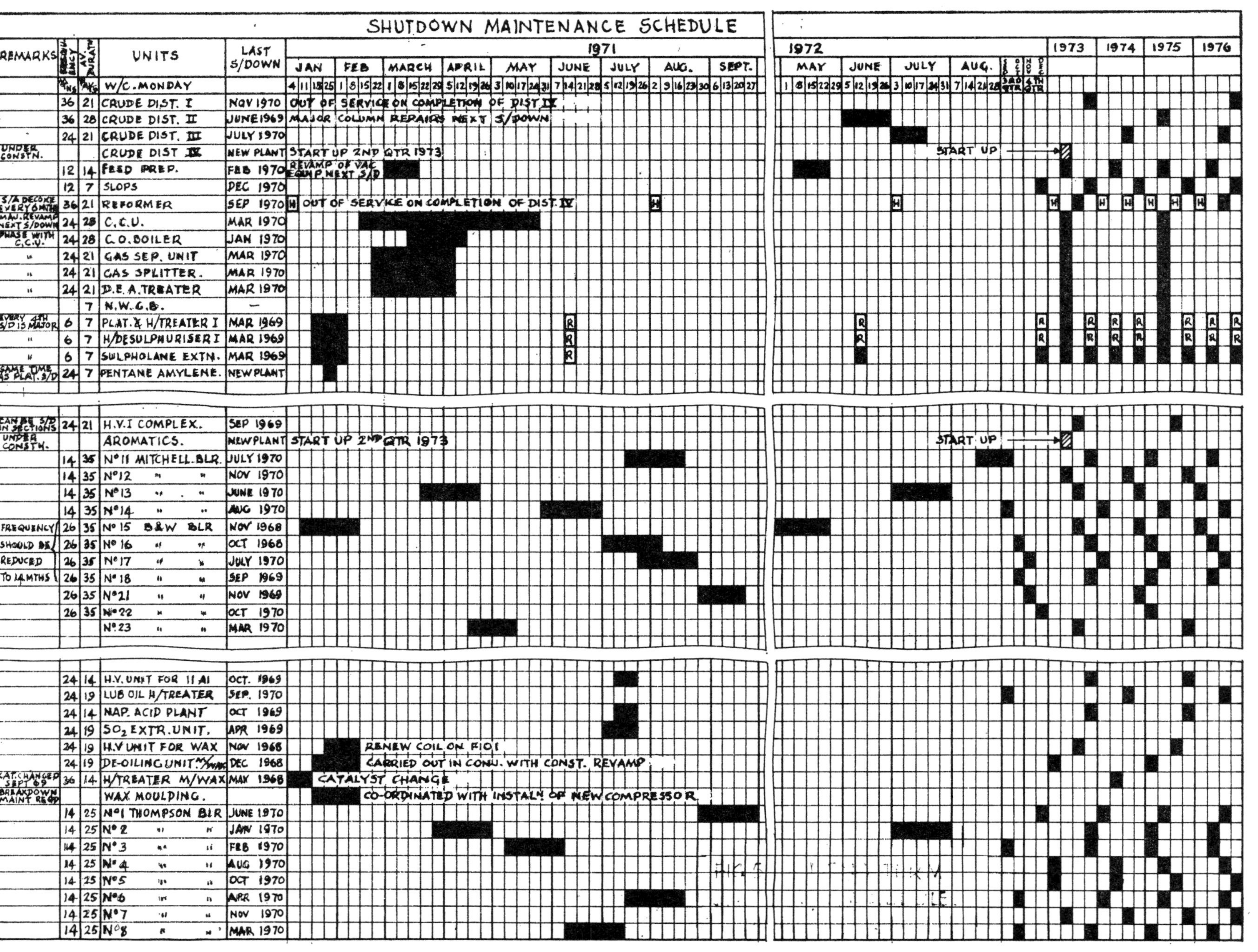

Fig. 53.5. Long-term shutdown schedule

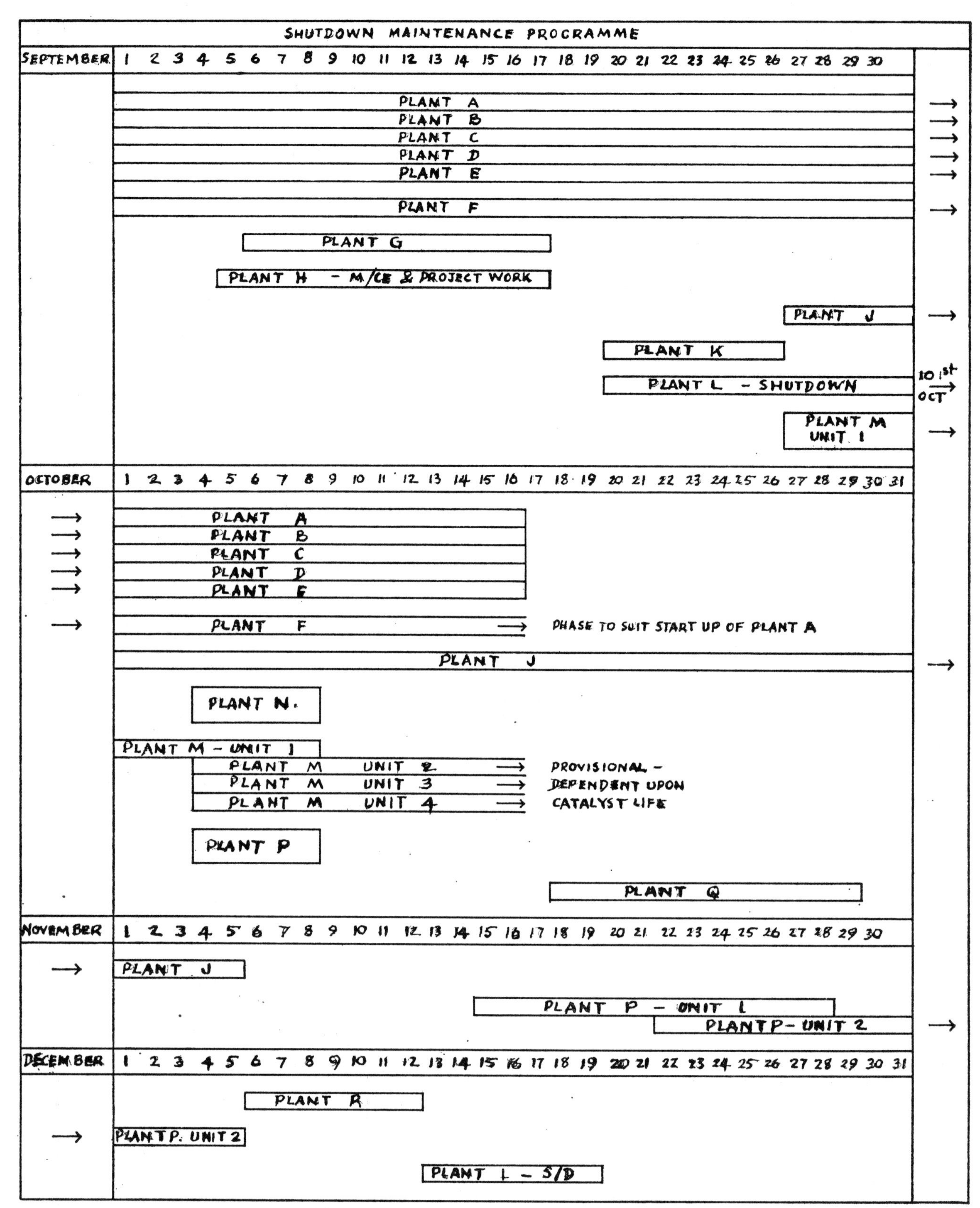

BOILER SHUTDOWN PROGRAMME

NO 16 BOILER - 21·6·71 - 10·3·71
NO 21 BOILER - 6·9·71 - 25·9·71
NO 17 BOILER - 13·9·71 - 8·10·71
NO 14 BOILER - 11·10·71 - 5·11·71
NO 12 BOILER - 8·11·71 - 3·12·71
NO 18 BOILER - EXPECTED BACK - OCT

Fig. 53.6. Short-term shutdown schedule

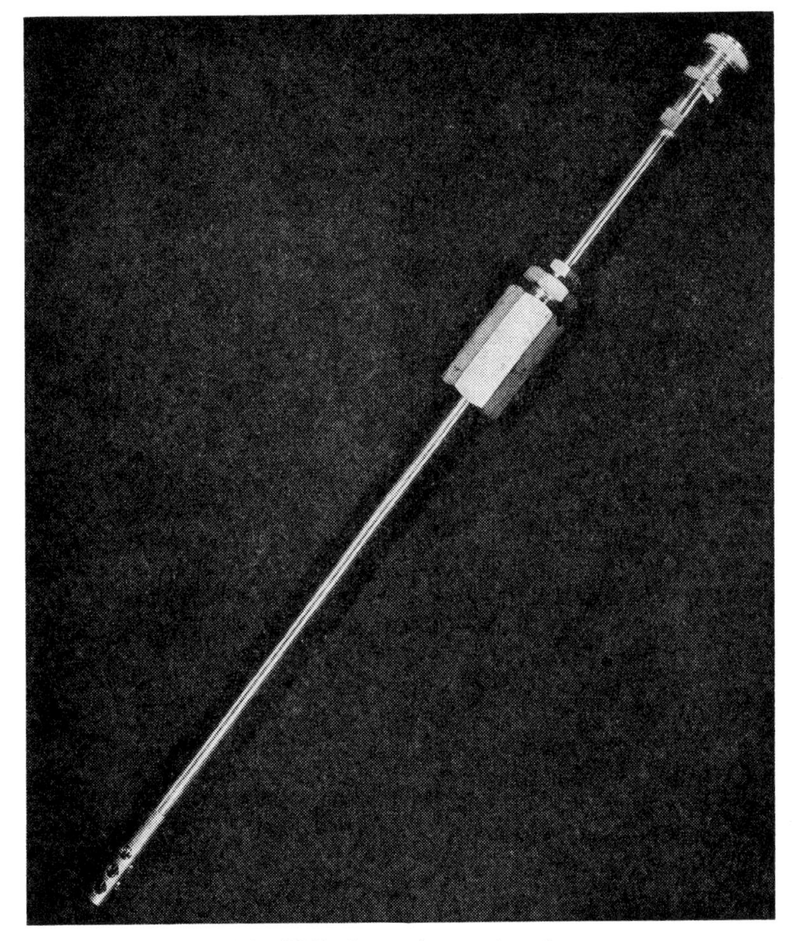

Fig. 53.7. Corrosion test probe

TOOLS, INSTRUMENTS, AND TECHNIQUES USED FOR INSPECTION

The experienced inspector with a sharp eye is the inspection department's most valuable asset. Without him, and the ability of the department to interpret and forecast the observations made, the use of sophisticated techniques is of little use and can, indeed, be misleading. Furthermore, the use of certain instruments calls for specialist operators if the readings are to be reliable and consistent, and the instruments themselves must always be kept in a good state of repair and regularly checked.

The following are the tools and techniques used:

Tensile testing and impact testing.

Hardness testing.

Material identification by spectrographic analysis.

Microscopic examination to determine the metal structure.

General visual inspection.

Hammer testing.

Radiography of welds and of vessel and nozzle walls to determine their thickness.

The checking of internal and external dimensions by tapes and rules, and measurement of wall thickness by calipers.

Ultrasonic flaw detection and wall thickness measurement.

Analysis of scale and corrosion products.

The use of periscopes, cameras, and TV in inaccessible places.

Optical methods for internal inspection of tubes.

Dye penetrant to detect surface flaws.

Magnetic particle detection of surface and subsurface flaws.

In situ hardness testing.

'Holiday' detection of internal linings.

Thickness measurement of paints and coatings.

INSPECTION RECORDS

A simple but effective recording system is essential if the results of periodic inspection are not to be lost. It must be possible to refer to the history of a pressure vessel quickly and know that the data recorded are accurate and reliable.

Record cards (Figs 53.8*a* and *b*) of each individual pressure vessel are kept up to date by the inspectors themselves. In addition to showing the original manufacturing data, initial dimensions and thickness, and design conditions, the cards are up-dated to show times and results of inspections carried out, the repairs effected, and modifications made. Copies of inspection and action reports are kept with the cards. These reports are issued to both operating and maintenance departments, a duplicate copy of the report being returned by maintenance department, giving details of the work done. Re-inspection is carried out to ensure that the repairs are of the right standard; this includes visual inspection, and radiography and pressure testing if necessary. Besides recording a history of the

PRESSURE VESSEL	**REGISTRATION No.** V·201

GENERAL INFORMATION

Purchase order No.	HP 0334/22/04 URS	Date	APRIL 1970
Maker	A.N.OTHER	Type	VERTICAL
Year of fabrication	1971	Serial No.	-
Design Book page No.	-	Size	5200°/D x $18,050^{O}$/L
Fabrication report No.	-	Mesc No.	-
Relief valve card No.		Mass	59480 kg
		Date in operation	1973

DATA ON
ENGINEERING

Constructed acc. to BS·1515 - PART 1 - 1965 UP TO AMEND. 3

		SHELL	COIL	JACKET
Max. allow. internal working press.	bars ga	-		
Max. allow. external working press.	bars ga	-		
Hydrostatic Testpressure (new and cold)	~~bars~~ ga	18·978 kg/m²		
Max. allow. working temperature	°C	-		
Volume of water for testing	dm³	257·85		
Inside diam.	mm	5148	-	-
Total length	mm	10500	TAN.TO Top End	TAN.TO Btm.End
Wall thickness shell/dished ends	mm	24/26	27	29
Corrosion allowance	mm	3	-	-
Stress relieved		~~yes~~/no	~~yes/no~~	~~yes/no~~
Radio graphic/ultrasonic tested 100%		yes/~~no~~	~~yes/no~~	~~yes/no~~
Special heat treatment		~~yes~~/no	~~yes/no~~	~~yes/no~~

Shell liner full/partly -thickness	None	mm
Dished ends lined ~~yes~~/no -thickness	No	mm
Insulated yes/~~no~~ 50 -thickness		mm
Heating surface coil/jacket		m²

DRAWING REFERENCE

General Arrangement No.	T.105,676-E
Requisition No.	T-1051-606-C

Fig. 53.8*a*. Record card (pressure vessel), L.H.S.

vessels, action reports and completed duplicates help to keep track of the work to be done and the job finished during a busy plant shutdown.

In addition to keeping records of individual vessels, reports are also made of all the work covered by a plant shutdown. These show the days run since the last shutdown, the condition of individual items of the plant, the repairs and modifications carried out, recommendations as to the length of time the plant can run till the next shutdown, the operating conditions that should be observed, and the materials and repairs that it is anticipated will be necessary at the next shutdown. These reports form the basis on which shutdown work is planned and they assist in the preparation of shutdown schedules.

Items of special interest highlighting particular problems and their resolution are the subject of separate reports, which are sent to the central office for circulation to other refineries.

LOCATION AND DUTY

No.	LOCATION	LOC. No.	DUTY	MEDIUM	DATE
1	UNIT 200V	201	PREFLASH VESSEL	CRUDE	1971
2					
3					
4					
5					
6					

DATA ON

DESIGN OPERATING CONDITIONS				ACTUAL OPERATING CONDITIONS 1	2	3	4	5	6
Working press. int./ext.	SHELL	8.2/ATMOS	kg/cm² ~~bars ga~~	5.5 Max 3.0 Min 4.3 Norm.					
Working temp.	SHELL		°C	180	203				
Working capacity	SHELL		m³/h						
Max. Working cap.	SHELL		m³/h						
Working press.	COIL		bars ga						
Working temp.	COIL		°C						
Heat Consumption	COIL		W						
Working press.	JACKET		bars ga						
Working temp.	JACKET		°C						
Heat consumption	JACKET		W						

MATERIALS

No.	PART	SIZE	SPECIFICATION	MESC No.
	Shell	10500 O/L x5200 O/D	BS 1501-151-26B	Semi Killed M/S
	Dished ends	5200 O/D x 27/22 THK	"	
	Liner shell	–	NONE	"
	Nozzles	30":24":4":2"	BS 3602, HFS GR23	
	Jacket	50 mm THK	INSULATION	LAGGING
	Plugs	–	–	
	Coil	–	–	
	Stud bolts		A 193GR B7 EXTERNAL	
	Nuts		A 194GR 2H EXTERNAL	
	Gaskets	CAF	BS 1832	

Fig. 53.8*b*. Record card (pressure vessel), R.H.S.

UNDERSTANDING THE PROCESS, FACTORS INFLUENCING CORROSION, AND EFFECTS OF PROCESS CHANGES

Process considerations

The design and construction of a pressure vessel are intimately bound up with the conditions to which it will be subjected. If, for example, corrosive fluids are to be handled, the materials of construction will be very different from those used under relatively non-corrosive conditions. This may be illustrated by hydrocarbon distillation units, where mild steel is generally considered to be satisfactory for the construction of columns, pipelines, and vessels. However, where sour crudes are processed, sulphur corrosion is introduced and alloy steels are necessary to counter the higher corrosion rate produced on mild steel at elevated temperatures.

This case emphasizes the need for an understanding of the process of which the vessels are a part. Particular attention must be given to the problems of this corrosion not only during the design stage but also during subsequent inspections. The reason for this is that the corrosion reaches a peak degree of severity between 370° and 425°C. Below 325° and above 450°C it is possible to use mild steel. Over 450°C, particular care must be taken with lagged vessels to avoid the insulation becoming defective; if this occurs, a local cool spot could result, which might lead to very rapid local corrosion, ending in a perforation.

Conversely, if a plant product stream were locally heated and brought into the active range, similar effects could arise. This could occur as a result of some operational upset or overheating. It is therefore very important that the inspector should be aware of the day-to-day incidents that occur on the plant, so that he can at the forthcoming shutdown period correlate the conditions under which equipment has actually been operating with his inspection requirements.

Materials considerations

A knowledge of the general properties of the material of construction of a pressure vessel is necessary in order to obtain a full picture of the conditions of the vessel, in relation to its suitability for duty, in a given service. For instance, a welded mild-steel vessel may be quite satisfactory in a 20 per cent caustic soda service at temperatures below about 70°C; but above this temperature, stress corrosion cracking can be expected. A careful watch must be kept for this type of defect when temperatures approach this critical value.

However, should the vessel be stress relieved, then the risk of cracking is reduced to a very low value. Thus, in this case, cracks would not be specifically looked for during inspections. In a similar manner, a vessel made of austenitic stainless steel would be inspected very carefully for cracks, if it were suspected that aqueous chloride solutions had been present. Such solutions promote stress corrosion cracking in this type of material at temperatures above about 70°C.

Summary

From the examples just quoted, it is clear that an understanding of the process and materials enables an inspector to acquire the following skills:

(1) To judge whether a vessel and all its branches and connections, etc., are, in fact, adequate for the services which they must perform.

(2) To anticipate the expected general condition of the vessel at the end of any particular period of service.

(3) To be able to judge when a vessel is reaching the end of its useful service.

(4) In the case of a failure on stream, to be able to know where and what to look for and to interpret any unexpected conditions found.

CONCLUSIONS

The inspection of pressure vessels, necessitating equipment and plant shutdowns, must be carried out at the longest intervals possible consistent with ensuring reliability and safety. On the one hand, leaks and unscheduled shutdowns must be avoided; and on the other, so must unnecessary depressuring, isolation, and opening up. The aim should be to inspect at intervals that, in the long term, result in the most economic level of repairs, and a minimum amount of unexpected work at each shutdown.

To achieve this aim, inspection and recording throughout the life of the vessels, from the design stage onwards, are essential, together with a full understanding of what is happening inside the vessels. This necessitates experienced keen inspectors and the use of specialized inspection tools and techniques.

ACKNOWLEDGEMENTS

The author wishes to thank members of the Engineering Services Department, Stanlow, for their assistance, and the Managers of Shell U.K. Ltd at Stanlow and London for permission to publish.

C54/72 THE DESIRABILITY OF PROOF TESTING REACTOR PRESSURE VESSELS PERIODICALLY

C. L. FORMBY*

The need for periodic proof testing is considered in relation to factors which can cause deterioration in service. Fatigue, irradiation damage, strain-ageing, and creep are examined and it is found that the safe working life is limited by them only when an extremely long crack is present initially. The lifetime guaranteed in these critical circumstances by a repeat proof test is impossible to predict with any accuracy, and the repeat proof test could itself shorten the life. The paper mentions some other ways of increasing confidence in reactor vessels, and the nature of these is such that it is nearly as easy to guarantee a vessel for its entire life as for the time between proof tests. The conclusions arrived at strongly oppose the use of repeat proof testing as a general safeguard, at least in simple, all-welded structures of the type examined.

INTRODUCTION

Fracture of a pressure vessel is a rare event, which involves unexpected conditions of material, flaws, and/or stresses. It follows that the probability of fracture depends upon the care taken to control these factors. Nevertheless, attempts have been made to estimate the failure probability of reactor vessels on the basis of reported failure rates of pressure vessels generally. If failure is defined as detection of a condition that is judged to require repair or replacement of the vessel, German experience (1)† indicates a failure rate of $\sim 10^{-4}$ per year. Phillips has examined U.K. information and the data generally and finds that 90 per cent of failures are due to cracks, and about two-thirds of these are discovered during visual examination (2). He estimates that the probability of catastrophic fracture during operation would be $\sim 10^{-5}$ per year for an inspected vessel (3). Since complete inspection is not possible in the case of a reactor vessel and repeat proof test is the procedure adopted when this is the case for conventional vessels, there are some grounds for considering the repeat proof testing of reactor pressure vessels. However, there is no statistical evidence of an effect of repeat proof testing on the probability of catastrophic fracture in service. Two attempts have been made to assess the usefulness of periodic proof testing using fracture mechanics approaches.

Irvine (4) used a phenomenological approach to estimate the fatigue lifetime of a vessel containing initially a full-thickness crack of the maximum possible length—i.e. one which was just insufficient to cause fast fracture during the pre-service proof test. He concluded that the lifetime under such circumstances was sufficiently short to warrant periodic pressure testing. The approach on which this belief is based can be criticized in that consideration is restricted to fatigue damage, in that it relies upon extrapolation from a few test results without consideration of all the important parameters (e.g. thickness and strain rate) and possible crack growth during the proof test is ignored. Yukawa (5) considers the question in a more general way, including the possibilities of embrittlement as well as fatigue and using the generally accepted fracture mechanics approach. Unfortunately the discussion lacks coherency, since the general review of fracture mechanics presented is not related explicitly to reactor vessels at all times. He concludes that periodic proof testing of nuclear vessels is neither warranted nor required and considers that the main objective of the initial proof test is to test for leakage, not to detect cracks. This report attempts to bridge the gap between the two views.

The case for repeat proof testing of steel reactor pressure vessels must be based upon an examination of the possibility of fracture, having regard to their particular material properties and operating conditions. Bearing in mind the fact that these vessels survived an initial proof test, it is only necessary to consider the changes that occur during the service life. These are fatigue, irradiation damage, strain-ageing, and creep, and each will be considered as it applies to the behaviour of a pre-existing crack.

CRACK GROWTH BY FATIGUE

Collected data show that the increment of crack growth per cycle of load (da/dN), when related to the stress intensity factor, ΔK, varies little from one low- or medium-strength ferritic steel to another (Fig. 54.1). K is defined by

$$K = Y\sigma\sqrt{a} \quad . \quad . \quad . \quad (54.1)$$

where σ is the applied stress, $2a$ is the crack length, and Y is a factor that depends upon geometry. This conclusion has been arrived at previously [see, for example, reference

The MS. of this paper was received at the Institution on 10th November 1971 and accepted for publication on 9th February 1972. 22

* *Central Electricity Generating Board, Berkeley Nuclear Laboratories, Berkeley, Gloucestershire.*

† *References are given in Appendix 54.1.*

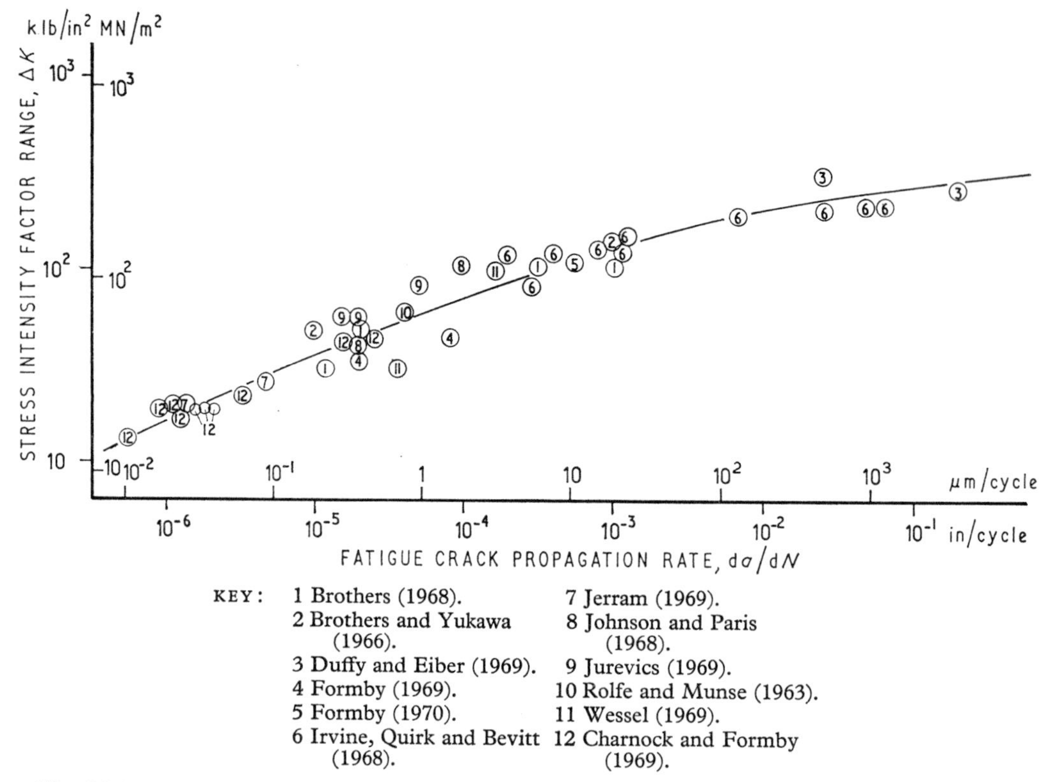

Fig. 54.1. Collected data for low- and medium-strength steels on the influence of stress intensity factor range on fatigue crack propagation rate

(6)], based upon small specimen results, and Fig. 54.1 extends the information to higher values of ΔK, incorporating results from vessels and pipes. The derivation of ΔK for such cases is dealt with later in the paper. Except at very high ΔK values the data approximates to

$$\frac{\mathrm{d}a}{\mathrm{d}N} = A\,\Delta K^4 \quad . \quad . \quad . \quad (54.2)$$

The number of full cycles of stress expected in the lifetime of a reactor pressure vessel is approximately 300. Part-load changes are accomplished by changing the gas flow rate, being accompanied by only small changes in vessel stresses and temperatures, and since $\mathrm{d}a/\mathrm{d}N$ is proportional to the fourth power of the stress [equations (54.1) and (54.2)], only the full-stress cycles resulting from full-load cycles need to be considered. Several cases will now be discussed, to cover the possible extreme conditions of crack size, material toughness, and stress.

Full thickness cracks in membrane regions, ductile material

Calculations have been performed for the specific case of cracks lying in the axial direction of a typical reactor pressure vessel [a cylinder 15·2 m (50 ft) in diameter, 76 mm (3 in) thick, with a hoop stress of 101 MN/m² (14·6 lb/in² × 10³) during operation]. The calculation of ΔK made allowance for bulging in the vicinity of the crack, using the work of Folias (7), which gives

$$\Delta K = \sigma\sqrt{\left[\pi a\left(1+\frac{1\cdot 6a^2}{Rt}\right)\right]}$$

where R is the radius and t the plate thickness. This expression was used in conjunction with the mean of growth rate data (solid line in Fig. 54.1) to predict the relationship between growth rate and crack length (Fig. 54.2). That the calculation of ΔK is sufficiently correct is evidenced by the fact that use of the same method in the compilation of Fig. 54.1 brought the vessel and pipe data into line with the flat plate data. The number of cycles to fracture,

$$N_F = \int_{a_i}^{a_c} \frac{\mathrm{d}N}{\mathrm{d}a}\,\mathrm{d}a$$

where a_i is the initial crack length and a_c is the critical crack length under service conditions. This is the area under the curve in Fig. 54.2, between the two limits. For example, a crack initially 254 mm (10 in) long will lead to rupture of the vessel after 77 000 cycles of load. The shape of the curve is such that N_F is insensitive to the value of the critical crack length but reduces very rapidly with increasing initial crack length.

It is possible to imagine a much longer crack being present if no weight is given to the non-destructive testing or to leak testing. The upper limit of crack size that can be present is then the critical crack length under the conditions of the proof test. The flow-stress method of Hahn and Sarrate (8) applied to this case predicts a critical crack length in the initial 1·6 times operating pressure proof test of 1·52 m (5 ft). The number of operating stress cycles required for rupture would be 340 if this length of crack were initially present (Fig. 54.3). The period of immunity conferred by a proof test is critically dependent on the level of overpressure. For example, Fig. 54.3 shows that an overpressure of 1·2 would guarantee only 23 cycles.

Whilst the above results represent the most probable situation on present knowledge, a number of assumptions and approximations are, of course, involved. The more important ones are now listed.

(1) Creep will occur in fact and be superimposed on the

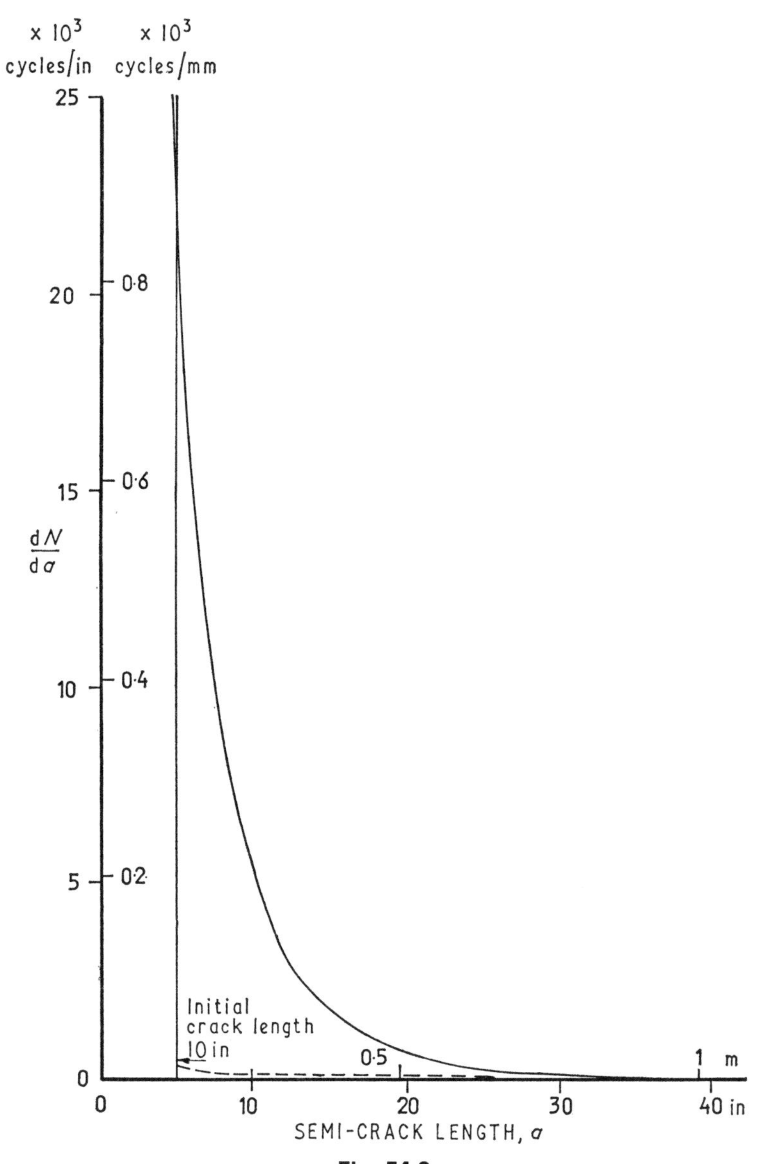

Fig. 54.2

fatigue. This will cause some increase of da/dN above that expected from Fig. 54.1 since da/dN is related to the amount of cyclic plastic deformation at the crack tip (**9**). It will be shown under 'Discussion', below, that the deformation can at most be doubled and hence da/dN increased by the factor 4. This is an upper limit of the effect of creep and it will be smaller than this in practice. Some of the data in Fig. 54.1 were, in fact, obtained at high temperature, although the hold times were short.

(2) Slow, ductile tearing at the ends of the initial crack is expected during the proof test since a very long, just subcritical crack has been postulated. There is insufficient information available to be very specific about this, but the most pessimistic estimate (see 'Discussion') would reduce N_F from the 340 cycles calculated above to 60. The calculations that assume a more modest initial crack length will be virtually unaffected since the amount of tearing is a sensitive function of crack length.

(3) Material of one sort or another (e.g. oxide) may occlude the crack completely or partly. This would reduce da/dN by limiting the amount of crack closure and hence the cyclic deformation at the crack tip.

(4) The variability of the test results in Fig. 54.1 also contributes to the uncertainty of the predictions. The worst possible case is attempted below under 'Full thickness cracks . . .', taking growth rates equal to the maximum of the scatter band and a high local stress.

(5) The proof test may itself influence the subsequent da/dN in service through the high strain at crack tips during it and the residual stress left after it. Little information is available, but there is some indication that the effect is extremely localized and of little importance (**10**).

Summary

In the case of the very long, through cracks discussed in this section, where it is clear that the number of operational cycles guaranteed by the initial proof test could be of the order of the number expected in service (about 300), there would be some merit in repeat proof testing. The judgement, however, hinges upon the magnitude of the damage done in this situation by crack growth during the proof test and upon the chance of cracks several feet in length being conceivable.

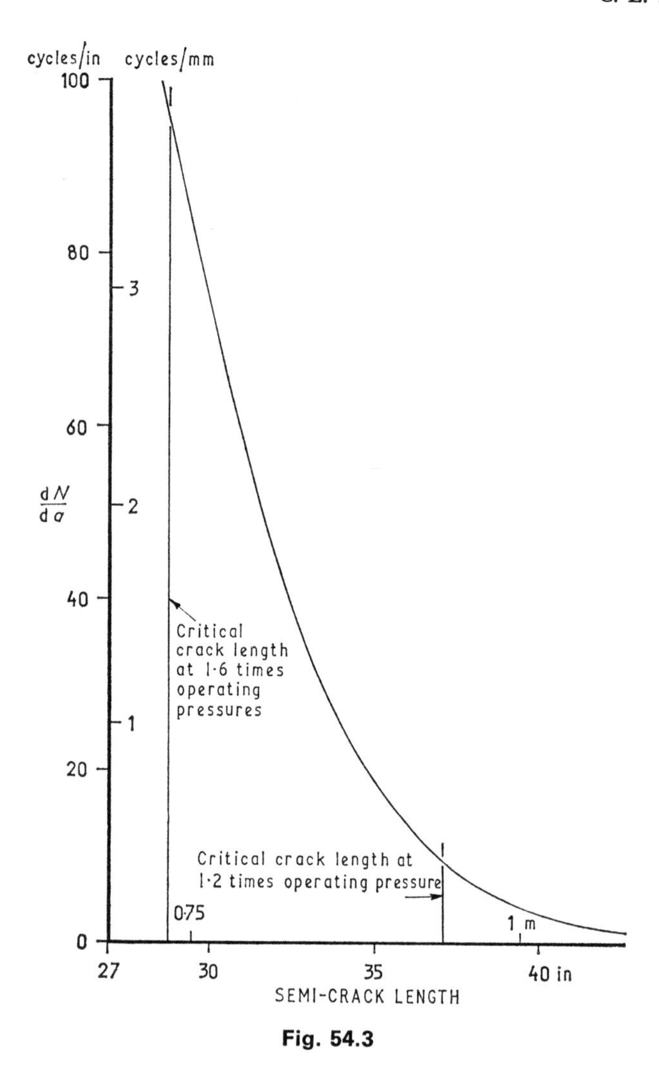

Fig. 54.3

Very long surface cracks in membrane regions, ductile material

If leakage can be said to rule out long, through cracks, the same cannot be said of partial penetration cracks. Complete snap-through would lead to fast fracture if the partial penetration crack extended over a length greater than the critical crack length of a full-thickness defect. Irvine (**4**) has shown that the criterion for failure of the remaining section beneath a partial penetration crack is attainment of an average stress across the section equal to the u.t.s. (ultimate tensile stress).

For a plate thickness of 89 mm (3·5 in), an operating stress of 108 MN/m² (16 lb/in² × 10³) with an initial overpressure of 1·6, and an u.t.s. of 430 MN/m² (63 lb/in² × 10³), the initial crack depth just insufficient to cause fracture at the proof test stress is 53 mm (2·1 in), and 13 mm (0·5 in) increase of depth is required to cause fracture at the operating stress. The value Y in equation (54.1) for this case (**11**) gives ΔK, and when coupled with the crack growth rate from Fig. 54.1 it becomes apparent that N_F is of the order 100.

This is clearly another limiting case in which repeat proof testing could be useful. However, the number of cycles to failure increases very rapidly as the initial depth of the crack is reduced. For example the number required would be of the order 10^4 if the starting crack depth were 25 mm (1 in). Again the judgement of the need for overpressure testing depends upon the chance of such very serious cracks being present and upon the amount of crack extension in a repeat test.

The presence of material of low toughness and of local stress concentrations

The limit approach will again be applied, namely that a crack of just subcritical length is present in the proof test, and it will be assumed that no deterioration in toughness occurs in service. It will also be assumed that the dependence of da/dN on toughness is not great (**6**). From equations (54.1) and (54.2) the number of cycles to failure,

$$N_F = \frac{1}{AY^4\sigma^4}\int_{a_0}^{a_c}\frac{da}{a^2} \quad . \quad . \quad (54.3)$$

a_o and a_c, the critical crack lengths at the proof test stress (σ_p) and the service stress (σ_s) respectively are related to the toughness (critical stress intensity factor), K_c, by

$$K_c = Y\sigma\sqrt{a_{o,c}}$$

Integrating and substituting into equation (54.3) gives

$$N_F = \frac{(\sigma_p/\sigma_s)^2-1}{AY^2K_c^2\sigma_s^2} \quad . \quad . \quad (54.4)$$

Thus the number of cycles to failure is greater the lower the fracture toughness. The point is illustrated in Table 54.1 where computations from equation (54.4) are presented. The value of A is determined from equation (54.2) and Fig. 54.1, σ_s is taken as 108 MN/m² (16 lb/in² × 10³), the initial degree of overpressure as 1·6 and Y as 4 (i.e. a stress concentration factor of 2). Two is the highest stress concentration factor present in most reactor pressure vessels, so that this represents a realistic condition at the shorter crack lengths but overestimates the stress at longer lengths because areas of high stress tend to be quite small. In fact, bulging would become the deformation mode at the longest of the crack lengths, as has been assumed in the previous calculations.

It can be concluded from Table 54.1 that the smallest values of N_F occur for long cracks in ductile material, the case discussed more particularly under 'Full thickness cracks . . . ' and 'Very long surface cracks . . .' above. However, it would be wrong to conclude that one should build a vessel out of material of low fracture toughness because most information on flaw sizes comes in practice from non-destructive testing, not from the proof test, and

Table 54.1. Computations

Fracture toughness, K_c (lb/in² × 10³ √in)	Critical semi-crack length in service, a_c (in)	Fatigue life guaranteed by the 1·6 times operating pressure proof test, N_F (service load cycles)
Computations from equation (54.4) (bulging ignored, S.C.F. = 2)		
50	0·6	4×10^4
100	2·5	$1{\cdot}0 \times 10^4$
200	10·2	$2{\cdot}5 \times 10^3$
300	23	$1{\cdot}1 \times 10^3$
400	41	640
Computation from Fig. 54.3 (with bulging and S.C.F. = 1)		
Fully ductile	56	340

the longer flaws tolerated in higher toughness material could be detected more reliably.

Full thickness cracks in regions of high stress, growth rate assumed to be the maximum of the scatter band, ductile material

This combination will give maximum growth rate. An extensive region of high stress is combined with a material whose fatigue crack growth characteristics fit to the high-rate end of the scatter band in Fig. 54.1. To allow for any possible local stresses, assume that the through crack exists in an extensive region where the enhanced stress is of yield stress magnitude σ_y (i.e. shake-down is only just achieved). The most pessimistic assumption is that the direct pressure stress and the extra local stress on the crack are additive, and since the service stress in reactor pressure vessels σ_s is about $\sigma_y/2$, the total stress can be separated into two equal parts of magnitude σ_s. Thus

$$\Delta K = \sigma_s \sqrt{\left[\pi a\left(1+\frac{1\cdot 6a^2}{Rt}\right)\right]} + \sigma_s\sqrt{(\pi a)}$$

Calculation of the $(\mathrm{d}N/\mathrm{d}a)v a$ curve by combining this expression with the $(\mathrm{d}a/\mathrm{d}N)v\,\Delta K$ relation at the extreme of the scatter band in Fig. 54.1 (at which the rates are five times greater than the mean line) gives the dashed line in Fig. 54.2. From this, initial crack length to give fracture after 300 cycles of stress is calculated to be 2 ft.

This result illustrates the extreme conditions required to give fracture by fatigue in 300 cycles, since 2 ft represents a very considerable crack notwithstanding the high degree of pessimism in the conditions chosen.

Thermal fatigue

One cause of fatigue, namely thermal stresses, is particularly difficult to treat because of the wide range of possible stress distributions that change with time in complex ways as well as the range of crack configurations. Such stresses are caused by temperature gradients developed during shut-down or start-up so that the number of cycles is about the same as the pressure cycles. Careful design will have limited the thermal strain to less than twice the yield so that the amount of cyclic deformation at the crack tip cannot exceed that in the last example. It follows that crack growth due to the thermal cycles of most reactor pressure vessels cannot be greater than that calculated for pressure cycles in the last example.

Fatigue as a mechanism of failure

The foregoing shows that crack growth rates in fatigue are so slow in general that 300 cycles will not damage the reactor vessels appreciably. The only exception occurs if no advantage is taken of the results of non-destructive testing, and it becomes necessary to postulate the presence of cracks in excess of 2 ft long and penetrating more than half-way through the wall.

STRAIN-AGEING EMBRITTLEMENT

Deformation of the C–Mn steel often used in reactor pressure vessels can lead to embrittlement if it is followed by thermal ageing or if it takes place in the temperature range 200 → 350°C (12). Most deformation at stress-raisers, such as cracks, will occur during the initial proof test, and smaller amounts of deformation will occur during subsequent changes of load in service at higher temperatures. This is exemplified in Fig. 54.4, where crack-opening displacement measurements made at the tips of a 300-mm (12-in) long crack in a 1·52-m (5-ft) diameter, 26-mm (1-in) thick cylindrical vessel are shown. The vessel was subjected to a sequence of loadings designed to simulate the service history of a reactor pressure

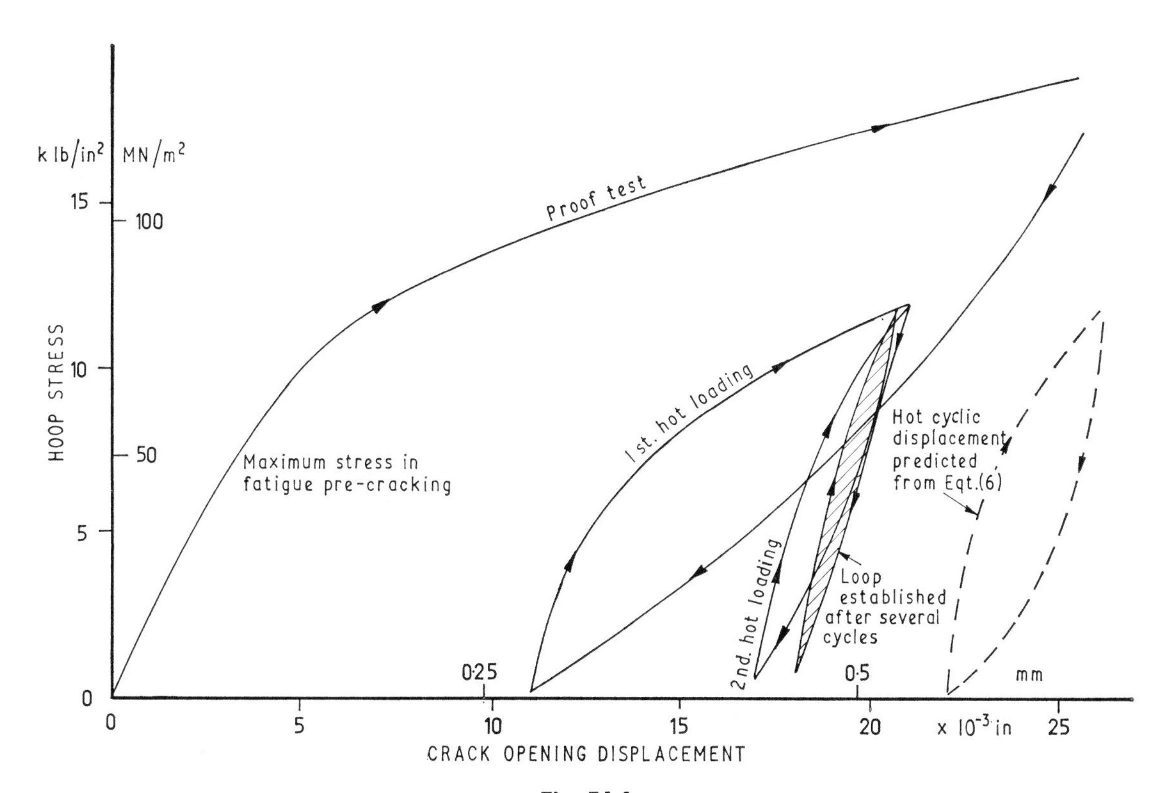

Fig. 54.4

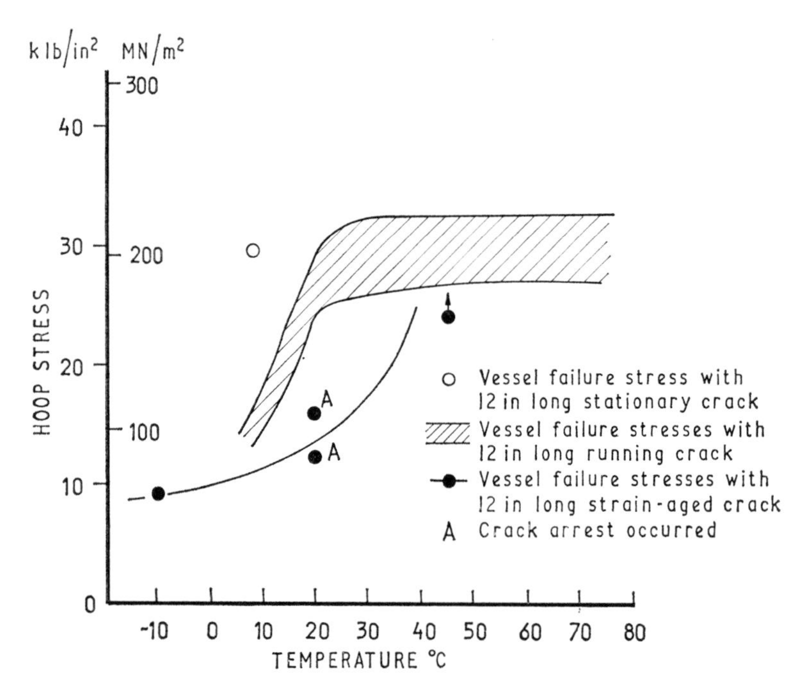

Fig. 54.5

vessel, consisting of repeated pressurization at 300°C, preceded by an initial pressurization at ambient temperature to 1·6 times the pressure applied hot. Final fracture of the vessels after this treatment showed that considerable embrittlement had been produced (Fig. 54.5). After having survived a proof stress of 131 MN/m² (19 lb/in² × 10³) the fracture stress was in one case as low as 62 MN/m² (9 lb/in² × 10³).

The loss in strength brought about by strain-ageing could be a cause of worry and might lead one to consider repeat proof testing, offsetting as it does arguments on hot pre-stressing as a protective influence, but factors operate to ameliorate the situation. Firstly, a transitional type of behaviour is exhibited, the fracture stress being very high above a certain temperature, which depends upon the crack length and the thickness. As an example, the vessel tests and their accompanying notched bend tests (**12**) showed that cracks 150 mm (6 in) long, or more specifically a stress-intensity factor, ΔK, during hot cyclic loading of 50 MN $m^{-3/2}$ raised the transition temperature locally above the ambient range (to approximately 70°C in reactor vessel plate) after 100 cycles of load. However, the effect will not be important in vessel operation above about 100°C since the transition temperature increase showed a saturation effect, no amount of hot deformation at the crack tip being sufficient to raise it above 40°C in the 26-mm (1-in) thick plate and above 90°C in 90-mm (3·5-in) thick plate studied.

Secondly, it was found that fracture of the strain-age-embrittled material at the crack tip did not necessarily lead to rupture of the vessel. Taking the case of a vessel containing a 300-mm (12-in) long crack with a hoop stress lower than that required to maintain a crack of that length in motion [determined in a previous set of tests (**13**)], the crack stopped after extending about 20 mm (0·8 in) at each end (see Fig. 54.5).

This only occurs in restricted situations; the stress required to maintain a crack running shows a transition from low values to very high values within a fairly narrow temperature range. This is approximately 5°–20°C in the case of the 26-mm (1-in) plate studies, compared with 40°C for the fracture toughness transition of excessively strain-aged material. Only at temperatures between these two transitions is it possible to encounter the above type of behaviour.

It can be concluded that the strength of a vessel at temperatures in excess of 100°C will not be affected by strain-ageing. At temperatures below this the safe level of pressurization is governed principally by the slope of the dynamic fracture toughness transition curve, which has been the subject of considerable study in these laboratories but will not be discussed here since it touches upon start-up procedures rather than repeat proof testing. Provided that scientifically based start-up procedures are adhered to, it is clear that strain-ageing does not present a hazard and therefore gives no grounds for repeat proof testing. It is equally clear, however, that the possibility of strain-ageing might limit the choice of temperature for such a test.

The repeat proof test temperature

It would on the face of it seem most natural to choose the lowest temperature at which the vessel is ever pressurized (say after a long shut-down) for the repeat proof test, since it is generally accepted that the risk of brittle fracture is greater the lower the temperature. Caution is required, however, since the stress applied in a proof test is much greater than the low stresses allowed during start-up from these temperatures. If it is accepted that initiation of cracking would be undesirable in such a test even if crack arrest occurred later, the fracture stresses of strain-aged vessels presented in Fig. 54.5 can be used to quantify the strength at these low temperatures. The results are only approximate because the number of tests performed has been rather small.

Fig. 54.5 shows that the fracture stress of the 1·52-m

diameter test vessels is approximately 62 MN/m^2 (9 lb/in$^2 \times 10^3$) at temperatures well below the transition temperature. This result is capable of more general interpretation if the critical crack opening displacement measured during the test is used. Expressions relating the crack opening displacement, δ, to the hoop stress, σ, for a cylinder containing a through, axial crack are (**9**)

$$\delta_1 = \frac{\pi\sigma^2 a}{\sigma_y E}\left(1+1{\cdot}6\frac{a^2}{Rt}\right) \quad . \quad . \quad (54.5)$$

in the initial proof test loading and

$$\delta_2 = \frac{\pi\sigma^2 a}{2\sigma_y E}\left(1+1{\cdot}6\frac{a^2}{Rt}\right) \quad . \quad . \quad (54.6)$$

in subsequent service loadings, where E is Young's modulus and σ_y is the yield stress.

Equations (54.5) and (54.6) can be used to estimate the fracture stress of a cool reactor pressure vessel after severe strain-ageing, assuming it to contain a long crack that was just short of the critical length in the proof test. Taking 1·5 mm (0·060 in) for the δ_1 in the initial proof test and 91 μm (0·0036 in) for the critical δ_2 after strain-ageing (measured in the test vessel) shows a fracture stress equal to one-third of the initial proof test stress. This is an underestimate, in fact, since any cracks present are likely to be much shorter than has been assumed. While the result may indicate the acceptability of applying a small stress to a reactor pressure vessel cold, it also indicates very strongly the inadvisability of repeat proof testing cold. In the presence of a crack of sufficient length [approximately 280 mm (11 in) in a 1·35 overpressure test, from equation (54.6)], such a procedure would cause a fracture that would never have occurred under service conditions.

It can be concluded that strain-ageing embrittlement does not give any support for repeat proof testing, as its worst consequences can be avoided by appropriate operating rules. However, if such tests are carried out for other reasons, it provides a minimum testing temperature which depends upon the material and the plate thickness and is 100°C in the case of the 9-mm (3·5-in) thick reactor vessel plate studied.

IRRADIATION EMBRITTLEMENT

Neutron irradiation has the effect of raising the fracture toughness transition temperature of ferritic steels. In the case of reactor vessels this is usually monitored by the periodic removal and testing of Charpy specimens. The amount of embrittlement will depend on neutron dose, material, and temperature of exposure, all of which are functions of the design. In reactors operated so far it is frequently the case that any real changes of vessel material transition temperature are largely induced by thermal or strain ageing and that radiation embrittlement has played little part. In such cases irradiation embrittlement is not an important factor in discussions on the value of repeat proof testing, but it might become so in certain other cases where the radiation dose is very high.

CREEP

Nuclear vessels are designed so that significant creep will not occur at the design temperature and creep-rupture is not a problem. Some creep is present inevitably, however. For instance, Wood (**14**) has shown that a C–Mn pressure vessel steel will creep approximately 1 per cent if strained 2·3 per cent and held at that stress for 60 min at 350°C. This information gives some idea of the amount of creep possible in regions of high stress, near crack tips for example; but at the much lower design stress the creep rate is negligible, of course.

There is a strong possibility that the very small plastic strain (about 0·1 per cent) required to relax out a field of residual stress could occur by creep below 400°C. As one of the benefits usually associated with proof testing is that it sets up a favourable pattern of residual stresses (**15**), it could be argued that periodic proof testing would re-establish these. The argument is incomplete, however. The benefits have not been quantified; creep could never relax the residual stress completely as it required a stress to drive it; the time scales involved are not known. It is likely that relaxation will occur either too rapidly or too slowly for periodic proof testing to be of value.

It is possible to be somewhat more specific with regard to the effects of residual stresses set up by the presence of a crack. They cause the crack-opening displacement on reloading to be smaller than that obtained on initial loading. For example, when reloading to the same stress, equations (54.5) and (54.6) show that $\delta_2 = \delta_{1/2}$. Assuming creep removes the residual stress completely, the amount of strain at each change of load will therefore be doubled and strain-age embrittlement will be more likely. This information is not of great concern, however, since the safeguard against strain-age embrittlement lies in temperatures being high enough for the toughness to be adequate when the full operating pressure is approached (see under 'Strain-ageing embrittlement').

On the grounds that the consequences of the limited creep strain are not serious and that it is not clear that the rate of stress relaxation is in the appropriate range, it can be concluded that creep provides no argument in support of repeat proof testing.

DISCUSSION

Examination of four processes that can affect the integrity of a pressure vessel (fatigue, irradiation, damage, strain-ageing, and creep) has shown that only fatigue gives conceivable grounds for repeat proof testing nuclear vessels. Fracture due to fatigue would require the presence of a crack initially several feet long and penetrating at least half the wall thickness.

Unfortunately, this is the situation in which a proof test could do damage by causing slow tearing at the ends of the postulated defect without inducing fast fracture. In the series of tests of 1·53-m diameter vessels used to compile Fig. 54.5, slow tearing at a 300-mm (12-in) long crack was first detected at 103 MN/m^2 (15 lb/in$^2 \times 10^3$) hoop stress, whereas final rupture did not occur until 199 MN/m^2 (29 lb/in$^2 \times 10^3$) was reached. The fracture surface showed signs of approximately 40 mm (1·5 in) of ductile growth prior to failure. A more thorough study of the phenomenon has been made by Fearnehough *et al.* in 1971 in relation to natural gas pipeline problems. For instance, a through crack 180 mm (7·9 in) long was found to grow to 220 mm (9·8 in) before fracture occurred, mostly in the last quarter of the stress rise. It can be concluded that a repeat proof test to an overpressure of say 1·35 could cause an increase in length of a just sub-critical crack of up to 20 per cent.

The critical value of a in a 1·35 overpressure is 850 mm (33·5 in) on the basis used in compiling Fig. 54.3. A crack just shorter than this might grow by 20 per cent to 1·0 m (40 in), and Fig. 54.3 shows that this is equivalent to the 'using up' of 85 service cycles of fatigue, leaving only a small and uncertain residue. It should not be concluded that a proof test will normally cause damage. At more moderate crack lengths the test can be likened in effect to a single cycle of fatigue only.

Summary

A repeat proof test would be a waste of effort unless a crack greater than 600 mm (2 ft) were present, in which case the test could do considerable damage.

Peripheral factors to be considered are: (*a*) the hazard during the repeat proof test itself, and (*b*) other ways of providing additional confidence in reactor vessel integrity. These two factors will be considered in order below.

(*a*) The risk of failure during a repeat test only operates as an argument against if the failure would not otherwise have occurred in service. This risk would naturally be limited by applying an overpressure short of the initial proof test.

It can be seen in Fig. 54.3, i.e. the case of ductile material, that fracture in a 1·35 overpressure test would occur only if the fatigue life remaining were less than 100 cycles, so that fracture would probably have occurred in service anyway. Since appreciable deterioration following the initial proof test is obtained only in the case of long cracks in ductile material, saving the case of strain-age embrittlement, it is concluded that the hazard in a repeat test is negligible. It has been seen that adoption of a minimum proof-testing temperature of 100°C would remove any risk from strain-age embrittlement.

(*b*) If the information on failure rates of pressure vessels generally (see under 'Introduction') indicates an unacceptably high failure rate when applied directly to reactor pressure vessels, periodic proof testing is not the only recourse. It is logical to take account of the features that mark out reactor vessels as members of a special class. Examination of the particular requirements for fracture provides some idea of the remoteness of the possibility, and also of the effects of periodic proof testing. This has been the task of the present paper.

However, there are several areas in which knowledge could be improved with advantage. For instance, while it is reasonable to assign a low probability to the very long partial penetration cracks necessary for fracture, insufficient is known about the possible origins of such cracks to put a number to their probability. Moreover, sufficient information about their rate of growth is not available. Again, the resistance to crack propagation falls off markedly in the lower ranges of vessel operating temperatures, a phenomenon that will presently be put on a sound fracture mechanics basis.

Additional areas of relative ignorance include the toughness of welds and the behaviour of cracks in situations of complex geometry. Further enquiry along these lines, coupled with improved non-destructive testing techniques, e.g. the detection of stress-wave emission from cracks, would seem to be a surer way than periodic proof testing to obtain assurance of the continuing safety of reactor pressure vessels.

CONCLUSIONS

(1) There is no statistical evidence upon which to judge the effectiveness of periodic proof testing, nor is there information on the effect of the degree of overpressure in the initial proof test on the probability of fracture. Fracture mechanics indicates that the fatigue lifetime guaranteed will depend sensitively on the level of overpressure applied.

(2) The fatigue lifetime guaranteed by a pressure test is so uncertain on present knowledge that survival of a proof test would not enhance confidence appreciably.

(3) Taking into account all the changes that occur in the service life of a reactor pressure vessel, fracture will not occur within 30 years unless a very large crack was present initially. It must be several feet long and penetrate the wall to a depth of at least half the thickness.

(4) If a crack approaching these dimensions were present, ductile tearing during a repeat proof test could lead to a fracture in later service that would not otherwise have occurred.

(5) On the basis of the previous conclusions, periodic proof testing would reveal nothing quantifiable about vessel integrity and might actually cause damage. This suggests periodic proof testing should not generally be carried out.

(6) Our assessment of the integrity of reactor pressure vessels can best be refined by continuing research into the mechanical and material requirements for fracture and by improving our techniques for non-destructive testing.

ACKNOWLEDGEMENT

This paper is published by permission of the Central Electricity Generating Board.

APPENDIX 54.1

REFERENCES

(1) Slopianaka, G. and Mieze, G. Inst. Reaktorsicherheit der Technischen Uberwachungs-vereine, E.V., Cologne, Rept No. IRS-134, 1968.

(2) Phillips, C. A. G. and Warwick, R. G. U.K.A.E.A. Rept AHSB(S) R. 162, 1968.

(3) Phillips, C. A. G. Personal communication, 1970.

(4) Irvine, W. H. U.K.A.E.A. Rept No. AHSB(S) R. 143, 1968.

(5) Yukawa, S. General Electric Rept No. HSSTP-TR-1, 1969; also *Proc. Symp. Tech. Pressure-retaining Steel Components*, *Nucl. Metallurgy* 1970 **16**, 250.

(6) Johnson, H. H. and Paris, P. C. *Engng Fracture Mech.* 1968 **1**, 3.

(7) Folias, E. S. *Int. J. Fracture Mech.* 1965 **1**, 104.

(8) Hahn, G. T. and Sarrate, M. *Proc. Symp. Fracture Toughness Concepts for Weldable Structural Steel* 1969 (Chapman and Hall, London).

(9) Edmondson, B., Formby, C. L., Jurevics, R. and Stagg, M. S. *Proc. Second International Conf. on Fracture* 1969 (Chapman and Hall, London).

(10) Jurevics, R. Personal communication, 1969.

(11) Brown, W. G. and Srawley, J. E. S.T.P. No. 410, 1966 (Am. Soc. Test. Mater., Philadelphia).

(12) Formby, C. L. and Charnock, W. Paper C1/71, *Conf. Practical Application of Fracture Mechanics to Pressure Vessel Technology* 1971 (3rd–5th May), 1 (Instn Mech. Engrs, London); also C.E.G.B. Rept No. RD/B/R1846.

(13) Edmondson, B., Formby, C. L. and Stagg, M. S. *Proc. Symp. Fracture Toughness Concepts for Weldable Structural Steel* 1969 (Chapman and Hall, London).

(14) Wood, D. S. *Weld. J.* 1966 **45**, 90-S.

(15) Nichols, R. W. *Br. Weld. J.* 1968 **15**, 21, 75.

BIBLIOGRAPHY

BROTHERS, A. J. General Electric Rept No. GEAP-5607, 1968.

BROTHERS, A. J. and YUKAWA, S. *J. Bas. Engng*, Paper 66-Met. 2, 1966 (Am. Soc. Mech. Engrs, New York).

CHARNOCK, W. and FORMBY, C. L. C.E.G.B. Rept No. RD/B/N1818, 1970.

COWAN, A. and KIRBY, N. *Proc. Symp. Fracture Toughness Concepts for Weldable Structural Steel* 1969 (Chapman and Hall, London).

DUFFY, A. R. and EIBER, R. J. *Proc. Symp. Fracture Toughness Concepts for Weldable Structural Steel* 1969 (Chapman and Hall, London).

FORMBY, C. L. C.E.G.B. Rept No. RD/B/N1219, 1969.

IRVINE, W. H., QUIRK, A. and BEVITT, E. U.K.A.E.A. Rept No. AHSB(S) R. 142, 1968.

IRVINE, W. H. Personal communication, 1971.

JERRAM, K. Personal communication, 1969.

ROLFE, S. T. and MUNSE, W. H. Ship Struct. Comm. Rept SSC-143, 1963.

WESSEL, E. Paper 72, *Proc. Second Int. Conf. on Fracture* 1969 (Chapman and Hall, London).

C55/72 PLANNING FOR IN-SERVICE INSPECTION OF NUCLEAR REACTOR COOLANT SYSTEMS IN THE U.S.A.

R. L. PHIPPS*

This paper reviews the A.S.M.E. Code requirements on which in-service inspection planning is based. Categories of inspection required by the code are described. The importance of design and accessibility reviews, the establishment of system boundaries, and the application of exclusion criteria with respect to inspection planning are discussed.

INTRODUCTION

PLANNING FOR IN-SERVICE INSPECTION is a continuing activity which is most effective when begun early in the plant design phase. (Section XI of the A.S.M.E. Boiler and Pressure Vessel Code is the source of the requirements on which the planning is based.) The intent of this paper is to discuss the code requirements on which the planning is based and the major steps in the planning activity.

GENERAL CODE REQUIREMENTS

The basic requirements of the code are for the performance of pre-service and scheduled periodic post-operational or in-service inspections of the plant. There are other general requirements which define the access, extent, and methods of inspection. Effective planning considers these requirements.

The first of these is the requirement to provide an inspectable plant. The most important aspect in this provision is that sufficient space be available around each inspection area to permit access by the inspector and his equipment. It also means that weld contours and surface finish must be set to avoid interference with the prescribed inspection. Space allowance for disassembly of equipment such as insulation is another requirement. Scaffolds, lighting, and handling equipment are also a desirable part of an inspectable plant.

The code stipulations as to the type of required inspection must be taken into account early in the planning activity. The more precisely the inspection technique can be defined, the more flexibility for other design goals such as economy. As a negative example, some past plants were designed to permit access for inspection of the reactor vessel from both the inside and outside of the vessel. This obviously resulted in increased plant costs over the early selection of the inside inspection method.

The code covers three categories of required inspection: these are visual, surface, and volumetric. The simplest of these and the least defined as to technique and rejection criteria is the visual method. Adequacy of the technique is based on the ability to discern a $\frac{1}{32}$-inch black line on a neutral grey card. This card is shown attached to the inspection area in Fig. 55.1. Direct or remote visual observation is allowable provided the required resolution standard can be met. Remote visual can involve the use of mirrors, television cameras, or boroscope. With this flexibility allowed it is obvious that early selection of the specific technique is desirable to ensure that the design provides proper access. As the pre-service or reference examination technique, we recommend colour photographs for critical areas, as shown in Fig. 55.1. Another simple example of effective planning is to take photographs of the pump interior after its final cleaning to preclude an additional disassembly.

The second category of inspection methods is surface inspection. The code specifically mentions liquid penetrant and magnetic particle examination as the allowed methods. These are both well-established methods used in other vessel and piping inspections. Since magnetic particle inspection will be effective for only ferromagnetic materials, not for stainless steel, its use in in-service inspection is limited. Therefore, in the interest of consistency, we recommend that the liquid penetrant method be used wherever surface inspection is required. This category, in general, is the least critical to the planning activity. There are, however, some required areas of inspection where remote application of the penetrant technique may be required for in-service inspection due to high radiation levels.

Most time-consuming of the inspection categories is the volumetric examination. For this category, the code requires either ultrasonic or radiographic examination, with allowance for the use of other newly developed techniques if they are equivalent in defect detection capability. In line with most of the inspection industry, and because radiography is not feasible for use in many inspection areas, ultrasonic inspection is recommended as the primary volumetric method. For many areas the ultrasonic examination can be applied either directly or

The MS. of this paper was received at the Institution on 26th October 1971 and accepted for publication on 11th February 1972. 44

* *Westinghouse Electric Corp., PWR Systems Div., P.O. Box 355, Pittsburgh, Pennsylvania 15230, U.S.A.*

Fig. 55.1

remotely by the operator. The selection depends on which will affect the detailed design of the plant. An example of the need for specific detailed selection of the ultrasonic procedure prior to final design will be given later.

The code defines system boundaries subject to inspection. There are three groups of components defined by these boundaries. The first of these are components outside the boundary which require no inspection by the code. The second group comprises equipment inside the boundaries which must be completely inspected in accordance with the examination categories defined in the applicable code tables. Equipment inside the system boundary which meets the exclusion criteria of the code is in the third category, being components which, in the event of failure, would not cause a loss of coolant which would prevent an orderly shutdown and cooldown of the reactor. These categories are illustrated by Fig. 55.2. Components in the third category are required only to be visually inspected during a hydrostatic test. With proper design this may not require removal of the insulation.

SETTING DESIGN CRITERIA

The first step in setting the design criteria is establishment of the system boundaries. Since the code is relatively new, it is still subject to varying interpretations and to changes. Therefore, a knowledge of the code history and, in some cases, the agreement of the code committee on certain interpretations is necessary. In defining the boundaries, knowledge of the design parameters and the operation of the power plant must be applied. In determining the exclusion criteria, for example, the following information is applicable:

(1) Capacity of the make-up system.

(2) Dynamic analysis of coolant loss as a function of break size.

(3) Closing time of automatically actuated valves. An example of the required information is shown in Fig. 55.3.

Once the boundary is initially designated it is examined for possible modifications which will decrease the amount

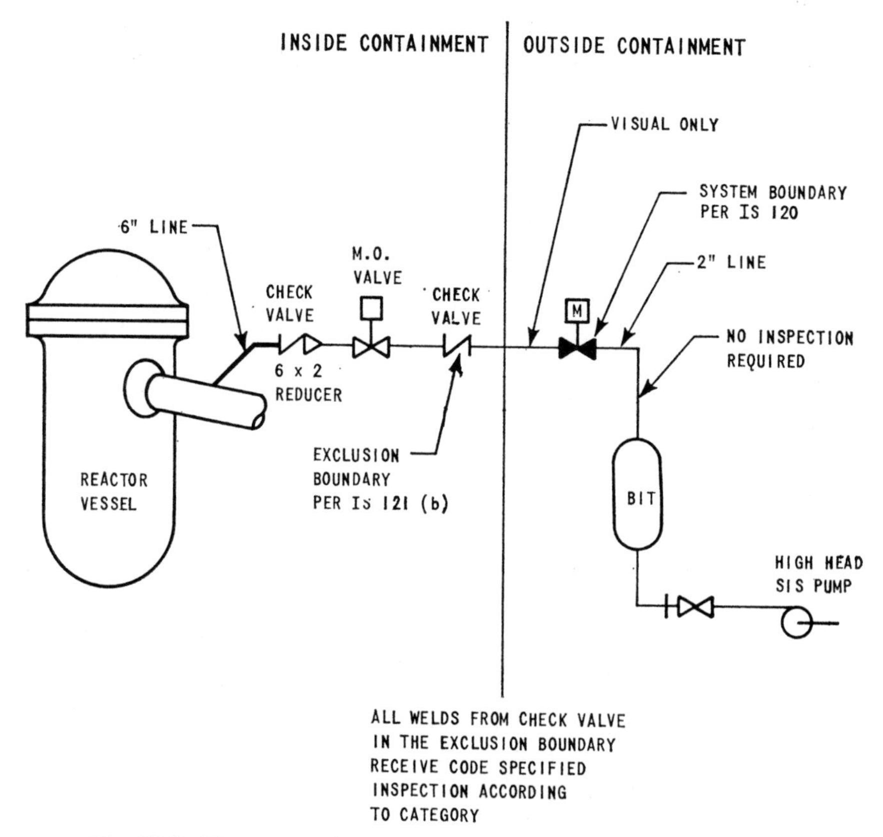

Fig. 55.2. Diagrammatic arrangement of boron injection system

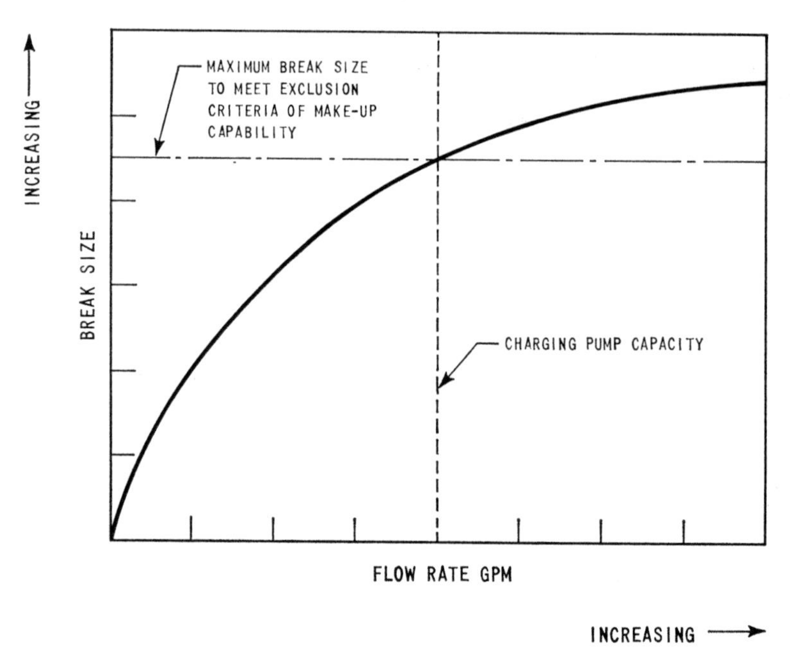

Fig. 55.3. Sample exclusion criteria analysis

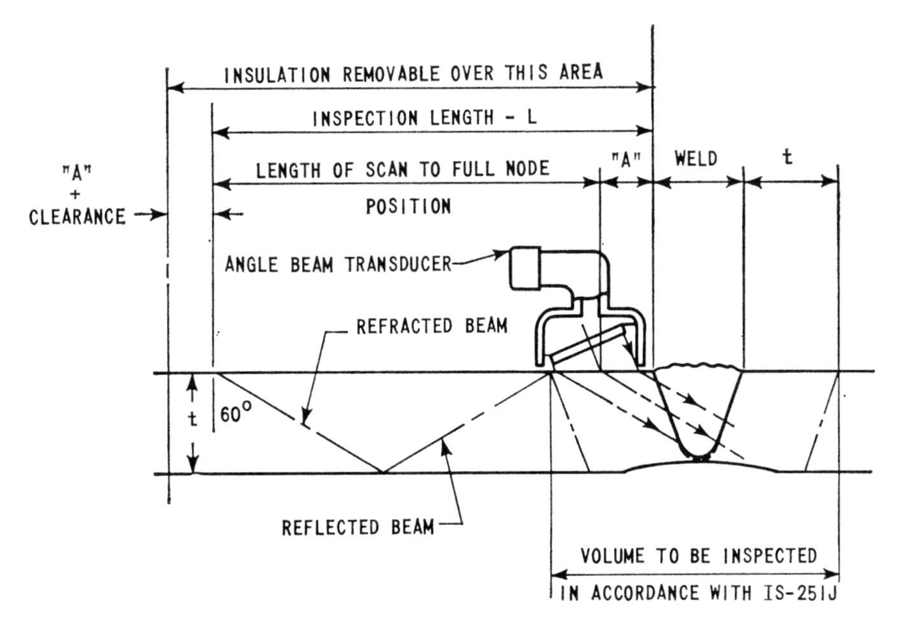

Note: Dimension 'A' will vary with transducer physical dimensions.

Fig. 55.4. Access requirement

of required inspection. A common example has been the physical relocation of valves to remove pipe welds from the inside to the outside of the boundary, without alteration of the flow diagram.

The second step in establishing design criteria is to select the specific inspection methods to be used. This selection is the basic factor in determining the free volume requirement about an inspection area. An example is the determination of the access dimension parallel to the pipe or component surface which, incidentally, is also used to determine removable insulation requirements. In setting this dimension, the pipe thickness, the weld detail, and the specific transducer angle to be used are pertinent factors. This is shown in Fig. 55.4. With the specific inspection method established, an access criterion is set for each area. When this criterion is set, the end result is a table showing access requirements for each weld by component or pipe size. A typical table is shown in Fig. 55.5. The access criterion when completed is to be used by the plant designers as a guide in producing drawings. It is also desirable to have the detailed drawings reviewed by personnel familiar with in-service inspection requirements.

PRE-SERVICE INSPECTION

Prior to performance of the pre-service inspection, final procedures for each type of inspection covering all inspection areas must be prepared and reviewed by the plant operator and the proper regulating agency. The procedures and a complete list of areas requiring inspection are then used to estimate the total inspection man-hours needed. This estimate, in conjunction with the plant construction or test schedule, is used to establish the pre-service inspection plan. The number of inspection personnel and

NOMINAL PIPE SIZE	PIPE SCHEDULE	NOMINAL O.D.	WALL THICKNESS	EACH SIDE OF WELD		
				SCAN LENGTH X	LENGTH L	REMOVABLE INSULATION LENGTH
14	160	14.0	1.406	5-5/8	6-3/8	10-1/2
12	160	12-3/4	1.312	5-1/8	5-7/8	10
10	140	10-3/4	1.000	3-3/4	4-1/2	8-1/2
10	40	10-3/4	0.365	7/8	1-5/8	5-3/4
8	40	8-5/8	0.322	3/4	1-1/2	5-1/2
6	160	6-5/8	0.718	2-1/2	3-1/4	4-3/4
6	40	6-5/8	0.280	1/2	1-1/4	2-3/4
4	120	4-1/2	0.438	1-1/4	2	3-1/2
3	160	3-1/2	0.438	1-1/4	2	3-1/2
3	40	3-1/2	0.216	1/4	1	2-1/2

Fig. 55.5. Typical access circumferential pipe welds

length of time on site are variables that can be set to best meet the plant schedular requirements. A sample schedule is shown in Fig. 55.6.

PLANNING THE IN-SERVICE INSPECTION

The pre-service inspection not only provides baseline data for subsequent in-service inspections, but should be considered as a rehearsal for in-service inspections. Difficulties noted during the pre-service examination should be listed, evaluated, corrected, if possible, and factored into the in-service inspection planning. Actual time used in performing pre-service examinations should be evaluated with predicted radiation levels.

The A.S.M.E. Code defines an in-service inspection interval as 10 years, and sets requirements for each one-third interval or three and one-third years. Generally, the code requires that at least 25 per cent, with credit for no more than $33\frac{1}{3}$ per cent, of the specified inspections be performed in each one-third interval. This is shown in Fig. 55.7. The amount of inspection required for an area varies according to the category, but is explicitly defined in the code. This is illustrated in Fig. 55.8. Certain examination categories are excluded from the intermediate

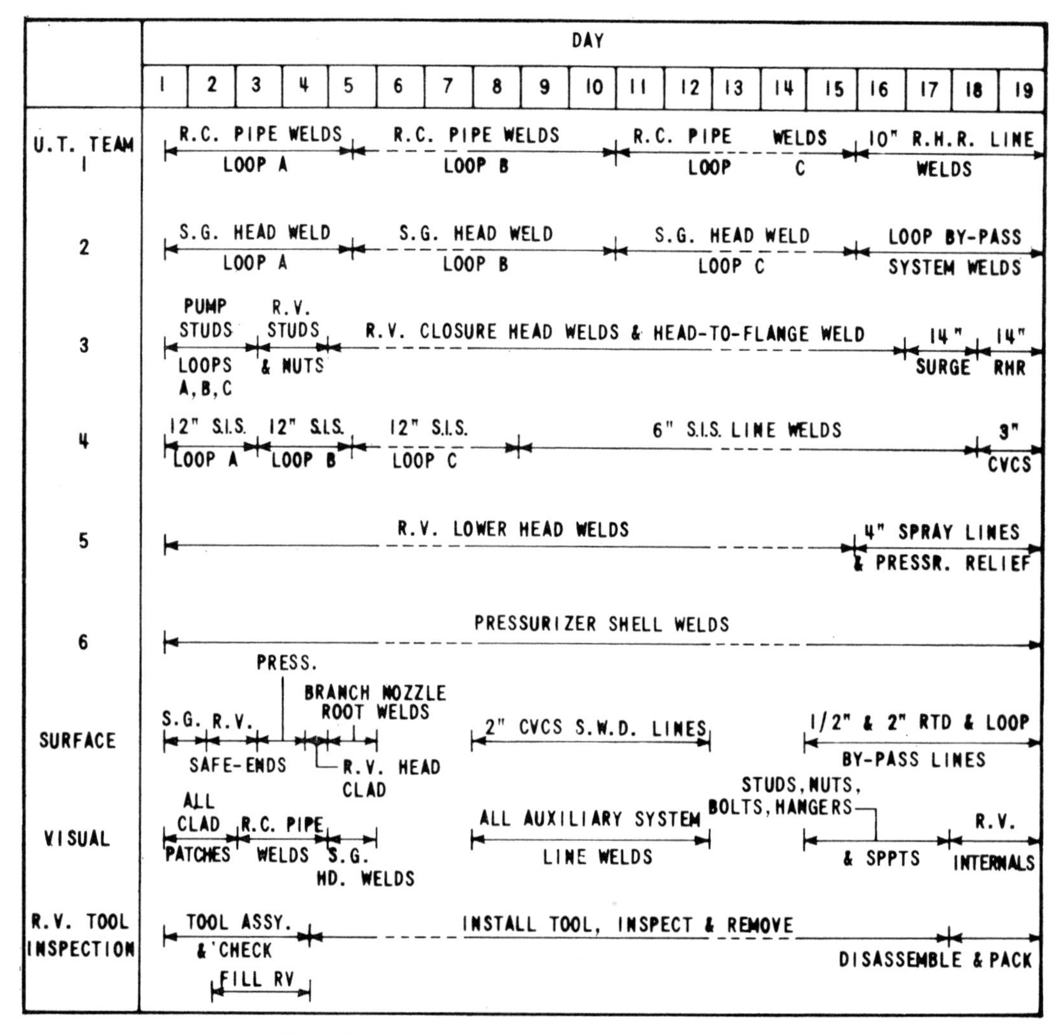

Fig. 55.6. Sample pre-service schedule outline

PRESERVICE	3-1/3 YEARS	6-2/3 YEARS	10 YEARS
100% +	AT LEAST 25% WITH CREDIT FOR NO MORE THAN 33-2/3%	AT LEAST 50% WITH CREDIT FOR NO MORE THAN 66-2/3%	100%

Fig. 55.7. Per cent of total 10-year inspection requirement

EXAMINATION CATEGORY	REQUIRED INSPECTION
B PRESSURE CONTAINING WELDS IN VESSELS	10% OF THE LENGTH, EACH LONGITUDINAL SHELL WELD AND MERIDIONAL HEAD WELD
	5% OF THE LENGTH OF EACH CIRCUMFERENTIAL SHELL WELD AND HEAD WELD
F PRESSURE CONTAINING DISSIMILAR METAL WELDS (SAFE-END WELDS)	100% OF THE CIRCUMFERENCE OF EACH SAFE-END WELD
J PRESSURE CONTAINING WELDS IN PIPING	100% OF EACH OF 25% OF THE CIRCUMFERENTIAL JOINTS

Fig. 55.8. Example of 10-year inspection requirements

inspection requirements and are allowed to be completely inspected at the end of the 10-year interval. These categories involve parts of the reactor vessel, welds in pump casings, and welds in valves. The code requirements mentioned above concerning the amount of an area to be examined, the requirements for each inspection interval, along with the experience from the pre-service inspection and predicted radiation levels are the basic input factors for the in-service inspection plan.

CONCLUSION

Following are the major points covered in this paper:

(1) Planning for in-service inspection is an important activity, most effective when initiated early in the plant design.

(2) Effective in-service inspection planning requires an intimate knowledge of the code and of the nuclear plant.

(3) Experience in performing the pre-service inspection should be factored into the in-service inspection plan.

C56/72 INTERNATIONAL REQUIREMENTS FOR THE PERIODIC INSPECTION OF PRESSURE VESSELS

J. J. WHENRAY*

This is an international review of the requirements for the periodic inspection of boilers, static and portable pressure vessels, and nuclear vessels, which are applied throughout the world. The review covers the requirements for external and internal examination, and hydraulic tests. Also included is information on the manufacturer's responsibility, exceptions, and inspection organizations.

INTRODUCTION

MOST COUNTRIES INSIST that an inspection of boilers and pressure vessels be carried out on a periodic basis to ensure that the vessel continues to be safe to operate. In general the tests consist of an internal or external examination, or an hydraulic test, or a combination of all three.

This paper contains tabulated summaries (Tables 56.1 and 56.2) of the periodic inspection requirements that are applied in 24 countries throughout the world. The information was obtained directly from the various inspection organizations. In the main, abbreviations and terms have been abstracted from the relevant *Technical Help to Exporters Technical Digest* (1)†. In some cases no information is listed owing to the fact that at the time of printing this paper the information had not been received from the inspection organizations. (It is hoped that by the time this paper is presented additional information will be available.)

Mention of the E.F.T.A. scheme refers to the 'Scheme for the reciprocal recognition of tests and inspections carried out on pressure vessels' (2). This scheme was implemented on 1st January 1971 and at the present time includes the E.F.T.A. countries of Denmark, Finland, Iceland, Norway, Portugal, Sweden, Switzerland, and the U.K.

As can be seen from the tables, all the countries listed have requirements for a periodic inspection of boilers and pressure vessels, but not all countries have inspection requirements for nuclear equipment. Many countries do not have nuclear equipment at present and therefore no requirements for regulations.

SUMMARY

In summarizing the information listed, the countries are divided into the areas of Europe, Australasia, and North America, and it is proposed to make a few comments with regard to these three areas.

The MS. of this paper was received at the Institution on 18th November 1971 and accepted for publication on 10th February 1972. 00

* *Technical Help to Exporters Service, British Standards Institution, Maylands Avenue, Hemel Hempstead, Herts.*

† *References are given in Appendix 56.1.*

EUROPE

All countries carry out periodic tests on some of the categories of pressure vessels, and these tests include the inspections plus the hydraulic test. The interval between periodic inspection varies from one year for boilers in Belgium, Sweden, and Italy to 10 years for pressure vessels in Portugal. The hydraulic test is required in all countries with the exception of Sweden. The country with the longest interval between tests is the Netherlands, where the interval is 25 years.

Manufacturer's responsibility

The manufacturer's responsibility ceases on acceptance of the vessel only in parts of Australia and in Canada, Czechoslovakia, the Netherlands, South Africa, Sweden, the U.K. and the U.S.A. In all the other countries given the responsibility continues indefinitely.

Inspection organizations

In general, organizations approved by the authorities are accepted for carrying out periodic inspections. Exceptions to this rule are France, Italy, and the Netherlands where the regulative authorities insist on carrying out their own inspection.

AUSTRALASIA

This area is only covered (at the time of writing) by information received from Australia and Japan. All the Australian states and Japan require that periodic tests are carried out, and these tests include inspection and hydraulic tests (with the exception of the Capital Territory of Australia where the hydraulic test is not required). The periodic inspection intervals are similar in all the Australian states from which information has been received, and for Japan. Boiler inspection is on an annual basis; the same applies for static pressure vessels, except in South Australia where the interval is two years. It would seem that, due to an absence of nuclear vessels in Australia, there are no requirements. In Japan the inspection of nuclear vessels is carried out annually.

Table 56.1. Periodic inspection of pressure vessels

Country	Interval between periodic inspection (years)				Inspection requirements	
	Boilers	Pressure vessels, static	Pressure vessels, portable	Nuclear	Internal and external inspection	Hydraulic test
AUSTRALIA						
Capital Territory	1	1	1	N/A	Yes	No
New South Wales	*	*	*	*	*	*
Northern Territory	1	1	1	N/A	Yes	Yes
Queensland	1	1	1	N/A	Yes	Yes
South Australia	1	2	2	N/A	Yes	Yes
Tasmania	1	1	1 to 4	N/A	No	Yes
Victoria	1	1	1 to 4	N/A	Yes	Yes
Western Australia	*	*	*	*	*	*
AUSTRIA	Internal 3 Hydraulic 6	Internal 3 Hydraulic 6	Internal and hydraulic 1 to 10	N/A	Yes	Yes
BELGIUM	1	5 or 10	2 or 10	1 or 3	Yes	Yes
CANADA						
Alberta	1	2	2	N/A	Yes	Yes
British Columbia	1	2	5	N/A	Yes	Yes
Manitoba	1	2	2	2	Yes	Yes
Newfoundland and Labrador	1	1	1	N/A	Yes	Yes
North West Territory	1	As required for safety	1	1	Yes	Yes, in some cases
Ontario	1	1	1	N/A	Inspector's decision	Inspector's decision
Prince Edward Island	1	1	1	*	Yes	Yes
Quebec	*	*	*	*	*	*
Saskatchewan	1	2	2	1	Yes	Yes
Yukon	1	1	1	N/A	Yes	Yes
CZECHOSLOVAKIA	Internal 5 External 3 Hydraulic 9	Internal 5 External 3 Hydraulic 9	Varies	In preparation	Yes	Yes
DENMARK	5	5	5	5	Yes	Yes
FEDERAL REPUBLIC OF GERMANY	Internal 3 External 1 Hydraulic 9	Internal 4 External 2 Hydraulic 8	Internal 4 External 2 Hydraulic 8	Part internal and external 1 Full internal and external 4 Hydraulic 8	Yes	Yes
FINLAND	2 or 8	4 or 8	2 or 10	N/A	Yes	Yes
FRANCE	10	10	5	1 or 2	Yes	Yes
GERMAN DEMOCRATIC REPUBLIC	Internal 3 External 1 Hydraulic 9	Internal 4 Hydraulic 8	Internal 4 Hydraulic 8	Internal regulations	Yes	Yes
INDIA	1	1	5	Varies	Yes	Yes
ITALY	1	1	5	Varies	Yes	Yes
JAPAN	1	1	5	Varies	Yes	Yes
NETHERLANDS	2	6	5	2	Yes	Every 25 years

Table 56.1.—*contd.*

Country	Interval between periodic inspection (years)				Inspection requirements	
	Boilers	Pressure vessels, static	Pressure vessels, portable	Nuclear	Internal and external inspection	Hydraulic test
NEW ZEALAND	*	*	*	*	*	*
NORWAY	5	5	5	5	Yes	Yes
PORTUGAL	Internal 3 External 1 Hydraulic 6	Internal 3 External 1 Hydraulic 6	General 5 Certain types 3	Not determined	Yes	Yes
POLAND	5	10	10	*	Yes	Yes
SPAIN	*	*	*	*	*	*
SWEDEN	1	1 to 8	1 to 8	1	Yes	No
SWITZERLAND	*	*	*	*	*	*
TURKEY	1	1	1	N/A	Yes	Yes
U.K.	14 or 26 months	Steam and air receivers 26 months	5	Where required; interval varies	Yes	Yes
U.S.A.	1	2	2	As required	Yes	Not known

* No information available at time of writing.

Table 56.2. Manufacturer's responsibility, exceptions, and inspection organizations

Country	Manufacturer's responsibility			Inspection organizations
	Ceases on hand-over of vessel	Continues indefinitely	Varies with equipment	
AUSTRALIA				
Capital Territory	Yes	No	No	Dept of Interior
New South Wales	*	*	*	*
Northern Territory	No	Yes	No	Organizations approved by Mines Branch
Queensland	Yes	No	No	Organizations approved by authorities
South Australia	Yes	No	No	Dept of Labour and Industry
Tasmania	No	Yes	No	Dept of Labour and Industry
Victoria	No	Yes	No	Dept of Labour and Industry
Western Australia	*	*	*	*
AUSTRIA	No	Yes	No	Organizations approved by authorities
BELGIUM	No	Yes	No	Organizations approved by authorities

Table 56.2.—*contd.*

Country	Manufacturer's responsibility			Inspection organizations
	Ceases on hand-over of vessel	Continues indefinitely	Varies with equipment	
CANADA				
Alberta	No	No	Yes	Dept of Labour
British Columbia	Yes	No	No	Dept of Public Works
Manitoba	No	No	No	Dept of Labour, M. & E. Division
Newfoundland and Labrador	No	Yes	No	Dept of Labour
North West Territory	No	No	Yes	Organizations; Boiler and Pressure Vessel Inspection Branch Dept of the Territorial Secretary
Ontario	Yes	No	No	Organizations approved by Dept of Labour
Prince Edward Island	No	No	Yes	Dept of Labour, Industry and Commerce
Quebec	*	*	*	*
Saskatchewan	Yes	No	No	Dept of Labour, Boiler and Pressure Vessel Branch
Yukon	Yes	No	No	Dept of Yukon Territory
CZECHOSLOVAKIA	Yes	No	No	Cesky Urad
DENMARK	No	Yes	No	Boiler Inspection Authority
FEDERAL REPUBLIC OF GERMANY	No	Yes	No	T.U.V.
FINLAND	No	No	Yes	General; approved organization, see E.F.T.A. scheme
FRANCE	No	Yes	No	Service des Mines
GERMAN DEMOCRATIC REPUBLIC	No	No	No	T.U. der D.D.R.
INDIA	*	*	*	*
ITALY	No	Yes	No	A.N.C.C.
JAPAN	No	Yes	No	Organizations approved by the government
NETHERLANDS	Yes	No	No	Dienst vor het Stoomwezen
NEW ZEALAND	*	*	*	*
NORWAY	No	Yes	No	Boiler inspection authority
POLAND	No	Yes	No	*
PORTUGAL	No	Yes	No	Organizations approved in E.F.T.A. scheme
SOUTH AFRICA	Yes	No	No	Dept of Labour
SPAIN	*	*	*	*
SWEDEN	Yes	No	No	(1) Swedish Steam Users' Association (2) Technical X-Ray Centre
SWITZERLAND	*	*	*	*
TURKEY	No	No; approx. 1 year	Yes	Turkish Lloyd
U.K.	Yes	No	No	Organizations approved by authorities
U.S.A.	Yes	No	No	State Authority

* No information available at time of writing.

Manufacturer's responsibility

In general the manufacturer's responsibility continues indefinitely, with the exception of the Capital Territories, Queensland, and South Australia.

Inspection organizations

In general, inspection is carried out by the regulative authorities, with the exception of the Northern Territories and Queensland, where organizations approved by the regulative authority are accepted.

NORTH AMERICA

We have obtained information from all the provinces of Canada, except Quebec, at the time of writing. The information on the U.S.A. is very generalized and was obtained from the A.S.M.E. organization. To produce information on each state in the U.S.A. would require a separate paper.

All the provinces of Canada and the states of the U.S.A. require periodic tests to be carried out. These tests include the inspections and hydraulic test. In all cases the periodic inspection of boilers is on an annual basis; other pressure vessels vary between one and two years. The inspection interval for nuclear vessels is between one and two years.

Manufacturer's responsibility

In British Columbia, Manitoba, Ontario, Saskatchewan, the Yukon, and the U.S.A., the manufacturer's responsibility ceases on acceptance of the vessel. In all the other provinces, excluding Newfoundland and Labrador, though the manufacturer's responsibility does not cease on acceptance of the vessel, neither does it continue indefinitely.

Inspection organizations

In most cases the inspections are carried out by the regulative authority. Exceptions are Ontario, North West Territory, and Prince Edward Island, where organizations approved by the Department of Labour are acceptable.

CONCLUSION

It is interesting to compare the intervals between periodic inspection across the three areas outlined above. Europe, being the oldest established industrial area, carries out less frequent inspections than the youngest industrial areas of North America and Australasia.

This could imply a number of different criteria, such as closer control of the inspection organizations, less stringent climatic conditions, and many other reasons on which we can only speculate.

APPENDIX 56.1

REFERENCES

(1) *Technical Help to Exporters Technical Digests*. These are documents published by B.S.I. on a country by country basis. Each digest gives detailed information on the regulations, codes of practice, approval organizations, and approval procedure of the particular country. For further details contact Technical Help to Exporters, B.S.I., Maylands Avenue, Hemel Hempstead, Herts. Telephone Hemel Hempstead 2341.

(2) *E.F.T.A. Scheme*: 'Scheme for the reciprocal recognition of tests and inspections carried out on pressure vessels.' Copies of the updated E.F.T.A. agreement can be obtained directly from E.F.T.A. Headquarters, 9–11 rue de Varembe, 1211 Geneva 20, Switzerland.

C57/72

MECHANIZED EQUIPMENT FOR IN-SERVICE INSPECTION OF NUCLEAR REACTORS

C. E. LAUTZENHEISER*

The paper describes some of the mechanized equipment and techniques developed for the inspection of light water moderated P.W.R. and B.W.R. systems and of heavy water moderated B.W.R. systems. Nozzle devices and vessel inspection devices are discussed, and other systems are considered. The current state of the art is reviewed, and a summary given of the position in January 1972.

INTRODUCTION

MOST CODES for in-service inspection of nuclear reactor systems require volumetric inspection of welds. Ultrasonic techniques are currently the only techniques available to examine volumetrically the base metal and welds of a reactor pressure vessel. Due to the problem of personnel exposure to radiation, mechanized equipment and techniques are usually required for in-service inspection of these vessels. The examination may be made from either the o.d. or i.d. (outside diameter or inside diameter) surface, depending upon the design of the reactor pressure vessel, the internals, the insulation, and the biological shield. This paper will describe some of the mechanized equipment and techniques that have been developed for the inspection of light water moderated P.W.R. (pressurized water reactor) and B.W.R. (boiling water reactor) systems and of heavy water moderated B.W.R. systems.

NOZZLE DEVICES

The first mechanized ultrasonic scanning equipment for inspection of a nuclear reactor was designed by Sw.R.I. (Southwest Research Institute) for use on the Elk River reactor pressure vessel. This equipment, two samples of which are shown in Figs 57.1 and 57.2, consisted of three assemblies for inspection of the 10-in steam outlet nozzles, the 8-in feedwater nozzles, and the 16-in force recirculation nozzles. Since the equipment was ordered after the system achieved criticality, these items were designed in accordance with the as-built drawings of the vessel. However, at the next shutdown, an inspection could not be made of all the nozzles because the as-built dimensions were considerably in error as to the actual i.d. of the nozzles, and the fact that the nozzle bore was elliptical in shape and not oriented normal to the centre of the vessel.

In the nozzles where an inspection could be made, the data generally agreed with that generated during the pre-service inspection. Based on these results, it is advisable to be sceptical of as-built data; this scepticism has been justified on other more recent reactor systems.

Two different types of devices have been developed for inspecting nozzles from the i.d. The device in Fig. 57.3 was developed for the inspection of nozzles that do not orient towards the centre of the vessel. The device can be inserted by almost any positioning system that can push the device into the nozzle bore. Once in the bore, the device is aligned along the nozzle central axis by self-contained feet, and can inspect the nozzle-to-shell welds as well as the nozzle-to-safe-end and pipe welds. The other type of device (Fig. 57.4) is similar to that which was used in Italy and in Sweden. This device requires a sophisticated manipulator as its orientation, and position in the nozzle is dependent upon the accuracy of the positioning system. Both these devices have been proved on pre-service inspections.

VESSEL INSPECTION DEVICES

For inspection of welds in reactor pressure vessel walls, there are two generalized types of inspection equipment: one type for inspecting from the vessel i.d. and the other for inspecting from the o.d. There are many variations of both these types, depending on the type of reactor and the access. One of the first types of positioning equipment was that used in Italy for inspecting the nozzles on the Garigliano and Trino Vercelles reactors. The first complete system for pre- and in-service inspection of a reactor coolant system was that for the Oskarshamn B.W.R. in Sweden.

Portions of this system were designed and fabricated by Sw.R.I., and others by T.R.C. (Tekniska Rontgen-centralen). The system consists of the following six major items.

Ring positioner

The equipment shown in Fig. 57.5 can position an ultrasonic module, a TV camera, nozzle inspection devices, or any other mechanism within the limits imposed by the reactor internals. This positioner is essentially a precision

The MS. of this paper was received at the Institution on 6th December 1971 and accepted for publication on 11th February 1972. 34

* *Southwest Research Institute, P.O. Box 28147, San Antonio, Texas 78228, U.S.A.*

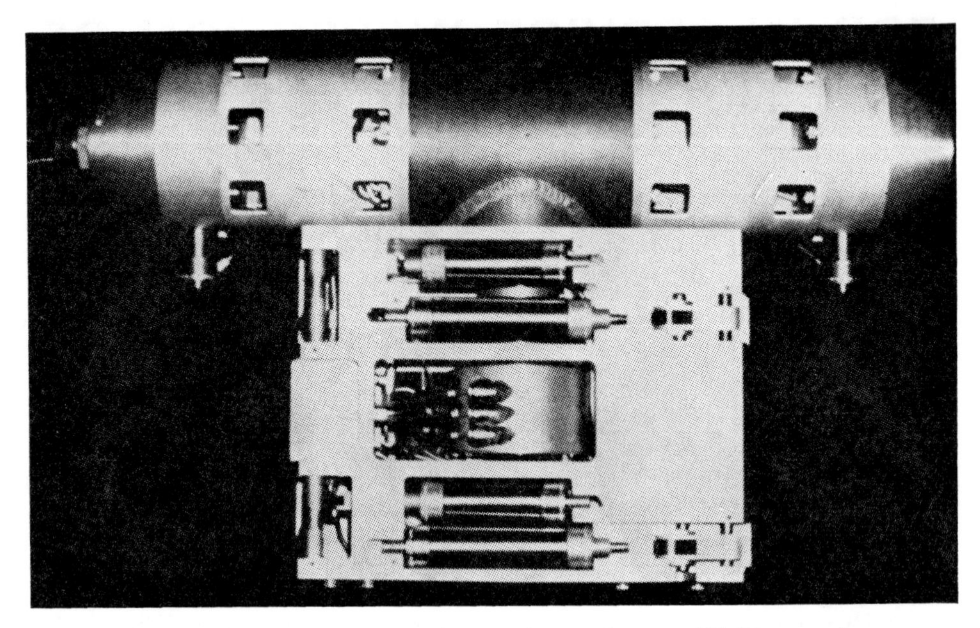

Fig. 57.1. Mechanized ultrasonic scanning equipment, Elk River project

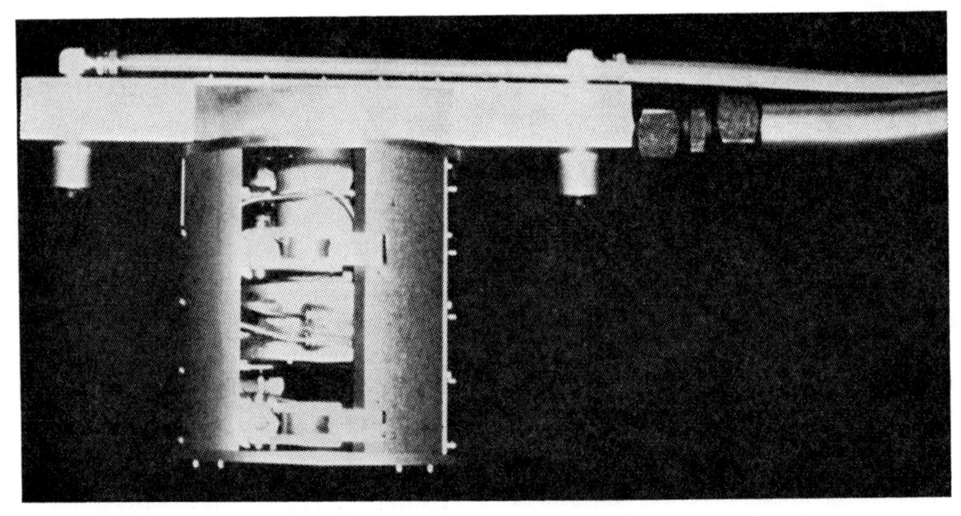

Fig. 57.2. Mechanized ultrasonic scanning equipment, Elk River project

Fig. 57.3. Nozzle interior inspection device

Fig. 57.4. Device developed for inspecting nozzles from the i.d.

Fig. 57.5. Ring positioner on Oskarshamn 1 reactor pressure vessel

Fig. 57.6. Ultrasonic module for inspection and defect analysis

Fig. 57.7. Gimbals for remote TV system

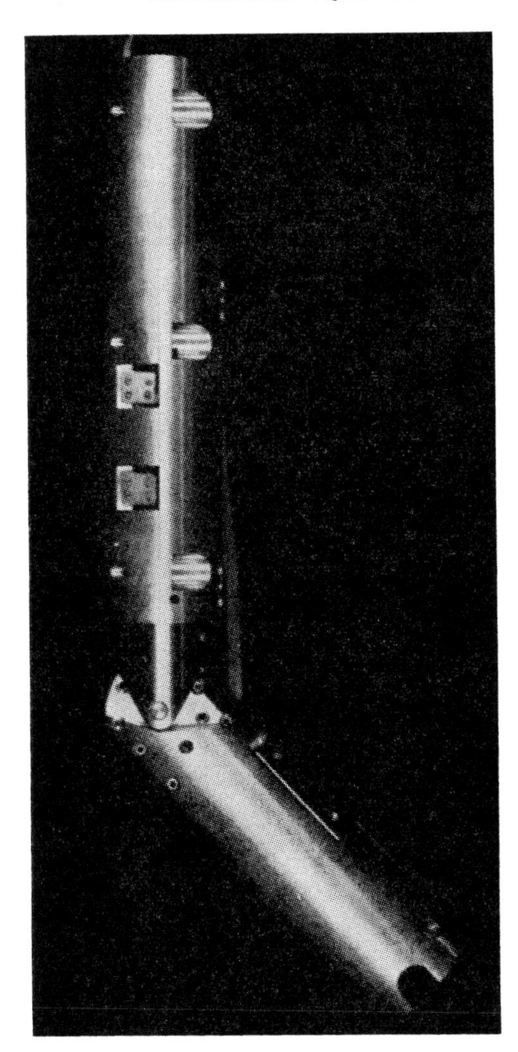

Fig. 57.8. Elbow for remote television system

rotary bridge crane, with rotational, translational, and vertical movements.

Ultrasonic module

The module shown in Fig. 57.6 was designed to pass through and operate within the 50-mm (2-in) annulus between the thermal shield and the vessel wall. In addition to the inspection, it was also essential that the module could be used for defect analysis. Thus, it contained a fixed longitudinal transducer, and two remotely variable angle transducers capable of propagating a surface wave in each direction. These variable-angle transducers can also be remotely moved towards and away from each other. This allows simultaneous inspection at two angles, the use of the pitch and catch, or use of the tandem technique. The entire module section can also be rotated 90° for inspection of longitudinal or circumferential welds.

Remote television system

This is supported by a mast passing through a gimbals located on the ring positioner (Fig. 57.7). The gimbals allows vertical movement, rotational movement, and up to 10° angulation of the mast. At the lower end of the mast is an elbow (Fig. 57.8), which allows the camera to be moved straight down to slightly above the horizontal plane. In front of the camera is a remotely rotatable mirror and lighting system (Fig. 57.9), so that the direction of the view may be varied as desired. The camera barrel is modified so that either a $\frac{1}{2}$-in or a 1-in lens may be used for normal viewing, or a 3-in lens, together with extension tubes, can be used for viewing up to 10Z magnification.

Device for inspection of steam outlet nozzles

This device (Fig. 57.10) moves in and out of the nozzles by the ring positioner, and provides rotational movement and remote transducer angulation to permit scanning of the nozzle-to-shell weld and the nozzle-to-piping welds.

Special nozzle device

This is used for the inspection of the forced recirculation nozzles. Changes were made in reactor design to allow use of this equipment. These changes were a pipe installed in the elbows of each of these nozzles, and a 'hot-tap' assembly used to insert the device, which can thus be placed in position with the reactor full of water. The small size of the insert pipe requires that the arms holding the transducers be remotely erected after insertion, and the complicated geometry necessitates sophisticated transducer angulation. The device inspects the nozzle-to-shell welds, the nozzle-to-piping welds, and the weld attaching the pipe to the nozzle elbow.

Fig. 57.9. Rotatable mirror and lighting system

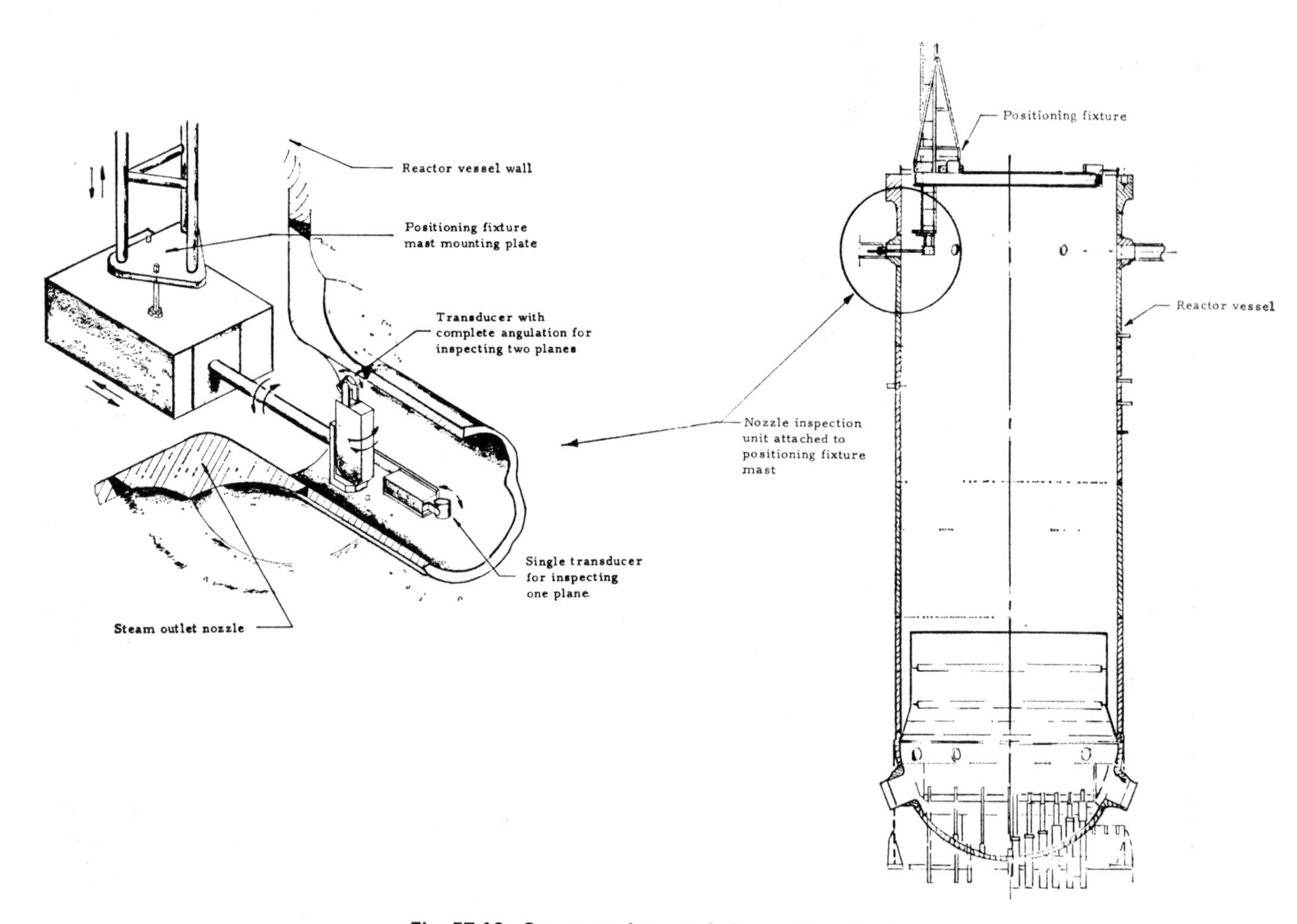

Fig. 57.10. Steam outlet nozzle inspection device

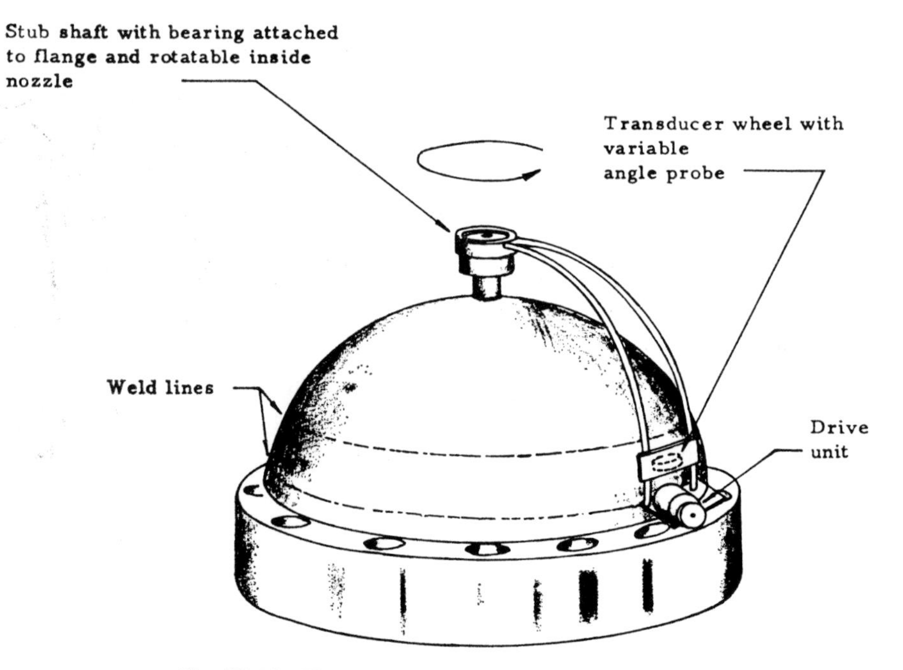

Fig. 57.11. Flange torus weld inspection device

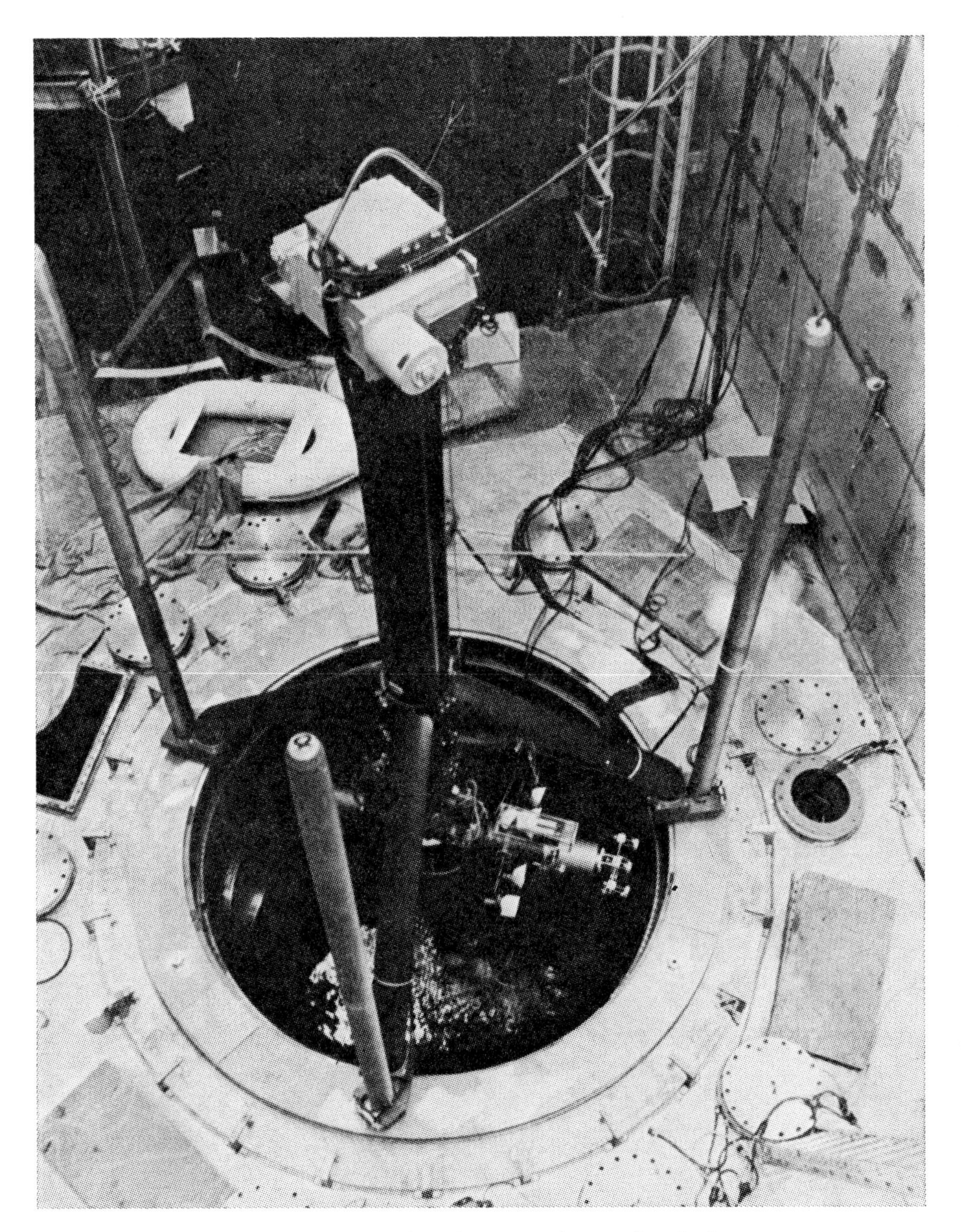

Fig. 57.12. Point Beach reactor inspection device

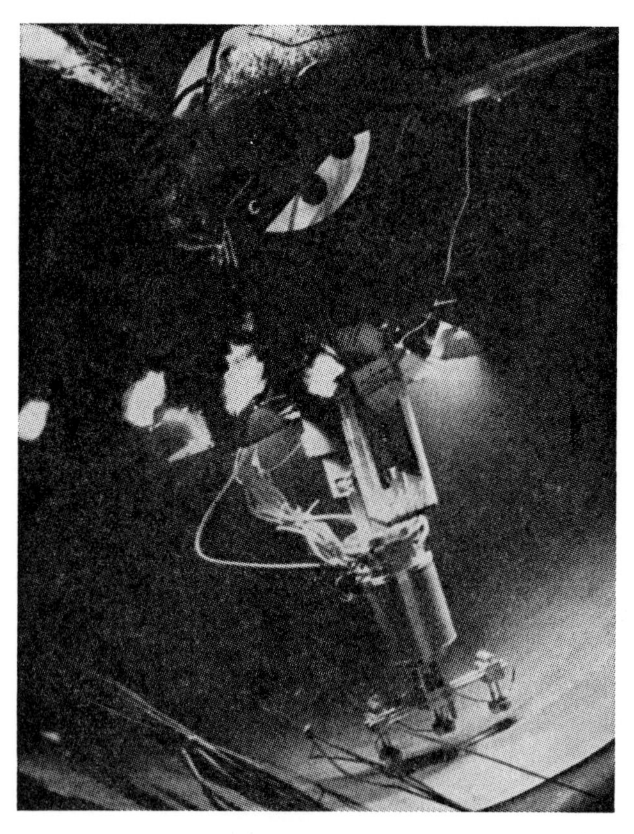

Fig. 57.13. Shell weld inspection module

Fig. 57.15. Atucha reactor pressure vessel o.d. inspection device

Fig. 57.14. Nozzle weld inspection module

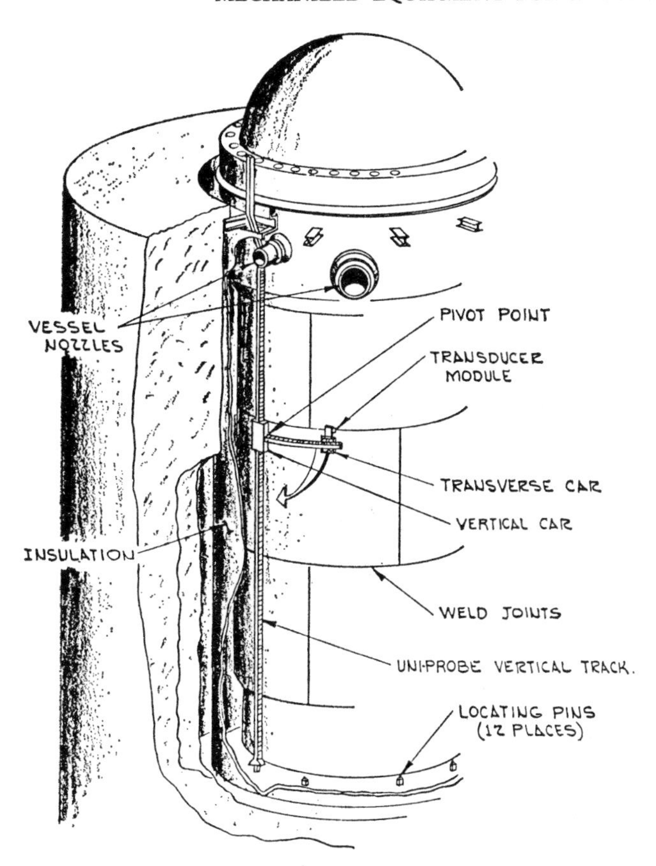

Fig. 57.16. Schematic of Unipole system

Device for examination of head flange-to-torus weld

This equipment (Fig. 57.11) is currently being modified to allow inspection of the longitudinal gore welds in the head. Experience has shown that such a ring positioner is the best type for inspection of a B.W.R. from the i.d. A similar system has been built for inspection of a P.W.R. from the i.d. However, a central mast system, such as that recently used for the pre-service inspection of Point Beach 2 reactor, may be more suitable for i.d. inspection of a P.W.R. The Point Beach device (Fig. 57.12) was designed and fabricated for the Wisconsin Electric Power Company by Pa.R. (Programmed and Remote Systems Corporation) with technical assistance from Sw.R.I.

This device, using ultrasonic modules engineered and fabricated by Sw.R.I. for each application, can inspect all the welds of the reactor pressure vessel, including those in the lower head, except as restricted by in-core instrumentation nozzles. The first time the device was used it required only nine days to assemble equipment within the containment, make measurements regarding alignment and reproducibility, and perform a complete inspection of the vessel. Figs 57.13 and 57.14 show, respectively, the module for inspection of the shell welds and the module for inspection of the nozzle welds. A similar device was designed and fabricated by M.A.N. (Maschinenfabrik Augsberg, Nuernberg), and used for a pre-service inspection of the Stade P.W.R.

OTHER SYSTEMS

Many more types of devices have been proposed for inspection of reactor vessels from the o.d., although to date none has been used for a pre-service inspection. A mock-up of a device for o.d. inspection of B.W.Rs was developed for the Empire State Atomic Development Association by Babcock & Wilcox. The device consisted of a fixed ring near the top and the bottom of the vessel, connected by a vertical member which rotated around the vessel. The ultrasonic module would ride up and down this vertical member, allowing inspection of the shell portion of the vessel. A similar device is currently being fabricated by Automation Industries for inspection of the Three Mile Island units.

Sw.R.I. currently has three systems under development for o.d. inspection. That shown in Fig. 57.15 is currently being installed for inspection of the Atucha heavy water reactor pressure vessel. This device consists of a rotating ring supporting a vertical member. The ultrasonic module and propulsion unit, remotely inserted from the top of the reactor, traverses up and down a track on the vertical assembly, permitting inspection of the welds in the shell section. An important feature of this system is that the rotating mechanism is designed for a 40-year life without maintenance.

The Unipole system (Fig. 57.16), together with a bottom head inspection arm (Fig. 57.17), is being fabricated for the inspection of three P.W.R. systems. The propulsion unit and the Unipoles provide vertical movement; and the ultrasonic module, traversing on a

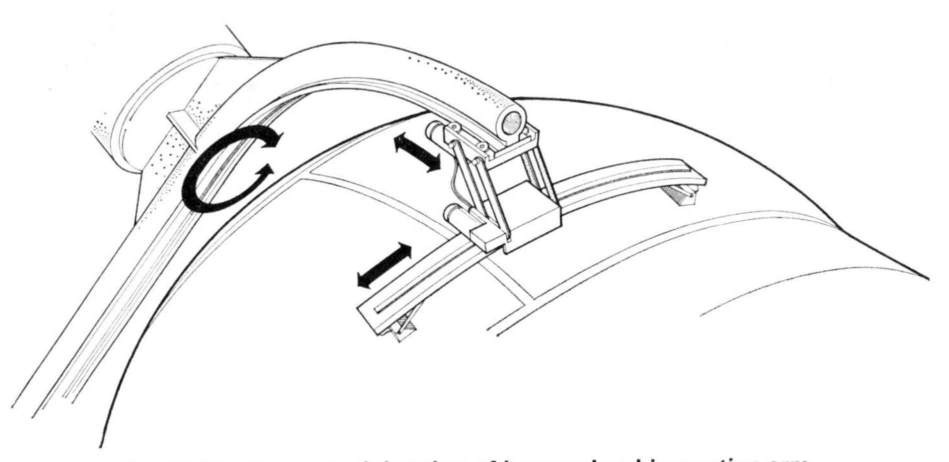

Fig. 57.17. Conceptual drawing of bottom head inspection arm

Fig. 57.18. Magnetic mouse with propulsion unit

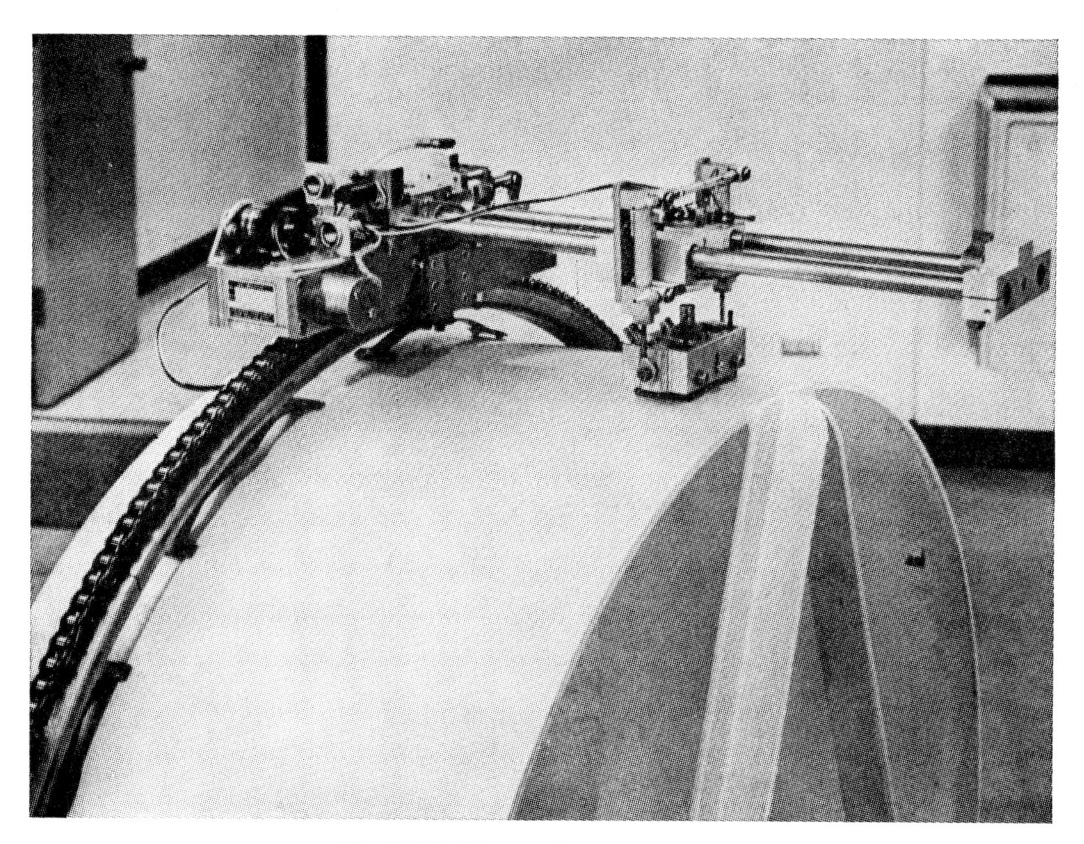

Fig. 57.19. Pipe car inspection device

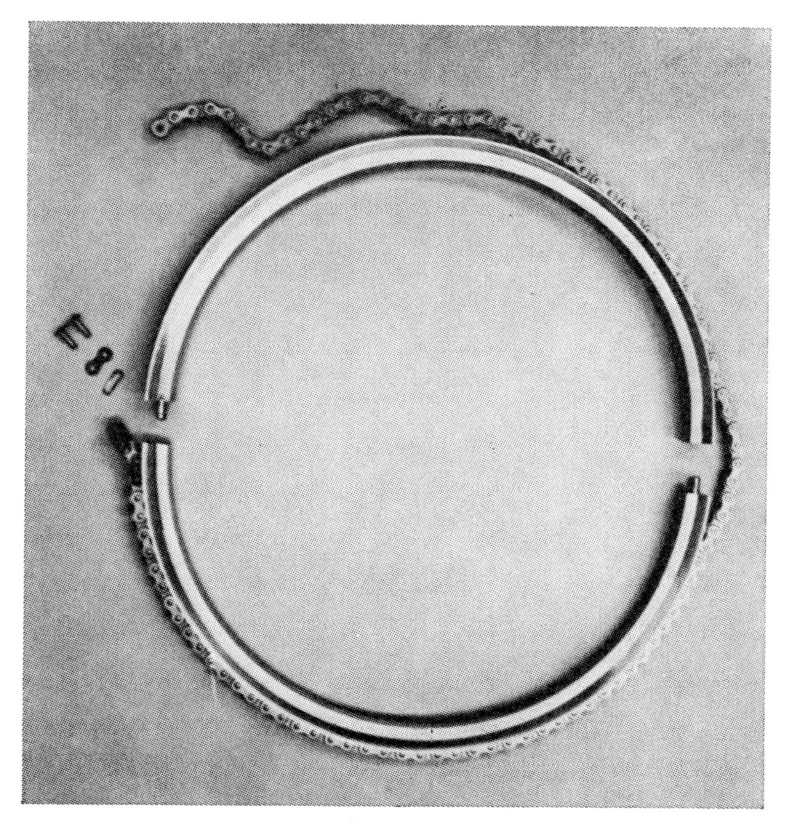

Fig. 57.20. Permanent pipe track

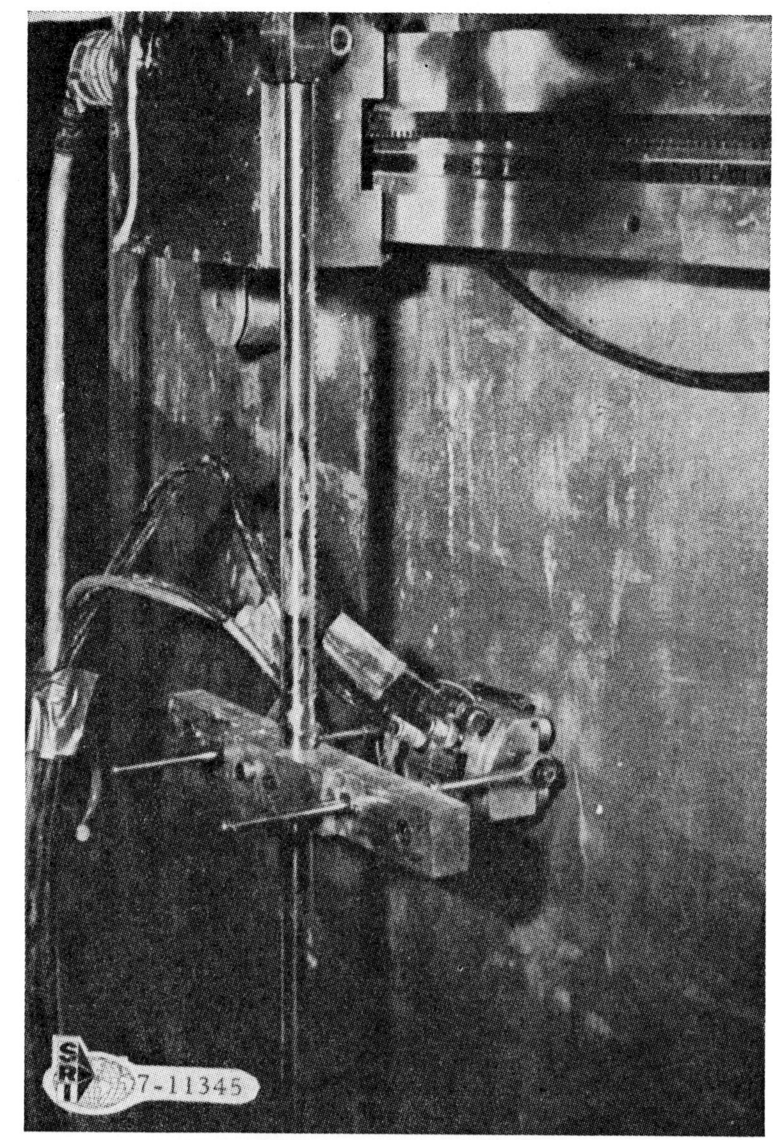

Fig. 57.21. Vertical seam scanner

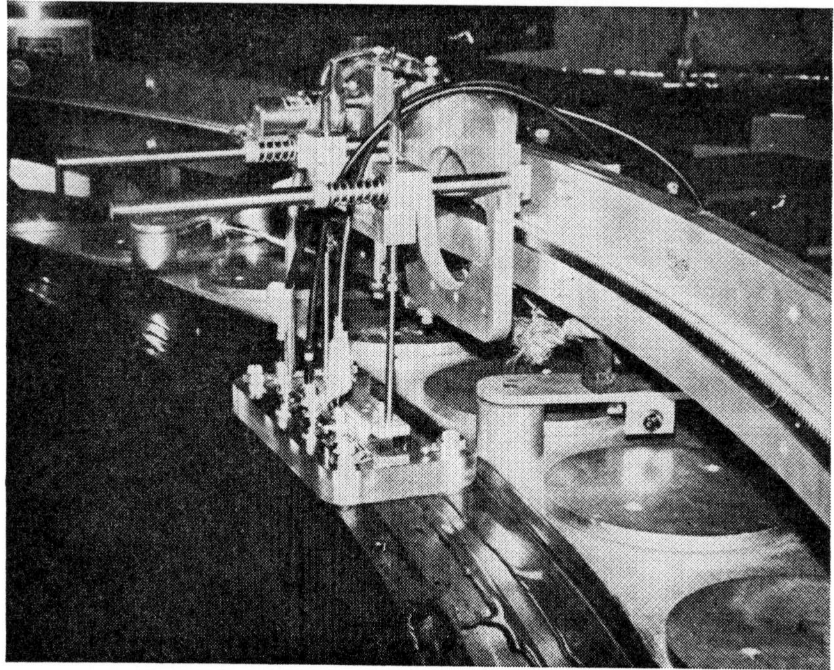

Fig. 57.22. Vessel shell-to-flange weld inspection

Fig. 57.23. Vessel shell-to-flange weld inspection

horizontal cross-arm, provides the circumferential movement for inspection of the welds in the vessel shell. For the lower head, the propulsion unit travels along the arm, and the arm rotates for complete inspection of this area.

The third system is a magnetic mouse, used primarily for inspection of B.W.Rs. This device is being designed to operate within a 100-mm (4-in) annulus, and to inspect all the reactor vessel shell welds. The propulsion unit of such a device has been fabricated (Fig. 57.18), but the problems of determining position and orientation have not been solved. Two positioning techniques are being investigated: one technique would use a taut wire for guidance; the other would utilize a laser beam for guidance and vertical position indication. A prototype of this device is scheduled for use on the Peach Bottom 2 unit of Philadelphia Electric Company in December 1971; use on pre-service inspection of the Fukushima 2 unit of Tokyo Electric Power Company is scheduled for April 1972.

The other items of o.d. inspection equipment developed by Sw.R.I. are the pipe car and the P.L.U.S. system. The pipe car (Fig. 57.19) can be used for inspection of pipe welds with a diameter of 250 mm (10 in) or larger. For high radiation areas, such as the nozzle-to-pipe welds, a permanent track is placed in position. The track (Fig. 57.20) is held in proper orientation by preloaded leaf springs. This system has been successfully used on the pre- and in-service inspection of many reactors. A recent adaptation to this unit is a side arm, to allow remote inspection of the nozzle-to-shell welds of B.W.Rs. It is hoped that this system can be extended to the inspection of these welds on P.W.Rs.

The P.L.U.S. system has been used successfully on the pre- and in-service inspection of many reactor systems. This is probably the most universal system developed by Sw.R.I., as it can operate on straight and curved track in areas with as low as 100-mm (4-in) clearances. The device has been used to inspect pressurizers (Fig. 57.21), vessel flange-to-shell welds (Figs 57.22 and 57.23), and similar inspections, such as head flange-to-torus welds. The P.L.U.S. is the basic propulsion system, with the ultrasonic module being engineered for a particular inspection.

ACCESS FOR INSPECTION

The main difficulty in designing equipment for inspection of nuclear reactor systems is that very little thought has been given to providing access and/or design for such inspection. This situation is improving, as is evidenced by the o.d. access being provided on most reactor pressure vessels. There always will be, however, certain areas where an inspection is required after system design has been finalized. An example of this is the requirement for inspection of the lower head-to-shell weld and support skirt-to-lower-head weld of several Swedish vessels. Access is limited and a system of curved tunnels is required for insertion and traverse along the welds. Even though this is a 'special' system, an attempt is being made to use standard assemblies and components to suit the requirements; in other words, to build up inspection equipment using modular concepts.

CURRENT STATE OF THE ART

Many other types of equipment have been under development around the world. Risley Laboratories, U.K.A.E.A. (United Kingdom Atomic Energy Authority), have developed equipment principally for gas-cooled reactors; M.A.N. is working on other types of equipment for B.W.R. systems; Mitsubishi Industries, Japan, and Westinghouse, United States, are developing their versions of an i.d. device for P.W.R. vessels; and General Electric, United States, is working on a magnetic bug for o.d. inspection. In addition, there is a large amount of activity in the areas of acoustic emission and acoustic holography. These are relatively new techniques that require special equipment for transducer placement and manipulation.

SUMMARY

Until January 1970 practically all the equipment for inspection of light water moderated reactors was built by Sw.R.I. and T.R.C. Since then—which is the date of issue of Section XI, 'In-service inspection of nuclear reactor coolant systems', of the A.S.M.E. code for boiler and pressure vessels—many other agencies have started to build equipment. However, from January 1972 these agencies' equipment, with the exception of that of M.A.N., has not been tried on a pre- or in-service inspection. A similar situation exists for gas-cooled reactors; the majority of the inspection equipment to date has been built by the U.K.A.E.A., with some equipment being built by Pa.R. It has become apparent that providing equipment for inspection of nuclear reactors is an important part of the project. The equipment is costly and development times are lengthy. It is incumbent on the utility to recognize the requirements early and to make the necessary decisions in time to provide equipment for thorough inspection prior to the pre-service inspection date.

C58/72 CONTINUOUS MONITORING OF NUCLEAR REACTOR PRESSURE VESSELS BY ACOUSTIC EMISSION TECHNIQUES

J. B. VETRANO* W. D. JOLLY* P. H. HUTTON*

The integrity of the primary circuit nuclear central station power plants that are now operating in the U.S.A. is of utmost concern. As part of an overall effort to resolve this concern, an association of investor-owned utilities in the U.S.A. is supporting research at Battelle. The purpose of this research is to develop a system that will continuously monitor nuclear reactor pressure vessels during operation to detect any incipient flaw growth. The system that has been developed is based on the acoustic emission phenomenon and is capable of detecting subcritical cracks in pressure vessel steels in the presence of temperatures, fluxes, and hydraulic flows that are characteristic of those in operating nuclear power plants.

INTRODUCTION

THE LAST DECADE has witnessed a marked growth in the number of nuclear central station power plants located near large population centres in the U.S.A. This has occurred simultaneously with an increasing awareness and concern for environmental preservation and public safety. The inevitable outcome of the interaction of these two forces has been an impetus to develop the technological know-how to design and inspect nuclear installations so that there is essentially no danger of uncontrolled release of radioactivity to the environment. The first line of physical defence against such occurrences (after the fuel cladding itself) is the nuclear reactor pressure vessel and primary loop piping. These provide the containment for the nuclear core, its associated subcomponents, and coolant circulated for steam generation. Attention has naturally been directed towards ensuring the integrity of this total system.

An increased understanding of fracture mechanics in heavy section steels is being sought to allow more realistic engineering design criteria. At the same time, the technology for periodic non-destructive evaluation of the more highly stressed zones on the pressure vessel is advancing, and interim code criteria are in force (1)†. Finally, an effort has been under way to non-destructively monitor the condition of the reactor primary pressure system continuously during operation. This last topic is the primary subject of this paper. Discussion is directed to surveillance of the pressure vessel only because this has been the focal point of development efforts. A successful system would, however, also be applicable to the primary piping.

Present-day nuclear fission reactors using light water coolant-moderators have reached single unit sizes of about 1200 MWe in commercially supplied boiling water reactors (BWRs) and pressurized water reactors (PWRs).

The MS. of this paper was received at the Institution on 26th October 1971 and accepted for publication on 21st February 1972. 33

* *Battelle, Pacific N.W. Laboratories, P.O. 999, Richland, Washington 99352, U.S.A.*

† *References are given in Appendix 58.1.*

Typical vessel design conditions are for $86{\cdot}2\times10^6$ dyn/cm^2 pressure at 300°C for the BWR and $172{\cdot}4\times10^6$ dyn/cm^2 pressure at 340°C for the PWR. Vessels are constructed at A.S.T.M. A302, Grade B, low alloy steel with about 3·2-mm thick austenitic stainless steel internal cladding. The vessel sizes range from 4·5 m outside diameter by 12 m high × 30 cm thick (PWR) to 6·7 m outside diameter by 21 m high × 15 cm thick (BWR). This is the pressure vessel that needs to be continuously monitored for any indications of subcritical size flaw initiation or growth. The method selected for development for this application is acoustic emission monitoring.

BACKGROUND

When a material deforms or fractures, part of the energy of the breaking bonds is released as elastic waves which propagate through the material until dampened out. This elastic wave is referred to as acoustic emission.

Acoustic emissions are quite easy to detect with various piezoelectric transducers which generate a voltage output proportional to the applied pressure. Thus, all the essentials of a non-destructive monitoring tool for complex structures are in hand: a failure mode which announces itself by the generation of acoustic emission, and a method of detecting the emission. This overall process is called acoustic emission monitoring.

Implicit in the above description of acoustic emission monitoring are two subtle points which deserve notice.

First, inherent in the method is the ability to determine the location of flaws as well as their presence. This can be done by using a multisensor array and appropriate timing circuits. By making measurements of the relative times that the transducers sense the wave front, and by knowing sonic velocity in the material and transducer spacing, the location of the origin of the acoustic emission can be calculated.

Second, in order for a flaw to be detected by the acoustic emission monitoring method, measurements must be taken during the actual time of flaw growth. This means that the material must be under some type of stress.

Acoustic emission monitoring is not a method for non-destructively locating flaws in unstressed parts. On the other hand, the flaw growth required to generate detectable acoustic emission is very small. Fatigue cracks only about 0·025 mm long have been detected by acoustic emission (2). In nuclear reactor pressure vessels, critical crack size is several centimetres (exact size depends on material, temperature, and stress), therefore, by continuous monitoring it is quite feasible to detect, locate, and follow the progress of subcritical-size flaws in nuclear reactor pressure vessels. From the acoustic emission history and from other confirming non-destructive evaluations made during reactor shutdown, a decision can be made regarding the seriousness of the flaw.

MAJOR PROBLEM AREAS

Three major problem areas were identified as crucial to the development of an on-line, continuous, acoustic emission monitoring system for nuclear reactor pressure vessels. These are: (1) how to cope with the high-temperature, high-radiation environment; (2) development of a data processing system for computing flaw location; and (3) how to detect acoustic emission in the presence of hydraulic flow noises.

Hostile environment problem

The temperature of the outer surface of a PWR or BWR pressure vessel is about 315°C. Thermal neutron fluxes can be as high as 1×10^5 n/cm^2 s. Access for maintenance is impossible during operation and severely limited during shutdown periods because of high residual radioactivity. This is the environment in which the acoustic emission monitoring system must operate for lengthy unattended periods.

Longitudinally poled PZT-5A operated in a cross-coupled mode is by far the most sensitive piezoelectric transducer for acoustic emission. However, its low Curie temperature (about 350°C) relative to reactor operating temperatures makes it unsuitable for long term use on the surface of a nuclear reactor pressure vessel. There are two approaches to this problem: one is the use of a higher temperature transducing technique; the other is to locate the PZT-5A out of the hostile environment and sonically couple it to the vessel with a waveguide. We selected the waveguide approach because of the reliability inherent in a purely mechanical interface with the pressure boundary.

The waveguide (Fig. 58.1) consists of a long, slender tube or rod of low-acoustic-loss material (such as Zircalloy or aluminium). One end is flared to a horn (single or two-piece construction) that can be welded or pressure coupled to a pad on the surface of the reactor pressure vessel. The PZT-5A crystal is cemented on the other end which, by virtue of its distance from the pressure vessel surface, is in a much more benign thermal and neutronic environment. The preamplifier is housed in a container which surrounds the transducer and waveguide end so that transmitting cables with their associated impedance losses are eliminated.

Some typical data obtained by a PZT-5A piezoelectric transducer mounted at the end of a 2·54-cm long waveguide coupler horn are compared to a similar transducer

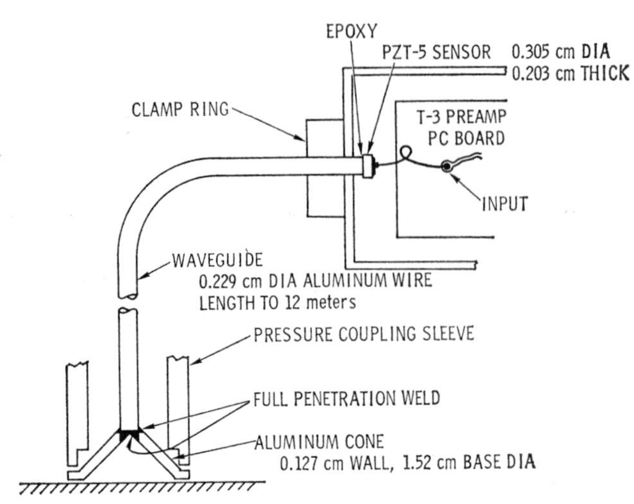

Fig. 58.1. Waveguide configuration

mounted on the vessel near the foot of the waveguide in Fig. 58.2. Some small acoustic losses in the coupler horn are more than compensated for by such factors as the orientation of the transducer face relative to the direction of the acoustic wave possible in this system, and the possibility of building the preamplifier right into the assembly so that impedance losses are eliminated. The net effect could be, in certain cases, a stronger signal than is possible with a piezoelectric material alone mounted directly on the sensing surface.

Waveguides as long as 6 m might be practical if it is necessary to get this much separation from the hot pressure vessel wall. Typical waveguide attenuation values measured in our laboratory are about −1·3 dB/m at 0·7 MHz.

By removing the sensor and electronic preamplifiers out of the hostile environment zone, it is feasible to use PZT-5 transducers for the acoustic emission system. Furthermore, by proper design of the acoustic coupler there should be little if any degradation of signal.

Source location problem

Basically, the source location problem is one of combining

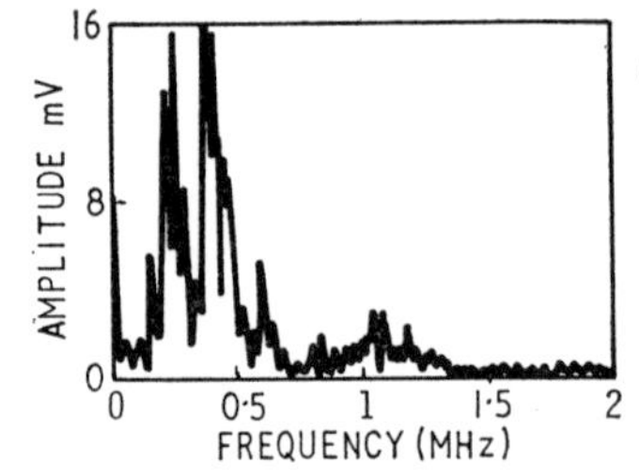

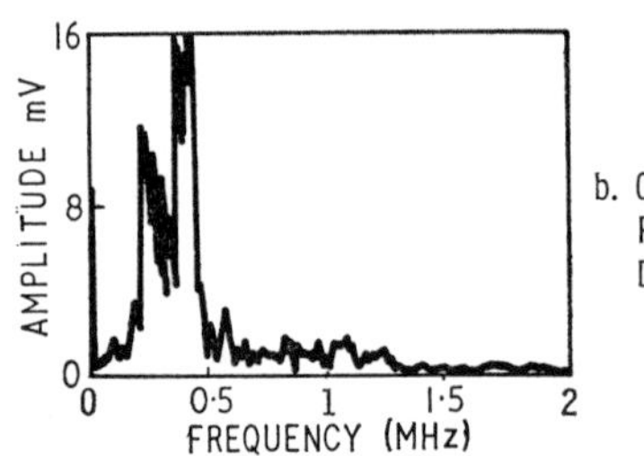

Fig. 58.2. Zircalloy waveguide response to simulated acoustic emission compared to direct response using the same sensor

the data of transducer spacing (relative to an imaginary co-ordinate or grid system), sonic velocity in the pressure vessel wall (accounting for propagation mode), and relative time-of-arrival of signals in the sensor array. This combination of data makes a seismic-like computation of the origin of the source relative to the imaginary grid system.

During a recent programme at Battelle many variations of transducer array patterns were evaluated. The following describes the pattern having the optimum benefits for this particular application.

A hexagonal transducer pattern was chosen because it offers advantages in simplification of hardware and reduction of calculation time. Using a hexagonal pattern it is possible to replace the computation of the intersection of two hyperbolas with a simple table relating relative arrival times at three transducers to the location of an emission source. As shown in Fig. 58.3, the source of emission is located by successive approximations. Each hexagon of the array is divided into six equilateral triangles. A triangle is identified by three transducers; for example, triangle ABC. Each triangle may be divided into six similar triangles or sectors by noting the order of response to an emission signal. An emission signal originating within the boundaries of sector (C, A, B) must first excite C, then A, then B transducers. Finally, a zone is determined from a table relating relative arrival times to signal origin. This table is represented by a hyperbolic grid on sector (C, A, B). Zone (2, 1) is located by entering the table with

$$(\Delta t_{CA})\left(\frac{v}{2}\right)K = 2 \quad \text{and} \quad (\Delta t_{AB})\left(\frac{v}{2}\right)K = 1$$

where K is a scale factor and v is the velocity of sound in the material. Because of the hexagonal symmetry, the zone table is the same for all sectors of the transducer array.

For this particular application, the transducer spacing is 1·5 m and a sector is divided into 10 zones, approximately 230 cm². Transducer spacing depends upon the effective range at the operating frequency. At least three transducers must be within effective range of any emission source. The hexagonal pattern has a minimum of overlap and conforms readily to cylindrical or spherical surfaces where the radius of curvature is large compared to sensor spacing.

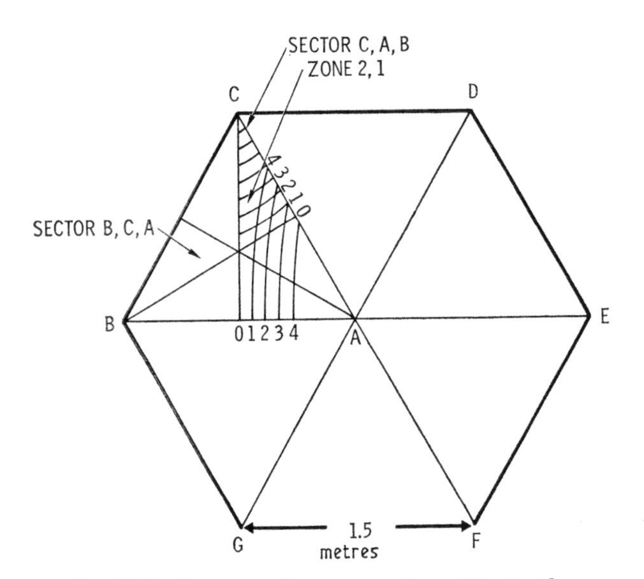

Fig. 58.3. Hexagonal sensor pattern illustrating method of locating acoustic emission

This method of data processing has many significant advantages for application to acoustic emission monitoring of reactor pressure vessels. First of all, using a 'table-look-up' logic rather than a digital computation, a much smaller computer memory is needed. This can result in a smaller capital investment in the final system.

Furthermore, the use of a floating origin rather than fixed origin grid system makes it much easier to physically locate the source of emissions on the vessel. Rather than go out so many metres in the x direction and so many metres in the y direction from a fixed origin point on the vessel to mark an active spot, one can simply lay a template between the three transducers around the indicated source and punch through to mark the vessel in the appropriate zone.

Noise problem

If nuclear reactor pressure vessels are to be monitored continuously by an acoustic emission method to detect the presence of growing flaws, one must be able to distinguish between the fracture-generated emissions in the vessel wall and the hydraulic noises generated by coolant flow. By measurement of the frequency and amplitude characteristics of flow noises in a modern PWR, it has been determined that the use of a combination of frequency-filtering and signal-enhancement techniques will enable flaw growth to be detected by acoustic emission during reactor operation.

Flow noise characteristics were measured on the head flange and the cold leg of the 'A' loop of the San Onofre reactor (San Clemente, California) during a restart cycle in November 1970 (3). Measurements were made under conditions of essentially no reactor coolant flow, full cold flow (80°C, zero power level), and full hot flow (282°C inlet, 298°C outlet, 417 MWe power level).

Several types of sensor were used to obtain the data: PZT-5A coupled with RTV116 silicone rubber or epoxy; stainless steel waveguides, both pressure coupled and epoxy coupled and terminated with PZT-5A chips; and epoxy coupled lithium niobate chips. The output of the sensors led to a 30 kHz–3 MHz signal conditioning and analysis system (Fig. 58.4).

Prior to installation on the San Onofre reactor, the frequency response to a broadband noise source was measured for all sensor and couplant combinations. This was done by making an amplitude–frequency response curve when the sensor was mounted to a steel test block that was being blasted with 50 μm alumina grit carried by compressed air. As the noise generated in this way is essentially white noise in this frequency range, non-uniformity in sensor response is due to peculiarities of the individual sensor, such as resonance peaks. Some point in the amplitude–frequency curve can then be selected as a reference, and corrections applied to other points to normalize the curve to a white noise response.

A correction factor for temperature variations was also applied to the measurements. This factor was determined experimentally in the laboratory for all the sensor–couplant combinations over the temperature range of measurement, and then applied to the raw data.

The corrected and averaged data for the spectral distribution of noise under full cold flow and full hot flow

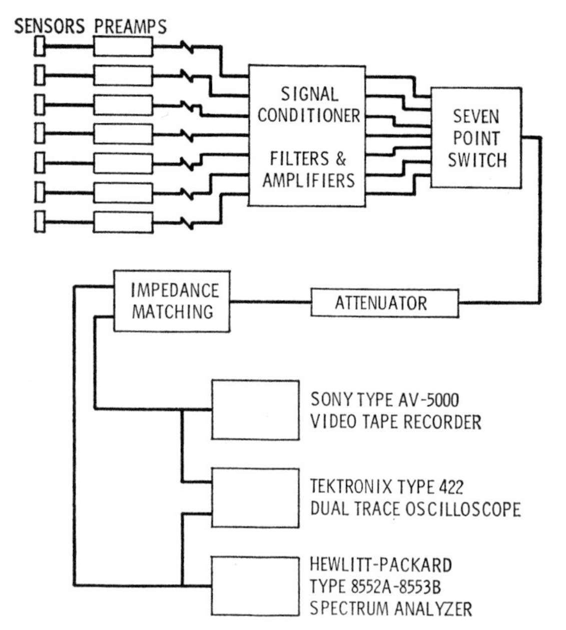

Fig. 58.4. Block diagram of the noise monitor system

are shown in Figs 58.5 and 58.6, respectively. The cross-hatched areas in these figures show typical acoustic emission signals as recorded from a distance of about 1·5 m with a similar measurement system.

These results are in general agreement with earlier measurements made by Battelle on the Dresden reactor (**4**). The evidence is growing stronger that previous measurements on the San Onfore reactor (**5**) which showed signal-to-noise ratios of about 1/120 at 100 kHz and 1/40 at 1 MHz are seriously in error.

If an acoustic emission system must operate with a signal-to-noise ratio of greater than about 1·2 to 1·5, it would need to use frequency filtering to eliminate signals below about 1 MHz for cold flow conditions or about 300 kHz for hot flow conditions. A signal enhancement method has been recently developed, however, which permits operation at signal-to-noise ratios of less than 1·0, and consequently allows use of a low-frequency system both under cold flow (reactor start-up) and hot flow (reactor operation) conditions. This method is described in the following discussion.

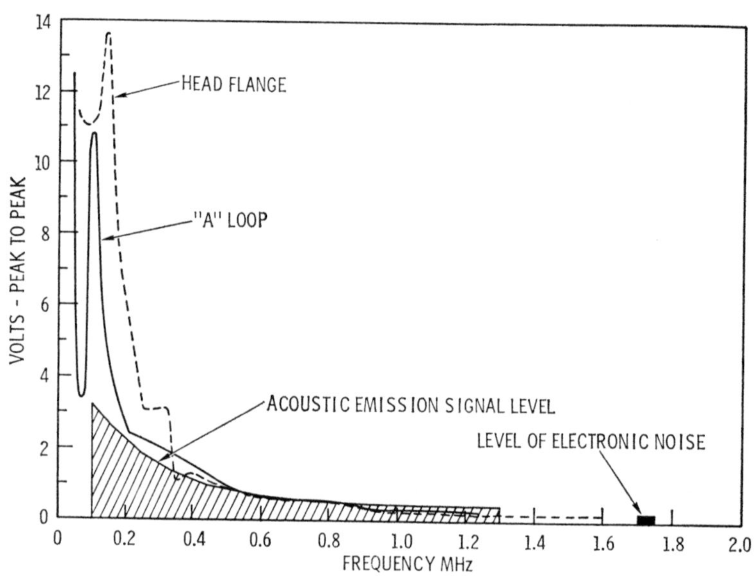

Fig. 58.5. Average noise spectrum—full cold flow, San Onofre nuclear power plant

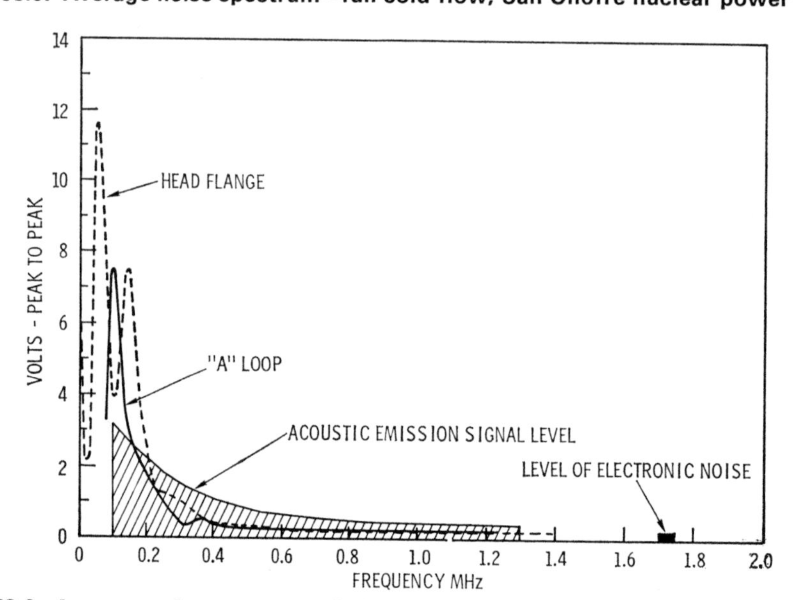

Fig. 58.6. Average noise spectrum—full hot flow, San Onofre nuclear power reactor

Signal enhancement

Signal enhancement may be effected if a signal possesses some unique characteristic which distinguishes the desired signal from undesired background noise. The background noise encountered at the pressure boundary of an operating nuclear reactor consists of turbulent flow noise and cavitation bubble collapse. Cavitation bubble collapse and acoustic emissions both represent energy released in randomly spaced impulses. Turbulent flow noise from a number of sources results in a randomly varying pressure wave at any given point on the pressure boundary. A transducer attached to the pressure boundary responds to the algebraic sum of all pressure waves passing the point of contact. Acoustic emission signals are difficult to distinguish from background noise unless the amplitude of the acoustic emission is significantly greater than the peak noise amplitude. The turbulent noise contribution may be reduced by frequency filtering but only at the expense of signal acquisition range. For example, the exclusion of frequencies below 1·0 MHz will almost eliminate the turbulent noise contribution, but the acoustic emission detection range is reduced to 1·5 m because of attenuation. The detection range extends to approximately 5 m when the cut-off frequency is 0·3 MHz.

Now the problem is to find a unique characteristic of the desired signal. Having given up the advantage of frequency filtering, we find that the acoustic emission amplitude is below the amplitude of random turbulent noise peaks and cavitation pulses and that the repetition rate of acoustic emissions is much less than the repetition rate of turbulent noise amplitude peaks and cavitation pulses. The distinguishing characteristic is not to be found in the signals received but in their respective points of origin.

Acoustic emissions from a given active flaw will arrive at a set of three transducers in a unique sequence, depending upon the distance from the flaw to each transducer. This is also true for discrete cavitation sources but not for the randomly varying turbulent noise peaks. Since the turbulent noise peaks are random at each of the three transducers, the apparent arrival sequence must also be random. Thus, if all received signals are assigned to physical locations on the basis of an arrival sequence, the turbulent noise will be uniformly distributed in all locations, while acoustic emission signals and cavitation signals will be assigned to specific locations. The result may be considered as a two-dimensional map of the monitored surface with the number of assigned signals in each location displayed in a third dimension (Fig. 58.7). Each discrete source appears as a prominence above the uniformly distributed turbulent noise count. Background subtraction may be used to enhance the prominences representing discrete sources. After background subtraction, a discrete source is detectable when the accumulated count at one location exceeds the spatial variation of the background count. Since the turbulent noise location assignment is random, the spatial variation of the background count diminishes with time while the count due to discrete sources at specific locations increases with time.

Once the individual signal sources are separated spatially, cavitation sources are readily distinguished from acoustic emission sources on the basis of count accumulation rate. Cavitation source count will increase continuously at a very high rate, while acoustic emission count from a flaw will increase slowly at a variable rate, depending on the rate of flaw growth at the crack tip.

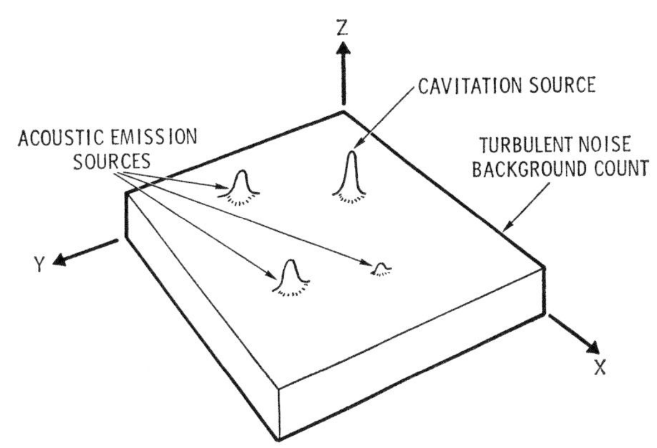

Fig. 58.7. Representation of the spatial distribution of signal sources obtained by assigning a location to all received signals

Spatial resolution of acoustic emission sources in the presence of background noise requires that all received signals be assigned to a spatial location on the basis of arrival sequence at a minimum of three transducers. This task is accomplished by means of incremental digital delay lines. The arrival sequence for a location or resolvable element is modified by a set of three time delays such that a signal originating at that location will cause a coincident output from the three time delays. Each resolvable element of the area to be monitored is identified in this manner. The number of coincidence counts is individually accumulated for each element. Thus, each resolvable element is monitored continuously. Background subtraction, generation of the spatial distribution map, and data analysis may be carried out in an on-line computer.

The coincident detection technique was verified experimentally using a single resolvable element. The experimental set-up is illustrated in Fig. 58.8. A steel plate, 1·2 m × 2·4 m × 0·1 m, provided with a turbulent noise source was used to simulate a portion of a reactor pressure boundary. An ultrasonic pulse generator located equidistant from three transducers represented an acoustic emission source. Coincidences were counted for conditions of signal only, signal and noise, and noise only.

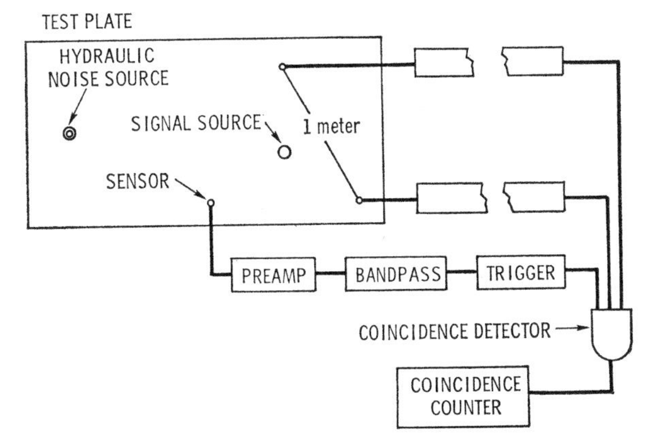

Fig. 58.8. Schematic of the experimental set-up to test the coincident detection–location method. Signal source was positioned for simultaneous arrivals at the three sensors. Signal amplitude was set to 0·5× noise amplitude in 0·3–0·6 MHz bandpass

When the signal-to-noise amplitude ratio was 0·5, 66 per cent of the transmitted signals were counted.

This one experiment with simplified boundary conditions clearly indicates the feasibility of the coincident detection location method for detection of relatively weak acoustic emissions in the presence of strong hydraulic flow noises. Further demonstration of this technique is now being performed, scanning the entire area enclosed by these transducers located in a triangular pattern.

CONCLUSIONS

As a result of the research efforts discussed above, solutions have been found to the principal problem areas for the application of an acoustic emission monitoring method for continuously monitoring modern BWRs and PWRs for flaw growth. On the basis of this groundwork it is reasonable to postulate that continuous on-line monitoring of all operating central station nuclear reactor plants could be a reality. This can be accomplished by demonstration type experiments in which controlled flaws in large vessels are monitored under simulated operating conditions. Such demonstrations are under way in the U.S.A. and are due for completion before June 1972.

ACKNOWLEDGEMENTS

This work was performed under subcontract to Southwest Research Institute, acting as programme manager for the Edison Electric Institute Research Project RP-79. The financial support and guidance of both are gratefully acknowledged.

APPENDIX 58.1

REFERENCES

(1) *A.S.M.E. boiler and pressure vessel code*, Section XI, 'Rules for in-service inspection of nuclear reactor coolant systems', 1968.

(2) Hutton, P. H. 'Acoustic emission applied to determination of structural integrity', BNWL-SA-3147, presented at the 11th Open Meeting of the Mechanical Failures Prevention Group, Williamsburg, Virginia, 1970 (7th–8th April).

(3) Hutton, P. H. 'Nuclear reactor background noise vs. flaw detection by acoustic emission', BNWL-SA-3820, presented at the 8th Symp. Nondestructive Evaluation, San Antonio, Texas, 1971 (April).

(4) Hutton, P. H. 'Nuclear reactor system noise analysis, Dresden-1 reactor, Commonwealth Edison Company', BNWL-933, 1968 (December).

(5) Parry, D. L. and Robinson, D. L. 'Incipient failure detection by acoustic emission, a development and status report', IN-1398, 1970 (August).

C59/72 REGISTRATION AND INSPECTION OF EQUIPMENT AND PROTECTIVE DEVICES IN THE CHEMICAL INDUSTRY

E. H. FRANK*

The requirement for the registration and inspection of steam-raising equipment and steam receivers is defined by statute. No comparable legislation exists for other equipment used on chemical plant where the risk of failure and the associated hazard may be greater. Traditionally registration and inspection of equipment has been centred round designated pressure vessels and their protection against overpressure. This may still form the bulk of equipment; however, it is now considered appropriate to include other items such as large-diameter piping systems and to introduce protection against overtemperature. A recommended practice has been drafted in an attempt to ensure that users of equipment operating under pressure and temperature within the chemical industry have an adequate system for the identification, inspection, and monitoring of the equipment so that it does not become unfit for its service duty.

INTRODUCTION

THE MATERIALS OF CONSTRUCTION and the detailed engineering design of equipment used in the chemical industry are chosen to make it suitable for the duty it has to perform. Basic protection against failure in service is part of the original design, and protective devices are added to prevent equipment being operated outside safe limits.

When equipment is taken into service and becomes part of a process plant it is exposed to the environment in which it operates. This environment, which may be similar to that in the original design or different to varying degrees, may react on the equipment in an unfavourable way. If the reaction is sufficiently unfavourable and is not detected soon enough, an on-line failure may result. There is a need, therefore, for an adequate system of identification, examination, and monitoring of equipment to ensure that its fitness for service is maintained.

The requirement for the registration and inspection of steam-raising equipment, steam and air receivers is defined by statute (Factories Act 1961). No comparable legislation exists for other equipment used on chemical plant.

Traditionally, the registration and inspection of equipment operating under pressure has been centred round designated pressure vessels. While these still form the bulk of the equipment at hazard, items such as large-diameter piping systems, piping systems in known or expected corrosive and erosive service, systems operating under reduced stresses at temperature, and vessels made out of standard piping components now form a significant part of modern plant. It is equally important for safe operation that such equipment be dealt with in the same way as pressure vessels, and the word 'equipment' be interpreted to include plant items where a failure might give rise to a significant hazard.

This paper sets out principles which it is suggested form the basis of an adequate system in so far as the user who follows them might be said to have taken reasonable precautions to prevent failure.

EQUIPMENT

Inventory

Items of equipment which require examination as a necessary part of continued safe operation should be identified and registered before the plant is taken into service. A unique distinctive equipment number should be ascribed to each item. All items should be collated into a plant inventory and, since from time to time equipment will be added to and subtracted from the plant, means should exist for keeping the inventory up to date.

Records

Each item of equipment subject to examination should have a file or other suitable record. This file should be the repository of all reports and information on the equipment, including initial reports during manufacture, material, and other test certificates. It should also contain an equipment specification sheet, and drawings or details of where these are filed, together with reference numbers and materials list. Initial inspection reports and all subsequent reports should be added to the file, preferably on a standard form, as should any repairs, modifications, or changes of duty. In cases where a particular type of deterioration is known to be possible, this should be recorded as a guide to future inspection.

On occasions it may be necessary to operate equipment outside the original design limitations requiring de-rating

The MS. of this paper was received at the Institution on 11th November 1971 and accepted for publication on 1st February 1972. 40

* *Imperial Chemical Industries Ltd, Petrochemicals Division, Billingham, Teesside.*

or up-rating. Such changes should be properly authorized and recorded with supporting evidence where appropriate.

The file should be readily accessible to those responsible for the operation and the inspection of the equipment.

Frequency of examination

As far as possible, all equipment should be examined on site at as late a stage as is practicable before it is taken into service, and the results should be recorded in the equipment file.

A frequency of examination should be ascribed to each item of equipment ideally before the equipment is put into use, and this frequency should be clearly stated in the equipment file. The assessment of the inspection frequency is a matter of experienced judgement, taking into account the details of the operational environment, corrosion allowance, experience with similar plant, etc. When, with apparently similar plant, experience over many years has shown that a long inspection interval is justified, it must not be too lightly assumed that the plant is so similar that it will behave in the same way. It is desirable that new equipment taken into service shall be examined within an operational period of two years, depending on previous experience, and that inspection intervals can be extended after demonstrable proof that the plant is behaving as expected. Experience shows that this initial inspection is not only required to confirm that the engineering design is correct in so far as the stresses imposed, vibration, erosion, etc. are not having a deleterious effect, but also to confirm that the materials of construction are behaving as expected. Minor changes in chemical composition of process fluid can have a very significant effect in the resistance of austenitic steels and the pH and oxygen content of the process fluid markedly affect the resistance of ferrous materials.

In some cases on-line thickness or corrosion monitoring is planned to monitor the equipment. In these cases it is desirable that full agreement be reached between the inspection and operations sections on the frequency of examination, that they review the data obtained from routine tests, and that the equipment record should detail threshold and retirement levels.

After the inspection interval is established it should not be extended without an authoritative review of the situation, nor should the inspection be overlooked or neglected.

PROTECTIVE DEVICES

Inventory

All devices which protect equipment against overpressure, overtemperature, or other factors that can lead to failure—e.g. corrosion, explosion—should be identified. An inventory should be drawn up for such devices and identification in the field with a unique distinctive number is desirable.

Records

A file or other suitable record should be kept for each device and should include a complete specification of duty, materials of construction, drawings, diagrams, etc. necessary to fully define its function and operation.

Any change made in the setting of a protective device should be properly authorized before it takes place, and recorded together with any supporting evidence.

The file should be readily accessible to those responsible for the operation and inspection of the protective device.

Protection is afforded by a variety of similar devices, some of which may at first sight look identical. It is desirable, therefore, to ensure that the correct device has been fitted to the equipment before it is taken into service, and that it will afford the protection that the designer intended.

Frequency of examination

A protective device requires periodic checking to ensure that it has not become inoperative. A recommended frequency of inspection should be ascribed to each device before it enters service and this information recorded in the file. The inspection interval should not be greater than the interval chosen for the inspection of the equipment that the device protects.

Once the inspection interval is established it should not be extended without an authoritative review of the situation, nor should the inspection be overlooked or neglected.

EXAMINATION

Organization

As the scale of manufacture increases so does the size of the equipment. The cost of a plant shutdown to examine equipment is increasingly measured in terms of lost production, the monetary value of which may far exceed the engineering cost of the examination. On some equipment—e.g. large distillation columns—visual examination of every welded seam where conditions are known to be non-corrosive is prohibitively expensive, time consuming, and unnecessary. Selective examination based on experience is in many cases adequate and in other cases the only practical way of checking the effects of the environment on the equipment. Such an approach requires a degree of 'professionalism' in the hands of those carrying out the examination knowing not only what to look for but also where to look for it.

It is considered that the best way to secure the necessary competence in the examination of equipment is to make examination the full-time responsibility of an inspection team, This team, or service, must clearly work in close contact with those responsible for production. However, ideally it should be separate from the production function in order that it may not be unduly influenced or biased in its work by production pressures.

In some cases it may be found convenient to use one of the engineering insurance companies to carry out the examinations. The larger operator may consider it more appropriate to undertake this service within his own organization. Should this be so, it is desirable that its independence from the production function should match that of the engineering insurance companies.

Competent persons

The competence of those carrying out the examinations must enable them to carry the responsibility placed upon them. There is no statutory definition of a 'competent person', though these people are referred to in the Factories Act 1961.

It is suggested that a 'competent person' is one who, by virtue of his training and experience, should be capable of conducting a thorough examination and test of the equipment or protective device, calling in, where necessary, specialists to advise on the equipment or device and its functions, and making a full report on its condition.

It is desirable that 'competent persons' should receive an amount of formal training and that, after their competence has been established, they be appointed in writing by a senior engineer in the company, perhaps the chief engineer.

Only appointed 'competent persons' should be authorized to sign inspection reports.

Reports

The Factories Act 1961 requires that the result of any examination of a statutory vessel on the prescribed form and containing the prescribed particulars shall, as soon as practicable, and in any case within 28 days, be entered in or attached to the general register (a repository of certain information required by law which the occupier of premises must keep on the premises). It further requires that the report shall be signed by the person making the examination, and if that person is an inspector of a boiler insurance company or association, countersigned by the chief engineer of the company or association.

A similar procedure is desirable for non-statutory equipment; the inspection report should be signed by the competent person who carried out the examination and/or test. Where a company has established its own internal inspection service, it is suggested that the report be countersigned by the works engineer on whose chemical works the equipment is located. This procedure emphasizes the close contact that is essential between inspection and operations, while acknowledging that the works engineer is ultimately responsible for the state of the equipment and its continued safe operation.

It is foreseeable that disagreement may arise between inspection and operations over the fitness of equipment for continued operations. Experience shows that a satisfactory solution can invariably be found. Should it be impossible to satisfactorily resolve any disagreement, the matter should be referred to senior management.

CHANGES IN DUTY

It is important to ensure that any significant changes in duty, modifications to equipment which affect its duty, or alterations to the setting or function of a protective device are properly evaluated and the change authorized. It is recommended that a formal system be used whereby these changes are approved in writing by the works engineer—acknowledging his responsibility for the overall safety of operations—and be notified to the inspection system so that they may consider the implications of the change on the equipment, advise on any alterations to the examination interval, carry out a more detailed examination at the next inspection, etc.

REPORTING OF FAILURES

The failure of any equipment or of a protective device in service, even though that failure may not result in a hazardous situation, is an indication that the inspection procedures have failed in some way to prevent that which they were designed to avoid. It is suggested that the lessons to be learnt from such failures are so important that each failure should be the subject of a report issued by the works engineer.

DISPENSATIONS

Most, if not all, companies have established recommended inspection intervals for equipment based on their experience and technology. Often the interval chosen is one of convenience arising from the shutdown of statutory equipment for compulsory examination.

After initial confirmatory evidence that equipment is not deteriorating as the result of exposure to its service conditions (and this may require more than one examination), it may be considered that a major extension to the inspection interval can be safely justified. Where the new interval recommended is in excess of the established and accepted interval, the extension should be authorized by the chief engineer. Requests for such extensions would be prepared by works engineers and have the support of the inspection service.

SUGGESTIONS AND CASES

In framing the principles outlined above, due account has been taken of experience with the operation of, and service failures in, modern chemical plant. The following examples are drawn from experience.

Records

Not only must there be an adequate record to fully describe the item of equipment and materials of construction, but the details need to be in a form which can be readily understood and interpreted by those who will have occasion to consult them.

Difficulties have been experienced with 'packaged' type equipment, particularly where a variety of items are purchased from others by the prime equipment vendor—for example, a package boiler. Difficulties can be manifold with this type of equipment bought from a European manufacturer; drawings and specifications may not be in English and the material specifications are difficult to trace to find U.K. equivalents.

In a recent case, the need for original documents was not recognized. A number of large control valves are now powered by POP (pressure operated piston) actuators instead of the conventional 3–15 lbf/in^2 air diaphragm motor. POP actuators require a reservoir of air at 80 lbf/in^2 to drive the valves to the open or closed position in the event of air failure. This reservoir, which can be of from 1 to 3 ft^3 capacity, is supplied by the valve manufacturer and is designed to air receiver codes such as B.S. 487 or 1099. In most cases no documents are sent out by the manufacturer and, since it is received against an instrument order to which P.V. codes do not apply, none are called for by the purchaser.

Increasing use is made of dimensioned sketches which cover a range of a type of equipment such as a heat exchanger. These sketches are often not to scale and may show features not included on a particular item bought. They do not enable a proper visualization of the equipment. Furthermore, separate parts or material lists are quite common, and differential cross-hatching for dissimilar materials has virtually disappeared. These factors may lead to novel design or unusual material features being overlooked, as the following example shows.

A tower for stripping light-ends out of cyclohexane had a reduced diameter top section into which were built two internal condensers, one using cooling water, the other a non-corrosive refrigerant. The vessel drawing showed identical material thicknesses and weld details for the condenser tube sheet-to-shell welds and quoted no material. Reference to a separate parts list would have revealed that the C.W. tube sheets were in aluminium bronze welded to the M.S. shell with a bimetallic weld, while the refrigerant ones were M.S. The bimetallic welds failed in service by cracking, releasing cyclohexane vapour and necessitating an emergency shutdown of the unit. The vessel had been inspected twice since entering service; through ignorance, no special attention had been paid to the tube sheet welds.

In another case, the channel end covers of three feed/product exchangers in a reformer circuit were of identical mechanical design as far as branches, thickness, and size were concerned, and were detailed on the one drawing. The hottest channel was, however, internally lined with 18/8 plug welded to the inside of the channel. During leak testing, using a detergent water solution, and quite by accident, a leak was found in the dome end of the clad bonnet. This proved to be in a capping weld over a screwed plug filling a hole used to vent and test the back of the lining during fabrication. This was found by X-raying the bonnet to determine the cause of the leak, which was subsequently confirmed when the detailed drawing was checked, though the plant staff were unaware of the type of construction used.

As far as possible, unusual or particular design features requiring special attention must be signalled to plant staff, which can be done by having design staff declare them in some formal way, e.g. on an equipment registration card.

Frequency of examination

While examination of equipment before it enters service is justified, one must ensure that it is built to the drawings, has correct internal fittings, is to the required standard of cleanliness, does not contain wrong materials of construction, etc., and the opportunity should be taken to establish bench marks against which its future performance will be judged. In known or anticipated corrosive situations these bench marks may be monitored by NDT techniques to confirm that the interval chosen between examinations is correct. Unfortunately, this is not an exact science, and where the corrosion pattern is difficult to forecast or where it changes with the process feedstock, as for example on a crude oil distillation unit, use has been made of sentinel holes in piping to increase the area of surveillance. On a unit like this, conflict often arises between the need to keep the plant on line for production reasons and to take it off for examination. After some visual evidence has been obtained, as a result of examination during shutdowns, confidence builds up, and areas selected for monitoring are those where attack is visibly greatest. If the process conditions are then monitored by means of corrosion probes and show a regular pattern, the degree of confidence that the worst place in the plant has been identified is so increased that it becomes possible to run the plant to the measured retirement thickness in the limiting place before shutdown.

Attention is drawn to the danger of assuming that new equipment of similar design and materials to existing plant on the same duty and known to have performed satisfactorily will behave in the same way. Initial inspections of low alloy and austenitic steel equipment after an interval of not greater than two years can, amongst other things, reveal the accidental inclusion of wrong material not apparent during the initial inspection, at which time the equipment had not been exposed to its process duty. In one case it was found that some internal support rings in a reformer reactor had been wrongly made from plain carbon steel, and as a result they had suffered extensive blistering from hydrogen attack. Fortunately, they did not form part of the pressure scantlings in this severe duty vessel.

More recently it was considered advisable to check a new plant on line which used a lot of low alloy steel in hot hydrogen service to ensure that it was built from the correct steels. A considerable number of errors were found, all of which had to be corrected and could well have resulted in an in-service failure during the first two years of operation. The errors included substandard material in welds, piping, flanges, pipe fittings, valves and one in material in a bought out heat exchanger. Arising from this experience, more attention is being paid to control over steel items during the construction phase, and on selected items the material composition is being proved before the plant goes on line. Errors cannot be allowed to reveal themselves in service.

Protective devices

The regular examination and inspection of protective devices is perhaps of greater importance than that of the equipment they are designed to protect, and it is often not possible to monitor their continued ability to give protection. A recent survey carried out on relief valves removed for overhaul and test after two years' service on one site showed that 5·6 per cent were inoperative for a variety of reasons.

It is difficult to provide adequate and effective protection against overtemperature on equipment operating above 350–400°C where such equipment is directly fired. Many vessels will safely withstand pressures up to 50 per cent higher than their design pressure without failure. The same cannot be said of 50 per cent overtemperature and more damage is done by overtemperature than by overpressure.

Examination

There is some evidence to show that where inspection is in the hands of a full-time inspection team, then not only is a more thorough job done but a certain 'nose for smelling trouble' develops. Full-time inspectors will tend to do more work and look and enquire further than local management would do so (perhaps at times too much) where there is adverse experience.

On a works where the local engineers were responsible for examinations, corrosion was occurring in the tubes of a vertical exchanger in a heat recovery system boiling water at a gauge pressure of 100 lbf/in^2. The boiler was circulated by a thermo-syphon; steam raised inside the tubes, together with unevaporated water, left the top of the exchanger through a 16-in pipe enlarging to a 20-in pipe run in the form of an inverted U to a drum where the steam and water separated. There had been many tube failures in just over two years' operation. The exchanger head, together with the 16-in exit pipe as well as the 20-in pipe inlet to the drum, had been visually inspected from the ends anticipating trouble; however, there was no evidence of corrosive attack away from the exchanger. At 03.30 one morning the 20-in pipe burst open below a tee branch arranged for a future connection, and caused an emergency shutdown of the plant. With hindsight it is easy to see that the impingement arising from the waterfall effect of separated water flowing over the sill of the tee would give rise to more arduous conditions than in the two-phase areas in the 16- and 20-in pipes. The expertise of knowing where to anticipate corrosive/erosive attack is of paramount importance.

Competent persons

Many inspection engineers have been recruited with seagoing experience and certificated. There has been a move away from steam driven ships towards diesel propulsion, and a decline in the number of ships in British ownership. The complex equipment and materials technology applied to modern chemical plant increasingly require detailed rather than general knowledge. Those who come into the industry from other areas now require considerable training, and it may be better to select most of the competent persons, in whose hands lies the responsibility for examinations, from those already within the chemical industry.

The complexity of some protective systems—for example, a high integrity trip system protecting a direct oxidation process—may be beyond the ability of the 'competent person' to examine and test. In a case like this, which is comparatively rare, the person can do no more than ensure that the examination and test have taken place as prescribed by others more competent than he.

Reporting of failures

We learn from our mistakes. Failures, even apparently trivial ones, must be satisfactorily explained and not repaired by first line maintenance crews without being reported. Catastrophic failures are fortunately rare.

A 14-in gas oil line had been in service for many years and was known to be unaffected by corrosion. Plant downstream was being affected where the temperature was higher, and it was decided to evaluate the effect of a corrosion inhibitor. This was in liquid form, supplied in M.S. drums and said by the manufacturer to be non-corrosive to M.S. Arrangements were made to inject the inhibitor through a valved 2-in drain branch on the bottom of the 14-in line. Four weeks after injection had started, a leak developed where the 2-in branch was welded into the 14-in line and the unit had to be shut down immediately for repair. A section of line carrying the branch was removed and examined on the spot. There was little evidence of penetration through the 2-in branch weld and it was concluded that it was a bad weld in the first place; vibration perhaps from the injection pump and extra stress from the line now attached to the drain had contributed to the failure. The removed section was replaced with an identical 2-in branch welded this time with proven good penetration, and the unit put back on stream. An identical failure took place 35 days later. Highly localized corrosion at the interface between the incoming inhibitor and the gas oil had removed the weld. In each case it was providential that the branch did not neck off with the subsequent release of gas oil at 320°C.

Dispensations

It is expected that the recommended inspection interval for equipment may be safely extended in many cases, but not so with its protective devices. If full advantage is to be taken of this situation, consideration must be given to duplicating protective devices or using other means whereby equipment may continue in service with adequate protection.

The following bases for extension of the usual inspection interval are suggested:

(1) Where a new or recently installed vessel performs a duty similar to that of an existing vessel for which authoritative records show satisfactory performance for at least six years, and provided that the following factors are substantially unaltered—

(*a*) materials of construction,
(*b*) detail of design,
(*c*) composition, pressure, and temperature limits of contents,

then the inspection interval for the newer vessel may be increased to a period not longer than that for the existing vessel.

Similarly, when in a plant one vessel forms part of a group of vessels on the same duty, sample vessels may be chosen in rotation for examination. If, on examination, evidence is found of unexpected deterioration in the condition of the vessel chosen for examination, then the position must be reviewed. In no case shall any individual vessel be operated for a period greater than 12 years without examination.

(2) When a vessel is used for service on a plant of whose process there is little practical experience or the conditions above cannot be satisfied, then it shall be subjected to an initial examination after two years. Thereafter, if all is in order, the interval may be extended to four years, and after six years' successful service can be considered under (1) above. Such a decision must, however, take into account evidence from thickness testing, corrosion monitoring (if applicable), and inferential evidence from surrounding equipment which can or has been examined more frequently.

(3) For vessels on a duty where the rate and amount of deterioration are unpredictable, no relaxation as outlined in (1) and (2) above shall be permitted.

ACKNOWLEDGEMENT

The author wishes to thank Imperial Chemical Industries Ltd for permission to publish this paper.

C60/72 AUTOMATED DATA HANDLING SYSTEMS FOR IN-SERVICE INSPECTION: ADDCOM SYSTEM

A. R. WHITING*   D. E. STOLLE*

This paper discusses a computer program that has been specifically designed to provide the means to analyse pre-service and in-service ultrasonic inspection. The program effectively reduces the costs of manpower that would otherwise be required. Extract reports from this program, together with descriptive information, are also presented.

INTRODUCTION

DURING THE PRE-SERVICE and subsequently during in-service inspection of plant components by use of the ultrasonic test method (whether automated or manual), test data can normally be presented and documented by a variety of means such as visual, strip chart, analog-tape, 35-mm film, video tape, etc. The documentation is either voluminous or minimal.

(*a*) *Voluminous documentation* requires awkward storage and retrieval methods. Such volume requires extensive manpower and cost to handle and analyse the acquired test data, and deliver meaningful analysis reports. The proper handling of such data to meet both licence compliance requirements and A.S.M.E. Section XI requirements presents a serious procedural problem.

(*b*) *Minimal documentation* has only minimal coverage, descriptive and comparison capability, particularly when such data are hand-recorded to meet A.S.M.E. Section XI requirements. Subsequent retest (in-service inspections) produce equally minimal documentation and render the necessary test comparison analysis between the pre-service and retest inspections extremely difficult and time-consuming.

In addition to test data acquisition, there are many other necessary operations at this time requiring handling, documentation, and analysis. A few of these are listed below:

(1) Test procedure planning, documentation, certification of compliance, and audit of performance.

(2) Pre-inspection engineering procedures, either pre-service or in-service, which must be available for recall in original form and delivered to inspection crews.

(3) Inspection procedure certification to ensure retest equivalency, and immediate comparative analysis between pre-service and retest data for defect propagation analysis.

(4) Documentation, storage, and retrieval requirements of all the above for the life of the plant.

Southwest Research Institute has evaluated the requirements in a systems approach involving the development of mechanical scanning and data handling equipment. This paper describes the data handling equipment and procedures.

SPECIAL COMPUTER PROGRAM

A computer program, designated E106N, was specifically designed to provide the effective means by which to analyse pre-service and in-service ultrasonic inspection. The program effectively reduces the costs and manpower expenditure otherwise required. Extract reports from this program are presented in the following sections, together with descriptive information.

ADDCOM SYSTEM

Two basic systems are incorporated into a combined package called 'ADDCOM', shown in Fig. 60.1, which simplifies the acquisition and handling of pre-service/in-service ultrasonic test data as it is generated. These systems are described below. Pre-test entry data for ultrasonic scanning are shown in Fig. 60.2.

ADDTAP system

The 'Analog to Digital Direct Tape Acquisition Package' (ADDTAP) support system is an electronic data handling instrument package utilizing a direct analog to digital data acquisition concept which produces a magnetic computer data file tape. The package allows the capability of automated ultrasonic scanning equipment control and defect identification on direct printer output. The ADDTAP package collects the ultrasonic output test data, converts it to digital format, and produces a computer tape of all test file data.

The data acquired by the ADDTAP system for direct computer analysis consists of:

(1) the X–Y co-ordinates locating the position of the recorded test data, and

The MS. of this paper was received at the Institution on 6th December 1971 and accepted for publication on 3rd February 1972. 40

* *Southwest Research Institute, P.O. Box 28147, San Antonio, Texas 78228, U.S.A.*

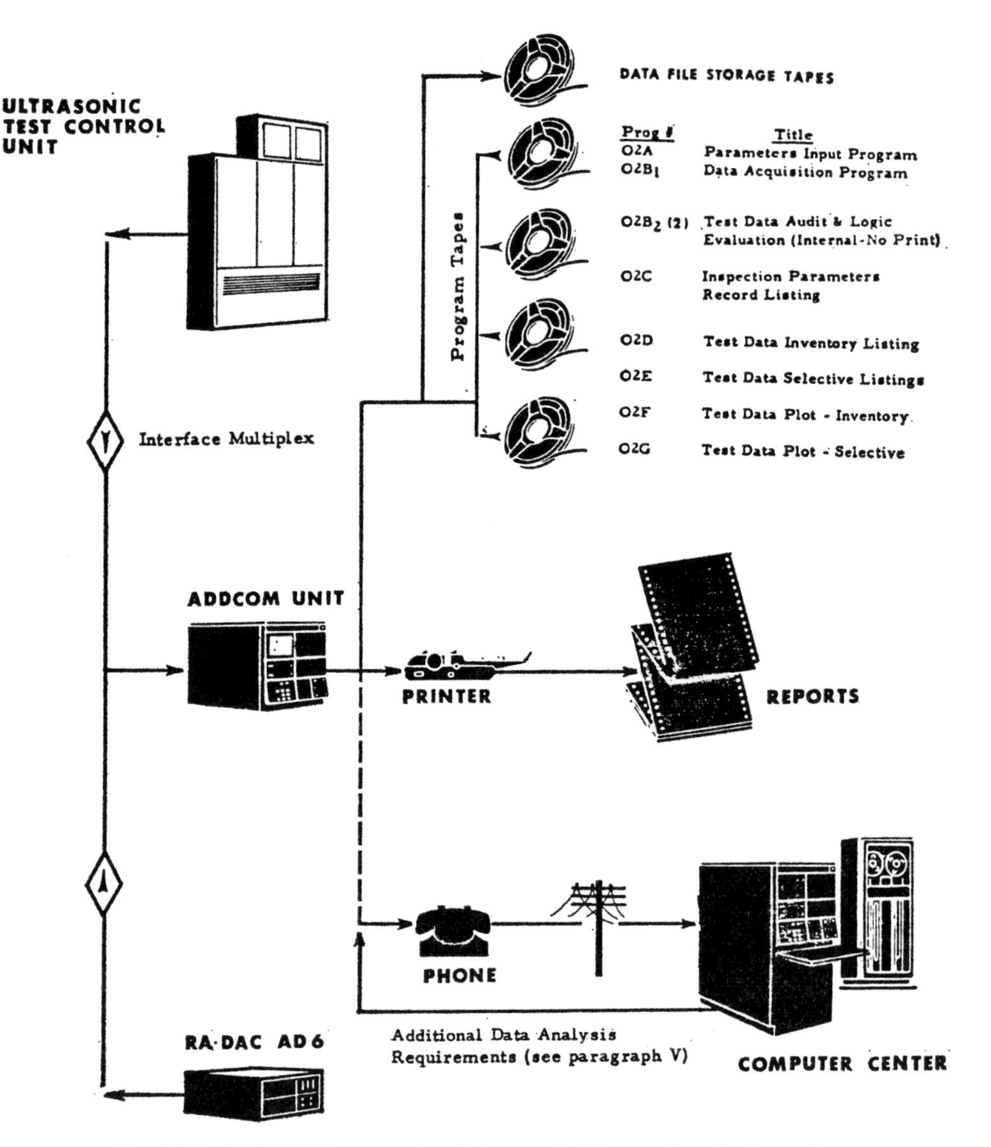

Fig. 60.1. ADDCOM computer data acquisition and analysis system

(2) the amplitude and time analog data from each ultrasonic transducer in operation.

COMCAP system

The 'Comparative Onsite Mini-Computer Analysis Package' (COMCAP) support system is a mini-computer module which ties into the ADDTAP package for on-site data analysis.

PROGRAM APPLICATION TO ULTRASONIC DATA ACQUISITION

A brief description of the various sub-programs is given below:

(*a*) Program 02A and $02B_1$ comprise the initial ADDCOM data acquisition programs and test identification parameters.

(*b*) Program $02B_2$ comprises the initial logic analysis of the data acquired and prepares it for the analysis programs.

(*c*) Program 02C provides a means of delivering a hard copy print file of the test identification and base test parameters for and used by the analysis programs. A sample of this hard copy is shown in Fig. 60.3.

(*d*) Program 02D provides the hard copy of a listing of all test data acquired for records and/or files retention. A sample of this is shown in Fig. 60.4.

(*e*) Program 02E is similar to Program 02D. However, this program is intended for files interrogation. Specific 'blocks' or areas of the test defined by low and high X and Y position limit are individually selected. Additionally, the specific nature of the data is also interrogated by selecting specific data whose amplitude and/or time analogs exceed pre-selected limits, or belonging to specific transducers.

(*f*) Program 02F provides a graphic display 'fingerprint' of all co-ordinate data generation, and displays data reduction in a format more easily identified and reviewed. A sample of this is shown in Fig. 60.5.

(*g*) Program 02G is similar to Program 02F. However, the similar selective limits, reduction, and data levels can be selected as described in Program 02E.

SYSTEMS SIMULATION

During mechanized pre-service and in-service testing operations, the ADDCOM is tied directly to the ultrasonic test control unit as shown in Fig. 60.1. This tie-in is

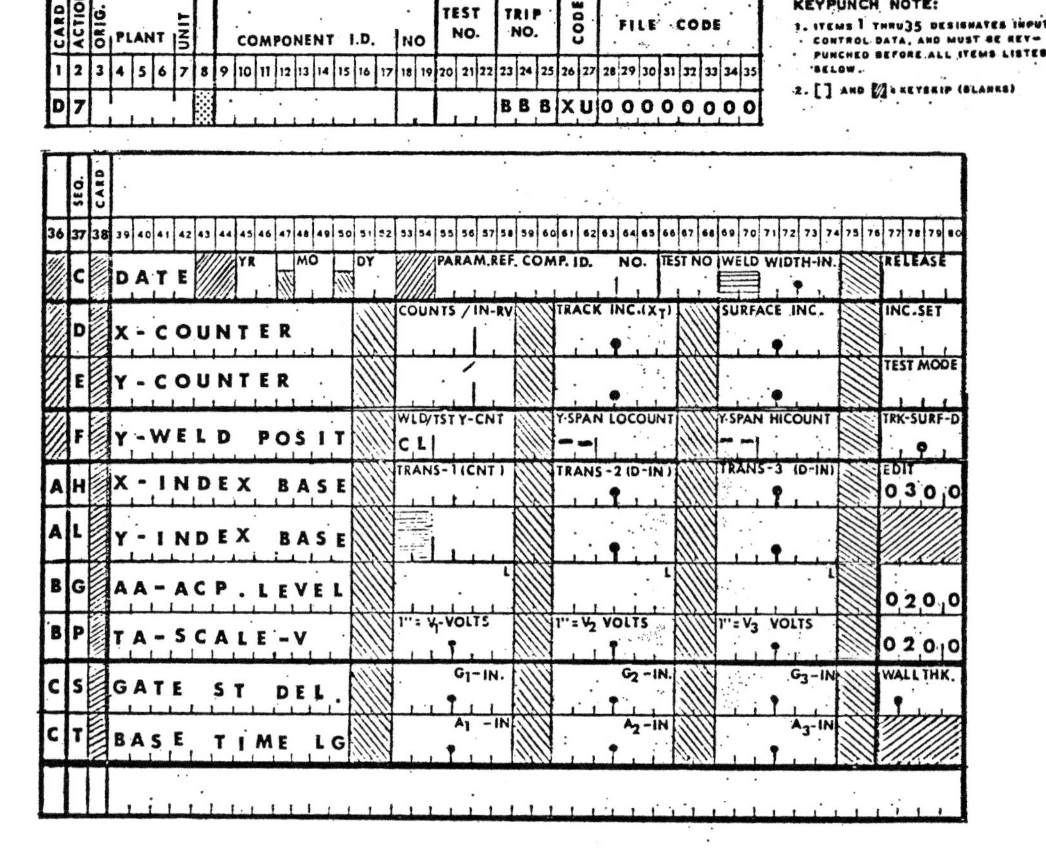

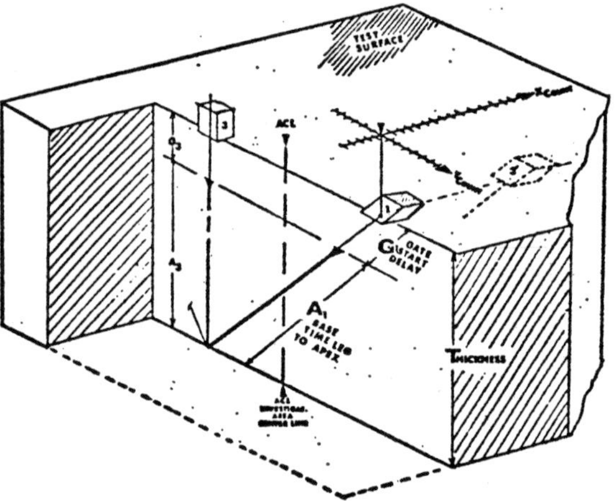

Fig. 60.2. Pre-test entry data for ultrasonic scanning

accomplished by the interface multiplexer. Tied directly into the ADDCOM computer is an on-line printer to deliver hard copy data upon request. Also the various report programs are available for inserting into the ADDCOM computer for computer function selection. Prior to actual operations, personnel must be trained for ADDCOM data acquisition and analysis operation. Also, the ADDCOM equipment must be checked out to ensure that all systems are functional. In order to test the equipment and to train or refresh technicians, a field pre-service or in-service test is simulated. To circumvent costly use of pre-service ultrasonic equipment and time, the RA-DAC-AD6 Systems Simulator is used to simulate a variety of field test operations and to deliver similar high volume data of the nature actually received during actual test. The RA-DAC unit can be selectively set to provide random data, or selected tests can be chosen to deliver simulated test operations and data acquisition.

TEST DATA ANALYSIS AND REPORTING

Depending on the extent necessary, full data analysis is available to provide a variety of detailed analyses. As listed above, ADDCOM data analysis programs are immediately available on-site to provide such selection. More detailed analysis may also be made by larger computers requiring more sophisticated mathematical and/or graphic requirements.

Figs 60.6 and 60.7 illustrate graphic data resolution analysis from the data generated by the ADDCOM system. Fig. 60.6 illustrates a preselected area of a pre-service test file, limited by preselected limits of the

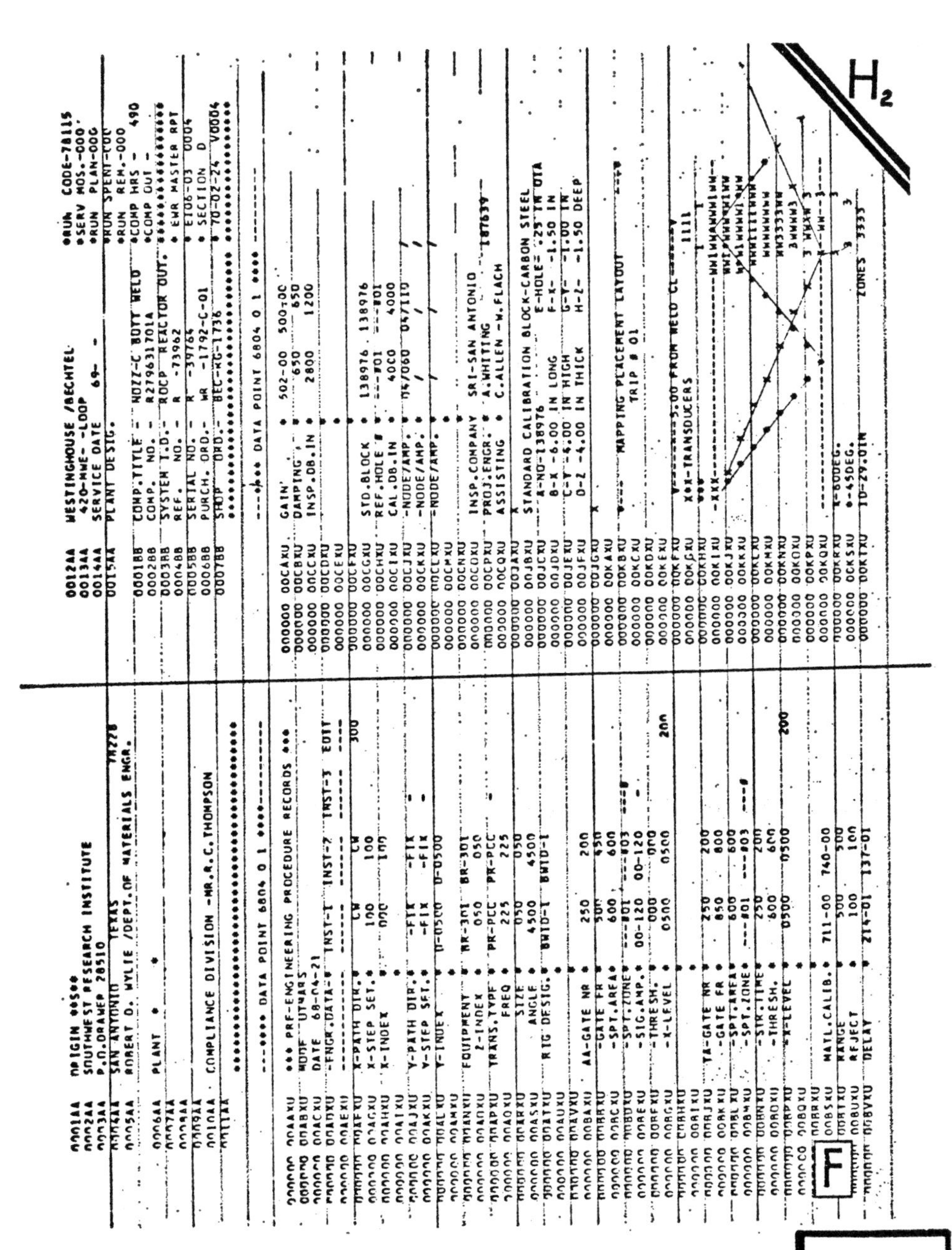

Fig. 60.3. Sample printout of test identification and base test parameters

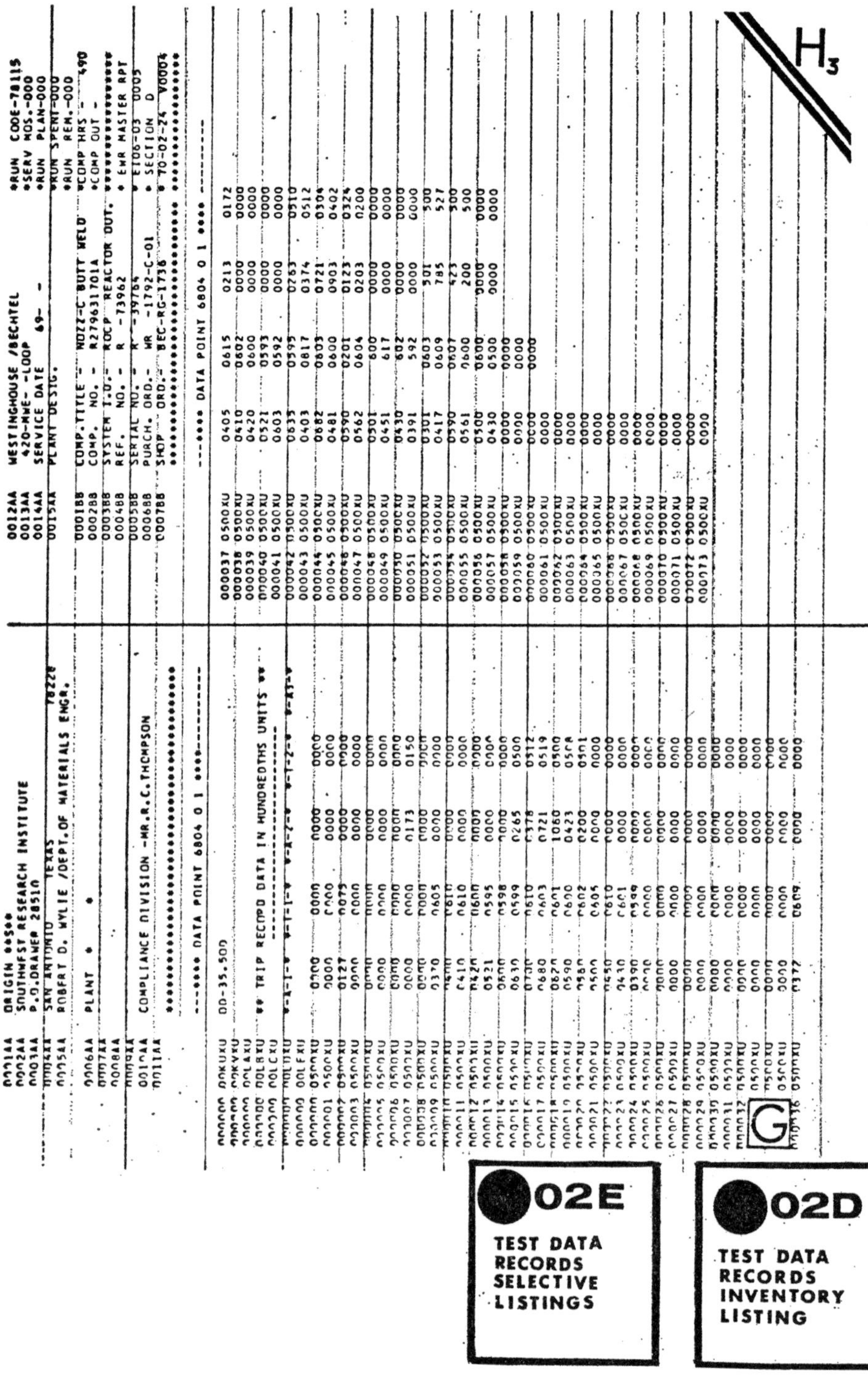

```
0001AA  ORIGIN **S**                                   0012AA  WESTINGHOUSE /BECHTEL              *RUN  CODE-78115
0002AA  SOUTHWEST RESEARCH INSTITUTE                   0013AA   420-MWE- -LOOP                    *SERV MOS.-000
0003AA  P.O.DRAWER 28510                               0014AA  SERVICE DATE    69-  -             *RUN  PLAN-000
0004AA  SAN ANTONIO      TEXAS           78228         0015AA  PLANT DESIG.                       *RUN  SPENT-000
0005AA  ROBERT D. WYLIE /DEPT.OF MATERIALS ENGR.                                                  *RUN  REM.-000
0006AA  PLANT  *    *                                  0001BB  COMP.TITLE  -  NOZZ-C BUTT WELD    *COMP HRS  -   490
0007AA                                                 0002BB  COMP.  NO.  -  R279631701A         *COMP OUT  -
0008AA                                                 0003BB  SYSTEM I.D.-  ROCP  REACTOR OUT.   ****************
0009AA                                                 0004BB  REF.   NO.  -  R  -73962           * EWR MASTER RPT
0010AA  COMPLIANCE DIVISION -MR.R.C.THOMPSON           0005BB  SERIAL NO.  -  R  -39764           * EI06-03  0005
0011AA                                                 0006BB  PURCH. ORD.-  WR  -1792-C-01       * SECTION  D
                                                       0007BB  SHOP   ORD.-  BEC-RG-1736          * 70-02-24  V0004
        ****************************************               ******************************  ****************

        ---**** DATA POINT 6804 0 1 ****----------             ---**** DATA POINT 6804 0 1 **** ---------

000000 00KUXU  DD-35.500                               000037 0500XU   0405   0615   0213   0172
000000 00KVXU                                          000038 0500XU   0410   0602   0000   0000
000000 00LAXU                                          000039 0500XU   0420   0600   0000   0000
000000 00LBXU  ** TRIP RECORD DATA IN HUNDREDTHS UNITS ** 000040 0500XU 0521  0593   0000   0000
000000 00LCXU  ---------------------------             000041 0500XU   0603   0592   0000   0000
000000 00LDXU  *-A-1-*  *-1-1-*  *-A-2-*  *-1-2-*  *-A3-*  000042 0500XU 0635 0595   0263   0510
000000 00LEXU                                          000043 0500XU   0403   0817   0374   0512
000000 0500XU   0000    0000    0000    0000           000044 0500XU   0682   0803   0721   0304
000001 0500XU   0000    0000    0000    0000           000045 0500XU   0481   0600   0903   0402
000002 0500XU   0127    0075    0000    0000           000046 0500XU   0590   0201   0123   0324
000003 0500XU   0000    0000    0000    0000           000047 0500XU   0562   0604   0203   0200
000004 0500XU   0000    0000    0000    0000           000048 0500XU   0501    600   0000   0000
000005 0500XU   0000    0000    0000    0000           000049 0500XU   0451    617   0000   0000
000006 0500XU   0000    0000    0000    0000           000050 0500XU   0430    602   0000   0000
000007 0500XU   0000    0000    0173    0150           000051 0500XU   0391    592   0000   0000
000008 0500XU   0000    0000    0000    0000           000052 0500XU   0301   0603    501    500
000009 0500XU   0370    0605    0000    0000           000053 0500XU   0417   0609    785    527
000010 0500XU   0400    0610    0000    0000           000054 0500XU   0590   0607    423    500
000011 0500XU   0410    0610    0000    0000           000055 0500XU   0561   0600    200    500
000012 0500XU   0420    0600    0000    0000           000056 0500XU   0500   0600   0000   0000
000013 0500XU   0521    0595    0000    0000           000057 0500XU   0430   0500   0000   0000
000014 0500XU   0600    0598    0000    0000           000058 0500XU   0000   0000
000015 0500XU   0630    0599    0265    0500           000059 0500XU   0000   0000
000016 0500XU   0730    0610    0378    0512           000060 0500XU   0000   0000
000017 0500XU   0680    0603    0721    0519           000061 0500XU   0000
000018 0500XU   0620    0601    1060    0500           000062 0500XU   0000
000019 0500XU   0590    0600    0423    0508           000063 0500XU   0000
000020 0500XU   0580    0602    0200    0501           000064 0500XU   0000
000021 0500XU   0500    0605    0000    0000           000065 0500XU   0000
000022 0500XU   0450    0610    0000    0000           000066 0500XU   0000
000023 0500XU   0430    0601    0000    0000           000067 0500XU   0000
000024 0500XU   0390    0599    0000    0000           000068 0500XU   0000
000025 0500XU   0000    0000    0000    0000           000069 0500XU   0000
000026 0500XU   0000    0000    0000    0000           000070 0500XU   0000
000027 0500XU   0000    0000    0000    0000           000071 0500XU   0000
000028 0500XU   0000    0000    0000    0000           000072 0500XU   0000
000029 0500XU   0000    0000    0000    0000           000073 0500XU   0000
000030 0500XU   0000    0000    0000    0000
000031 0500XU   0000    0000    0000    0000
000032 0500XU   0000    0000    0000    0000
0000?? 0500XU   0000    0000    0000    0000
0000?? 0500XU   0000    0000    0000    0000
0000?? 0500XU   0000    0000    0000    0000
000036 0500XU   0372    0609    0000    0000
```

02E
TEST DATA
RECORDS
SELECTIVE
LISTINGS

02D
TEST DATA
RECORDS
INVENTORY
LISTING

G

H_3

Fig. 60.4. Sample printout of test data for records or files retention

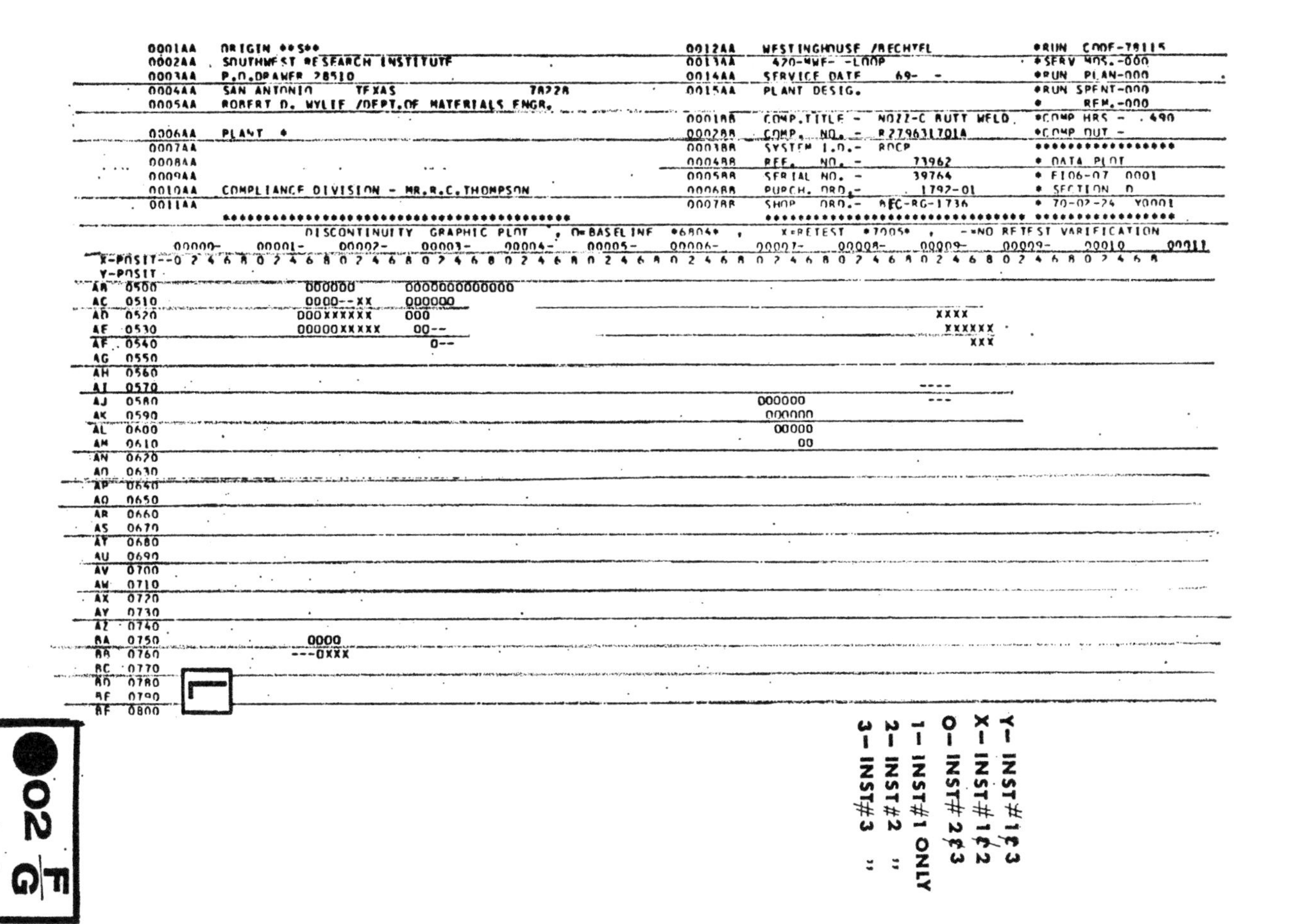

Fig. 60.5. Graphic display of test data exceeding preset limits

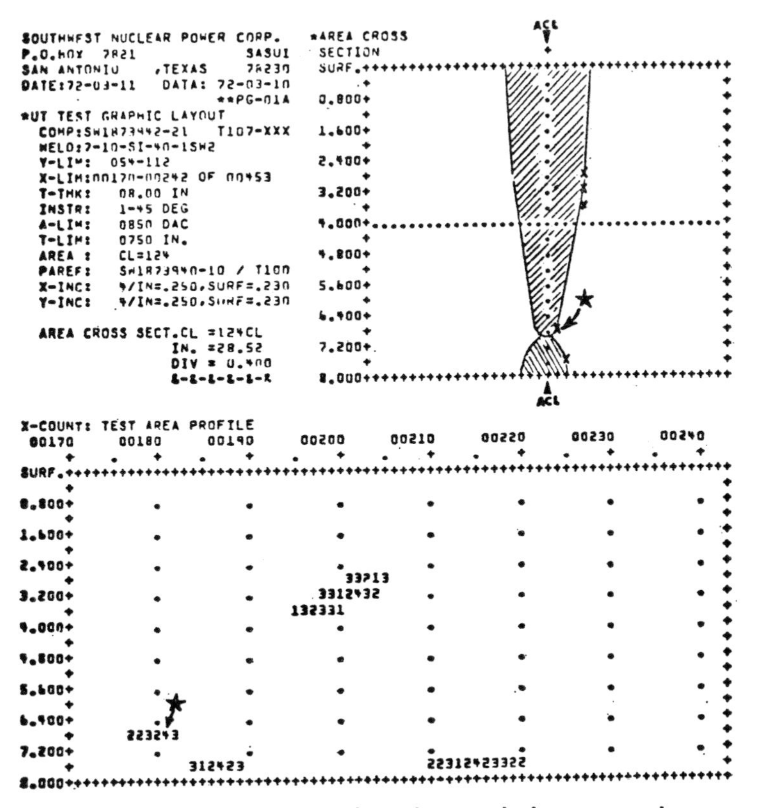

Fig. 60.6. Profile and envelope data analysis presentation

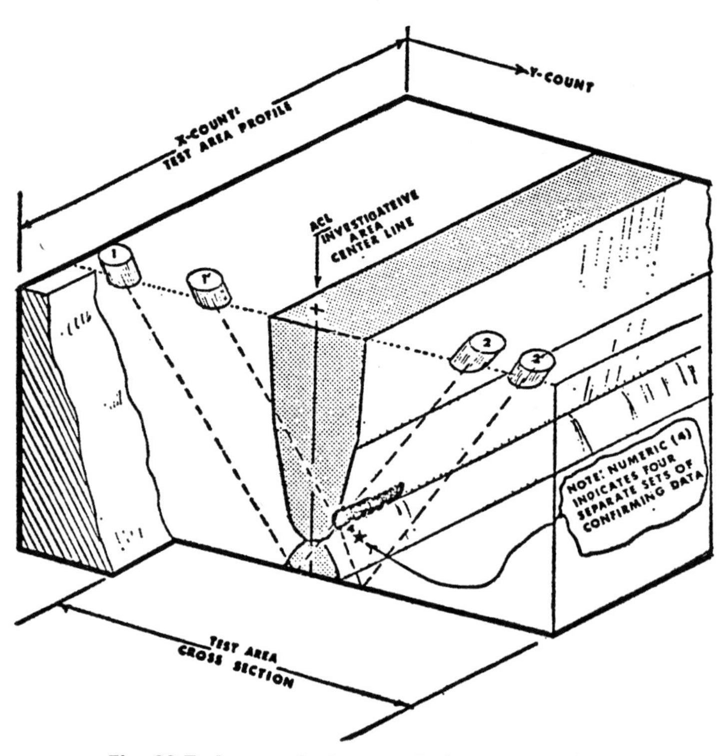

Fig. 60.7. Isometric data analysis presentation

amplitude and time analog data. This graphic analysis resolves the data to show a cross-section and profile view of the test area in question. Fig. 60.7 illustrates an isometric view of the reported analysis.

This programmed analysis is capable of analysing pre-service and in-service data, as well as a comparative differential analysis between selected pre-service and in-service inspection records.

The ADDCOM system also serves as an 'on-site' computer terminal to communicate directly via teleprocessing with larger computer hardware when required. This capability will allow 'on-site' detailed data analysis for the more complex analysis requirements.

Manual analysis of this type of data, particularly when defect propagation analysis is desired, becomes impractical. The type of analysis presented in Fig. 60.6 displays the acquired data in a meaningful manner which permits the evaluation to be presented with far more realism.

Further automatic data reduction permits more complete and on-time reporting which is so necessary in decision-making regarding return to power generation with safety. In addition, it provides a means of checking the accuracy of the inspection with a built-in quality control system by comparing indexing, location of previous indications, and landmarks when available.

SUMMARY AND CONCLUSIONS

The data handling system presented herein is designed to speed up the process of reporting. In its ultimate form, no written report would be required. Inspection results from the pre-service to all in-service intervals in the life of the plant would be stored for ready comparison. The system is also versatile enough to store quality control data from fabricators for recall. The ready availability of this information can aid rapid, accurate, and safe decision-making.

Discussion

K. P. Q. Appleton Member

E. H. Frank states in paper C59 that the determination of the frequency of examination of non-statutory pressure vessels is a matter of experienced judgement, taking into account plant history, etc. But he omits one method which might appear obvious to the layman. Why not use the frequency defined by statute for the relatively innocuous duties of steam and air as a yardstick for defining the frequency of inspection of chemical pressure vessels? If this were the basis, would it not be unlikely that periods between inspection of five years or more would be allowable?

H. Baschek Baden, Switzerland

S. M. Bush, paper C26, reported about the system hydrostatic test as required by the A.S.M.E. Section XI Code. He mentioned that at present it is under discussion to clearly specify a test pressure of 10 per cent above *operating* pressure and a test temperature of about 100 R above the applicable nil ductility temperature. In my view, such a test, performed at pressures considerably below 1·25 the *design pressure* times should be classified as a leak test and not as a hydrostatic test.

Based on experiences and on many discussions with the supplier of our plants and with others, we have concluded that the best method for leak-testing is to heat up the system in accordance with the normal operating procedure and to inspect the system carefully during and after this heat up period. The system pressure may be increased up to a value marginable to the set point of the safety valves, but we think this is unnecessary.

From the contributors to session 4, I would like to have an opinion on what critical flaw sizes should be expected in the nozzle regions, and whether these critical defects can be detected by the in-service inspection equipment. For example, I have seen published or unpublished critical values in the range between about 4 and 15 mm for the nozzle region. I believe that it is essential that the people who sell in-service inspection equipment and the fracture mechanics people should talk to each other more frequently.

In the last session, I tried to explain that, according to my understanding, the values for critical flaw sizes which have been calculated by fracture mechanics and are based on K_{1C} values, are not applicable to vessels while they are actually subjected to higher temperatures. Can any of the authors comment on this?

D. Birchon Fellow

In the verbal discussion of Session 5, W. March, whilst accepting the exciting promise of stress wave emission, voiced some very proper doubts about the advisability of commiting oneself to a developing technology, such as stress wave emission, in an advanced engineering project. But he has also spoken of the difficulties already encountered with long established non-destructive testing techniques, which have sometimes either falsely indicated defects, causing damaging or unnecessary repairs to be conducted, or have missed altogether large defects which have sometimes even caused the structure to fail.

Since stress wave emission is a dynamic method of testing, it should enable us to learn about the presence of any defect which is growing, irrespective of its size or orientation, provided that in growing the defect makes a noise which can be heard above the background noise; this means that we must have correlative information for different materials and defect extension processes, since it is now quite clear that slow ductile tearing in some tough materials can be a quiet process.

At the Admiralty Materials Laboratory we have been working on stress wave emission for some time, and, as was reported to the Institute of Physics on 14 March 1972, are just completing our first computer on line defect location system, with a 20 sensor input, for testing pressure vessels and other structures.

In the early days of this research, we were keen to see whether we could detect growing defects in a real structure, and we therefore conducted a number of service trials, albeit with rather elementary techniques.

It is not often that one can actually 'wreck' a structure which has been observed to possess a defect which is considered to be 'unacceptable'. Therefore, when a pressure vessel containing a defect was to be tested to destruction, we monitored the test, using very simple stress wave emission equipment. The pressure vessel was thought (as a result of ultrasonic inspection) to contain a defect running circumferentially around a thread root in one of the screwed end caps. During the pressure test, we were able to observe a slight noise at about 22·5 MN/m^2. The ultrasonic inspection at this stage indicated that the crack had not got any longer but might have grown a little more radially. As the test continued, a leak developed, and since stress wave equipment is such an effective leak detector, we were entirely swamped by noise and could hear nothing until the vessel burst at about 52 MN/m^2.

Two things were of interest. One was that the vessel burst from a longitudinal split, in no way related to the defect which caused it to be rejected from service! The second was that, on taking a section through the original defect and breaking it open, the fracture appearance was as shown in Fig. D1. The lower dark area was the original fracture (thereby confirming the ultrasonic indication) and the central area was an increment of crack extension which we succeeded in detecting through the background noise, while the light area indicated by the arrow is the final fracture induced to permit examination of the broken surface.

This experience gave us considerable confidence in the

Fig. D1. Macrograph of part of fracture of pressure vessel

ability to use stress wave emission in practical situations to detect discontinuous subcritical crack extensions.

F. D. Chaplin Member

I appreciate that a tabular presentation is often very convenient, but as the author of paper C56, J. J. Whenray, pointed out at some length in his introductory comments, there are many exemptions to Table 56.1, and it may be that this table is simplified to the point of being misleading. Thinking particularly of the continuous operation process industry, a requirement for annual internal and external inspection with hydraulic testing would greatly interfere with reasonable operation, and experience has shown it not to be justified.

Considering one country in the table as an example, I understand that in Italy the inspection requirements for boilers and static pressure vessels include inspection at manufacturers' works and an on-site internal inspection and hydraulic test before commissioning. In-service inspection requirements comprise an internal inspection after the first year of operation and thereafter, alternately, a check for correct operational control and internal inspection at yearly intervals, until the tenth year of equipment life, when a hydrostatic test is required with internal inspection and a check of correct operational control. This implies a shutdown every two years, for which a two month period of grace is allowed. The ten year cycle is then repeated. For unfired pressure vessels in oil or chemical plant operation of a continuous nature, which after one year of operation are found free of corrosion or appreciable surface defects, the period between internal inspection may be extended to six years. For l.p.g. storage vessels manufactured in accordance with stringent welding procedures, the first in-service internal examination may be delayed until the sixth year with re-inspection intervals also of six years. Where a hydrostatic test is due, e.g. at tenth year of equipment life, it is combined with an internal examination. For pressure vessels of less than 1 m^3 capacity, the internal examination and pressure tests are not required during the first installation examination. Further exemptions are granted when dealing with low pressure/capacity boilers, de-superheaters, steam vessels, air vessels, vessels for refrigerating plants, etc.

Regarding portable vessels, l.p.g. cylinders up to 80 litres have to be tested every tenth year, according to a specified procedure.

Thus the situation is seen to be very different from that inferred by reading Table 56.1, in particular for the continuous operation oil and chemical industries. I should be interested to know how far similar amplifications or exemptions are applicable to the other countries listed.

G. P. Fagence Graduate

Whilst accepting the statistical approach of the authors of paper C48, we do not consider they make sufficient correction to the conventional data to draw the most realistic conclusions on the safety of nuclear vessels. We cannot agree that the Class I vessels reviewed in reference (**4**) of the paper were built and operated to standards comparable with nuclear plant. Requirements for the latter are far more stringent than those of the vessels which gave the 100 000 vessel-years experience. Although individual vessels may approach nuclear standards, their significance is not reflected in the statistical population.

The contrast in quality is exemplified in the fact that a formal stress report based on extensive design and stress analysis is required for a nuclear vessel and submitted for the client's approval. From our experience, such effort is not matched in conventional vessel design, as indicated in the following example.

Vessel	Design effort, man-hours	Computer analysis, hours
Complex conventional vessel for chemical plant . .	821	12
Light water reactor . .	9600	135

Nuclear vessels are also assessed for specific operating and transient conditions. Operation of nuclear plant within these conditions is ensured by extensive control and instrumentation. Predetermined operating conditions of this form are not features of general Class I vessels.

Non-destructive testing of nuclear vessels during manufacture is far more extensive than that performed on the sample vessels. For a water reactor, a client would probably specify three comprehensive inspections; before and after stress relief and after hydrotest.

Considering the numerical values listed in Table 48.3 in the context of nuclear plant meeting the requirements of A.S.M.E. Section III Pressure Vessel Code rather than general 'Class I standards', we would propose the following modifications:

P_D $2{\cdot}5\times10^{-5}$ based on nuclear design procedures;

P_C 1×10^{-4} Paper C48 uses three catastrophic failures from Table 48.2, but only two are applicable to nuclear vessels.

P_{US} 10^{-1}–10^{-2} 10^{-2}–10^{-3} Wessel (1) has shown that for a light water reactor a 'significant' defect is one over 3 in deep. While we do not propose such a gross basis of significance, we consider ultrasonic reliability to be very high for these defects, and propose a further 10^{-1} factor.

Revised Table 48.3

Item	Assessed probability of failure per five year cycle	Selected values for equation (48.1)
P_D	$2{\cdot}5\times10^{-5}$	$2{\cdot}5\times10^{-5}$
P_M	5×10^{-5}	5×10^{-5}
P_C	1×10^{-4}	1×10^{-4}
P_{PT}	$1\times0{\cdot}985$	1·0
P_{AT}	As in paper	
P_{US}	10^{-1}–10^{-2} (first inspection)	0·5 for design and material defects
	10^{-2}–10^{-3} (repeat inspection)	10^{-2} for constructional defects
P_{VE}, P_L	As paper	

Using the revised table, equation (48.1) gives a mean vessel failure rate of 5×10^{-7} per vessel-year, which satisfies the requirements of Table 48.1, as does the authors' own value.

The authors find themselves unable to demonstrate that the highest standards available meet those required. We cannot see how this conclusion follows from the authors' text. In any event, making appropriate allowance when applying Class I data to nuclear vessels, we consider that the highest standards available are more than adequate.

The questions are asked requiring affirmative answers but, having posed the question early in the text, the authors do not return to them. Questions of this nature concern the future of the nuclear industry and require specific replies. Based on this contribution, our answers are categorically in the affirmative and we see no limitation on the use of steel vessels within the industry. This conclusion is subject to the requirement of the nuclear codes being met by competent personnel.

REFERENCE

(1) Wessel, E. T. and Mager, T. R. 'Fracture mechanics technology as applied to thick walled nuclear pressure vessels', *Conf. Practical Application of Fracture Mechanics to Pressure-Vessel Technology* 1971 (May), 17 (Instn Mech. Engrs, London).

R. Haas Essen

Referring to the NDT procedure requirements for in-service inspection of water cooled reactor pressure vessels in Germany, the recommendations given by the Reactor Pressure Vessels subcommittee of the German Reactor Safety Commission include the following requirements.

The wall of the entire vessel must be inspected visually at the inner and the outer surface. As far as necessary optical or TV equipment may be applied.

In the present state of testing techniques, the additional performance of an ultrasonic inspection is mandatory. If different techniques can be introduced which are at least as informative as the ultrasonic tests, this recommendation will be amended correspondingly. Because the ultrasonic tests are given preference, this technique is emphasized in the following items without denying the potential for developing further techniques, for instance eddy current, magnetic flux or liquid penetrants for surface cracks.

The design of the pressure vessel and the testing apparatus must be suitable for the ultrasonic examination of the entire thickness of the complete wall. In particular, the condition of weld cladding must not prevent testing.

As the dose rate of radiation in the vicinity of the pressure vessel at the time of in-service inspections cannot be predicted with sufficient accuracy, the ultrasonic inspection should be carried out by a fairly mechanized system. For regions where the theoretical dose rate of radiation will be low, and where therefore tests are usually performed by hand, it must be demonstrated how mechanized remote control testing systems can be made feasible in the case of an unexpected high dose rate prohibiting the presence of testing personnel.

The maximum of testing feasibilities is required, because, with regard to the youth of the technology of the nuclear power industry, operational experience is still not sufficient to provide the essential information for suitable selection of the vessel parts to be tested. Nevertheless, it is intended to apply ultrasonic testing procedures to those vessel parts which appear to be important and representative according to the present state of knowledge. The areas to be tested could be extended, if future experience should make it desirable.

The process of selecting the parts to be tested in a preliminary state has been based upon the possibilities of hidden defects created by manufacturing or of operational stress. Thus, the following parts are to be tested as a minimum: all butt and nozzle welds, ligaments between the holes for the control rod penetrating the head (P.W.R.) or the bottom (B.W.R.), inner edges of the coolant nozzle holes, bolts and nuts of the flange connection between the head and the vessel.

Ultrasonic tests can be conducted by directing the sonic beam from the inner as well as from the outer vessel surface. Because the testing equipment which is provided for any of these cases can fail, and because it may become necessary to carry out the ultrasonic test from a complementary surface in order to check or interpret indications which might be found within the applied system, it must be demonstrated in what manner such supplementary tests can be made possible at all times after the vessel has been put into service.

The testing procedure has to be suitable, in order to be able to detect free crack surfaces in the interior of the vessel wall which are oriented vertically to the essential stress directions. This means, for a large part of the region to be tested, that the ultrasonic beam has to be directed perpendicularly to these crack faces or that a tandem or pitch and catch technique has to be applied. In order to find cracks which are located close to the vessel surface, the use of effect of angular reflections may be applied. Testing with a single probe technique which cannot fulfil one of these conditions for the beam directions is only acceptable if a more suitable technique is not feasible.

The transfer losses in the entire testing area and for each sonic beam path are to be determined and accounted for as much as is feasible. If the coupling is not secured by immersion technique, it must be monitored continuously.

The sensitivity calibration depends upon the methods which are applied in the individual case and the conditions in the testing area. Therefore, requirements for the testing sensitivity have to be defined on a case by case basis. However, the following conditions and rules should be met.

If the ultrasonic beam can be directed vertically to the expected crack surfaces or if the tandem or pitch and catch technique is applied with regard to the expected crack orientation, the level for flaw registration has to be set 6 dB lower than the echo level of an equally oriented circular reflector of 10 mm diameter. These conditions have to be met at each point of the beam cross-section which is being applied.

If only a single probe technique is applied, the effect of angular reflection cannot be used, and the direction of the beam is not perpendicular to the surface of the expected defect, a higher sensitivity must be calibrated. Circular reflectors with a diameter of 3 mm which may be oriented vertically to the main beam determine the registration limit to potential flaw echoes. This condition must be met at each point of the beam cross section which is used in the tests.

For different techniques, the required sensitivity calibration has to be determined by experiments. If in this manner only tests with a relatively low sensitivity can be conducted, then the intervals between these tests have to be correspondingly small.

H.-J. Hantke Cologne

I am surprised that only common steel vessels are mentioned, although the conference covers periodic inspection of all kinds of pressure vessels. Thus I would like some comments on multilayer vessels and prestressed vessels, for instance, P.C.P.V. In this respect I think it will not be easy to apply the same practices for recurring inspections as for common steel vessels. For instance, I draw attention to the requirements in Section XI of the A.S.M.E. Boiler and Pressure Vessel Code which distinguishes between the three categories of required inspections, visual, superficial and volumetric. The interesting liner, representing in the P.C.R.V. the leak-tight pressure shell, is covered with a multilayer insulation. The cables, which compensate for the pressure, are grouted within the concrete wall; hence, visual surface and volumetric investigations cannot be applied in the same way. On the other hand, the failure mode of a P.C.P.V. will be quite different, and the safety factor of the design and layout will be such that a catastrophic failure can be excluded. There are other possibilities of controlling the behaviour and the integrity of the P.C.R.V., such as strain gauges and thermocouples in the concrete, strain measurement devices on the cable anchors, etc., which continuously measure the load conditions.

Also, some day, acoustic emission measurements for the occurrence of overload cracks could be taken into account, but so far only experiments have been carried out in this field, from which no final conclusions or results can be drawn. Therefore it might be possible to restrict the applications of the periodic inspection only to the parts outside the concrete, i.e. penetrations and steel closures. I should therefore be very pleased to hear the opinions of the experts, and especially to learn about the practice which is applied to the P.C.R.V. in the U.K. and France.

P. H. Jones Member

With reference to paper C48, by R. O'Neil and G. M. Jordan, it is extremely difficult for nuclear safety authorities to assess the reliability of reactor equipment, when the justification presented inherently involves an element of value judgement on materials, construction and quality control, and to a lesser extent, design. This is compounded when the justification is for continuing operation following an in-service inspection of selected parts of the plant. However, in their understandable enthusiasm to escape from this position, I believe the authors have reached conclusions not justified by the facts available. They appear to be constructing a statistical framework for safety and reliability assessment, in a situation where there is an almost total absence of relevant statistical data. I am concerned that in our anxiety to develop a valid statistical framework, we might allow this underlying weakness to be driven beneath the surface, with dangerous consequences.

The authors suggest that there might be a gap of 10^{-1} to 10^{-2} between vessel reliability requirements and best current achievement, and conclude that, unless this can demonstrably be bridged, '... there may be a serious limitation on the use of steel vessels in the nuclear industry'. Examination of the deductive reasoning and data used leads me to conclude that the uncertainties in the underlying assumptions cumulatively far outweigh the claimed gap. Examples of the debatable points within the assumptions include:

The nuclear safety policy is based on the Farmer criterion, which, although described numerically, must involve much subjective judgement in its derivation. The very least such a criterion must achieve before application is acceptance internationally by nuclear and medical authorities. To what extent has this been achieved?

Under the heading **'REQUIREMENTS FOR NUCLEAR PRESSURE VESSEL INTEGRITY'** the following statements appear: '. . . pressure vessel failure is likely to be the dominating mode . . .' How can this statement be justified? How do the authors draw the boundary of the pressure vessel in this context for a multi-loop reactor plant containing several large individual items of pressurized equipment?

In **'Category A failures'**, the authors say '. . . it is possible to reasonably postulate that not more than about 10^7 Ci will be released . . .' They also suggest that engineered safeguards and containment might reasonably be assumed to fail completely at a rate of 10^{-1} to 10^{-2} per demand. In **'Category B failures'** the authors suggest '. . . a further decontamination factor (d.f.) of 10^{-2} from these sources might be anticipated . . .' How can these statements be justified? To whom are they reasonable, or capable of anticipation, as the case may be, and on what authority can they be accepted?

There are several other such examples, but I think these are sufficient to make the point on this aspect.

One further worry on this section of the paper concerns the reasoning leading to identical allowable failure rates for Category A and Category B failures. This seems to stem from assuming, for both failures, a release of 10^7 Ci of I-131 to the containment area, and futher assuming that, for Category B failures, I-131 is always released to the atmosphere, while with Category A failure, the probability of release is 10^{-1} to 10^{-2}. These assumptions puzzle me, and I should be grateful if the authors would explain them in a little more detail.

With regard to the statistical evidence of nuclear pressure vessel integrity, I agree with the authors that there has been no known nuclear pressure vessel failure, and that the population from which to draw the statistics of conventional vessels is very small. I cannot agree, however, with the use to which the available data on conventional vessels from the Phillips and Warwick paper has been put in this paper.

Of the seven catastrophic failures in Table 48.2, the authors state that four were due to operational defects, i.e. lack of water in the boiler, oil on heat transfer surfaces, etc. Reference to Phillips and Warwick, C48 (**4**), shows that, of the remaining three, one was defective due to laminated material, one was due to a defect on the inner surface of a pipe, and the last was due to a crack in a header inspection cap.

How can the authors possibly justify relating this kind of evidence to a catastrophic failure of a nuclear pressure vessel? To do so seems to ignore completely the quality environment surrounding such equipment. It is simply not correct to imply, as the authors have done, that terms like 'highest possible standards, Grade I construction', etc. are used to define standards of design manufacture, testing and inspection of nuclear pressure vessels. Specifications are much more explicit than that, requiring defined standards to be met at each stage of the process. It is simply not acceptable to organize it any other way.

Furthermore, the authors ignore the intensity of engineering skills, technological support and management attention that a nuclear reactor project demands—from the preliminary designers' drawing board, right through to product support, including in-service inspection. This 'quality' environment is essential, and is expensive to set up and maintain; to ignore the pay-off from this, as the authors appear to have done, seems highly unreasonable to me on all counts (including that of engineering economics!).

On the question of the probability of vessel failure, equation (48.1) essentially is:

$$\text{probability of failure} = (\text{probability of vessel being defective}) \times (\text{probability that a defective vessel passes all tests and inspection})$$

The small number of one kind in existence, the relatively low production rate, and the advance of their technology suggest that the probability of any one nuclear pressure vessel being defective cannot be assessed meaningfully on a statistical basis.

The probability that a defective vessel passes inspection is assessed by the authors on the basis of engineering judgement and the data given in Table 48.2. In my opinion, that data cannot be used for this purpose. For example, two defects out of 132 defects (i.e. 1·5 per cent) were found by pressure tests, suggesting that the pressure test is of little value. However, this assumes that the 130 defective vessels which passed the test were defective at the time of testing. Phillips and Warwick, C48 (**4**), state that, generally, vessels were only pressure tested during commissioning. A large number of the defects might conceivably appear in service caused by fatigue, corrosion, etc., i.e. after commissioning. Consequently, the above assumption is wrong, and I believe pressure testing has a value in its own right unassociated with other non-destructive examination methods. The value of pressure testing is too complex to discuss here, and is a worthy subject for further debate.

The usefulness of visual examination and, to a lesser degree, leakage in finding defects is stressed. Whilst not disagreeing with this conclusion, it does not follow from Table 48.2 as suggested. The fallacy in the argument is similar to that for pressure tests. We can presume that a defect found by visual examination will not be subjected to other tests (e.g. non-destructive testing). Even if it is, this will be reported as being found by visual examination, even though it could also have been detected by the other tests. If the vessel is visually examined first, this will reduce the number of defects attributed to the other inspections, and hence diminish their apparent usefulness.

With regard to the individual failure probabilities, Table 48.3 gives probabilities for substitution into equation (48.1). The points made above throw some doubt on the validity of some of these numbers.

My final point concerns the importance of so-called design faults. I accept that such faults may not be capable of being perceived at the pre-service validation stage. However, I would take issue with the authors' deduced assertion that '. . . a complete and independent assessment of the design is a necessary part of the overall quality control of a nuclear pressure vessel'. Such a suggestion flies in the face of experience, generally, in the quality control field, where there is now a strong body of opinion saying the best way to ensure quality is to put the responsibility for it firmly and explicitly on the supplier. I would personally go further and suggest that the existence of an independent inspecting authority for design may well detract from the overall quality achieved, because it smears responsibility and blurs accountability. Surely, independent bodies can only be advisory and constrained to ensuring the supply organization is working to acceptable codes of practice.

Within this area of independent design reviews, the authors suggest that an assessment would require a success rate of 90 per cent to give an improvement of 10^{-1} in overall vessel reliability. Frankly, I find it difficult to imagine how, in practice, one can define 90 per cent of design faults; even assuming one could list such faults, surely their type and consequence come into it? Again, ignoring these practical difficulties, I do not understand the manipulation of the statistics. Equation (48.1) is evaluated, using Table 48.3, to be the probability of vessel failure $= P_F = 4 \times 10^{-6}$. If 90 per cent of the list of 'equal' design faults were eliminated, then $P_F = 1{\cdot}3 \times 10^{-6}$. If all the design faults were eliminated, then $P_F = 1 \times 10^{-6}$. These figures do not support the authors' stated conclusions.

M. Kelly Fellow

In Session 3 I replied to a question from H.-J. Hantke concerning in-service testing and inspection carried out on concrete pressure vessels in Great Britain.

Oldbury-on-Severn power station was the first British nuclear power station to use concrete pressure vessels; within these vessels were the reactor cores and the boilers.

The Oldbury-on-Severn pressure vessel is a vertical cylinder of prestressed concrete, internal diameter 23·5 m, internal height 18·3 m, and with a 4·57-m thick wall and 6·70-m thick slab ends. It is constructed of high strength concrete prestressed by layers of cables within the wall in a helical pattern and in flat layers in the top and bottom slabs. The 4400 cables in each vessel are anchored on the outside face of the vessel with anchorages of the Freyssinet type.

The inner surface of the concrete is lined with a steel plate for gas tightness and maintained at a safe temperature by the water cooling system in the concrete. The liner is separated from the reactor gas by stainless steel foil insulation protected by steel cover plates.

Reactor I was commissioned in November 1967 and Reactor II in January 1968. Since that time, in-service inspections of the vessels have been carried out by frequent operational monitoring of temperatures of the liner, concrete and cooling system and regular monthly assessment of the embedded strain gauges. All these measurements have shown that the vessels are in a satisfactory state and the strains are less than predicted.

Average concrete temperature has been 28–32°C and maximum concrete temperature about 50°C.

In addition, inspection of cables and exterior surfaces are carried out annually on each vessel. The cable checks are: a load check for relaxation on 1 per cent sample, a gross slippage check for anchorage defect on 1 per cent sample, and a corrosion check of three cables, which are detensioned and withdrawn. One strand from each cable is replaced and the sample removed is cut up to carry out metallurgical checks. Examination of the concrete surface for any crack developments, particularly adjacent to cable anchorages, are carried out annually.

The results of all these inspections to date have shown that the cables are satisfactory with relaxation as predicted. None of the results are considered to be of any structural significance.

It should also be noted that in Great Britain there is close consultation between the government, Atomic Energy Authority and the Central Electricity Generating Board. The nuclear power stations operate under a licence issued under the Nuclear Installations Act. This licence includes the requirements for inspection and maintenance of the plant.

D. S. Lawson Member

I would like to comment on paper C48 by O'Neil and Jordan, which is a most interesting paper and makes a useful contribution by providing a consistent language within which a quantitative assessment of safety can cover design, manufacturing and inspection. The paper is a general paper covering steel vessels and bases the statistical numbers on past experience on a wide range of vessels defined as Class 1. Within Class 1 vessels, one can get a wide variation of design, manufacturing and inspection, and I would suggest that a more specific approach would be useful. If one considers the standardized reactor vessels being built for light water reactors (L.W.Rs.) today, then the numbers produced and short operating experience to date are such that operational statistics of the level required are not available. What can be said with certainty is that the standards of manufacture and inspection are higher than for traditional Class 1 vessels. This statement in itself is not quantifiable, although the standard of safety can be assessed by breaking the problem into predictable areas. This can be done by taking the probabilities of failure in design, manufacture and construction and assessing these on the basis of the specific task.

The design of such vessels is now being repeated by about 7 different major manufacturers, many of whom have now produced stress reports on more than 10 vessels. Comparison between these can rule out gross design errors, and one is left to assess the accuracy of the calculations predicting the operating stress levels. In the highest stressed region around the nozzles, finite element techniques are used, and some simplifying assumptions have to be made. This will introduce some errors which at least on the surface (from which failure is most likely to start) have been measured by extensive strain gauging during hydro tests and found to be small. Hence the statistical distribution for design should be able to be estimated with a small deviation.

As the materials for L.W.R. vessels have been standardized and frequent sampling is demanded during manufacture, a statistical distribution of the relevant material properties exists, a possible exception being the long term irradiated material properties, which will therefore require a wider deviation to cover these doubts.

The manufacture of L.W.R. vessels is covered by frequent inspection and NDT steps, as shown in Fig. 25.2 for initial manufacture and paper C26 for in-service. The probability of inspectors missing faults of a given type has been roughly assessed. Fracture mechanics, with some assumptions, can predict the effect of flaws in the fabricated vessel. Combining these, an estimate can be made of the probability of a given size flaw existing after inspection and this related to the design stress and material properties, and hence to the probability of vessel failure.

Fig. D2. Types of cladding

The above approach is, in fact, the more usual design approach and follows the statistical consideration given to determining factors of safety described in many texts, e.g. (2) (3). In assessing problems of this type I find it useful to examine them from different angles and compare the results. Therefore I suggest the above approach is complementary to that given in the paper. This duplication should give added reliability/safety to the conclusions!

REFERENCES

(2) JOHNSON, R. C. *Optimum design of mechanical elements* 1961 (John Wiley, London and New York).

(3) MCCRORY, R. J. 'Mechanical reliability concepts', *A.S.M.E. Design Engineering Conf.*, New York, 1965.

H. J. Meyer Nuremberg

Up to the present, the Maschinenfabrik Augsburg-Nürnberg AG (M.A.N.) has carried out in-service inspections and baseline inspections in six nuclear power stations in Europe (three of the B.W.R. type and three P.W.R. types). Two of these nuclear stations had already been commissioned, and what was termed 'primary inspection' had to be carried out under hot nuclear conditions. For the baseline inspections, the equipment was designed in such a manner that the same rigs and probe systems can in the future be used for in-service inspections under hot conditions.

While the experience which we have been able to accumulate during these inspections suggests a certain measure of success and the practicability of the approach adopted, I feel it would be appropriate also to point out the various difficulties which the inspector faces in tackling such jobs.

Let me first of all mention the influence of cladding on the amount of information obtainable with ultrasonic techniques. To protect the ferritic pressure vessel material against corrosion, the surfaces in contact with the coolant are covered by deposit-welded austenitic cladding. The type and method of cladding varies widely and, in some instances, different techniques are applied to one and the same pressure vessel.

Fig. D2 (left-hand side) shows three typical types of cladding (from top to bottom these are manual cladding with electrodes, strip cladding and oscillating wire cladding). These specimens show the typical surface structure obtained after surface welding. On the right-hand side of Fig. D2 we have the same types of cladding viewed from the underside after etching away of the ferritic parent material. It can be clearly seen how much the so-called interface can differ. The outer surface of the cladding can be dressed down to conform with certain standards of smoothness or a certain degree of uniformity can be specified in the light of ultrasonic inspection experience. With a view to obtaining reproducible results, requirements for a smooth surface tend to be most exacting. The structure of the cladding, and especially the structure of the interface, cannot be changed on the pressure vessel and their influence has to be allowed for, or the inspector can try to compensate for it by suitable ultrasonic techniques. The first requirement is that adequate information should be available on the various influences involved.

Transverse waves of 2 MHz on penetrating austenitic welds are subject to a marked degree of attenuation. Independent tests of different laboratories have shown a value of 1·60 dB/mm to be typical. Compared to this, attenuation of this type of wave in ferritic steel is as low as 0·03 dB/mm.

Fig. D3 shows different attenuation patterns in a typical 5 mm thick cladding, if the beam angle is varied between 35° and 80°. Since the time the beam travels through the cladding is a function of the beam angle, attenuation tends to increase steeply for large angles. For instance, in scanning the same target, the difference between a 45° beam angle and a 70° beam angle with the same travel distance is 14 dB. If quantitative information is required on the target sizes, this difference figures as a very important factor in the accuracy of definition. As will be realized

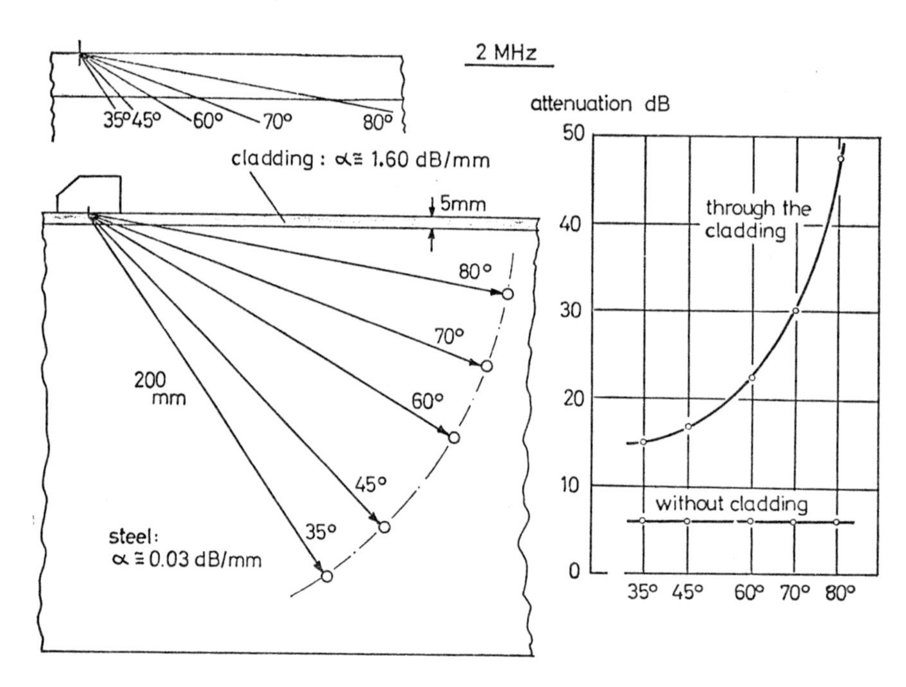

Fig. D3. Influence of cladding on angle beam penetration

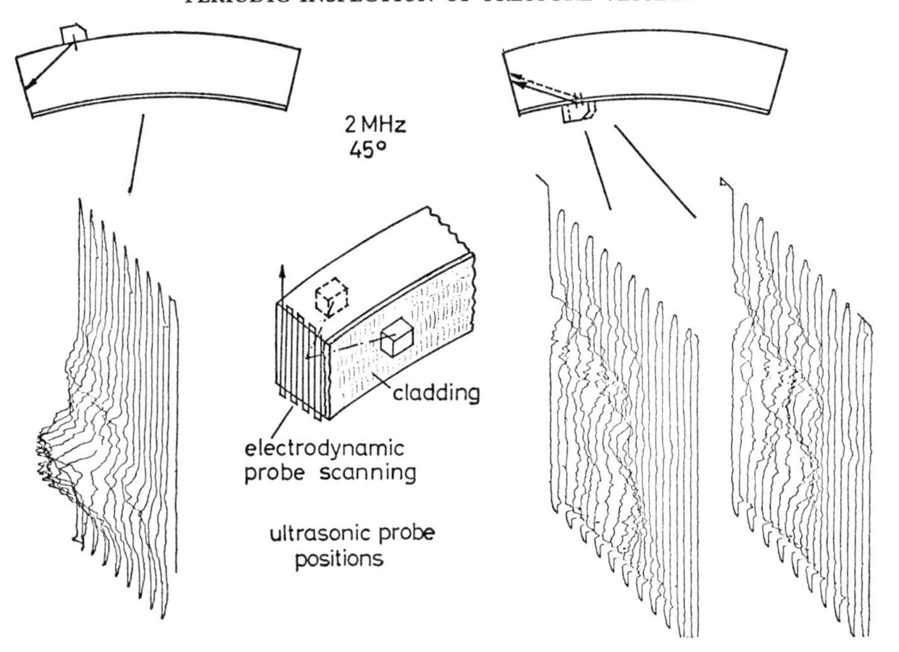

Fig. D4. Distortion of sound-energy distribution due to cladding influence

from this example, variations of sound attenuation result not only from the beam angle but also from different thicknesses of the cladding. A variation in wall thickness of, say, 4 mm with a 45° angle would cause a change of 10 dB but, with a beam angle of 70°, the change would be as much as 20 dB.

It follows that it should be borne in mind in selecting an appropriate beam angle that small angles (say 45°) will be least affected under conditions of varying wall thicknesses.

As extensive tests on specimens and clad pressure vessel walls have shown, variations in attenuation behaviour will occur not only as a result of changes in angles and wall thicknesses but, even when the probe is traversed at short increments, the nature of the interface influences the sound field pattern and local attenuation variations. As an example of the distortion of the sound energy distribution pattern, Fig. D4 shows the results of tests with an electrodynamic probe applied to the end of a specimen. Whereas in examining the wall from the unclad side, a clearly defined concentric sound pressure distribution was measured, the pattern became 'mashed' as the beam passed through the cladding and there was no longer any correlation between the main sound beam and quantitative flaw size evaluation.

If the surface of a clad pressure vessel is scanned with a rigid transmitter/receiver system (Fig. D5), the so-called 'transfer measurement' will result in local variations which, under adverse conditions, may result in variations as high as ±10 dB when the probe is traversed only a few millimetres. Tests have shown that, depending on the type of cladding, these variations may occur both when making transfer measurements from the outside as well as when these are made from the inside. Again, the cause of this phenomenon appears to be largely in the structure of the interface.

In order to eliminate these phenomena effectively during practical inspections or, at least, in order to be able to measure their effect, our test probe system incorporates probes that are designed to establish the transfer conditions. The values obtained can either be recorded synchronously or applied to control the sensitivity of the probes.

Fig. D6 illustrates a typical probe system of this type for testing pressure vessel walls from the outside. Probes 1 and 7 are used to monitor the transfer conditions. The local influence of the cladding at the deflection point is thus established, and allowance can be made in selecting the gain level of the individual probe groups or individual probes in the electronic scanning of the individual probe groups.

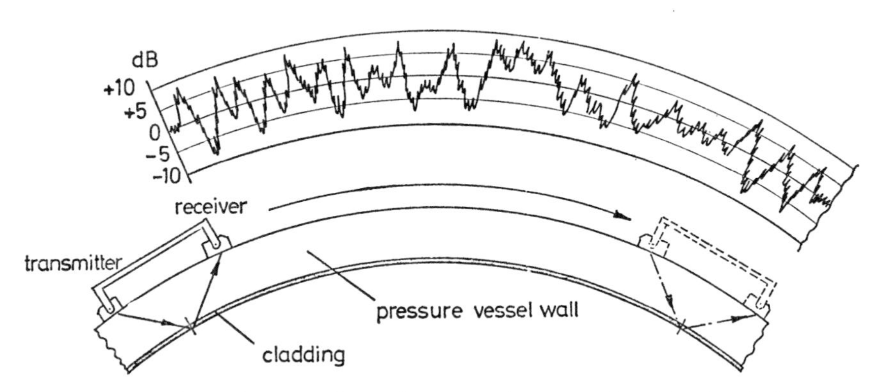

Fig. D5. Transfer recording of attenuation behaviour due to cladding

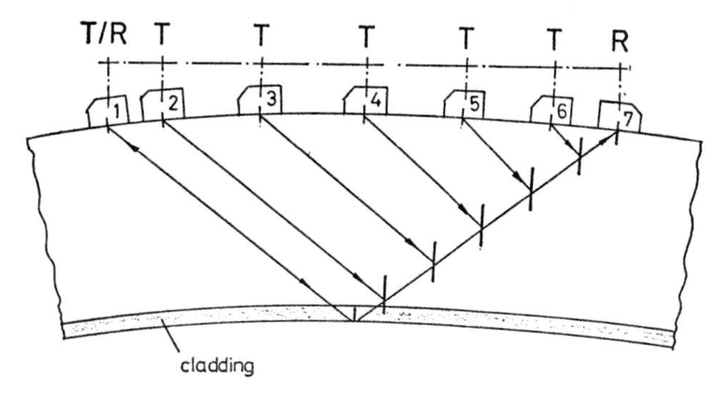

Fig. D6. Multiple probe system ('harp') with transfer control

The task of in-service inspection is to detect any defect that may arise in the course of reactor operation and to provide information on location and size. Only quantitative information will enable our understanding of fracture mechanics to be applied or, in other words, the growth of a defect to be compared from test to test. The orientation and size of a defect with respect to ultrasonic wavelength is an important consideration. Although, in view of the more or less rough surface of a crack arising in service, a certain amount of reflection is conceivable in all directions, the most accurate flaw size evaluation will be possible if the sound beam is applied perpendicular to a flat defect or a flat defect lying in mirror position for a tandem probe configuration. If a reflector is sloped to this ideal position ($\Delta\delta$ in Fig. D7), the actual size relations will be distorted. A large reflector with ideal mirroring surfaces may escape discovery if the slope is only a few degrees, whereas smaller reflectors, while producing indications only with higher gain levels, will be amenable to detection also if the slope is greater, due to their reflection properties having a more spherical characteristic (Fig. D7). Since cracks in the pressure vessel walls due to service stresses are unlikely to be oriented other than more or less perpendicular to its surface, the tandem technique appears to be preferable, in view of conceivable separations in the walls under these conditions.

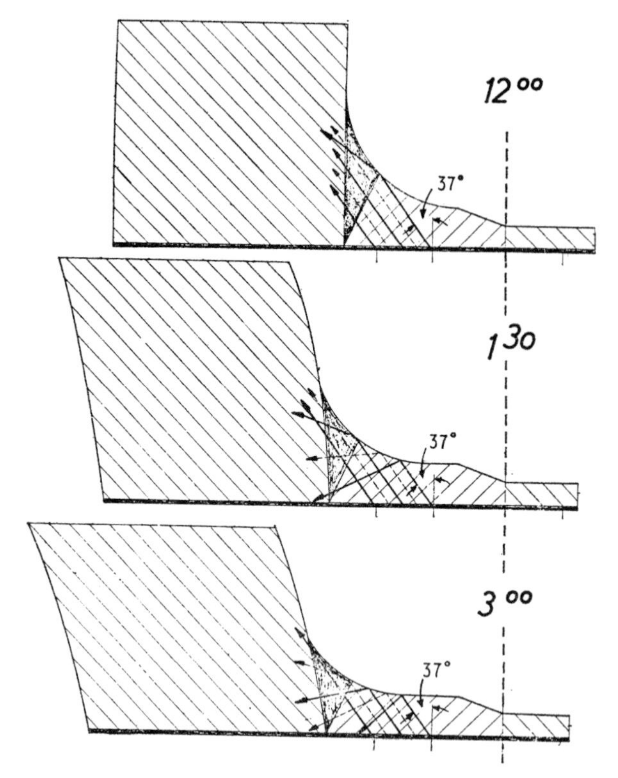

Fig. D8. Variation of sound beam geometry on weld seam built up supports

In examining nozzle welds, conical connections of flange rings to cylinders and the edges of nozzle openings, an understanding and evaluation of the sound beam geometry are important factors.

An example is shown in Fig. D8. Due to the saddle curve of the intersection between the pressure vessel cylinder and the nozzle cylinder, the sound beam pattern

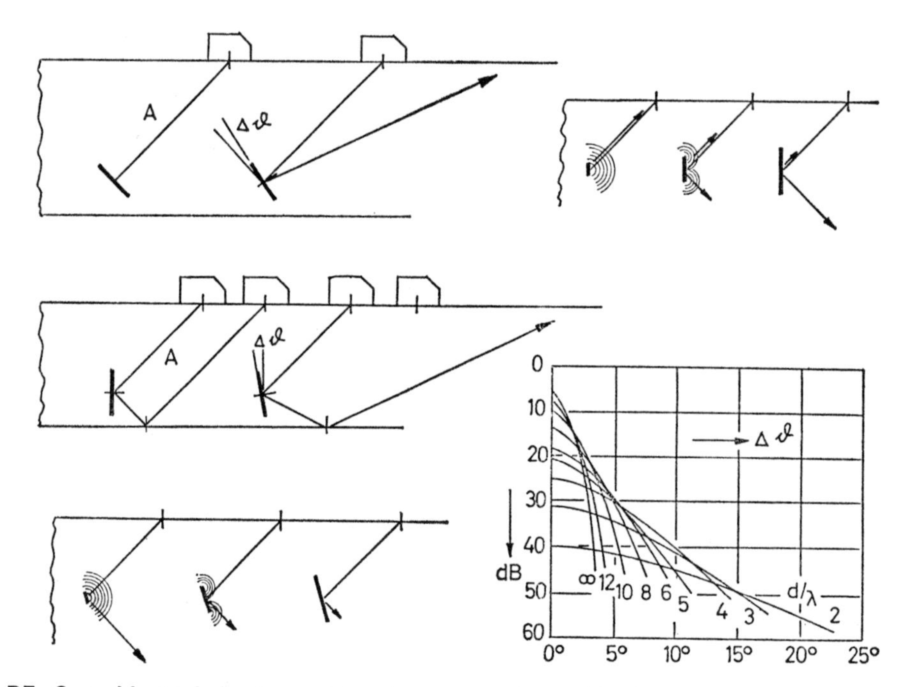

Fig. D7. Sound beam behaviour with reference to size and inclination of a plane reflector

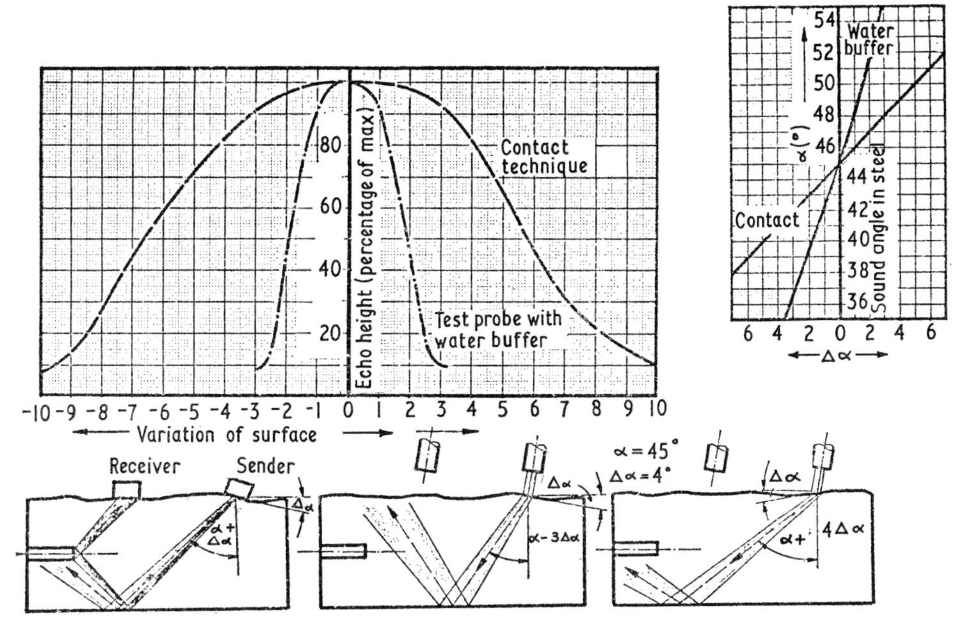

Fig. D9. Sound angle variation of wave surface

is different for every clock position within each peripheral quadrant. Small changes of the contour of the outer fillet or local ground spots are liable to create conditions that interfere with the locating of any reflections or make this impossible. The results of tests at such points therefore have to be very carefully analysed, possibly by the use of computers or auxiliary devices (perspex models with light beams).

As initially mentioned in connection with the influence of the cladding, the structure of the surface to which the probes are applied plays an important role, and considerably influences quantitative determination of flaw size and flaw position. In this context, the question is frequently raised as to whether it would not be preferable in view of the difficulties presented by wavy surfaces to adopt immersion techniques with an adequate water depth interposed between the probe and the surface instead of contact techniques (with the probe positively guided by the surface to be examined).

Fig. D9 illustrates the conditions obtaining with the contact and immersion techniques. Due to the different refraction conditions between perspex and steel with probe to metal contact and probe–water–steel coupling in the immersion method, the change of beam angle in the steel is roughly twice that in the contact technique with the same angle change at the surface. Conversely, this physical interrelationship means that a target exposed to maximum beam energy will cause less variation of indication if the surface angle changes in the contact technique than the same target would with the same surface angle and the immersion technique.

Another problem which appears to have been ignored in the planning and performance of in-service inspections on reactor pressure vessels is that of beam angle changes due to the elevated temperature of the pressure vessel and the coolant. An analysis of the conditions arising due to the change of sonic velocity in water and perspex, respectively, yielded the results reproduced in Fig. D10. Using the immersion technique with 45° probe angle, an increase of as little as from 20°C to 50°C was found to cause a change in beam angle of 2·8°, whereas with a 70° probe angle the beam angle change in steel was 6·2°. The distortion found with the direct contact method using a 45° probe angle was 1·3° and with a 70° probe angle the distortion was 3·7°. These results, too, advocate the use of the direct contact method whenever quantitative evaluation of flaw size and location is intended.

The aspects I have touched upon are typical examples of the problems arising in applying ultrasonic techniques to the in-service inspection of reactor pressure vessels.

In spite of all existing problems, ultrasonics have continued to represent the only volumetric inspection method capable of giving the type of information needed to evaluate the condition of a pressure vessel. Its use calls for a full appreciation of the existing difficulties and a systematic effort to minimize the factors that tend to distort the quantitative results.

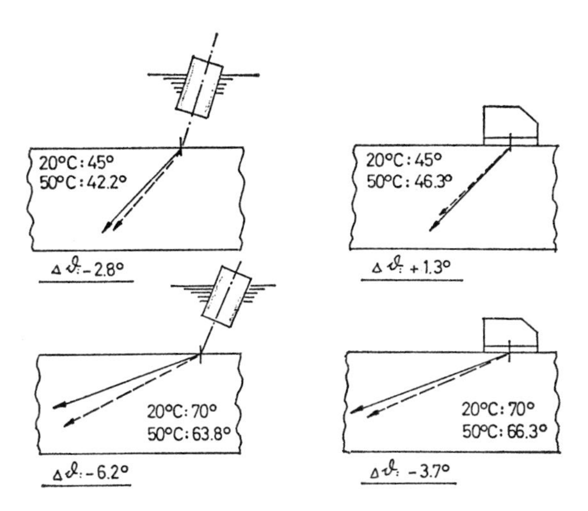

Fig. D10. Influence of inspection temperature on beam angle

D. H. Njo Würenlingen, Switzerland

There exist differences of opinion among specialists in the field of ultrasonic testing regarding the applicability of this method for the detection of defects in austenitic materials. I put this matter forward for discussion during the conference, and an interesting discussion took place. Several speakers during this discussion confirmed the differences in opinion, but were rather unclear and inconclusive.

Since we know that austenitic materials have been used, are used and will be used in the foreseeable future, as far as I can judge, in the primary systems of nuclear power plants. I think it is very important and high time to know where we stand today regarding the inspectability of austenitic materials and to know what further research is needed in this field. Furthermore, since cracks have been reported to occur in austenitic materials used in the primary systems of nuclear power plants, I do not think we can afford to 'forget' austenitic materials during the periodic in-service inspections. Therefore I would like to put a general question concerning this subject to the authors of several papers who have done in-service inspections and no doubt are well versed in the field of ultrasonic testing, e.g. B. Watkins, H. Jackson, S. H. Bush, Y. Ando, A. Tietze *et al.*, H. W. Keller, R. D. Wylie, C. E. Lautzenheiser, etc. and to other persons who are working in this field, and ask for their answers and/or comments.

The question is: is it scientifically and technically possible (and proven) to use ultrasonic testing to inspect austenitic materials (with reference to ferritic materials) conclusively?

If the answer is positive, how does one achieve this (what special technique is required, what kind and size of defects can one find, etc.)?

If it is a conditional answer, under what circumstances would one get a positive reply or a negative reply and why?

If the answer is negative, what are the scientific and/or technical reasons for the negative reply?

A. A. Pollock Cambridge

The speakers in session 5 have indicated that there are two ways in which acoustic emission can be applied to the long-term surveillance of structures. First, one can attempt to monitor a structure continuously during its working life, and to detect the growth of flaws as it takes place in service. Second, and alternatively, one can monitor during periodic proof tests to gain information about flaw growth that has taken place during the preceding period in service.

The second approach, monitoring during periodic proof testing, is one which we are currently assessing in two quite separate applications. The basis of this approach (due to Dunegan) is the following:

Imagine a structure such as a pressure vessel subject to successive proof tests at pressure P_P (Fig. D11). Suppose that the emission during the first test of the structure follows the curve OA. On a second test, due to the Kaiser effect, much less emission will be detected (curve OB). Now let the structure be put under working conditions for an extended period at a pressure P_W, and then retested at P_P. If flaws have grown during the working period, then the stress intensity in the pressure range P_W–P_P will be greater than they have previously encountered and emission activity should be renewed (curve OC). Thus emission at these levels should indicate that flaws have grown and source location devices should enable them to be found. We are working to see whether this concept can be successfully applied to practical structures.

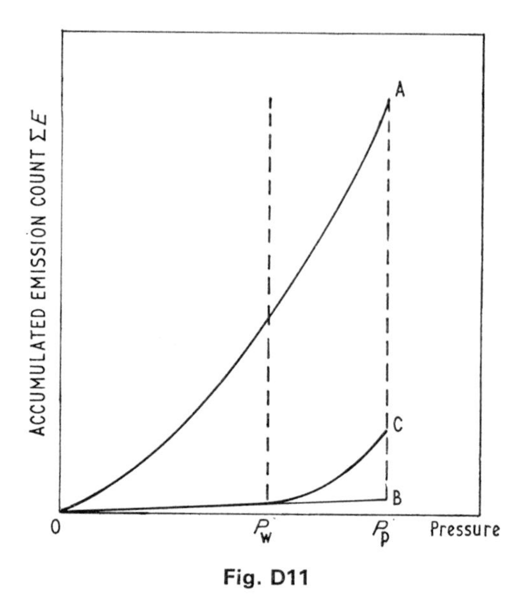

Fig. D11

Base line data for these measurements should ideally be obtained during the initial proof test of the structure. I am afraid I do not yet have such data for a pressure vessel, but I would like to show some of the information we collected during the proof testing of a military bridge at the Military Vehicles and Engineering Establishment, backed by laboratory experiments. The proof load was applied 11 times and we had 7 transducers on the bridge. Fig. D12 shows cumulative emission as a function of load, during the first load application. The load was applied in 10 ton increments, and the arrows on the graph indicate the points at which loading ceased. Note that at high loads, emission continued for a period after load application ceased; the relative amount of this 'hold emission' increased as the test proceeded. This behaviour suggested that the structure was getting near to its working limit and other measurements tended to confirm this. During these tests we also gained information on amplitude distributions and source locations.

The base line data provides a reference point for subsequent trials of the bridge. We are concerned with the reliability and reproducibility of emission data and our approach to the reliability problem is based on pulse injection/transmission measurements which we carry out concurrently with emission testing. The purpose of these measurements is twofold; to measure the transmission properties of the structure, and to check transducer mounting and instrumentation performance.

Assuming that spurious signals are not a source of error, the reliability of acoustic emission measurements is as good as the reproducibility of pulse transmission measurements.

Time is often a limiting factor in field applications, and we are developing standard operating procedures so that these calibration measurements can be made efficiently in the time available. We have recently been through our first exercise in operating our equipment from a mobile

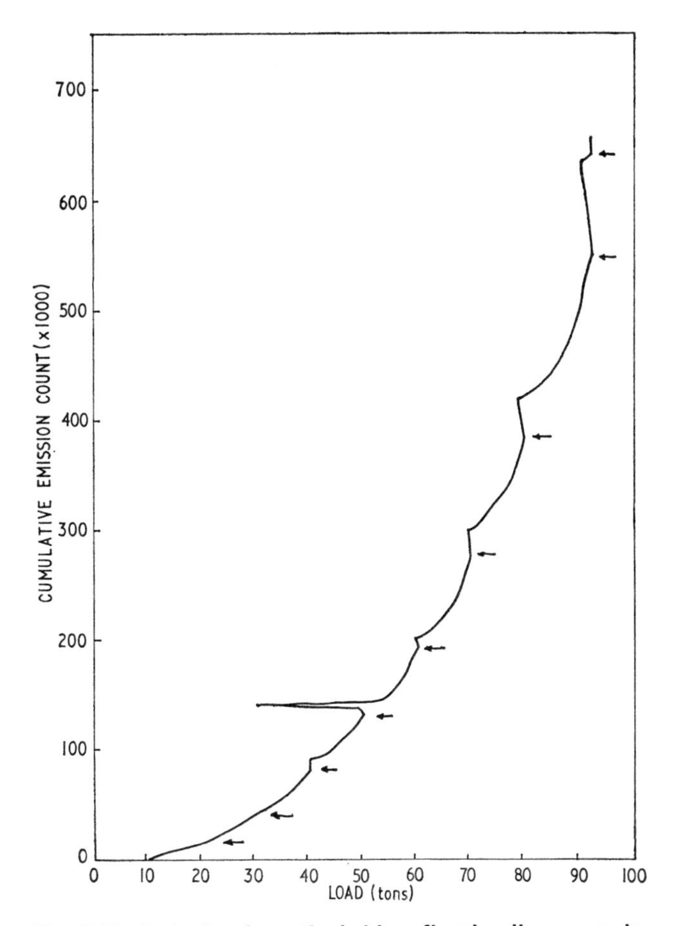

Fig. D12. Emission from the bridge, first loading, cumulative count (× 1000) versus load, gain 89·5 dB. Arrows mark the points where the gantry was stopped

van; this greatly reduces the time needed to set up and take down the system.

I would like to ask the speakers a question about source location. It has been mentioned in one of the written papers that waves of several types, compression, shear, surface and others, are present in the emission signal and travel with different velocities. The velocities of the various types are typically spread over a 2:1 ratio. I would ask, which components are used for source location, how are the different components separated, and how do we know, when performing source location computations on real data, that we are using the appropriate velocity?

The material on military bridging is subject to Crown copyright.

J. C. Quinn Barberton, Ohio

Paper C48 presents methods of calculating the probability of the failure of nuclear pressure vessels as a result of inadequacies related to design, material, construction, leak testing and non-destructive examination. The statistics used in the calculations have been compiled almost entirely from experience related to 'Class 1 Conventional Vessels' since, as stated, extensive operating experience is not yet available in the nuclear power field.

We do not believe that it is logical to predict the rate of failure of nuclear vessels based entirely upon experience with conventional vessels, particularly in view of the fact that the design, materials, construction, testing, and non-destructive examination of nuclear vessels are governed by rules more stringent than the rules governing conventional vessels. These more stringent rules, established by the American Society of Mechanical Engineers, the United States Atomic Energy Commission and British agencies, enhance the safety of nuclear vessels. Certainly, the application of nuclear codes with significantly greater controls should ensure fewer failures of nuclear vessels than conventional vessels designed and fabricated to non-nuclear codes. In particular, there is emphasis in the nuclear field on the prevention of catastrophic failures postulated by the authors and identified by them as Categories A and B. These are the failures which could result in the penetration of the reactor vessel containment by fragmentation of the reactor in a gross way or small amount respectively.

We believe that the only failure of nuclear vessels that warrants serious consideration is 'Category C'. The authors have defined these as failures resulting in limited leakage, but not likely to compromise the containment structure. Thus, the elimination of two of the three types of failures described by the authors significantly reduces the probability of failure of nuclear vessels.

One of the authors' conclusions is that the requirements of safety are best satisfied by an acoustic emission device which continuously monitors the vessel with respect to structural integrity. This device would provide data to follow the progress of any flaw that is found to be growing in size, thereby permitting vessels to be removed from service before rupture or leakage occurs. It is agreed that such an instrument, if developed for use, would provide considerable assurance of safe operation to the power plant personnel. However, any failure causing a power plant outage, even when found in time to prevent compromising safety, is very costly to owners. Therefore, with or without the acoustic emission equipment, every effort must be made to prevent any failure by continued control of design, materials, construction, testing, non-destructive examination, and in-service inspection.

W. Rath Cologne, W. Germany

My questions regarding the contribution by B. Watkins and H. Jackson 'Technique for inspection of light water reactor pressure vessels' (paper C27, session 4) are as follows.

How does one overcome the cladding influence with respect to change of inspection sensitivity? We have found sensitivity variations at about 20 dB with sound angles of 45° and 2 MHz transverse waves.

How is one able to find cracks in the austenitic cladding and cracks starting into the base material at the transient zone between the cladding and the base material?

How does one take into account the phenomena that the farther the reflector is away from the crystal, the longer it is stroked by a sound pulse while moving the inspection assembly over the specimen, which means that an equal sized reflector seems to be larger on a *B* scan recording when it is far away from the crystal.

Could the authors give details on data comparison of test results received from inspections which were done at different times, in order to recognize the growth of a reflector?

Summing up

R. W. Nichols Fellow

It is impossible in the space available to do justice to the many valuable papers and contributions made to this enthusiastically-attended conference. Every paper had something of value to say, but to refer to each by name would convert this review into a catalogue. What follows then is not a summary; rather a number of personal jottings on aspects which may prompt further thought or future work.

Session 1 drew attention to the legal requirements that often provide a background to what in-service inspection is done. In particular, referring to the position in the U.K., J. T. Toogood said that there was an absolute requirement on certain operators to provide efficient maintenance, which he interpreted as requiring inspection; his paper indicated the U.K. requirements for certain types of nuclear vessels. While at present the period of inspection, where such was required, was fixed, he hoped to see more flexibility in this aspect in the future, a viewpoint echoed by several contributors to the discussion. The subjective nature of decisions on this aspect of inspection period was indicated by the marked variation between the requirements in different countries, as summarized by J. J. Whenray. An important contribution from S. H. Bush gave some of the background to the A.S.M.E. Section XI Code which plays such a central role in the periodic inspection of steel pressure vessels for light-water moderated, nuclear reactors. He said that the issue of extensions to this Code to cover Class 2 items should be made in Autumn 1972, and there would be other modifications later in the year. Working groups had started preparing draft requirements for inspection of prestressed concrete nuclear pressure vessels and on sodium cooled fast breeder reactor components. He described an approach to 'inspection acceptance requirements' which could eventually form part of the A.S.M.E. Section XI Code. The paper by R. O'Neil and G. M. Jordan was novel and constructive in its approach to providing new methods of assessing failure hazard; the vigorous discussion showed it had proved intellectually stimulating to many of the audience. Discussion suggested the need for international recording and analysis of data on defects discovered both in service and in manufacture. R. W. Nichols pointed out that the approach could be extended by use of fracture mechanics to convert the problem into one of defining the probability of different sizes of defect arising, a more readily obtainable statistic than that of number of failures. There were also vigorous exchanges on the value of the over pressure test, and on the need for setting different acceptance standards in manufacturing stages (based on quality control requirements) to those applying to the in-service inspection (based on fitness for purpose).

Session 2 described the wealth of experience arising from repeated inspections of pressurized components in a wide variety of industries, including boilers and steam drums in the power industries, and marine applications, pressure vessels of various types in chemical and petrochemical plant, transportable pressure vessels and gas containers. In some cases in-service inspection had alerted the authorities to the evasion of safety rules; in others to the difficulty of predicting operating conditions and corrosion behaviour. Of the 239 reported boiler explosions in the period 1964–68, 162 were due to overheating, shortage of water, overpressure or water hammer and 73 to corrosion or erosion failure at normal working pressure, the remaining four being failure of a man-hole door or joint. Results obtained by the Allied Offices Technical Committee over years of statutory and insurance inspection emphasized the important protection given by in-service inspection in such cases. These aspects, and the economic gains that can be achieved by in-service inspection were also stressed in a paper by A. E. Lines and C. J. Holman on naval practice. J. E. Macadam gave valuable details of the organization and recording of such inspections, based on oil industry experience, while E. H. Frank covered similar ground for the chemical industry. W. Dean described the requirements for transportable containers, and prompted discussion on the value of the 'stretch' test, perhaps unique to this type of vessel. Other discussion stressed the importance of recording present defects to guide future practice.

With Session 3, the Conference turned to the in-service inspection of pressurized components used in nuclear power production. The paper by J. M. Carson and F. Turner described techniques applied to U.K. gas cooled and steam generating reactors, techniques somewhat similar to those described in Session 2. Photographs taken inside the Calder Hall vessels after 15 years' operation showed that the condition of the inner surfaces was excellent. As well as visual and ultrasonic inspection, strain gauges and dimensional checks can provide valuable information for in-service inspection. Monitoring for changes in material properties by means of special specimens and similar activities to measure the extent of corrosion, were also important aspects that should be considered as part of the repeat inspection assessment. Y. Ando, in a comprehensive paper covering Japanese experience, stressed similar points, and described an ingenious electrical resistance probe which had been found useful for in-service inspection where cracks were known to be present near the plate surface. The Japanese work also stressed the importance of taking account of the real reactor environment when making simulative tests, a point supported in discussion. Experience in Germany and the U.S.A. in the inspection of steel vessels for light water reactors was outlined in other papers to this session, these papers giving valuable information on operational aspects;

showing irradiation doses for inspection staff to be small. However, the German speakers considered that such inspections had been found to be less effective than those for conventional pressure vessels and steam boilers, stressing the need for the over-pressure test as a repeat inspection tool, and the importance of an ultrasonic 'fingerprint' using comparable techniques. These points were elaborated in an important discussion where future requirements by the German surveying body T.U.V. were outlined. This put a requirement that the vessel design and manufacture should be such as to allow inspection; the cladding technique could be important in this aspect. They considered that the ultrasonic techniques used should include searches at different angles to cover the possibility of defects of different orientations. These comments led to an important discussion contribution from H. J. Meyer covering sources of error in sizing of defects by ultrasonics and the effect of cladding. The latter, and work described by French workers, indicated some of the increased difficulties of applying ultrasonic techniques to austenitic steel components. Another important contribution was from the C.E.G.B. describing repeat inspection techniques applied to prestressed concrete pressure vessels at Hinkley Point.

The discussion on remote inspection of steel pressure vessels for water cooled reactors was taken at greater length in Session 4, based upon several papers from organizations that had developed special skills and facilities for this purpose. Amongst the large number of ingenious pieces of equipment described, perhaps the most novel features mentioned were the permanent fittings for external inspection of an Argentinian plant; a multi-probe, multi-angle search tool; an adaptable device for locating the inspection equipment at any chosen position inside the vessel; devices for inspecting nozzles; a variable angle scanning probe which presented its results as a cross-section view of the component under inspection (B-scan). In discussion it was mentioned that one problem in the design of such equipment was that reactor components as built were not always according to the drawing; this led to the requirement of adaptability of the devices. The heavy costs of prolonged down-time made it important to have reliable equipment, planned operation and easy interpretation of examination results. The importance of calibration and reproducibility at intervals of many years was emphasized, discussion again touching on the effect of component, weld and defect geometries, position in depth, influence of cladding and of possible cracks in the cladding and just below it. It was pointed out that whilst careful selection of techniques permitted many austenitic steels to be inspected ultrasonically, there were still cases which were virtually impossible. In such cases radiography could be appropriate, and had been done effectively even under conditions of considerable residual radioactivity. With regard to selection of techniques and of inspection schedules, and particularly with regard to the interpretation of results, the conference highlighted the importance of putting such inspections into a framework based on fracture mechanics/material properties/stress level.

The discussion on potential future developments in Session 5 again centred rather on the nuclear pressure vessel application, although several of the techniques could have nore general application. C. E. Lautzenheiser described a massive U.S.A. co-operative programme from which several of the later papers arose. A paper on residual stress determination stressed the point made in earlier sessions that in-service inspection was not limited to the aspect of defect detection, even though the emphasis in the discussion tended to concentrate on that aspect. The use of holographic techniques applied to ultrasonic examination was promising, giving an ability to record on one film the whole information covering a variety of depths in the plate. However, most discussion centred on the three papers dealing with the application of the acoustic emission technique as a repeat inspection tool. These papers indicated that whilst its use as a permanent monitor or as a failure warning device was not yet possible, its development had reached a practical stage for the single time overall survey during a repeat pressure test. This would overcome many of the objections based on the difficulties of ensuring a fully comprehensive inspection of operating vessels. E. J. Burton *et al.* showed how such tests could make use of sensors fitted to existing permanent attachments as well as the vessel itself, and in this and other papers the aspect of identifying defect signals in the presence of other noise was discussed in detail, showing in most cases such problems could be overcome, at least under pressure-test conditions. Much further work was needed before this effective search tool could be used to define defect size and shape, before one understood the behaviour of different materials or before it could be used as a continuous failure monitor. Nevertheless the discussion indicated that acoustic emission was a very promising technique with important applications possible even now.

The final session centred on other development aspects—automated data handling systems, the effect of in-service inspection on pressure vessel design, and the use of the over-pressure test as an in-service inspection tool. Continued discussion on this last point indicated widely differing views on the purpose and effectiveness of the over-pressure test. The view was expressed that, unless such a test was supported by acoustic emission measurements, in many cases it added little that could be quantified to over confidence.

This and several other aspects were obvious points to which discussion will return. Indeed the vigorous discussion and continued full attendance throughout the conference showed the widespread interest whilst the content of the papers and discussion showed that there was rapid development in many of the areas covered. Indeed, there appeared to be a case for a further conference on a similar topic in, say, two years' time. This very successful conclusion is of course the result of much hard work by the organizing committee, authors, contributors, chairmen and Institution staff.

May I close with a short parody indicating the qualities needed by anyone aiming to carry out such in-service inspections.

(With apologies to Rudyard Kipling)

If you can make your test when all about you
There's lots of 'r' but very little room,
If you have built your masts and probes and boxes
And TV scopes which pan and tilt and zoom,
If you have fixed your waveguides and your sensors
To listen to each creak and groan and thud,
If you can do all this and keep your senses
To tell their meaning, you can join the club!

Authors' Replies

E. J. Burton

A. A. Pollock has shown an interesting application of acoustic emission to the proof testing of a bridge, and one would expect this to be typical of the kind of applications that can be made in general engineering practice. In reply to his questions, I would say that transverse wave propagation is used as a basis of source location. The smaller contribution that arises from longitudinal waves can be rejected by suitable electronic processing. This approach has been confirmed by a number of tests of vessels and other structures. This has included the use of artificial sources as well as true acoustic emission, and it is found that location can be achieved within a few inches. Location is more complicated by vessel penetrations and by weld lines, and these in our experience are more likely to give rise to errors than to any from uncertainty in the wave velocity.

D. Birchon's experience would emphasize the value of source location for interpretation of acoustic emission data. The location capability is a powerful aid in distinguishing spurious signals from water leaks.

S. H. Bush

Several modifications to the current A.S.M.E. Section XI Code are being considered by the code committee. These include a new section dealing with the operational testing of pumps and valves. This section is nearly complete. A new section, covering the in-service inspection of components defined in A.S.M.E. Section III as Class 2, has been completed and awaits final approval by the various code committees. A section covering the in-service inspection of components defined in A.S.M.E. Section III as Class 3 is in draft form.

A working group has been established to develop in-service inspection codes for (a) gas-cooled reactors, prestressed concrete reactor vessels, etc.; (b) concrete containment structures. A working group has been established to develop an in-service inspection code for liquid metal fast breeder reactors.

The Section XI sub-group is working on four major modifications to the present In-service Inspection Code. These are the development of a mandatory appendix covering the ultrasonic examination of Class 1 components and the development of a separate set of acceptance standards for the evaluation of indications and/or defects. A basic assumption in this approach is that the system has been code stamped as meeting Section III requirements, so that these standards control for Section XI inspections. An example of possible standards is given in Fig. D14. Please do not consider that any of the values represent a definite opinion of the code committee; the numbers are just an attempt at a suggested position. The development of a non-mandatory appendix covering the analysis of defects by linear elastic fracture mechanics (l.e.f.m.) to establish their acceptability, need for more rigorous inspection, or need for repair. Fig. D13, abstracted from reference (4), represents a graphical presentation of the three possible positions. Again, I wish to emphasize that the code committee has taken no action as to acceptable values based on l.e.f.m. Consideration as to the feasibility of developing modified rules for repair of defects in an irradiated vessel utilizing a technique such as weld bead self annealing rather than a mandatory stress relief. Whether such an approach is possible depends on the quality and quantity of data available.

The Section XI Committee has been reviewing and modifying the conditions controlling the overpressure test required by the code at least once during every inspection period (10 years). The latest modification considers the stresses throughout the entire primary system. These suggest limiting the overpressure to about $1{\cdot}1P_o$, where P_o is operating pressure. The committee intends to review the results of acoustic emission tests on the Experimental Beryllium Oxide Reactor (E.B.O.R.) vessel to determine the value of acoustic emission to complement the overpressure test.

REFERENCE

(4) CLARK, W. G., jun. 'Fracture mechanics and nondestructive testing of brittle materials', A.S.M.E. Paper 71-PVP-4.

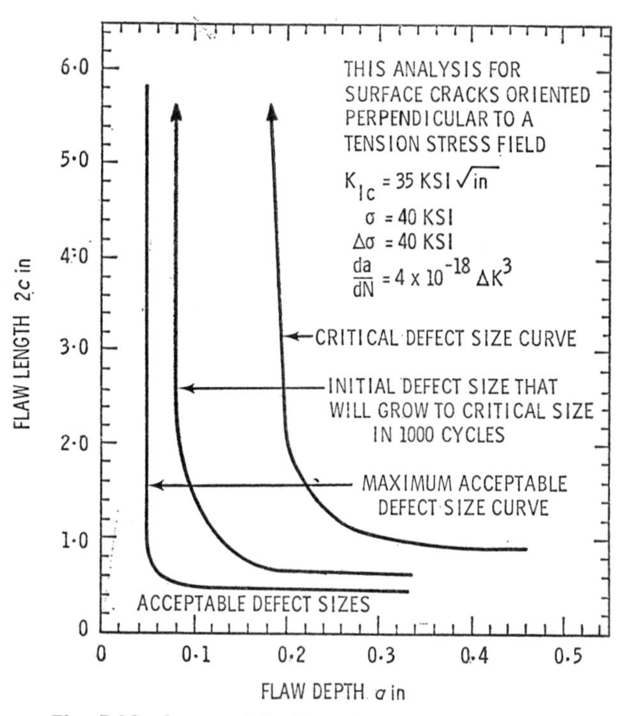

Fig. D13. Acceptable flaw size and shape curve for the example problem

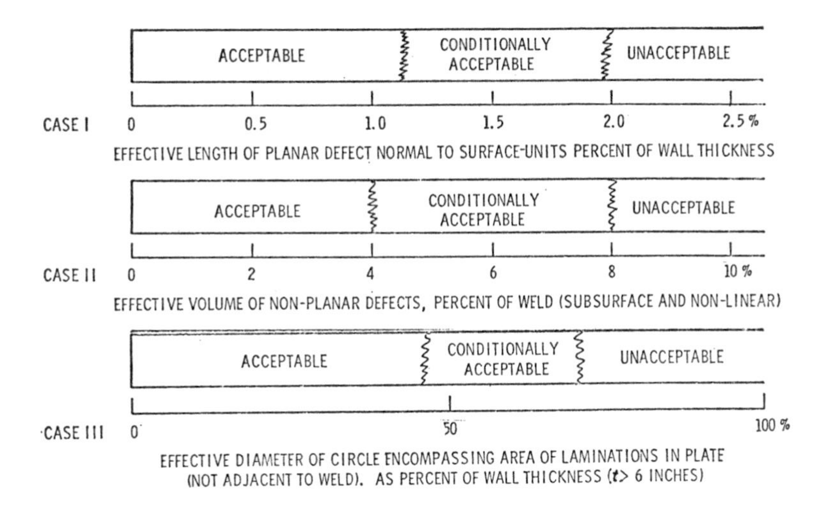

Fig. D14. Spectrum of types and sizes of defects considered acceptable, conditionally acceptable based on l.e.f.m. analysis, and unacceptable. Size based on assumption that U.T. procedure is rigidly controlled so that magnification factors are known and indication corrected to true size

G. J. Dau and W. P. Jolly

H. Baschek's request for comments on critical flaw size detection sensitivity gets to the heart of the matter. However, discussions on equipment sensitivities for minimum sized flaws are not very meaningful until some agreement is reached on what constitutes a critically sized flaw. Thus we cannot add anything to the discussion at this time.

In reply to A. A. Pollock, the Rayleigh or surface wave component is used for source location because the compression wave component is rapidly dispersed, loses energy due to mode conversion and does not refract efficiently at large angles of incidence; the shear wave component suffers from the same losses. The surface wave component is usually established within one wall thickness or less, depending upon the depth of the source, and propagates at the Rayleigh group velocity thereafter. No separation of modes is necessary. The exact velocity of the detected emission may be computed if a minimum of four receivers are used. Triangulation using two sets of three receivers will produce coincident source locations only for the correct velocity.

E. H. Frank

In reply to K. P. Q. Appleton, the frequency of inspection of steam and air vessels was established in the last century and has remained substantially unaltered since. The periods of 14 and 26 months have, by and large, been proven to be adequate, and are applied to equipment containing steam or air irrespective of the type of works on which it is installed and operated and where, in some cases, technical and practical engineering surveillance may not be available. It seems that the inspection interval is set to ensure that even the poorer operator with little or no technical support is largely protected and policed by compulsory examination of his steam and air equipment by competent people from outside his operation. Much of this equipment is used to provide a service, for example in garages, laundries, mills, general engineering and textile works where the main business is the manufacture of something else.

Steam and air vessels form only a small percentage of the vessels in a chemical plant, perhaps 10 per cent, but are part of an operation which by its nature has the other 90 per cent of the equipment exposed to conditions which may be equally or more severe, with the complication of chemical attack, than those in the steam and air equipment. It is axiomatic, therefore, that in chemical plant operation the operations are closely controlled, continuously monitored and take place under the surveillance of suitable technically and practically qualified staff. With this level of competence built into the operating and maintenance teams it would not be sensible to use the frequency defined by statute for steam and air vessels as a yardstick for the chemical plant. Unlike steam and air, where the physical and corrosive properties are well known and within reason easily controlled, chemical plant may operate under very aggressive and variable conditions. It is not always possible to exercise complete control over these nor to select materials which are fully resistant. Under these circumstances more frequent inspection than is defined for statutory vessels may be prudent and does in fact take place. Conversely, for example, on pure product distillation trains, where air is excluded from the process and the products are known to be non-corrosive, satisfactory service may be confidently predicted and inspections at intervals considerably longer than those prescribed for statutory vessels safely established.

It is to be hoped that the discretion at present allowed the operators of non-statutory equipment in the chemical industry to set the frequency of examination of their equipment in the light of plant history and experienced judgement can be retained. If a maximum statutory interval were to be prescribed, this would have to be related to the vessels with greatest need and would result in a lot of unnecessary work on many vessels known to be perfectly satisfactory, as well as giving rise to additional plant outage to carry out the inspections.

H. W. Keller, D. C. Burns and **T. R. Murray**

In reply to H. Baschek, the size of the defect that must be detected by the in-service inspection equipment developed by Westinghouse is set in accordance with the requirements of the A.S.M.E. Code Section XI. The critical flaw will obviously vary, depending on the location in the vessel, as well as on the flaw size, shape, orientation, and the vessel material, operating history and type loading conditions. Although the analytical tools to perform a rigorous analysis are not fully developed, a conservative analysis indicates that the critical flaw size exceeds the required sensitivity for the reactor vessels the Westinghouse tool was designed to inspect.

In reply to D. H. Njo, Westinghouse has been able to inspect austenitic materials as demonstrated by both laboratory and field measurements. We achieved this by use of optimum transducer size and angles with special attention to surface condition. The size defect that can be detected complies with the requirements of the A.S.M.E. Code Section XI.

In general, it has been found that forged austenitic materials are almost always inspectable, and cast austenitic materials, including welds, are usually inspectable, but this must be determined on a case-to-case basis.

Referring to R. Haas, in a recent reactor pressure vessel case in the United States, a mid-plane 18 mm through wall flaw could not be detected in all cases by the tandem technique, but could be detected in all cases by single probe pulse echo techniques. The problem with the tandem technique was that the reflection had to pass through the cladding on the i.d. of the reactor and the sound was either totally dispersed or directed so that it was not received by the receiving transducer. In this case the crack was extremely tight and could not be detected by radiography. In our opinion, this crack (and other similar cracks) demonstrates that the pulse echo technique is more definitive and reliable than a tandem technique in which the sound must pass through the cladding.

C. E. Lautzenheiser

H. Baschek's comments relating to critical flaw sizes should really be sent to a fracture mechanics expert. I can comment that a finite element fracture mechanics analysis was made on a defect in a nozzle to shell weld of an American reactor pressure vessel. The analysis was for a 30 mm through wall flaw located in the central portion of an 180 mm thick vessel wall. This flaw was well below critical flaw size for this location. For the same situation, an analysis was made for a 30 mm through wall flaw located at the vessel i.d.; this analysis determined that the flaw was slightly below critical size. Flaws of this magnitude can be readily detected if they are at or near a vessel surface. They are not so readily detected if they are in the central portion of the plate, whether one uses standard pulse echo techniques or the tandem technique. The flaw in question is a crack in the central portion of the vessel (the actual crack size is 18 mm through wall). The amplitude of this flaw bore no relation to the actual flaw dimension and, in addition, certain areas of the flaw could not be detected by the tandem technique. This is probably due to the fact that the sound must pass through the cladding when using the tandem technique, and in many areas, transmission from the clad interface or through the cladding is not possible.

Based on our experience, I would estimate that a flaw 5 mm deep by 10 mm long located on the vessel i.d. can be reliably detected when inspection is from the o.d. I have no factual data, but I would estimate that the flaw would have to be at least twice this size on the o.d. surface to be readily detected when inspection is made from the i.d. (clad surface).

In reply to D. H. Njo, it is scientifically and technically possible (and proven) to use ultrasonic testing methods to inspect austenitic materials. However, this cannot conclusively be stated to apply to all austenitic materials. We at Southwest Research Institute use attenuation measurements, and pitch-and-catch methods across the weld, to prove that we are getting sound transmission and to adjust the sensitivity level of the inspection to agree with that on the standard. There have also been some improvements in techniques, such as using a special narrow band ultrasonic pulser, special split transducers, and refracted longitudinal waves to improve inspection of austenitic materials.

R. O'Neil and **G. M. Jordan**

Arising as they do from a wide experience in the design of quality products, P. H. Jones' views have been studied by us with great interest.

If one accepts a definition of statistics as that of a mathematical treatment related to the collection and processing of *directly applicable* numerical data, then statistics has little place in our paper since its main ingredient—directly applicable data—does not exist. Statistics are not created by edict and it was certainly not our intention to 'construct a *statistical* framework for safety and reliability assessment'. Such an exercise would not only be meaningless, but would involve the dangers that P. H. Jones so rightly emphasized.

In the face of such difficulties, it is tempting to a safety authority to merely lay down the required safety objectives, not concerning itself with the means by which compliance with these objectives is to be demonstrated. Such an approach is sterile. We have attempted to generate that dialogue between designer and safety authority which is considered to be essential to effective safety justification and assessment.

In order to have such a dialogue, a 'language' is required. For large volume products (e.g. electronic components) statistics is the language and objective interpretation is possible. This is not possible for nuclear pressure vessels, and a considerable degree of engineering judgement is required. Despite this limitation, it is our view that this judgement should be expressed in *quantitative* terms. 'Quantified judgement' is the language. Wherever possible relevant statistical evidence should be used to guide and test this judgement and this is the approach used in the paper.

We associate ourselves with the view of Freudenthal who stated (**5**) that 'With respect to the use of probabilities, the frequently voiced objections against the operation with probability figures that are beyond the range of statistical observations are irrelevant. Reliability analysis of large structures or complex systems that can never be tested in sufficient numbers to provide an acceptable statistical formulation can only be based on probabilistic physical arguments, not on statistical theory'.

The quantified judgements of paper C48 represent our attempts to provide the 'probabilistic physical arguments' of Freudenthal.

With regard to the assessed gap of 10^{-1} to 10^{-2} between demonstrable achievement and objective, we contend that this is the most appropriate safety conclusion from our survey. The qualification of *demonstrable* achievement is important, and while it is obvious that the uncertainty in the underlying assumptions may be greater than the gap, we have adopted the usual safety practice of a 'lower bound' estimate of demonstrable achievement. As stated in paper C48, the true situation may well be better but, in the absence of justification of this, the present interpretation is believed to be the only one permissible for safety.

As defined in paper C48, the Farmer criterion was the basis for evaluation. Discussion of the many considerations leading to this criterion was considered to be outside the scope of the paper, but attention is drawn to Appendix 48.1.

The precise Farmer risk/consequence relationship (Fig. 48.1) has not been adopted internationally. However, we cannot agree that such acceptance is a prerequisite to application of a specific safety criterion. Indeed, if such were the case, it is doubtful whether any single criterion would ever become applicable. National considerations will always lead to differences in the precise functional relationship between risk and consequence. It is perhaps more meaningful to examine the adoption of a risk/consequence philosophy. There are firm indications of a growing acceptance of such an approach in nuclearly sophisticated countries.

In an interim statement of general policy and procedure for implementation of the National Environmental Policy Act 1969, the U.S. Atomic Energy Commission state (**6**): 'In the consideration of the environmental risks associated with the postulated accidents, the probabilities of their occurrence and their consequences must both be taken into account. Since it is not practicable to consider all possible accidents, the spectrum of accidents, ranging in severity from trivial to very serious, is divided into classes.

Each class can be characterized by an occurrence rate and a set of consequences.

In Japan, the Committee on Reactor Safety Evaluation and Reliability of Engineered Safeguards have developed probabilistic risk/consequence criteria for application to safety assessments (**7**).

P. H. Jones is also referred to the Proceedings of a Symposium on Safety and Siting held by B.N.E.S. (**8**). In his opening address, Sir Owen Saunders, Chairman of the Nuclear Safety Advisory Committee, advocated the pursuance of a probabilistic approach despite the difficulties that still have to be overcome.

Main pressure vessel failure is considered to be the dominating mode for major release of fission products to the atmosphere, since such failures have the potential for exceeding the capacity of all engineered safeguards. Consideration of the reliability of engineered safeguards for other pressure-containing components was stated to be outside the scope of the paper, but would, of course, form part of an overall safety appraisal.

Similarly, consideration of the details of fission product transport and leakage leading to the assessed release rates was not considered appropriate to the subject matter of paper C48. P. H. Jones is referred to the literature for this aspect.

The comparison between failure rates for Category A and B failures which puzzle P. H. Jones arises from their stated definitions. For a Category A failure, involving containment puncture by fragments, no significant attenuation of the released activity is obtained from the containment.

Thus, for such failures, the presence of the containment does not affect the 'consequence' element of the hazard. On the other hand, for Category B failures the content remains substantially intact and the resultant hold-up of fission products reduces the activity release. Thus whilst P. H. Jones' interpretation of the 'risk' argument is correct, his neglect of the 'consequence' aspect has led to his confusion.

In the very difficult task of application of conventional vessel failure data to nuclear vessels, we were cognisant of the 'quality' environment which surrounds the production of nuclear pressure vessels. However, if 'quality' is synonymous with 'fitness for purpose', then quality is present in no small measure for the power and chemical plant vessels which formed a large part of the Phillips and Warwick, C48 (**4**), and the Kellerman and Tietzle, C48 (**6**), surveys. Furthermore, there is one aspect of nuclear vessel technology which has hitherto fallen short of conventional practice—that of in-service inspection. It was in recognition of the fact that the 'quality' of nuclear vessels fails to fully compensate for this shortfall that in-service inspection codes such as A.S.M.E. XI were produced.

It is acknowledged that such observations are largely qualitative. The main difficulty faced by us was to quantify such effects, and it is regretted that P. H. Jones' opinion on the methods and results of such quantification have not been presented in his contribution. It is clear that such opinions must have been formulated, since without them, consideration of the 'engineering economics' of the expensive quality practices would not appear to be possible. One major objective of paper C48 was to evoke expert opinion in positive (and preferably quantitative) terms, in an attempt to refine judgement of such difficult aspects.

We acknowledge that for the assessment of reliability of inspection processes, a direct application of Table 48.2 data is uncertain, since such data will be biased by the relative degree of usage of such processes in conventional vessel inspection. In general, it was considered that such usage must reflect to some extent the industry's confidence in the various processes.

The source data for Table 48.2 related to 132 failure cases and the mean age at failure was 7·2 years. Thus, it was considered that the sample should be representative of time-dependent effects such as fatigue, corrosion, etc.

With regard to the failure rate of pressure testing (P_{PT}), it was considered that a case by case analysis, to evaluate the risk of a pressure test failing to reveal a pre-existing defect, was not possible with the data available. A 'lower bound' safety assumption was therefore made with regard to the frequency and efficiency of such testing. A broad engineering evaluation of pressure testing (see p. 142) tended to support the assessed high value of P_{PT}. The relative simplicity of pressure testing makes it economically attractive in the nuclear environment, and we agree with P. H. Jones that further consideration should be given to it. However, its potential danger must not be overlooked (see Formby, paper C54).

A similar approach was used in the assessment of visual

examination (P_{VE}), except that in this case 'lower bound' safety assumptions required a significant reduction in the efficacy indicated in Table 48.2 for conventional vessels.

We find P. H. Jones' hypothesis, that an independent design review may lead to lower quality, totally unacceptable. Responsibility is not divisible, and the existence of an independent assessment in no way diminishes the responsibility of those assessed. (See, for example, Section NA3351 of A.S.M.E. III.) Indeed a willingness so to be assessed is surely indicative of a responsible attitude, cf. 'second opinion' in the medical profession.

Our philosophy on this aspect can be illustrated by simple probability theory. If A assesses B (and vice versa), then the probability of a deficiency escaping both is $P_A \times P_B$, which for all real situations is less than the respective 'failure' rates P_A, P_B. In pressure vessel assessment, as in many other judgement processes, two heads are better than one.

The above considerations were used to illustrate the assessment reliability necessary to achieve a tenfold reduction in the design failure rate (P_D).

We are indebted to P. H. Jones for pointing out an error in the text, under DISCUSSION (p. 145, r.h. column, first paragraph, last line) the reference to 'overall vessel reliability' should, of course, refer to 'reliability against design deficiencies' only. The design deficiencies are, as stated, only those judged likely to lead to vessel failure.

We are greatly encouraged by the positive contribution made by C. P. Fagence. In view of the fact that some of his points coincide with those of P. H. Jones, we will avoid repetition by addressing ourselves to the additional points made—in particular those relating to the proposed revision of Table 48.3.

First, the general agreement between the assessed values is encouraging. Differences exist between values for P_D, P_C and P_{US}. Of these, the first two differences lie within the range of specialist opinion, and these perhaps reflect our philosophy of 'lower bound' estimation, compared to the 'best estimate' which may have formed the basis for C. P. Fagence's proposals. We would welcome the opportunity of further discussion of this aspect.

The failure rate for ultrasonic testing (P_{US}) shows a greater divergence of opinion, and while P_{US} (construction faults) is not a dominant factor in the probability chain, it is important to explain the reason for the divergence. This undoubtedly appears to be due to the respective definitions of significant defect size (1 cm versus up to 3 in). Since many ultrasonic techniques are 'flaw area dependent', the failure rate must be grossly affected by size assumptions. The reliability of flaw detection by normal ultrasonics is not as high as one might expect. C. P. Fagence is referred to the investigatory work reported briefly by S. H. Bush in paper C26. With regard to defect significance, our own investigations lead us to the view that small defects can have a significance to the integrity of the structure and reliable detection of these demands very high levels of ultrasonic technology. The initial 'fingerprint' type of inspection is of value in this respect.

J. C. Quinn's main point relates to the effect of different standards of design and construction between nuclear and conventional vessels. However, the fact that he, like P. H. Jones, has not quantified his judgement makes it difficult to evaluate.

J. C. Quinn's point with regard to acoustic emission is important. It is anticipated that a probabilistic assessment of availability, similar to the safety analysis presented, would also emphasize the fundamental importance of the control of design, materials, construction testing, NDT and in-service inspection.

We are in very close agreement with the proposals outlined in D. S. Lawson's contribution. Indeed, we are, in association with other organizations, pursuing all of the aspects noted, with a view to refining and extending the present assessment. These activities can best be summarized by reference to Fig. 48.2, and include acquisition and processing of operational history (load changes, start-up and shut-down, hydrotest, etc.); evaluation of present and projected methods of stress analysis, including finite element methods, in order to deduce reliability of prediction and identify potentially unreliable areas; and examination of variance of fracture mechanics predictions, including l.e.f.m., stress concentration, C.O.D. and contour integral methods, collection and processing of material property data, and evaluation of capability and reliability of inspection techniques and procedures.

REFERENCES

(5) FREUDENTHAL, A. 'Fatigue sensitivity and reliability of mechanical systems, especially aircraft structures'. S.A.E. Paper 459A, 1962 (Jan.).

(6) Amendments to Appendix D of Regulation 10 CFR 50, 1971, U.S.A.E.C. Federal Register **36** (No. 129).

(7) HATTORY and TAKEMURA. Meeting of specialists on the reliability of mechanical components and systems for nuclear reactor safety, 1969, E.N.E.A. (RISO report 214, Roskilde, Denmark).

(8) Symposium on Safety and Siting, 1969, B.N.E.S. (Inst. Civil Eng.–British Nuclear Energy Society).

G. J. Posakony

In reply to the questions raised by D. H. Njo, there is nothing characteristic about austenitic materials, as a class of materials, that precludes the use or applicability of the ultrasonic method of testing. Technically, the limit of ultrasonic wave propagation in any material is related to the boundary condition of and within the material. Fine grain austenitic materials can be and are readily tested. The boundary condition that causes the greatest concern in testing austenitic materials (or any other material) is the metallurgical grain size. Certain processes produce a product with very large grains. Large grains can disturb the normal wave propagation pattern of the ultrasonic beam. The net result is a diffusion or dispersion of the beam, and an inordinate attenuation of the ultrasonic energy.

One partial solution is to use a lower test frequency, so that the relative size of the grain represents a very small portion of a wavelength at the test frequency. Another is to use the longitudinal wave, rather than the shear wave, to interrogate the structure. A third is the use of multiple crystal techniques.

Without knowing more about the particular problem (e.g. weldment, rolled plate, tubing, etc.) there is no general response that can be given to the other questions posed by D. H. Njo. It is scientifically and technically possible and practical to use standard ultrasonic test techniques to inspect fine grain austenitic materials. Coarse grain or columnar grain structures can also be inspected; however, the ultrasonic spectral transmissivity of the material must be established before the test frequency and the test procedure can be established.

F. Turner

The following addenda to paper C31 should be made. Figs D15 to D26 inclusive.

REFERENCES

(9) CROOK, C. E. and NEWCOMBE, A. R. 'Acoustic excitation of the Calder reactor pressure circuits', Report 75W (British Nuclear Fuels Ltd).

(10) CROOK, C. E., NEWCOMBE, A. R. and STEWART, G. 'Operation of Windscale A.G.R. superheater headers in the creep range', Report 76W (British Nuclear Fuels Ltd).

The following corrections should be noted.

P. 73, l.h. col., **Present position and future procedures,** last line should read '. . . pressure of 285 lb/in^2 gauge at a temperature of 170°C'.

P. 74, r.h. col., *Emergency cooling water vessel,* line 6; for S/P, read 'standpipe seals'. 9 lines from bottom of page; for '1961' read '1971'.

J. B. Vetrano, W. D. Jolly and P. H. Hutton

About the time that the deadline time for submittal of paper C58, a very significant field evaluation phase started for the programme described. The purpose of this addendum is to present some of the more significant data that was recently obtained.

The reactor surveillance system evaluation project conducted by Battelle Northwest and Southwest Research Institute under sponsorship of the Edison Electric Institute has resulted in direct evidence of the detection, location and correlation of acoustic emission from a growing fatigue crack in the presence of simulated reactor full power operating noise. The data illustrated in Fig. D27 show the acoustic emission event accumulation in relation to the ultrasonic indication of crack propagation as the fatigue load on the crack is decreased and then increased again.

The fatigue crack was started from a 'D' slot normal to the vessel outer surface and parallel to the edge of a longitudinal seam weld in the 15 cm cylindrical section of the EBOR reactor pressure vessel. The fatigue crack was grown by hydraulic pressurization of the 'D' slot at a rate of 10 cycles/min. The crack depth at centre of the 'D' slot was approximately 3·81–4·32 cm when the data illustrated was recorded.

The acoustic emission events recorded were large burst

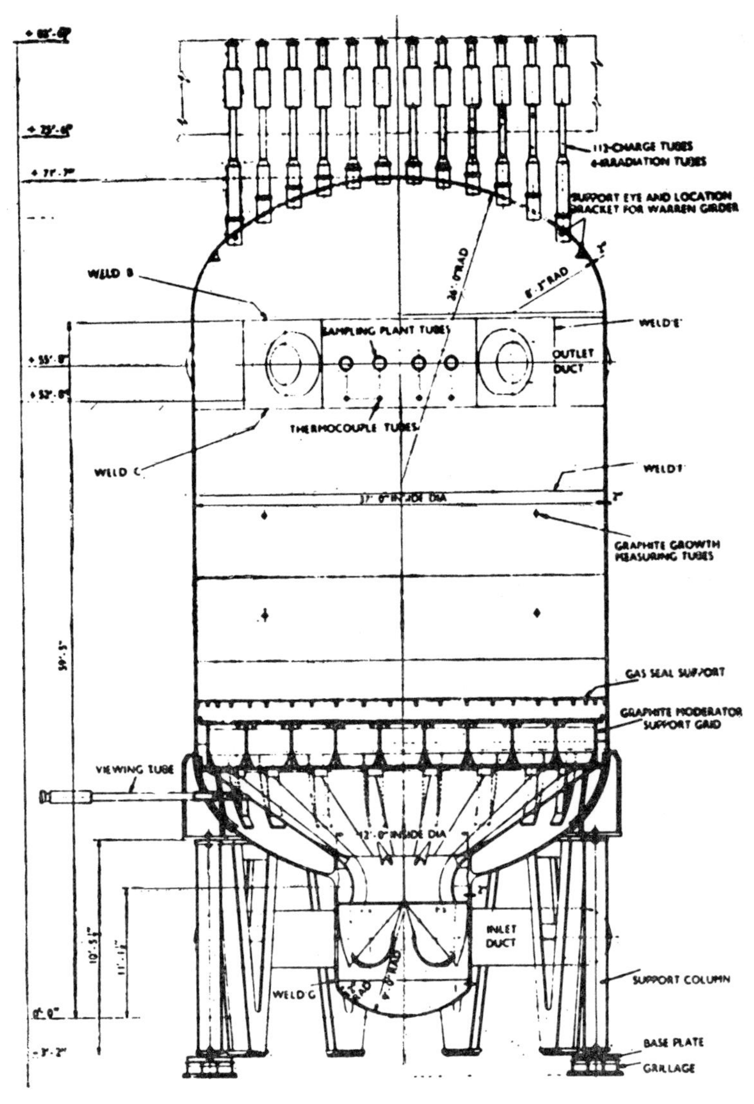

Fig. D15. Calder Hall pressure vessel

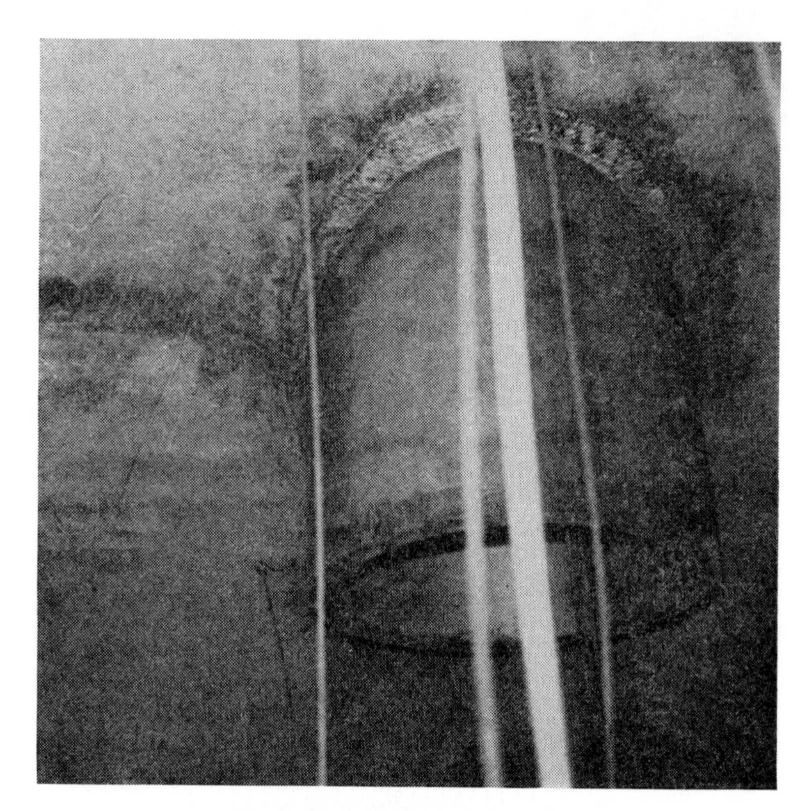

Fig. D16. Top head nozzle weld

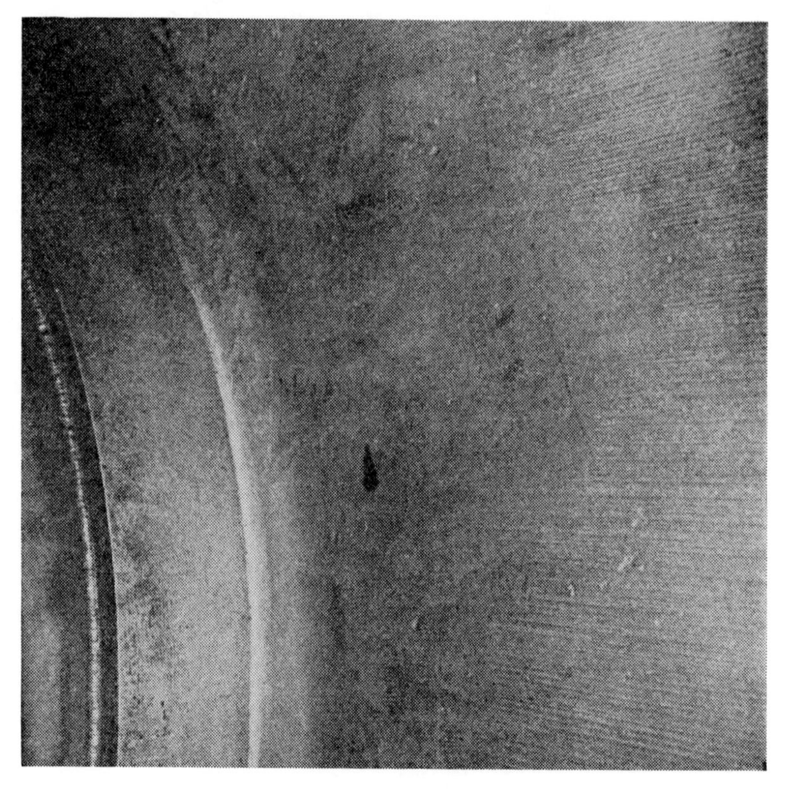

Fig. D17. B.C.D. nozzle weld

Fig. D18. Hanger attachment weld

Fig. D19. Bellows fairing

Fig. D20. B.C.D. tubes

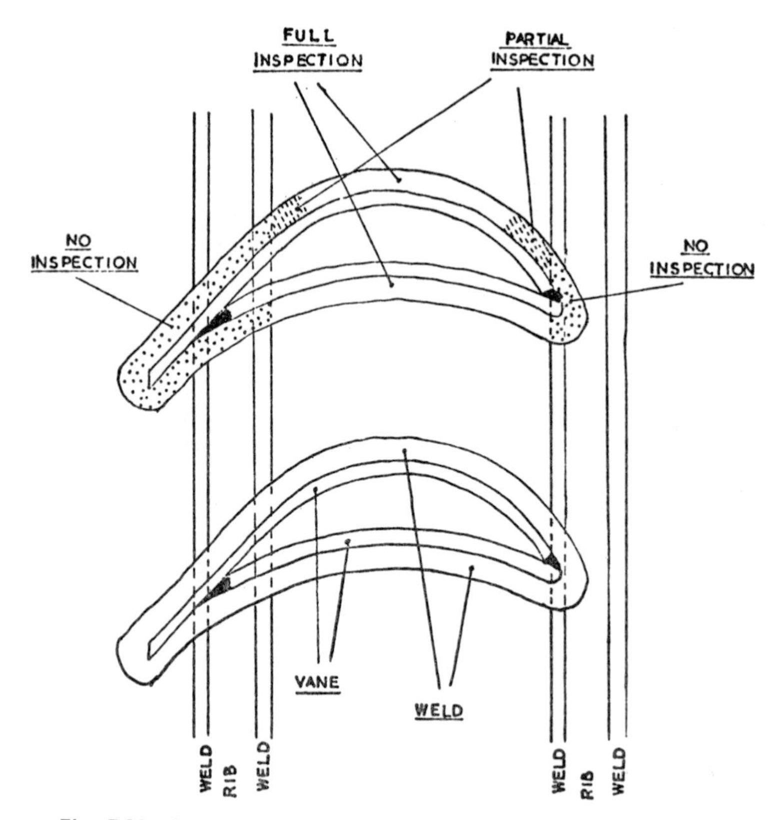

Fig. D21. Cascade corner: relative position of vanes and ribs, showing weld regions inspected

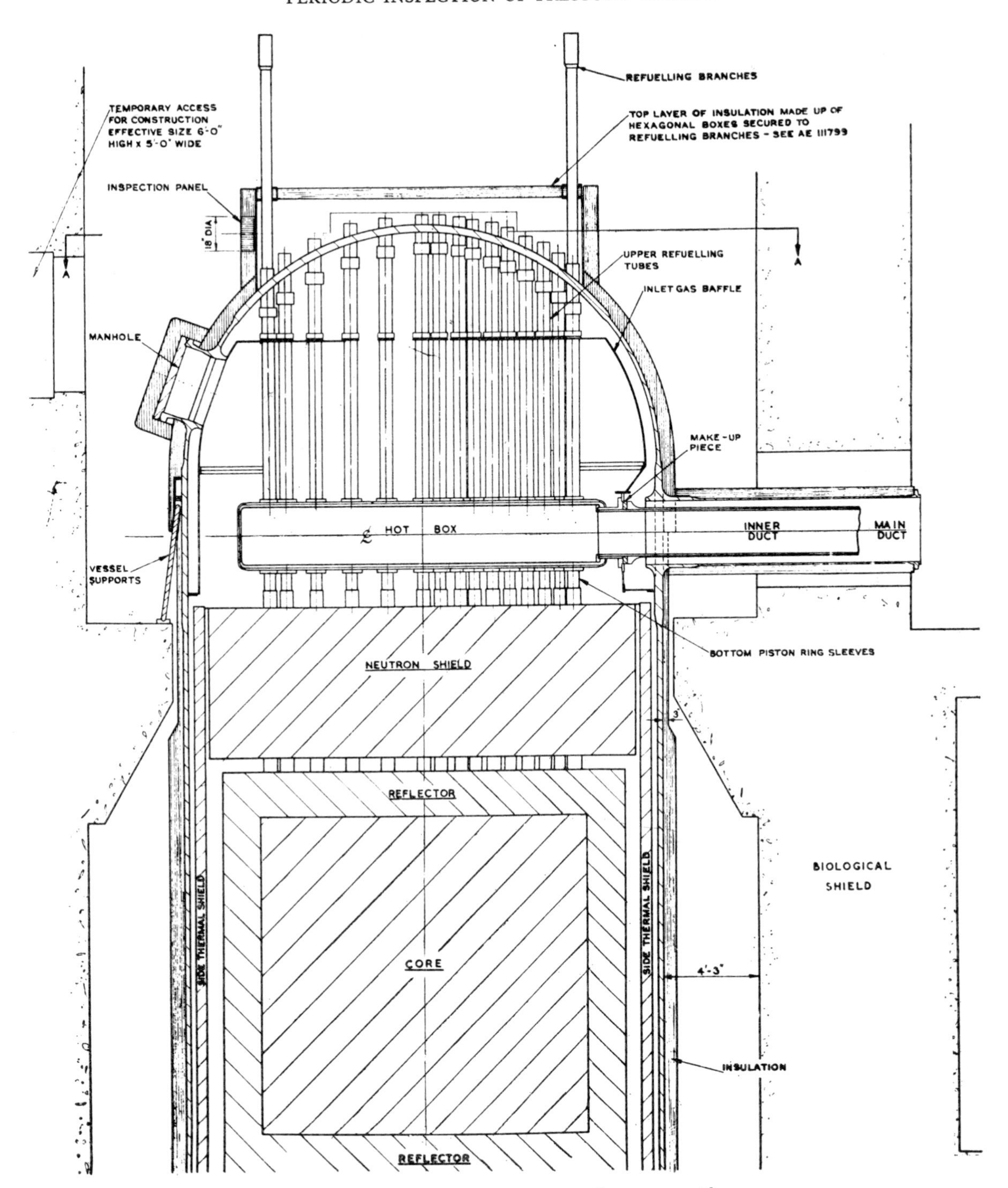

Fig. D22. W.A.G.R. pressure vessel, upper section

type pulses representative of rapid extension of the crack front. These burst pulses were found to be of sufficiently greater amplitude than the simulated reactor full power noise to permit detection and triangulation to the location of the fatigue crack.

The data were recorded on two different source location systems, which are described in paper C58. The prototype reactor surveillance system used aluminum waveguides to couple signals from the vessel surface to remote transducers and monitored a frequency band of 300 kHz centred at 600 kHz, while the coincident detection–location system monitored a broad frequency band between 200 kHz and 1·0 MHz.

The ultrasonic indication of crack growth was obtained by a pitch–catch arrangement of angle beam ultrasonic transducers. Crack growth was indicated by the reduction of received signal as the crack edge intercepted a greater fraction of the cross-section of the ultrasonic beam.

The indicated change of crack growth rate due to changing the fatigue load from 36 000 lb/in^2 to 30 000 lb/in^2 and back to 36 000 lb/in^2 is reflected in the change of acoustic emission event accumulation rate. These field data indicate that a continuous on-line acoustic emission monitor system for reactor pressure boundary surveillance system is feasible.

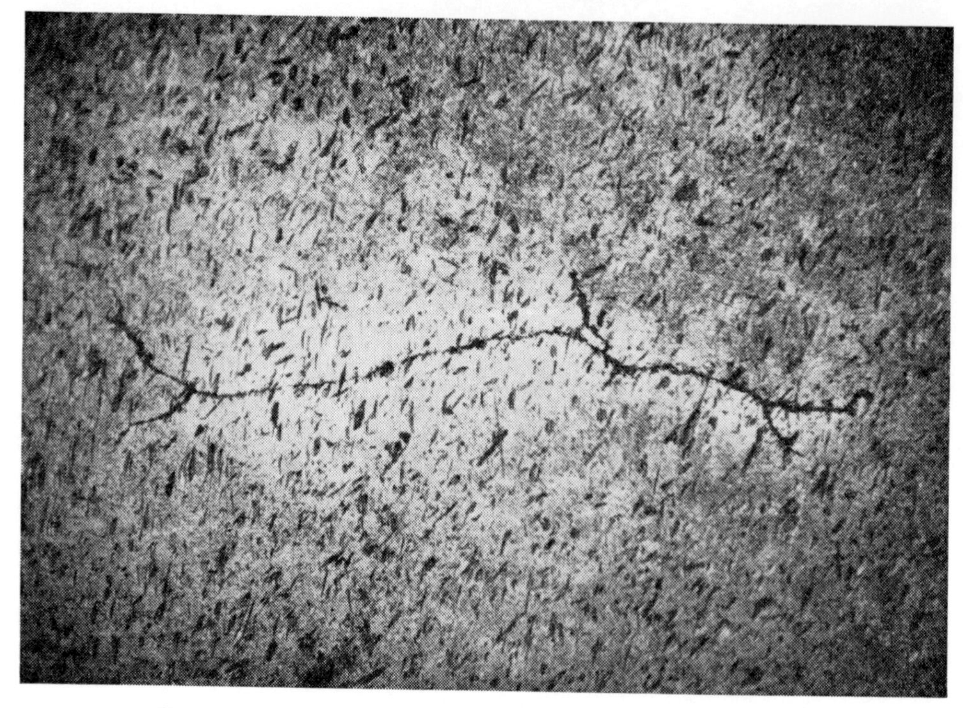

Fig. D23. Crack in support ball of W.A.G.R. heat exchanger

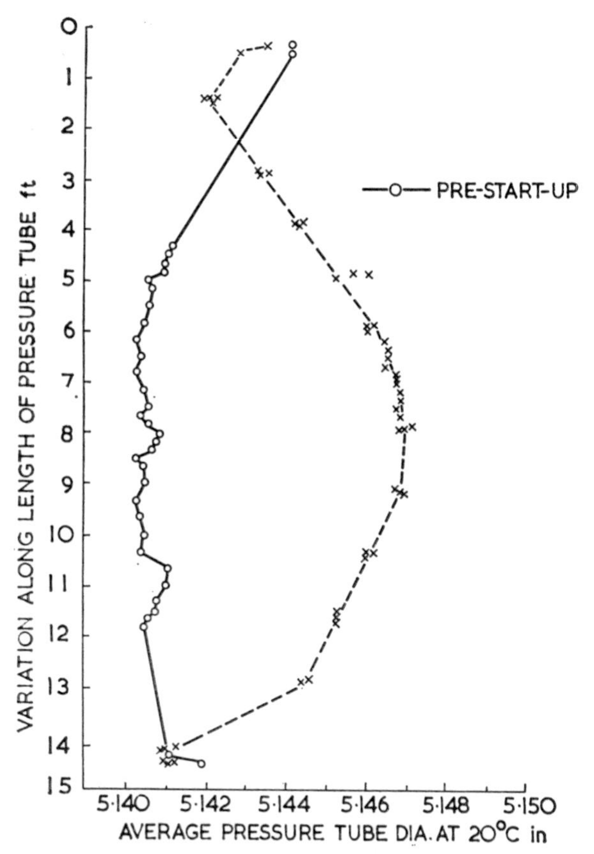

Fig. D24. S.G.H.W., variation in Zircaloy 2 pressure tube channel diameter

B. Watkins and **H. Jackson**

We thank R. Haas for his statement on German recommendations. It should be noted that virtually any standard of inspection can be achieved, provided that appropriate access can be made; the question then becomes one of balancing cost against the returns from particular requirements.

In reply to H. Baschek's query, we presume that the questioner is asking about defects at the nozzle leading edge or radius. There is now a system that will show the presence of surface or slightly sub-surface cracks, but is unable to measure the depth of the defect. The manipulation required to scan these complex curves in a satisfactory way can be very difficult, expensive and time consuming. We would agree with the implication of H. Baschek that some of the techniques presently being offered are not capable of resolving defects in such situations.

In regard to paper C27, we cannot overcome completely the effect of cladding on sensitivity. The figures quoted by W. Rath are realistic for stationary probes; however, the B-scan approach has advantages in this respect, but scanning probes build up a dynamic picture, and the small areas which are difficult to penetrate at one angle are frequently favourable at another angle.

As an analogy, take the case of a fence with evenly spaced palings. If one stands behind a plank, one sees nothing, but if one moves along, one gets a perfect picture of the scene behind the fence. I would refer W. Rath to the work undertaken by B.A.M. Berlin and R.T.D. Rotterdam on this subject. There is no difficulty in finding the cracks mentioned using modern techniques, and an evaluation of their length can be made. They cannot, however, have depth measurements made with any accuracy. Since the B scan is merely a manipulative and presentation system processing information from conventional A scan probes, the problem is only that which is common to all systems.

Since the B scan is calibrated against standard blocks, the size of a defect at any range can be compared with a standard.

Since each system is calibrated against special reference blocks which are permanently retained, then inspections at any time should be comparable.

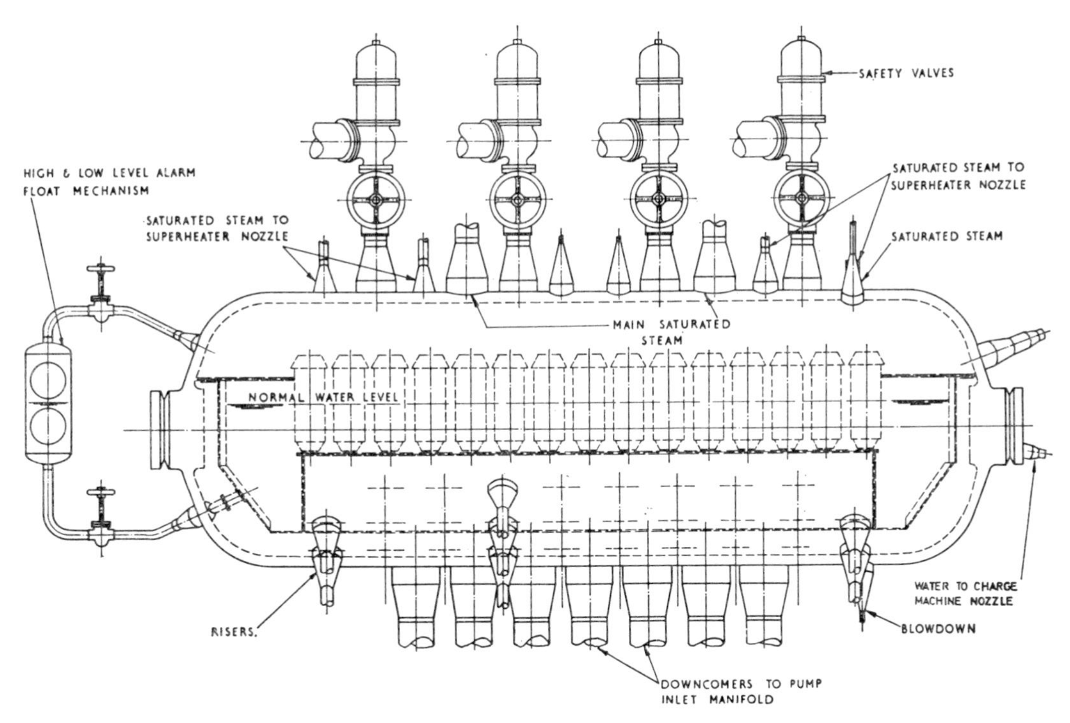

Fig. D25. Steam drum of S.G.H.W.

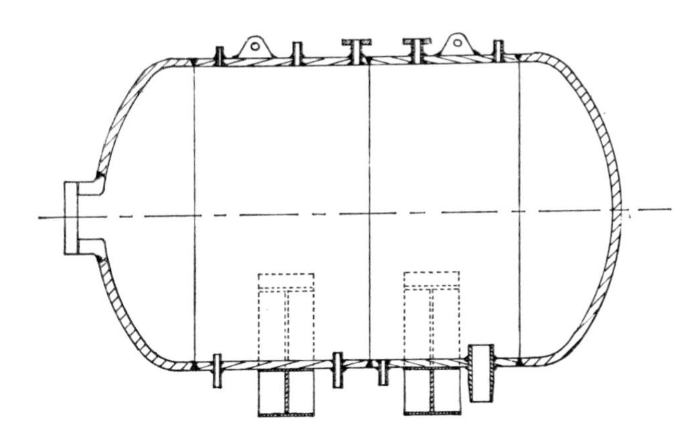

Fig. D26. Emergency cooling water vessel

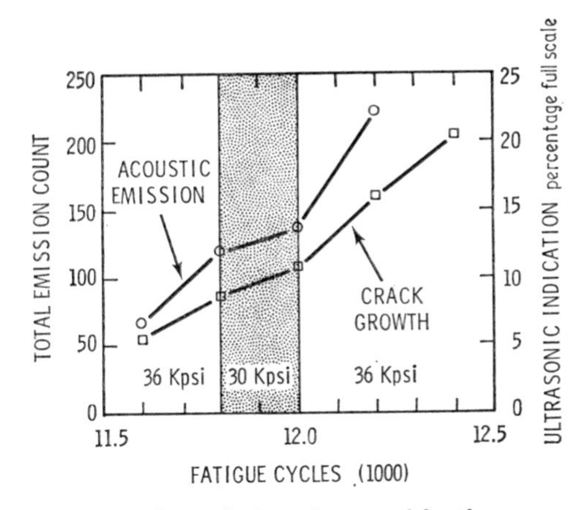

Fig. D27. Acoustic emission detected in the presence of simulated reactor full power background noise, and compared to crack growth

In reply to the query on paper C31 regarding the repetitive measurement of defects, this depends on the accuracy of the index system, the accuracy of the recovery position and the accuracy of calibration of the equipment. Our system can be read to one millimetre in a radial or axial direction. In addition, for nozzle work we measure to the same tolerance in radial or axial displacement.

The recovery position on dead reckoning is ± 5 mm. These tolerances are not so important with the B scan approach since movements are not by discreet steps but by continual movement, and hence, since the unit will encompass a defect in its scan, only the last read-out is critical. With a stationary probe, the recovery situation is much more critical. With regard to calibration, we would recommend that this be done on each occasion. We calibrate against full scale reference specimens of comparable geometry and fully clad. The attenuation of such specimens would be matched to the section of the reactor under test.

J. J. Whenray

In answer to F. D. Chaplin's question on the accuracy of paper C56, with regard to the continuous operation process industry, I have the following observations to make.

My paper is of a general nature, and due to the space and time available, there was no possibility to investigate the special requirements of various industries using pressure vessels. The paper is for manufacturers of boilers and pressure vessels rather than for users.

The information on Italy which F. D. Chaplin has selected is interesting, and I have studied this section carefully. My conclusions are that my information is accurate but less detailed than his.

With regard to the other countries, the information

given should be accurate as it was obtained directly from the approval organizations.

In concluding I would advise F. D. Chaplin that under no circumstances when checking regulations should he accept information from any document other than that issued by the approval organization, and even then there may be unstated exceptions. For detailed information he should write to the approval organization on every occasion.

The following corrections to the paper should be noted.

P. 201, Table 56.1, line 2, New South Wales: insert in columns, l. to r., 1, 1, 1, 1, Yes, No.

P. 202, line 1, NEW ZEALAND, 1, 1, 1, *, Yes, At surveyor's discretion; line 5, SPAIN, 5, 10, 2 to 10, 10, Yes, Yes; line 7, SWITZERLAND, 2, 2 to 4, 2 to 5, *, Yes, Yes. Transpose lines 3 and 4, POLAND and PORTUGAL. *Add* SOUTH AFRICA: 1, 2, 2, *, Yes, Yes. Table 56.2, line 2, NEW SOUTH WALES, insert No, No, Yes, Inspectors approved by Dept. of Labour.

P. 203, NEW ZEALAND; No, No, Yes, Dept. of Marine. POLAND; Technical Supervision Office. SPAIN; Yes, No, No, Ministry of Industry. SWITZERLAND; No, No, Yes, (a) Static, Society of Switzerland B.O., (b) Portable, Federal Materials Testing.

R. D. Wylie

The discussion by D. H. Njo asked about the inspectability of austenitic materials. We have found a considerable spectrum of ultrasonic attenuation in austenitic materials of the same basic composition. Generally, this is the result of varying grain size. Our test work in the laboratory indicates that it may be possible with heat input and ferrite control to achieve more uniform ultrasonic properties. We believe that this work in progress may lead to a method of control of inspectability which will be potentially available for procedure qualification requirements.

T. Yamaguchi, Y. Fukushima, S. Kihara, T. Endo and Y. Yoshida

We agree with H. Baschek's opinion that the people who are engaged in development of in-service inspection equipment and the researchers on fracture machines should co-operate closely with each other.

We are now developing a method for estimating critical flaw size in nuclear reactor vessels, using fracture mechanics. In order to establish the method, we must investigate various technological problems, such as estimation of occurrences of various degree of stress levels during vessel lifetime, analysis of local stresses, estimation of radiation embrittlement of materials, theoretical and experimental assumptions on the propagation speed of cracks, etc., so that it will not be an easy task to calculate actual critical flaw sizes.

As to the critical flaw size in the nozzle region, we suppose that though radiation embrittlement is rather small, the stress level is at its highest, therefore the critical flaw size would be smallest in the nozzle region. According to our preliminary study, and the opinions of other researchers, the critical flaw size in the nozzle region is about half an inch.

Our experience on ultrasonic testing, using a reactor vessel mock-up, shows that, though the subsurface flaws in the nozzle radius section can be detected when they are as small as 10 mm, surface cracks as deep as 20 mm were not detectable.

A few years ago, we did an in-service inspection on a Japanese reactor vessel (J.P.D.R.) using a 'Smeck gauge' and found that surface cracks in the nozzle radius section as shallow as 1 mm deep could be detected (11).

Regarding the question raised by D. H. Njo, we think that it will be conditionally possible to use the ultrasonic-testing method for austenitic materials.

If the grain structures of the austenitic weld metal can be made finer than normal, then it will be possible to inspect austenitic material.

If the grain structures of austenitic weld metal are comparable to that of the welded joint fabricated by ordinary welding procedures, then it will be rather difficult to achieve good flaw detectability, even if a frequency as low as 0·5 MHz and a transducer 2–3 times as large as conventional ones are used.

REFERENCE

(11) YAMAGUCHI, T. 'Inspection technique using the electric resistance probe method on a nuclear reactor vessel', *First Int. Conf. on Pressure Vessel Technology* 1969 (A.S.M.E., New York).

The following corrections to the paper should be noted.

P. 2, Fig. 25.1, centre figure: add height '12 366'.

Table 25.1. The titles 'C.R.T. Display' refer to the lower illustrations; suffixes of B_1, B_2 are unnecessary; delete vertical line between two right-hand sections of 'Remarks' column.

P. 3, Table 25.2, third column, for '2*x*' read '2×' at each mention; delete brackets round 'wall thickness'.

P. 9, Fig. 25.9, top diagram: delete 'plate' after 'forging'.

List of Delegates

AHONEN, T. — Imatran Voima Osakeyhtio, Helsinki
AINSWORTH, S. I. — C.E.G.B., London
AL-DAMALUJI, S. S. — Iraq Petroleum Company, London
ALDREW, A. R. — I.C.I. Agricultural Division, Billingham
ANDERSON, D. — British Nuclear Fuels Ltd, Chapelcross, Annan
ANDO, Y. — University of Tokyo, Bunkyoku, Japan
ANDREAS, L. R. — Dow Chemical N.V., Terneuzen, Holland
APPLETON, K. P. Q. — BP Chemicals International Ltd, Hull
ARCHER, J. — SOGERCA, Le Plessis Robinson, France
ATTFIELD, R. A. — Gas Council, London
BARBIER, R. — Iberduero S.A., Bilbao, Spain
BARDRAM, E. G. — AEK RISØ, Roskilde, Denmark
BARNES, S. A. — C.E.G.B., S.E. Region, London
BARRASS, A. — I.C.I. Organics Division, Manchester
BARTLE, P. M. — Welding Institute, Cambridge
BARTON, J. W. — Ministry of Defence, Bath
BASCHEK, H. — NOK Atomkraftwerk Beznau, Dottingen, Switzerland
BAUZA, J. Q. — Spanish Atomic Energy Commission, Madrid, Spain
BAWDEN, D. C. — British Oxygen Co. Ltd, Brentford
BEARE, J. W. — Atomic Energy Control Board, Ottawa, Canada
BEETHAM, G. H. — C.E.G.B., London
BEIE, D. — Kernkraftwerk Lingen GmbH, Lingen, W. Germany
BENNETT, J. C. — Clarke Chapman John Thompson Ltd, Gateshead
BENNETT, R. W. — I.C.I. Ltd, Billingham
BENTLEY, P. G. — U.K.A.E.A., Risley, Warrington
BERNARD, A. M. — FRAMATOME, Courvevoie, France
BERG, H. B. — BBR, Mannheim, W. Germany
BERTOLOTTI, G. B. — Union Electrica S.A., Madrid, Spain
BEUKES, H. J. — South African Atomic Energy Board, Pretoria, South Africa
BEVITT, E. — U.K.A.E.A., Risley, Warrington
BILLINGS, G. McK. — Gas Council, Solihull
BIRCHON, D. — Admiralty Materials Laboratory, Poole, Dorset
BLACK, W. S. A. — U.K.A.E.A., Winfrith, Dorset
BOEDECKER, K. W. — Kernkraftwerk Obrigheim GmbH, W. Germany
BOERSTOEL, B. M. — Metals Institute TNO, Delft, The Netherlands
BOULTON, H. — C.E.G.B., London
BOWLES, L. F. — G.E.C. Reactor Equipment Ltd, Leicester
BROEKHOVEN, M. J. G. — University of Technology, Delft, The Netherlands
BROMKAMP, K. H. — Hochtemperatur-Reaktorbau GmbH, Mannheim, W. Germany
BROOKES, A. — Nuclear Installations Inspectorate, Bootle
BROOME, T. — C.E.G.B., London
BROWN, H. A. — BP Chemicals International Ltd, Hull
BROWN, J. C. — British Engine Boiler and Electrical Insurance Ltd, Manchester
BROWN, R. — Kodak Ltd, Harrow
BUERGO, L. A. — Spanish Atomic Energy Commission, Madrid
BURDEN, P. J. — I.C.I. Agricultural Division, Billingham
BURMAN, A. — U.K.A.E.A., Risley, Warrington
BURNUP, T. E. — U.K.A.E.A., Risley, Warrington
BURT, G. A. — C.E.G.B., S.E. Region, London
BURTON, E. J. — U.K.A.E.A., R.E.M.L., Risley, Warrington
BUSH, S. H. — U.S. Atomic Energy Commission, Battelle-Northwest, Washington, U.S.A.
BUTLER, R. — Engineering Services, Shell U.K. Ltd, Ellesmere Port
CAMERON, J. — Gas Council, London
CARR, C. A. — Shell Chemical Co. U.K. Ltd, Manchester
CARRUTHERS, H. M. — Associated Nuclear Services, London
CARSON, J. M. — Lloyd's Register Industrial Services, Croydon
CAULFIELD, J. C. — British Nuclear Design and Construction Ltd, Whetstone, Leicester
CERECEDA, M. C. — TECNATOM, Madrid
CERQUEIRA LOPES, J. J. — Instituto de Soldadura, Lisbon, Portugal
CHAPLIN, F. D. — Shell U.K. Ltd, London
CHOCKIE, L. J. — General Electric Company, San Jose, U.S.A.
CLUCAS, J. — C.E.G.B., Bristol
COLLINS, E. W. — Quality Control Dept, Babcock and Wilcox (Operations) Ltd, Renfrew
COWAN, A. — U.K.A.E.A., R.E.M.L., Risley, Warrington
CRISCI, J. R. — Naval Ship Research and Development Centre, Maryland, U.S.A.
CROCKETT, C. B. — South of Scotland Electricity Board, Glasgow
CROOK, C. E. — British Nuclear Fuels Ltd, Sellafield
CROSSLEY, D. P. — British Engine Boiler and Electrical Insurance Company Ltd, Manchester
D'ANNA, C. — ENEL, Rome, Italy
DARE, D. W. F. — The Nuclear Power Group, Knutsford
DARLASTON, B. J. L. — C.E.G.B., Nuclear Laboratories, Berkeley, Glos.
DAU, G. J. — Battelle-Northwest, Washington, U.S.A.
DAVENPORT, L. H. — Cominco Ltd, Trail, B.C., Canada
DAVIS, J. H. — South of Scotland Electricity Board, Glasgow
DEAN, W. — I.C.I. Mond Division, Runcorn, Cheshire
DEBONO, J. — Gas Council, London
DE JONG, J. J. R. — N.V. K.E.M.A., Arnhem, The Netherlands
DEL BUONO, A. — C.N.E.N., Rome, Italy
DENNING, I. F. V. — Foster Wheeler John Brown Boilers Ltd, London
DE RAAD, J. A. — Rontgen Technische Dienst N.V., Rotterdam, The Netherlands
DICKENSON, S. W. — Plant Inspection and Standards Engineering Dept, BP Trading Ltd, London
DICKIE, D. — Bridge Engineering Ltd, Motherwell, Scotland
DICKINSON, K. — U.K.A.E.A., Risley, Warrington
DJURDJEK, I. — Elektroprivreda, Zagreb, Yugoslavia
DOUGLAS, J. S. D. — Scottish Boiler and General Insurance Ltd, Glasgow
DUMOUSSEAU, P. F. — C.E.T.I.M., Senlis, France
DURY, T. V. — Babcock and Wilcox (Operations) Ltd, Renfrew, Scotland
EDMONDSON, B. — C.E.R.L., Leatherhead
EDWARDS, J. G. P. — C.E.G.B., Bristol
EHRENTREICH, J. W. — Commission of the European Communities, Brussels, Belgium
EISENBLAETTER, J. — Battelle Institut e.V., Frankfurt
EISFELDER, H. J. — AEG-Telefunken, Frankfurt, W. Germany
ELSON, A. W. — A.E.E., Winfrith, Dorset
ENGSTROM, J. A. — State Power Board, Vallingby, Sweden
ERDMANN, J. — Technischer Uberwachungs Verein Rheinland e.V., Cologne, W. Germany
ESTER, P. — Rotterdam Dockyard Company, The Netherlands
ESTRUCH, B. — I.C.I. Ltd, Agricultural Division, Middlesbrough
EVANS, U. L. — Rolls-Royce and Associates, Derby
FAGENCE, G. P. — Babcock and Wilcox (Operations) Ltd, Renfrew, Scotland
FATIGAROV, G. A. — International Atomic Energy Agency, Vienna, Austria
FELDMANN, J. — Allianz Versicherungs-AG, Munich, W. Germany
FERRIE, J. G. — Ministry of Defence, Bath
FIELDER, E. — U.K.A.E.A., Risley, Warrington
FIGLHUBER, D. — Siemens AG, Erlangen, W. Germany
FOREST, G. H. F. — Electricité de France, Paris
FORMBY, C. L. — C.E.G.B., Nuclear Laboratories, Berkeley, Glos.
FOTHERINGHAM, I. K. — I.C.I. Ltd, Agricultural Division, Billingham
FOWLING, A. V. — Ministry of Defence, Foxhill, Bath
FRANK, E. H. — I.C.I. Petrochemicals Division, Billingham
FREIJ, B. — South Swedish Power Co., Malmo, Sweden

FRY, R. M. Australian High Commission, London
FRYER, D. R. H. Nuclear Inspectorate, Dept of Trade and Industry, London
FUCHS, I. J. United States Testing Inc., New Jersey, U.S.A.
FUIJIMUSA, T. F. Japan Atomic Energy Research Institute
FUKUSHIMA, Y. Kobe Technical Institute, Kobe, Japan
FURSTE, W. O. A. Krupp Kerntechnik, Essen, W. Germany
FURUNO, K. F. Nuclear Plant Dept, Kawasaki Heavy Industries Ltd, Kobe, Japan
GALLIZIOLI, G. E.N.E.N., Rome, Italy
GOETHALS, J. Association Vincotte, Rhode St Genese, France
GOW, R. A. Operations Department, C.E.G.B., London
GRANT, A. E. J. G.E.G.B., Bristol
GREEN, E. U.K.A.E.A., Winfrith, Dorset
GRIFFITH, W. B. C.E.G.B., Merioneth
GRONOW, W. S. Nuclear Inspectorate, Dept of Trade and Industry, London
GROSS, L. B. Babcock and Wilcox, Lynchburg, U.S.A.
GRUGNI, G. G. CISE, Milan, Italy
GUENOT, R. L'Air Liquide, Centre de Recherche Claude-Delorme, Jouy-en-Josas, France
GUTHRIE, J. E. Esso Engineering Services Ltd, New Malden, Surrey
HAAS, R. Technischer Uberwachungs-Verein e.V., Essen, W. Germany
HABERFIELD, G. D. Ministry of Defence, Bath
HAHNEL, G. Institut fur Reaktorsicherheit der Technischen, Uberwachungs-Verein e.V., Cologne, W. Germany
HAIRE, T. P. C.E.G.B., London
HAKE, G. Nuclear Safety Engineering, Atomic Energy of Canada Ltd, Ontario
HAMILTON, J. G. BP Chemicals International Ltd, Grangemouth
HARRIS, F. R. S.S.E.B., West Kilbride
HASLAM, G. H. National Engineering Laboratory, East Kilbride
HASLEHURST, F. H. Rolls-Royce and Associates, Derby
HAUPTFLEISCH, D. K. Jersey Nuclear Company, Diegem, Belgium
HAY, E. Vessel and Heat Transfer Group, C. J. B. Ltd, Portsmouth
HAYDEN, R. L. J. Foster Wheeler John Brown Boilers Ltd, London
HEATON, J. Clark Chapman–John Thompson Pipework and Pressure Vessel Division, Wolverhampton
HEDDEN, O. F. Combustion Engineering Inc., Windsor, U.S.A.
HERNALSTEEN, P. Traction et Electricite, Brussels, Belgium
HEYES, R. Kennedy and Donkin, Manchester
HIBBERT, J. Nuclear Power Group Ltd, Knutsford
HIBBERT, N. S. Reactor Plant Inspection Service, U.K. A.E.A., R.E.M.L., Risley, Warrington
HILBORN, J. W. Atomic Energy of Canada, Chalk River, Ontario, Canada
HILDEBRANDT, H. BASF Energie-Abteilung, TLK/80, Ludwigshafen, W. Germany
HILL, G. J. G.E.C. Turbine Generators, Whetstone, Leics.
HODGES, N. W. C.E.G.B., London
HOLMAN, C. J. Ministry of Defence, Bath
HOLMES, P. ICI Organics Division, Huddersfield
HITCHINS, C. A. Esso Petroleum Co., Southampton
IDASHIMOTO, J. Tohoku Electric Power Co., Sendai, Japan
IDE, A. Nuclear Power Group Ltd, Knutsford
IMATOMI, S. Mitsubishi Corporation, Fukuoka, Japan
IRVINE, W. H. U.K.A.E.A., A.S.R.D., Risley, Warrington
ISHIYAMA, T. Showa Measuring Instrument Ltd, Tokyo, Japan
JACKSON, H. U.K.A.E.A., Risley, Warrington
JAMES, D. P. Gas Council, Newcastle upon Tyne
JAX, P. Battelle Institut e.V., Frankfurt, W. Germany
JENNINGS, A. R. T. C.E.G.B., London
JOHNSON, A. L. Department of Trade and Industry, London
JONES, A. W. Rolls-Royce and Associates Ltd, Derby
JONES, P. H. Rolls-Royce and Associates Ltd, Derby
JORDAN, G. M. U.K.A.E.A., Risley, Warrington
JOYCE, L. Automation Industries, Rotterdam, Holland
JUNGHEM, A. Tekniska Rontgencentralen AB, Stockholm
KAFKA, P. Laboratorium fur Reaktorregeung und Anlagensicherung, Garching, W. Germany
KARNOWSKI, K. Technischer Uberwachungs-Verein e.V., Stuttgart, W. Germany
KASHIMOTO, S. Shikopu Electric Power Company, Tokamatsu, Japan
KATZ, L. R. Westinghouse Electric Corp., Pittsburgh, U.S.A.
KELLER, H. W. Westinghouse Electric Corp., Pittsburgh, U.S.A.
KELLY, M. C.E.G.B., Bristol
KEMP, M. J. British Standards Institution, Hemel Hempstead
KERR, C. E. National Vulcan Engineering Insurance Group Ltd, Manchester
KIRBY, N. U.K.A.E.A., Risley, Warrington
KLAUSNITZER, E. Siemens AG, Erlangen, W. Germany
KOBAYASHI, A. M. Kansai Electric Power Co., Osaka, Japan
KORFF, B. Nuclear Power Station Dept, Dienst voor het Stoomwezen, The Hague
KOVAN, R. W. *Nuclear Engineering International*, IPC Business Press Ltd, London
KRAMMER, J. H. Gemeinschaftskernkraftwerk Tullnerfeld GmbH, Vienna, Austria
LALOY, P. A. L. Babcock Atlantique, Paris, France
LANGSER, B. L. ASEA-ATOM, Vasteras, Sweden
LAURENT, N. SOCIA, Courbevoie, France
LAUTZENHEISER, C. E. Southwest Research Institute, San Antonio, U.S.A.
LAWSON, D. S. G.E.C. Reactor Equipment, Whetstone, Leics.
LAWSON, D. W. U.K.A.E.A., Risley, Warrington
LAZZERI, L. CNEN, Rome, Italy
LEDER, H. INTERATOM, Cologne, W. Germany
LEGG, G. G. Atomic Energy of Canada, Ontario, Canada
LEHNHOFF, H. Maschinenfabrik Augsburg-Nuernberg AG, Abholfach, W. Germany
LE ROUX, J. L. R. G.A.A.A., Le Plessis Robinson, France
LINES, A. E. Ministry of Defence, Bath
LUCIA, A. EURATOM C.C.R., Ispra, Italy
LUMB, R. F. Gas Council, Newcastle upon Tyne
MADSEN, A. G. ELSAM, Aabenraa, Denmark
MARCH, W. Rolls-Royce and Associates Ltd, Derby
MARIQUE, L. A. Westinghouse Electric Nuclear Energy Systems, Brussels, Belgium
MATTHEWS, R. R. C.E.G.B., London
MAURER, H. Commission of the European Communities, Brussels, Belgium
MAXWELL, I. E. C.E.G.B., Bristol
MAZUR, H. Technischer Uberwachungs-Verein, Hanover, W. Germany
MECKEL, N. T. Southwest Research Institute, San Antonio, Texas, U.S.A.
MEIRIK, A. Koninklijke Machinefabriek Stork N.V., Hengelo, The Netherlands
MEYER, H. J. Maschinenfabrik Augsberg-Nuernberg AG, Abholfach, W. Germany
MINAMOTO, S. Quality Control Section, Mitsubishi Heavy Industries Ltd, Kobe, Japan
MIYOSHI, S. M. Fuji Electric Co. Ltd, Tokyo, Japan
MORGENSTERN, F. H. INTERATOM, Bensberg, W. Germany
MORI, R. Nuclear Power Section, Hokuriku Electric Power Co., Toyama, Japan
MUKAI, J. Atomic Energy Bureau, Tokyo, Japan
MURRAY, R. Babcock and Wilcox (Operations) Ltd, Renfrew, Scotland
MCARTHUR, M. Lennig Chemicals Ltd, Middlesbrough
MCHUGH, B. Associated Offices Technical Committee, Manchester
MCKENNA, A. Engineering Services, BP Chemicals International Ltd, London
MCNALLY, J. Automation Industries UK, West Lothian, Scotland
NAKAGAWA, K. Ministry of International Trade and Industry, Tokyo, Japan
NAKAMURA, K. Tokyo Electric Power Inc., Tokyo, Japan
NAKATA, T. N. Krautkramer Japan Ltd, Tokyo, Japan
NAKAZAWA, N. F. Power Reactor and Nuclear Fuel Co., Tokyo, Japan
NICHOLS, R. W. U.K.A.E.A., Risley, Warrington
NICKELS, P. J. Commercial Union Assurance Group, London
NJO, D. H. Eidgenossisches Amt fur Energiesirtschaft, Berne, Switzerland
OATES, E. C.E.G.B., Manchester

OCHI, E. Japan Electric Association, Tokyo
O'CONNOR, J. P. Electricity Supply Board, Dublin
OGUSHI, A. Hitachi Ltd, Tokyo, Japan
OIZUMI, H. Kobe Steel Ltd, Kobe, Japan
ONO, M. Electric Power Development Ltd, Tokyo
O'NEILL, R. Reactor Assessment Section, U.K.A.E.A., Risley, Warrington
ORME, J. M. International Combustion Ltd, Derby
OWEN, P. U.K.A.E.A., Caithness
PASQUAL, A. Iberduera S.A., Bilbao, Spain
PATTERSON, M. M. General Electric Co., San Jose, U.S.A.
PAUL, A. C.E.G.B., New Romney
PHILLIPS, C. A. G. Department of Trade and Industry, London
PHILLIPS, N. A. BP Chemicals Ltd, London
PHIPPS, R. L. Westinghouse Electric Corp., Pittsburgh, U.S.A.
PICKEL, E. Kernkraftwerk Obrigheim, W. Germany
POLLOCK, A. A. Cambridge Consultants Ltd, Cambridge
POMIE, P. E. C.F.A. C.E.N. Cadarache, St Paul lez Duranie, France
POORT, J. Stress Analysis Group, Rotterdam Dockyard Co., The Netherlands
POSSA, G. CISE, Segrate, Italy
POUCHET, P. M. Babcock Atlantique, Paris, France
POUNDER, H. R. Commercial Union Assurance Co., London
PRANTL, G. Swiss Federal Institute for Reactor Research, Wurenlingen, Switzerland
PROT, A. C. CEN, Saclay, France
PUGH, B. E. C.E.G.B., S.W. Region, Portishead
PULL, D. J. Cremer and Warner, London
QUIRK, A. U.K.A.E.A., Risley, Warrington
RADHAKRISHNAN, A. T. Esso Engineering Services Ltd, London
RAES, H. D. B. R. PNF Pressure Systems N.V., St Niklaas, Belgium
RAIKES, J. L. Operations Department, C.E.G.B., London
RAO, S. Abatomenergi, Sweden
RAPETTI, A. A. Consolidated Energy Services Inc., New York
RASMUSSEN, I. Danish Atomic Energy Commission, Roskilde, Denmark
RASTOIN, J. Commissariat a l'Energie Atomique, Gif sur Yvette, France
RATH, W. Dr U.u.H. Krautkramer, Cologne, W. Germany
RAWOE, E. Association des Industriels de Belgique, Brussels, Belgium
READ, G. B. Associated Offices Technical Committee, Manchester
REGGIORI, A. Istituto di Ricerche, Breda, Milan, Italy
REPKE, W. Kraftwerk Union AG, Frankfurt, W. Germany
RICHARDS, R. W. S. Gas Council Engineering Research Station, Newcastle upon Tyne
RICHARDSON, J. Scottish Boiler and General Insurance Ltd, Glasgow
RISEBOROUGH, W. A. Unit Inspection Co., Swansea
ROCHE, R. Commissariat a l'Energie Atomique, Paris, France
ROCKENHAUSER, W. Westinghouse Nuclear Energy, Brussels
ROQUEFORT, P. I. Electricité de France, Paris, France
ROWLEY, T. C.E.G.B., Marchwood, Southampton
RUSSELL, J. L. Humphreys and Glasgow, Redhill, Surrey
RUSSELL, D. Conoco Ltd, Immingham, Lincs.
RUSSELL, J. Motherwell Bridge Eng. Ltd, Motherwell
SAGLIO, R. S. CEN Saclay, Gif sur Yvette, France
SALTER, W. B. W.E.N.E.S.E., Brussels, Belgium
SAMMAN, J. Babcock Atlantique, St Nazaire, France
SANDBERG, E. O. Oskarshamnsverkets Kraftgrupp AB, Stockholm, Sweden
SANDONA, E. NOK Atomkraftwerk Beznau, Dottingen, Switzerland
SAWAFUJI, K. Ishikawajima-Hasima Heavy Industries Co., Yokohama, Japan
SCHENK, H. Kernkraftwerk Obrigheim GmbH, Obrigheim, W. Germany
SCHIEL, L. R. Gebr. Sulzer AG, Winterthur, Switzerland
SCHIPPER, P. G. Institute of Metals, Delft, Holland
SCHLOSSER, K. Osterreichische Studienges für Atomenergie, Vienna, Austria
SCHOFIELD, B. H. Teledyne Materials Research, Waltham, U.S.A.
SCOTT, A. V. Non-destructive Testing, Babcock and Wilcox, Renfrew, Scotland
SHIMIZU, M. Japan Nuclear Ship Development Agency, Tokyo, Japan
SHIPP, W. J. C.E.G.B., Marchwood, Southampton
SIDLER, E. E. Swiss Association of Steam Boiler Owners, Zurich, Switzerland
SILK, M. G. A.E.R.E., N.D.T. Centre, Harwell, Didcot
SLAWSON, D. C. J. Foster Wheeler Ltd, London
SMEDLEY, G. P. Lloyd's Register of Shipping, London
SMETS, J. Association Vincotte, Rhode St Genese, France
SMITH, G. T. Technical Services, Shell UK Ltd, Stanford le Hope
SMITH, T. A. U.K.A.E.A., Risley, Warrington
SNEE, M. J. Electricity Supply Board, Dublin
SOAR, B. R. C.E.G.B., Trawsfynydd, Merioneth
SOUCH, A. E. Applied Physics Division, C.E.G.B., Berkeley, Glos.
SOUTER, W. D. Eagle Star Group Engineering Division, Cheltenham
SPANDICK, W. Fried. Krupp GmbH, Essen, W. Germany
SPENCER, R. M. Inspection Section, Humphreys and Glasgow, Redhill
STAQUET, M. Electricité de France, Clamart, France
STEWART, G. British Nuclear Fuels Ltd, Sellafield
STONES, C. H. U.K.A.E.A., Risley, Warrington
SVAHN, B. Tekniska Rontgencentralen AB, Stockholm, Sweden
SWINDEN, S. A. U.K.A.E.A., Risley, Warrington
SWITHENBANK, T. Reactor Plant Inspection Service, U.K.A.E.A., Risley, Warrington
SYS, A. C.R.I.F., Brussels, Belgium
TAK, A. G. M. Central Laboratory TNO, Delft, Holland
TAMIMURA, K. Japan Atomic Power Co., Tokyo, Japan
THEIRETZBACHER, H. Technischer Uberwachungs-Verein, Vienna, Austria
THOMAS, I. H. Shell UK Ltd, Ellesmere Port
THOMPSON, L. Guardian Royal Exchange Assurance Group, Ipswich
TIETZE, A. TUV Rheinland, Cologne, Germany
TILER, W. P. W.E.N.E.S.E., Brussels, Belgium
TOKIEDA, T. T. ASEA-ATOM/STAL Group, Kobe, Japan
TOOGOOD, J. T. Inspectorate of Factories, Department of Employment, London
TRUSCOTT, J. M. I.C.I. Agricultural Division, Billingham
TURNER, D. B. North of Scotland Hydro-Electric Board, Edinburgh
TURNER, F. U.K.A.E.A., Risley, Warrington
TUSA, J. M. S. Finnish Atomic Energy Commission, Helsinki, Finland
UEMURA, T. Nuclear Construction Section, Kyusyu Electric Power Co. Ltd, Fukuoka, Japan
UNDERWOOD, P. C. Foster Wheeler Ltd, London
VESTERDAL, J. K. ELSAM, Esbjerg, Denmark
VETTERLEIN, P. Kraftwerk Union AG, Frankfurt, W. Germany
WAKAYAMA, P. E. Chobu Electric Power Inc., Nagoya, Japan
WALD, F. W. Electricité de France, Clamart, France
WALKER, C. W. C.E.G.B., London
WALKER, J. B. U.K.A.E.A., Thurso, Caithness, Scotland
WALLACE, R. G. Shell Chemicals UK Ltd, Manchester
WALLIS, R. M. I.C.I. Ltd, Northwich
WALTER, J. C. Det Norske Veritas, Oslo, Norway
WARNE, R. Rolls-Royce and Associates, Derby
WARWICK, R. G. Associated Offices Technical Committee, Manchester
WASLEY, B. A. C.E.G.B., Bristol
WATANABE, S. Japan Steel Works Ltd, Tokyo, Japan
WATERS, J. R. U.K.A.E.A., Risley, Warrington
WATKINS, B. U.K.A.E.A., Risley, Warrington
WATSON, G. M. U.K.A.E.A., Thurso, Caithness, Scotland
WERBER, W. TUV Baden, Mannheim, W. Germany
WHENRAY, J. J. Technical Help to Exporters Service, B.S.I., Hemel Hempstead
WHITAKER, M. I.C.I. Agricultural Division, Billingham
WHITE, C. M. Whessoe Ltd, Darlington, Co. Durham
WILLIAMS, H. D. C.E.G.B., Midlands Region, Nottingham
WINTERMARK, H. S. Det Norske Veritas, Oslo, Norway
WOLF, M. R.W.E. AG, Essen, W. Germany
WOLLHAG, R. E. G. State Power Board, Vallingby, Sweden
WORLTON, D. C. Jersey Nuclear Co., Washington, U.S.A.
WUTSCHIG, R. TUV Stuttgart e.V., Stuttgart, W. Germany
YARDLEY, J. N. Atomic Power Constructions Ltd, Sutton
YATES, D. A. J. I.C.I. Plastics Division, Welwyn Garden City
YOSHIDA, T. Chugoku Electric Power Co., Hiroshima, Japan
ZETTERWALL, N. E. Swedish Steam Users Association, Stockholm, Sweden

Index to Authors and Participants

Names of authors and numbers of pages on which papers begin are in bold type.

Subject Index

Titles of papers are in capital letters.

MADE AND PRINTED IN GREAT BRITAIN BY WILLIAM CLOWES & SONS, LIMITED, LONDON, BECCLES AND COLCHESTER

STRATHCLYDE UNIVERSITY LIBRARY
30125 00075876 2

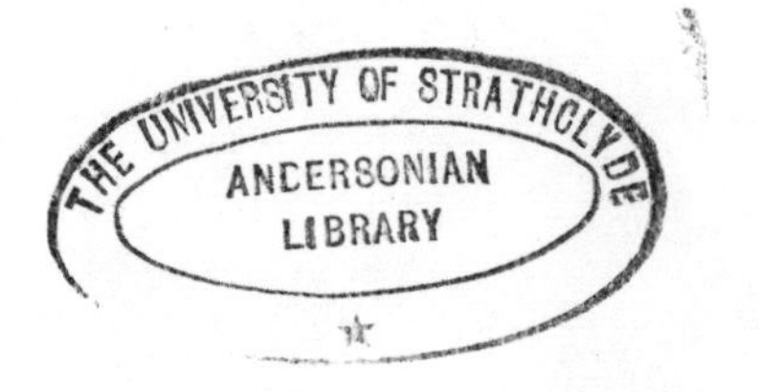
THE UNIVERSITY OF STRATHCLYDE
ANDERSONIAN
LIBRARY

ANDERSONIAN LIBRARY
WITHDRAWN
FROM
LIBRARY
STOCK
UNIVERSITY OF STRATHCLYDE